ANIMAL CREATIVITY AND INNOVATION

Explorations in Creativity Research

Series Editor

JAMES C. KAUFMAN

ANIMAL CREATIVITY AND INNOVATION

Edited by

ALLISON B. KAUFMAN
University of Connecticut, CT, USA

JAMES C. KAUFMAN
Neag School of Education
University of Connecticut, CT, USA

AMSTERDAM • BOSTON • HEIDELBERG • LONDON
NEW YORK • OXFORD • PARIS • SAN DIEGO
SAN FRANCISCO • SINGAPORE • SYDNEY • TOKYO

Academic Press is an imprint of Elsevier

Academic Press is an imprint of Elsevier
125, London Wall, EC2Y 5AS.
525 B Street, Suite 1800, San Diego, CA 92101-4495, USA
225 Wyman Street, Waltham, MA 02451, USA
The Boulevard, Langford Lane, Kidlington, Oxford OX5 1GB, UK

ISBN: 978-0-12-800648-1

British Library Cataloguing-in-Publication Data
A catalogue record for this book is available from the British Library

Library of Congress Cataloging-in-Publication Data
A catalog record for this book is available from the Library of Congress

For information on all Academic Press publications
visit our website at http://store.elsevier.com/

Typeset by MPS Limited, Chennai, India
www.adi-mps.com

Dedication

For James—ABK

For Allison—JCK

Are we allowed to do this?—ABK

I don't think so—JCK

Okay. Then we'd like to both dedicate this book to the tireless, wonderful folks who help rescue abused, neglected, and abandoned animals. Our own lives have been greatly enriched by our shelter dogs—love and eternal thanks to Angela Gschwind of Carolina Loving Hound Rescue (for Sweeney) and Diane Hendrix of New Leash on Life (for Pandora)—ABK & JCK

Contents

I

EVIDENCES OF CREATIVITY

II

REQUIREMENTS FOR CREATIVITY

III

THE STRUGGLE FOR CREATIVITY

IV

PUSHING THE BOUNDARIES OF CREATIVITY

List of Contributors

Ian A. Apperly School of Psychology, University of Birmingham, UK

Alice M.I. Auersperg Department of Cognitive Biology, University of Vienna, Austria

John Baer Rider University, Education Department, Lawrence Township, NJ, USA

Sarah R. Beck School of Psychology, University of Birmingham, UK

Ronald A. Beghetto Department of Educational Psychology, University of Connecticut, Mansfield, Storrs, CT, USA

Mathias Benedek Department of Psychology, University of Graz, Graz, Austria

Robert M. Bilder Department of Psychiatry and Biobehavioral Sciences, David Geffen School of Medicine at UCLA, Los Angeles, CA, USA; Department of Psychology, David Geffen School of Medicine at UCLA, Los Angeles, CA, USA; Semel Institute for Neuroscience and Human Behavior, University of California, Los Angeles, CA, USA

Gordon M. Burghardt Departments of Psychology and Ecology & Evolutionary Biology, University of Tennessee, Knoxville, TN, USA

Josep Call Max Planck Institute for Evolutionary Anthropology, Leipzig, Germany; School of Psychology and Neuroscience, University of St. Andrews, St. Andrews, UK

Jackie Chappell School of Biosciences, University of Birmingham, Birmingham, UK

David H. Cropley Department of Engineering, Defence and Systems Institute (DASI), University of South Australia, Mawson Lakes, SA, Australia

Nicola Cutting School of Psychology, University of Birmingham, UK

Debora Cutuli Department of Psychology, Medicine and Psychology Faculty, University Sapienza of Rome, Rome, Italy; Laboratory of Experimental and Behavioral Neurophysiology, IRCCS Fondazione Santa Lucia, Rome, Italy

Paola De Bartolo Department of Sociological and Psychopedagogical Studies, Faculty of Formation Sciences, University "Guglielmo Marconi", Rome, Italy; Laboratory of Experimental and Behavioral Neurophysiology, IRCCS Fondazione Santa Lucia, Rome, Italy

Robert Epstein American Institute for Behavioral Research and Technology, Vista, CA, USA

Agnes Ferisa Orangutan Kutai Project, Kutai National Park, E. Kalimantan, Indonesia

Marie J.C. Forgeard Department of Psychology, University of Pennsylvania, Philadelphia, PA, USA; McLean Hospital, Harvard Medical School, Boston, MA, USA

Liane Gabora Department of Psychology, University of British Columbia, Vancouver, BC, Canada

Vlad Petre Glăveanu Aalborg University, Department of Communication and Psychology, Aalborg, Denmark

Beth A. Hennessey Wellesley College, Psychology Department, Wellesley, MA, USA

Jessica Hoffmann Yale Center for Emotional Intelligence, Yale University, New Haven, CT, USA

Samuel T. Hunter Department of Psychology, Pennsylvania State University, State College, PA, USA

Sarah Jaumann Department of Ecology, Evolution and Behavior, University of Minnesota — Twin Cities, St. Paul, MN, USA

Eranda Jayawickreme Department of Psychology, Wake Forest University, Winston-Salem, NC, USA

David J. Jentsch Molecular, Cellular and Integrative Physiology Interdepartmental Program, University of California, Los Angeles, CA, USA; Semel Institute for Neuroscience and Human Behavior, University of California, Los Angeles, CA, USA

Allison B. Kaufman Department of Ecology and Evolutionary Biology, University of Connecticut, Field Technician, Storrs, CT, USA

David S. Kaufman Department of Psychology, Saint Louis University, Saint Louis, MO, USA

James C. Kaufman Department of Ecology and Evolutionary Biology, University of Connecticut, Field Technician, Storrs, CT, USA

Kendra S. Knudsen Semel Institute for Neuroscience and Human Behavior, University of California, Los Angeles, CA, USA

Purwo Kuncoro Orangutan Kutai Project, Kutai National Park, E. Kalimantan, Indonesia

Kevin Laland University of St. Andrews, Behavioural and Evolutionary Biology, St. Andrews, Fife, Scotland, UK

Daniela Laricchiuta Department of Psychology, Medicine and Psychology Faculty, University Sapienza of Rome, Rome, Italy; Laboratory of Experimental and Behavioral Neurophysiology, IRCCS Fondazione Santa Lucia, Rome, Italy

Phyllis C. Lee Behaviour and Evolution Research Group, Psychology, School of Natural Sciences, University of Stirling, Stirling, UK

Janet Mann Department of Biology, Georgetown University, Washington DC, USA; Department of Psychology, Georgetown University, Washington DC, USA

Robert W. Mitchell Department of Psychology, Eastern Kentucky University, Richmond, KY, USA

Antonio C. de A. Moura Department of Engineering and Environment, Federal University of Paraiba, Rio Tinto, PB, Brazil

Ana Navarrete University of St. Andrews, Behavioural and Evolutionary Biology, St. Andrews, Fife, Scotland, UK

Weihua Niu Pace University, Dyson College of Arts and Sciences, New York, NY, USA

William J. O'Hearn Department of Ecology and Evolutionary Biology, University of Connecticut, Field Technician, Storrs, CT, USA

Eric M. Patterson Department of Biology, Georgetown University, Washington DC, USA

Irene M. Pepperberg Department of Psychology, Harvard University, Cambridge, MA, USA

Laura Petrosini Department of Psychology, Medicine and Psychology Faculty, University Sapienza of Rome, Rome, Italy; Laboratory of Experimental and Behavioral Neurophysiology, IRCCS Fondazione Santa Lucia, Rome, Italy

Karen Pryor Karen Pryor Academy, Watertown, MA, USA

Apara Ranjan Department of Psychology, University of British Columbia, Vancouver, BC, Canada

Roni Reiter-Palmon Department of Psychology, University of Nebraska Omaha, NE, USA

Sandra W. Russ Case Western Reserve University, Department of Psychological Sciences, Cleveland, OH, USA

Anne E. Russon Psychology Department, Glendon College, York University, Toronto, ON, Canada

Alcino J. Silva Department of Neurobiology, David Geffen School of Medicine at UCLA, Los Angeles, CA, USA; Department of Psychiatry and Biobehavioral Sciences, David Geffen School of Medicine at UCLA, Los Angeles, CA, USA; Department of Integrative Center for Learning and Memory, David Geffen School of Medicine at UCLA, Los Angeles, CA, USA

Dean Keith Simonton Department of Psychology, University of California, Davis, Davis, CA, USA

Emilie Snell-Rood Department of Ecology, Evolution and Behavior, University of Minnesota – Twin Cities, St. Paul, MN, USA

Daniel Sol CREAF (Centre for Ecological Research and Applied Forestries), CSIC (Spanish National Research Council), Bellaterra, Catalonia, Spain

John H. Stathis Connecticut College, Biology Department, New London, CT, USA

Eli Swanson Department of Ecology, Evolution and Behavior, University of Minnesota – Twin Cities, St. Paul, MN, USA

Emma C. Tecwyn School of Biosciences, University of Birmingham, Birmingham, UK; Department of Psychology, University of Toronto, CA

Susannah K.S. Thorpe School of Biosciences, University of Birmingham, Birmingham, UK

Oshin Vartanian University of Toronto Scarborough, Toronto, ON, Canada

Thomas B. Ward University of Alabama, Department of Psychology, Tuscaloosa, AL, USA

Stephanie A. White Department of Integrative Biology and Physiology, University of California, Los Angeles, CA, USA; Molecular, Cellular and Integrative Physiology Interdepartmental Program, University of California, Los Angeles, CA, USA

Thomas R. Zentall Department of Psychology, University of Kentucky, Lexington, KY, USA

Foreword

Within the field of creativity research, psychologists tend to stick to the study of humans. Biologists or ethologists usually either focus on animal problem solving or else consider creativity to be an evolutionary adaptation. However, much of human creativity theory can be applied to animal behavior. Creativity and innovation found in the animal kingdom appear to follow similar rules, constraints, and models to those in humans. There has been one outstanding edited book on the topic—Animal Innovation by Simon Reader and Kevin Laland (Oxford University Press, 2003)—but it was, with some exceptions, focused on work by animal researchers. Our vision was for a dialog between those who study creativity in animals and those who study the same topic in humans. We sought a wide variety of contributors and commenters, both in terms of fields (we span psychology, biology, neuroscience, engineering, business, ecology, and education) and location (the contributors come from 12 countries). There has not been a synthesized collection and exchange of ideas between the two communities until now, however, despite the clear benefits to understanding the benefits of creativity in both an evolutionary and cognitive sense.

The goal of this edited book is not only to gather the most cutting edge research on innovation in animals, but to also allow chapter authors some degree of freedom to extrapolate their findings toward what might be considered "anthropomorphism." Traditionally, this type of extrapolation in scholarly literature can bring academic scorn; however, researchers have recently begun to see some value in small amounts of personification of their animal subjects. As part of the crossover effort, we feature chapters by leading scholars in animal cognition and then commentaries by top creativity researchers. We are delighted to be able to present chapters and commentaries by some of our very favorite researchers in both areas.

We start with a section detailing exciting examples of creativity in the animal world. Irene Pepperberg discusses the accomplishments of Alex, the late and legendary African Gray Parrot, whose abilities led Ronald Beghetto to rethink much of the way he looks at creativity from a human perspective. Next, Robert Mitchell describes an excellent example of creative play between a person and a dog, in which Jessica Hoffman finds many parallels to creative play in children. Alice Auersperg discusses creative innovations in parrots and corvids, and

Beth Hennessey and John Stathis examine how compatible these are with human definitions of innovation and creativity. Lastly in this section, Eric Patterson and Janet Mann discussion innovation in cetaceans, which Vlad Glăveanuthen uses as a springboard to propose "proto-c" creativity.

The second section focuses on requirements for creativity. Gordon Burghardt begins with a discussion of how play leads to creativity, and Sandra Russ expands on these ideas. Daniel Sol, followed by Liane Gabora and Apara Ranjan, discuss possible evolutionary models of innovation in animals. Finally, two different chapters discuss neurological constructs of creativity—Laura Petrosini, Debora Cutuli, Paola De Bartolo, and Daniela Laricchiuta tackle the cerebellum, with a commentary by Mathias Benedek; and Kendra S. Knudsen, David S. Kaufman, Stephanie A. White, Alcino J. Silva, David J. Jentsch, and Robert M. Bilder take a lab based, cross-species approach, followed by thoughts from Oshin Vartanian.

Our third section, The Struggle for Creativity, discusses barriers to creativity and innovation which animals may encounter. The idea relationship between brain size and creativity is addressed by Ana Navarrete and Kevin Laland, with a commentary by Thomas Ward. Two teams, Jackie Chappell, Nicola Cutting, Emma C. Tecwyn, Ian A. Apperly, Sarah R. Beck, and Susannah K. S. Thorpe; and Phyllis C. Lee and Antonio C. de A. Moura, discuss how basic resources impact creative abilities and innovative products. Roni Reiter-Palmon and Marie J. C. Forgeard and Eranda Jayawickreme (respectively), then discuss how many of the same issues are present when humans engage in creativity and innovation. Thomas Zentall discusses the relationship between social learning and creativity; which is also evident in humans, according to John Baer. Lastly, Robert Epstein proposes generativity theory as an approach to understanding creativity, which Dean Keith Simonton then comments on.

The final section, Pushing the Boundaries of Creativity, begins with a discussion by Josep Call of how tradition and creativity clash; an issue all too common in humans as well, according to Weihua Niu. Anne E. Russon, Purwo Kuncoro, and Agnes Ferisa discuss how their definition of creative tool use continues to expand upon finding more examples among the population of orangutans they study, and then David Cropley discusses how new tools are similarly viewed among people. Emilie Snell-Rood, Eli Swanson, and Sarah Jaumann discuss creativity in species not normally considered to be creative, and Samuel Hunter applies the same theories to business and group dynamics. Finally, Karen Pryor discusses how we can teach and train creativity in animals and James C. Kaufman expands this to how we can do the same for people.

Acknowledgments

We would like to thank Nikki Levy for her support and encouragement throughout the process of writing this book. We would also like to thank our children, Jacob and Asher, and our menagerie of animals: Pandora, Kirby, Sweeney, Septimus, and Eliza.

PART I

EVIDENCES OF CREATIVITY

1

Creativity and Innovation in the Grey Parrot (*Psittacus erithacus*)

Irene M. Pepperberg

Department of Psychology, Harvard University, Cambridge, MA, USA

Commentary on Chapter 1: What Can Creativity Researchers Learn from Grey Parrots?

Ronald A. Beghetto

Department of Educational Psychology, University of Connecticut, Mansfield, Storrs, CT, USA

INTRODUCTION

Creativity and innovation imply the capacity to form something that, in general, has never, in any way, previously existed. Such a definition is, however, at the extreme end of the continuum of the meaning of these terms, which can also be applied to describe the amalgamation of existing forms into something novel, to novel ways of using existent forms, or even of inferring novel solutions to problems based on circumstantial information. The terms can, of course, refer to behavior patterns in art, music, science, policy—any topic imaginable. The focus of this book, however, is on nonhuman creativity and innovation; thus this chapter describes these abilities in Grey parrots (*Psittacus erithacus*).

Animal Creativity and Innovation.
DOI: http://dx.doi.org/10.1016/B978-0-12-800648-1.00001-2

3

The topics under consideration will involve both vocal and cognitive processes: novel combinations of human labels to describe novel items, issues of sound play, and aspects of insightful and inferential learning and behavior.

CREATIVITY AND INNOVATION IN THE VOCAL MODE

Considerable data exist to demonstrate that Grey parrots use elements of English speech referentially (Pepperberg, 1999). The question has been raised, however, as to whether these birds can, like young children (e.g., de Boysson-Bardies, 1999; Greenfield, 1991; Marschark, Everhart, Martin, & West, 1987; Peperkamp, 2003; Tomasello, 2003; note Gillen, 2007), intentionally demonstrate creative or innovate use of such speech. Such examples may involve using or recombining existing utterances in novel ways to describe novel objects, engaging in sound play to create novel utterances that could relate to items in the environment, transferring of concepts across domains, and segmentation—the ability to divide and recombine parts of existing labels to create novel utterances for novel situations. Such examples may involve novel ways of replicating models, but primarily entail demonstrating some level of newness. Although the data are not extensive, several examples exist to demonstrate that at least one Grey parrot was capable of several of these forms of creativity and innovation.

Sound Play: Combinations, Recombinations, Simple Extensions in the Presence of Trainers

The bird in question, Alex, had been trained via a modeling procedure (M/R training, Pepperberg, 1981; Todt, 1975) to identify several objects, colors, shapes, and common foods. The amount of practice and creativity involved in acquisition of his initial labels is unknown, as his solitary sound play was not taped until a decade later (Pepperberg, Brese, & Harris, 1991). Early on, however, he produced novel utterances—utterances he had not heard from humans—in his trainers' presence.

Some of these were quite simple combinations of existing labels, and were triggered by the presence of only slightly novel objects (Pepperberg, 1981). Thus, after being trained to use the labels "key," "wood," and "hide" to identify, respectively, a silvery key, uncolored tongue depressors and pieces of rawhide, and to combine these with the label "green" to refer to their colored counterparts, he was shown a green clothes pin. His label for clothes pins was "peg wood," and he immediately stated "green wood, peg wood." Similarly, after being trained on the label "rose hide" for red hide, he transferred to "rose paper" and then other red items without training. Green corks were also identified immediately, without training, as were blue hide and blue clothes pins after "blue" was acquired. The same was true after other color labels were learned. After training to label triangular and square pieces of maple as "3-corner wood" and "4-corner wood," respectively, he promptly labeled their paper and rawhide counterparts, although he needed to chew the items first, to identify their material. He later similarly transferred other shape labels. After learning the label "rock" by repeatedly querying his trainers about a lava-stone beak conditioner, he began to call the dried corn kernels in his food "rock corn," presumably to differentiate it from the fresh, soft variety he was also fed. He used "rock nut" to refer to a Brazil nut that he could not crack on his own, but his lack of sustained interest in the item led to the disappearance of the utterance. He labeled an unshelled almond "cork," possibly because the shell looked like one, but accepted our term "cork nut" to distinguish between the two objects (Pepperberg, 1990). Other examples of labels that were initially applied to objects of little interest and did not remain in his repertoire are discussed in Pepperberg (1990).

In a similar vein, other parrots in my laboratory creatively and spontaneously extended the use of labels trained with respect to specific exemplars to additional items. Thus my birds, somewhat like children (e.g., Brown, 1973), tested meanings of newly acquired labels; possibly they saw humans as a means of providing additional referential information (Pepperberg, 2002). For example, a younger subject, Arthur, in ways similar to Alex, uttered one of his newly acquired labels, used previously in a very specific context, in a novel situation—here, "wool," trained as a label for a woolen pompon, subsequently uttered while pulling at a trainer's sweater. He seemed to be testing the situation, and our responses—"Yes, WOOL!" given with high positive affect—stimulated him further, revealing the potential power of an utterance and encouraging his early categorization attempts. Notably, such behavior was not trained in any direct manner.

Creativity with other labels seemed to involve more complicated processes. At one point, we tried to train Alex to identify an apple; at the time, he knew "banana," "grape," and "cherry" for the other fruits he was fed. After undergoing considerable training via our modeling procedure, Alex ignored the targeted label and said "Banerry...I want banerry." He not only repeatedly used this term, but also slowed production and sharpened his elocution ("ban-err-eee"), much as trainers do when introducing a new label (Pepperberg, 1990). No one had ever used the term in his presence, but it did make some semantic sense, apples tasting a bit like a banana and looking like a very large cherry.

Alex also often produced new vocalizations in what appeared to be vocal play by recombining other existing label parts in their corresponding orders (Pepperberg, 1990). His spontaneous novel phonemic combinations would occur outside of testing or training, appearing in contexts and forms reminiscent of children's play (NB: other juvenile Greys behave similarly; reviewed in Pepperberg, 2004a; see also Pepperberg & Shive, 2001). These utterances were rarely if ever used by trainers, but sometimes resembled both existing labels and separate human vocalizations. Notably, these recombinations suggest that Alex, somewhat like children (see Weir, 1962; note discussion in Treiman, 1985), abstracted rules for utterances' beginnings and endings. In over 22,000 vocalizations, he never made backward combinations (e.g., never said "percup" instead of "cupper/copper"; Pepperberg et al., 1991). When we *referentially mapped* these spontaneous utterances—providing relevant objects to which they could refer—Alex rapidly integrated these labels into his repertoire. He acquired "grey," for example, after seeing himself in a mirror and querying "What color?"; he then produced sound variants (e.g., "grape," "grate," "grain," "chain," "cane") that we mapped to appropriate referents (respectively, fruit, a nutmeg grater, seeds, a paper-clip ring, sugar cane; see Pepperberg, 1990, 1999). With the exception of "grape," which trainers had consistently labeled, these labels had never been produced in his presence. He promptly began to use them consistently to refer to the items. In contrast, he abandoned sounds whose combinations we could not map (e.g., "shane," "cheenut"), or for which mapped referents were not of interest (e.g., the dried banana chips we used for "banacker"; Pepperberg, 1990). Thus, Alex's spontaneous utterances that may initially have lacked communicative, symbolic value could, as they do for children, acquire this value if caretakers interpreted them as such (Pepperberg, 1990): that is, although his initial combinations could have simply been "babble-luck" (Thorndike, 1943), he behaved as if our interactions "conventionalized" both the sound patterns and sound-meaning connections in the direction of standard communication. Likely, after the first such instance, the excitement of the trainers and the introduction of a novel item may have impelled Alex to attempt more innovative vocal play.

Alex created very few action-related phrases, but the few he did form bear mentioning. To fashion most action-related variants, Alex excerpted part of a standard phrase, which he then used as a request; just as in referential mapping for labels, trainers responded *as though* he had made an intentional demand (Pepperberg, 1990). Thus, during a long training session, Alex uttered "go chair"—a portion of the (by then) commonplace phrase "wanna go chair" (see the section "Novel Combinations in the Service of Vocal Learning"). The trainer sat, temporarily discontinuing the session. Alex subsequently used this utterance to call a halt to whatever was happening. He also excerpted "go away," which he used to break contact with trainers. This phrase was closely related to two situations in which it was first routinely used in his presence. The first situation occurred when he refused to comply with our requests for object labeling: he routinely said "no," turned or walked away, and began to preen. We responded with a brief time-out, accompanied by the phrase "I'm gonna go away!" He subsequently asked ("go away") that the trainer depart. The second situation involved scheduled departures of trainers (e.g., at lunch or the end of the day). At those times, we added "I'm gonna go eat lunch/dinner," and placed Alex inside his cage. Interestingly, during long or difficult sessions, Alex began to request students to "go away," or occasionally stated "I'm gonna go away," after which he climbed off his training chair. Generally, attempts to continue training after such a declaration were fruitless.

Novel Combinations in the Service of Vocal Learning

My students and I had, by eavesdropping, found that Alex sometimes engaged in monologue speech in the absence of his trainers (Pepperberg & Matias, unpublished observations), seemingly playing with the sounds in his repertoire while he was in the process of acquiring new labels, as though he was practicing in private before attempting the labels in public. Note that Alex always experienced negative consequences for erring in his trainers' presence: trainers scolded him and removed training objects from sight. Such negative events did not occur during his private monologues. Might he, like children (e.g., Weir, 1962), have deliberately engaged in innovative practice and play, in a consequence-free environment? Might the absence of negative feedback or the need to use correct semantic, syntactic and pragmatic forms during monologues have encouraged practice, creativity, innovation and accelerate learning? Some researchers argue that such is indeed the case for children (Kogel, Dyer, & Bell, 1987; Krashen, 1976; Rice, 1991; note Salmon, Rowan, & Mitchell, 1998; see also Winsler, Feder, Way, & Manfra, 2006).

My students and I (Pepperberg et al., 1991) thus began to audiotape Alex surreptitiously, before trainers arrived in the morning and after they left at night, often during periods of either natural or artificial low light ("dawn" and "dusk"). We found that Alex indeed engaged in a variety of sound play related to labels that were being trained. So, for example, during training of "nail", he produced utterances such as "banail," "benail," "blail," "chail," "jemail," "lemail," "lobanail," "loobanail," "mail," "shail," "wgail." Although these combinations of phonemes did not seem to be a deliberate attempt to create a new label from specific sound patterns that resembled the target, but rather deliberate play within a range of existent patterns in an attempt to hit on a correct pairing that matched some remembered template, the behavior did demonstrate creativity. That is, Alex's behavior demonstrated an understanding of the combinatory nature of his utterances, but did not necessarily show that he understood, at least at that time, how to segment the novel targeted vocalization exactly, then match its components to those in his repertoire in order to create a trained label.

Such deliberate behavior did exist, however, many years later, in two forms. Both were the outcome of a study on phonemes, undertaken as a consequence of the previous data that suggested that Alex did in fact understand that labels are made of some kinds of subunits that can be used in different ways. Both also involved very few examples, but were nonetheless important additions to our understanding of Alex's innovative behavior.

My students and I began by training Alex to sound out phonemes to see the extent to which he understood that his labels were indeed made up of specific sounds that can be combined in different ways to make new labels (Pepperberg, 2007). We used wooden or plastic refrigerator letters, each a different color for each trial; Alex was taught the sounds of the different letters or letter combinations. He would be shown a tray on which several differently colored letters and letter combinations were placed and be asked, for example, "What color is 'SH'?" or "What sound is purple?" His reward for a correct answer was the chance to chew apart the letter or request some other reward. He had become quite proficient, and my students and I were asked to demonstrate his competence for a number of visiting dignitaries at the MIT Media Lab, where we were in residence at the time. I showed Alex a tray of his letters, asking several questions like those used in training. Alex was correct and consistently stated "Wanna nut" for his reward. We were under strict time limitations, so I refused to provide the nuts. Each of Alex's requests became a bit more emphatic until he finally said, slowly and seemingly deliberately, "Want a nut. Nnn…uh…tuh!" One important point was that he had been trained on "N" and "T," but not on "U"; he therefore had figured out how to parse the label without any training and, again, without any training, produce the appropriate phoneme. The second point was that he had seemingly assumed that, to receive the reward at this time, he would have to relate his request to the phonemic task at hand.

Alex also demonstrated unexpected abilities with respect to sounds and labels after our younger bird, Arthur, had acquired the label "spool" to refer to plastic and wooden bobbins (Pepperberg, 2007). Interestingly, given the difficulty of producing /p/ sans lips, Arthur used a whistle-like sound for the first part of the label (for sonagrams, see Pepperberg, 2007). Arthur did, however, follow the usual form of acquisition: labels usually appeared in sessions initially as rudimentary patterns—first a vocal contour, then with vowels, finally with consonants (Patterson & Pepperberg, 1994, 1998) and Arthur's production indeed began with his uttering /u:/ ("ooo") and ending with a distinct, fully-formed "spool" (/spu:l/; see figure 1A in Pepperberg, 2007; the IPA transcription is approximate because of the whistle for /p/).

Unlike Arthur, and unlike this usual form of acquisition, Alex, during training after watching Arthur playing with the object, began using a combination of existing phonemes and labels to identify the object: /s/ (trained independently in conjunction with the physical letter, S) and wool, to form "s" (pause) "wool" ("s-wool"; /s-pause-wUl/; figure 2 in Pepperberg, 2007). The pause seemed to provide space for the absent (and difficult) /p/ (possibly to preserve the number of syllables or prosodic rhythm cf the targeted vocalization; see Leonard, 2001; Peters, 2001). Alex's repertoire did not include any labels that contained /sp/, nor could he utter "pool" or "pull," or any other label that included /Ul/; he did, however, know "paper," "peach," "parrot," "pick," and so forth. He knew /u/ from labels such as "two" and "blue" (Pepperberg, 1999, 2007). Note that both Alex's and my /p/, when analyzed for VCT (voice onset time), fall solidly into the voiceless category, distinct from the voiced /b/; analyses suggested that Alex produced a viable /p/ via a form of esophageal speech (Patterson & Pepperberg, 1998), and /sp/ may have been even more difficult. He retained this "s-wool" formulation for almost a year of M/R training, with no change whatsoever in the form of his production, although normally only about 20—25 modeling sessions (at most, several weeks of training) are sufficient for learning a new label (Pepperberg, 1999).

At the end of this year-long period, Alex spontaneously produced "spool," perfectly formed (/spul/; see figure 3 in Pepperberg, 2007). Thus, Alex added the sound—which humans heard, sonagraphically viewed, and transcribed, as—/p/ and also shifted the vowel toward the appropriate /u/ sound; that is, his vowel also changed (for sonographic evidence, see Pepperberg, 2007). Alex's utterance sounded distinctly human, differed from Arthur's whistled version, and clearly resembled mine (Pepperberg, 2007), even though students had performed 90% of the training.

The pattern of acquisition was not unique to "spool"; Alex exhibited a similar pattern for the vocalization "seven" (first in reference to the Arabic numeral, then to an object set; Pepperberg & Carey, 2012). His first production of the label could best be described as "s...n", a bracketing using the phonemes /s/ and /n/; he then quickly progressed to "s-one" (Pepperberg, 2009; /s/-pause-/wǝn/) which looked sonagraphically quite different from my "seven," but followed the form of "s-pause-wool." Eventually, he replaced "s-one" with something sounding to the human ear like "sebun," much closer to my "seven" (Pepperberg, 2009).

Alex's data thus demonstrate a functional understanding that his existent labels were comprised of individual units that could intentionally be recombined in novel ways to create referential, novel vocalizations (Pepperberg, 2007, 2009). Although he seemingly generated novel meaningful labels from a finite set of elements, the rule system he demonstrated was relatively limited. Nevertheless, the data add another intriguing parallel between Alex's and young children's early label acquisition.

Inferential Vocal Learning, Including Transfer Across Domains

Alex also inferred how he could use existing vocalizations in novel ways, both to exert some control over his environment and to demonstrate an expanded understanding of the concepts on which he was being trained. The former, lexical substitution in sentence frames and extension of labels across categories, could be seen as a relatively simple innovation (e.g., Peters & Boggs, 1986). The latter often involved transferring the use of labels from one concept to another, a considerably more complex and creative cognitive task.

We found that after learning the simple, but utilitarian phrases, "I want X" and "Wanna go Y" with respect to a limited number of items—paper and cork for the former, gym and back (to his cage) for the latter—Alex spontaneously replaced these Xs and Ys soon after learning new labels for objects or locations (Pepperberg, 1988a). Thus, for example, we never trained "Wanna go shoulder" or "Wanna go knee"; my students and I simply trained the labels for those locations. Similarly, after learning the label of any novel desirable item, he would be capable of requesting it for play or ingestion. Clearly Alex had figured out how novel labels could be combined with existing phrases to enable him to communicate his needs and wants to his trainers.

Alex's abilities to transfer labels to demonstrate his understanding of novel concepts was considerably more advanced. So, for example, Alex had been trained to view two objects and respond to questions of "What's same?" and "What's different?" with the labels "color," "shape," or "mah-mah" (matter), including pairs whose colors, shapes, and materials he could not label (Pepperberg, 1987). He was then taught to use the label "none" if nothing within the pairs was same or different; that is, if they were totally identical or completely different (Pepperberg, 1988b). He was then taught concepts of bigger-smaller: to view a pair of objects and respond to the queries of "What color bigger?" or "What color smaller?" (Pepperberg & Brezinsky, 1991). When, for the first time, he was asked about a pair that was equal in size, he transferred, without any training, his use of "none" from the study on same-different in order to answer correctly (Pepperberg & Brezinsky, 1991). Such ability to transfer is a mark not only of complex cognitive processing (see Rozin, 1976) but also of innovation and creativity.

He also transferred his use of "none" in another, quite striking, manner: a spontaneous, untrained transfer to describe a nonexistent numerical set of objects; that is, a zero-like concept. The behavior occurred during a study of number comprehension (Pepperberg & Gordon, 2005), in which he was shown heterogeneous sets of objects, either three differently colored, different numerical sets of the same object (e.g., three red jelly beans, four green jelly beans, and five yellow jelly beans, of different sizes) or three different objects of the same color (e.g., two red corks, five red keys, and six red wooden sticks), all scattered in an intermingled manner randomly on a tray and asked, for example, respectively, "What color (is the set of) four?" or "What object (is the set of) six?"

On the 10th trial within the first dozen, Alex was asked "What color three?" to a set of two, three, and six objects. He replied "five." Obviously, no set of five items existed, and the questioner asked him the original question twice more and each time he replied "five." The questioner finally said "OK, Alex, tell me, what color five?" Alex immediately responded "none" (Pepperberg & Gordon, 2005). As noted above, Alex had been taught to respond "none" with respect to same or different (Pepperberg, 1988b), and he had never been taught the concept of absence of quantity or to respond to absence of an exemplar. Nevertheless, he had not only provided the correct response but also set up the question himself. We had not trained the conventional term, *zero*, to indicate the absence of quantity; Alex's use of "none" for this purpose was unexpected. Students and I thus repeated the question randomly throughout other trials with respect to each possible number, to ensure that this situation was not an odd happenstance. On these *none* trials, Alex's accuracy was 5/6 (83.3%; $p < 0.01$, binomial test, chance $=1/4$ [the three relevant color labels plus *none*]). His one error, interestingly, was to label a color not on the tray (Pepperberg & Gordon, 2005).

His use of "none" to denote absence of a particular quantity was impressive for at least four reasons. First, labeling a null set, whatever the means, has an intriguing human history: Bialystok and Codd (2000, p. 119) question whether early human cultures could even represent zero as a quantity. That zero was represented in some way by a parrot, with a walnut-sized brain whose last common ancestor with humans was likely 280 million years ago (e.g., Petkov & Jarvis, 2012) is striking. Second, the notion of *none*, even if already associated with other concepts (Pepperberg, 1988b; Pepperberg & Brezinsky, 1991), is abstract and relies on violation of an expectation of presence (Pepperberg, 1988b); Alex's transference of the notion to quantity, without training or prompting by humans, was unexpected. Third, children's comprehension of zero—none generally lags behind comprehension of other small numbers (Wellman & Miller, 1986), and one might expect parrots to behavior similarly; nevertheless, Alex deliberately used "none" in a number comprehension task. Finally, and likely most important, we did not initiate the topic. He insisted on stating "five" when asked about "three"; when asked about the non-existent "five," he responded appropriately (Pepperberg & Gordon, 2005).

CREATIVITY AND INNOVATION IN CONCEPTUAL LEARNING

Grey parrots can demonstrate innovation and creativity in other ways that are not directly involved in their vocal abilities; as with those involved in vocal learning, they can entail various forms of inference. Some were related to numerical competence and were also a part of the studies with Alex, but others were in totally different domains and with other Grey parrots.

Inference in Numerical Concepts

Alex had, as noted earlier, been trained on concepts of bigger-smaller (Pepperberg & Brezinsky, 1991) and had also learned to identify numerical sets of up to six items, even if they were part of heterogeneous collections (Pepperberg, 1994). He had separately been trained to identify Arabic numerals 1–6 with the same vocal English labels but not to associate Arabic numbers with their relevant physical quantities. Subsequently, without any additional training, he was shown (i) pairs of Arabic numbers or (ii) an Arabic numeral and a set of objects and was asked for the color of whatever was bigger *or* smaller (Pepperberg, 2006b). He succeeded on the first task, demonstrating that he (i) understood number symbols as abstract representations of real-world collections, (ii) deduced that an Arabic symbol has the same numerical value as its *vocal label*, and thus the relationship between the Arabic number and the quantity, and (iii) had also figured out, again without any training, the ordinal relationship of his numbers from their cardinality. To succeed, he specifically had to compare a *representation* of quantity for which the label stands, infer rank ordering based on these representations, then state the result *orally*. Importantly, specific stimuli within pairs would not, unlike in other nonhuman studies, have been associated with reward of the corresponding number of items (Wynne, 1992). He also succeeded on the second task (when the comparison involved a single Arabic numeral and a set of physical objects), failing only if the set of objects involved a single item; in that case he consistently answered "none." If given two Arabic numerals representing the same quantity and asked about their relative values, he responded "none." Interestingly, he also inferred, without any training or differential reward, that he should respond "none" when given Arabic numerals representing identical quantities but of different sizes (e.g., "2" vs. "2"); thus he eventually answered on the basis of the representation of the symbols and not on their physical dimensions. Such behavior clearly involved various levels of innovation and creative thinking.

Another experiment, involving the summation of quantities of total value of up to six, was unplanned, but instituted after Alex demonstrated innovative behavior (Pepperberg, 2006a). My students and I had begun a sequential auditory number session (training to respond to, for example, three computer-generated clicks with the vocal label "three") with another Grey parrot, Griffin, by saying "Listen," clicking (this time, twice), and then asking "Griffin, how many?" Because Griffin refused to answer, we replicated the trial. Alex, who often interrupted Griffin's sessions with phrases such as "Talk clearly" or who occasionally gave the answer even though he was not part of the procedure, said "four." I told him to be quiet, assuming his vocalization was not intentional. We then replicated the trial yet again with Griffin, who remained silent; Alex now said "six." He had never been trained to sum auditory sets. Given the issues involved in controlling cuing in tests for auditory summation (Pepperberg, 2006a), I decided to replicate the object addition study that Boysen and Berntson (1989) had performed with chimpanzees. Alex was presented with a tray on which various sets of small objects (e.g., differently sized jelly beans or pieces of crackers) had been covered with identical plastic cups. The experimenter brought the tray up to Alex's face, lifted the cup on Alex's left, showed him what was under the cup for a few seconds, and then replaced the cup over the quantity; the procedure was replicated for the cup on Alex's right. The experimenter made eye contact with Alex, who was then asked, vocally, and without any training, to respond to questions, such as "How many beans total?" No addends were visible during questioning and Alex was required to answer with a vocal English number label. To respond correctly, he had to remember the quantity under each cup, perform some combinatorial process, and then produce a label for the total amount. His accuracy suggested that his addition abilities were comparable to those of nonhuman primates and young children. Again, the entire experiment was based on an innovative behavior, and no training was involved.

Subsequently, he was able to transfer, without any training, to the summation of two Arabic numerals or three sets of objects hidden under cups, up to a total value of eight (Pepperberg, 2012). In a few trials in the Arabic numeral addition study, he was also shown variously colored Arabic numerals, placed above the cups on the tray, while the addends were hidden after their brief presentation, and asked "What color number (is the) total?" Without training, he demonstrated that he could use his knowledge of the referents of two different Arabic numerals, sum them mentally, and then represent the sum as either a vocal numerical label or the color of an appropriate numerical symbol.

Alex was also capable of inferring the cardinality of novel numbers from their ordinal placement on a number line (Pepperberg & Carey, 2012), a task so far not replicated with other nonhumans. Here he was trained to label vocally the Arabic numerals 7 and 8 in the absence of any sets of objects corresponding to their value and then to order these Arabic numerals with respect to the numeral 6, whose value was already known (Pepperberg, 1994, 2006b). He subsequently inferred the ordinality of 7 and 8 with respect to the smaller numerals and, critically, although 7 and 8 might have represented any of the larger integers (e.g., he was not given any reason not to infer they meant, say, 10 and 20), he inferred use of the appropriate label for the cardinal values of 7 and 8 items. These data suggest that he constructed the cardinal meanings of "seven" and "eight" from his knowledge of the cardinal meanings of one through six, together with the place of "seven" and "eight" in the ordered count list, inferring that each number on the list represents one more than the previous number, much as do normal young children (Carey, 2009).

Innovation in the Physical Realm

The ability, without training, to obtain food suspended by a string by reaching down, pulling up a loop of string onto a perch, stepping on the loop to secure it, and repeating the sequence several times (e.g., to demonstrate an understanding of intentional means-end behavior; see review in Willatts, 1999) has often been used to assess "insight" in avian species; simply reaching down for the food is not sufficient (Funk, 2002; Heinrich, 1995). Though some researchers (e.g., Taylor et al., 2010) argue for an alternative interpretation based on a positive perceptual-motor feedback cycle—that is, a form of associative learning, based on the reward coming closer as the string is being pulled—the point is that the initial act, which has no antecedents, still must involve some form of innovation. Two Grey parrots, Kyaaro and Arthur, when given the standard task of a suspended desired treat, immediately performed the targeted action, without any form of trial-and-error learning or training (Pepperberg, 2004b). Arthur, when given a choice between two suspended items, one a very desirable food and the other a toy in which he had never shown much interest, immediately went for the food. Two other parrots, the aforementioned Alex and Griffin, however, refused even to try, instead responding by looking at the trainers and saying "Want nut." Interestingly, neither Arthur nor Kyaaro had more than a couple of English labels in their repertoire, whereas Alex had an extensive vocabulary (Pepperberg, 1999) and Griffin had at least a dozen labels and the same commands as Alex (Pepperberg, Gardiner, & Luttrell, 1999; Pepperberg, Sandefer, Noel, & Ellesworth, 2000; Pepperberg & Shive, 2001). Kyaaro and Arthur, lacking any alternatives, solved the problem physically. Although Alex and Griffin seemingly failed the test of insight, one might argue that their attempt to direct humans to assist them in achieving their goals was equally innovative (Pepperberg, 2004b).

Inferential Abilities: Exclusion

Inference by exclusion is a fairly advanced cognitive process: the individual must derive knowledge about a situation by excluding other alternative explanations. It can be tested in the linguistic domain, whereby one decides that the novel of two items on the table must be a "blurgle" simply because it is not the well-known "ball" (Carey, 1978). Similarly, exclusion can be tested in the physical domain, whereby one decides that a treat must be hidden in one of two containers because the other has been shown to be empty (e.g., Premack & Premack, 1994). Although these tasks might seem somewhat trivial, and arguments can be made that exclusion is not innovative but rather simply learning to avoid something in the absence of an appropriate cue (i.e., the correct label or the presence of a treat), such tasks can be solved by children only when they reach an appropriate developmental age (Hill, Collier-Baker, & Suddendorf, 2012). Interestingly, the Grey parrot Griffin demonstrated a simpler form of linguistic exclusion (Pepperberg & Wilcox, 2000), and he as well as Arthur and several other Grey parrots showed understanding of the physical form (Mikolasch, Kotrschal, & Schloegl, 2011; Pepperberg, Koepke, Livingston, Girard, & Hartsfield, 2013; Schloegl, Schmidt, Boeckle, Weiss, & Kotrschal, 2012). I thus will describe such abilities even if the level of innovation may be questionable.

Griffin's linguistic form involved his acquisition of color terms. Children learn such terms not only by having their caretakers label various hues independent of objects, but also by exclusion; that is, by being asked to "Bring me the *red* ball; no, not the *blue* one, the *red* one." Supposedly, a child is thereby led to analyze the object for which it has a label—ball—and additional aspects of the environment and input for another property to which the additional, unknown label could be applied, for example, color or shape (Markman, 1990). Such analysis thus not only allows second labels to be associated with an item, but also drives acquisition of the concept of attribute labels. To distinguish this latter analysis from mutual exclusivity proper, I have labeled it *attribute deductivity* (Pepperberg & Wilcox, 2000).

Griffin's training on colors was quite different from that of Alex (Pepperberg & Wilcox, 2000). Alex learned his color labels in a manner unlike that of most children. Unlike children, most of the objects for which he initially acquired labels were uncolored—e.g., white paper, wooden sticks, untinted rawhide, silvery keys. When it came time to teach colors, we used two objects of a single color, and introduced a compound label, for example, "green wood" and "green key." Thus, we structured training so Alex was unlikely to assume that "green" was an *alternate* or second label for the same object, but rather that it was an *additional* label for an item that had some additional quality, and he was required to produce both the object and color label when queried "What's this?" After he learned "green" we introduced other color labels ("rose," "blue") in the same manner. As noted above, he quickly transferred his color labels to other items without training. Griffin, however, was exposed to objects of various colors during his initial label acquisition; he therefore had to recognize that "key" was a common term to describe variously colored and shaped keys, in and of itself a deductive process. Then, when it was time to teach colors, Griffin was first trained on color labels ("rose," "green," "yellow," "orange," "blue") by humans using a computer display (colored circles on a white background), a process totally unlike that used for Alex. Only after Griffin could produce the color labels did we query him about various real-world objects. Initially, Griffin reacted as do children who, having learned, for example, to refer to their pet as a *dog*, will insist, "It's not an *animal*, it's a *dog*" (Merriman, 1991)—that is, who believe that the terms *dog* and *animal* are mutually exclusive. Thus, when Griffin was asked "What color?" for an item for which he already had a label (*wool*), he at first ignored the query, rejected the color label, and responded "wool." Over time, he learned the relationship between the hierarchical term "color" and the specific color labels, for "shape" and his shape labels, and how they related to real-world objects. Although he could quickly make one-to-one associations between novel objects and novel labels, he initially persisted in using his first (object) label when queried about anything with respect to an object, and had to engage in a subsequent deductive process to realize that specific types of queries and the associated second labels involved object attributes (Pepperberg & Wilcox, 2000).

Grey parrots are also proficient in physical exclusion tasks. In a study by Mikolasch et al. (2011) that tightly controlled for issues such as avoidance and simple associational learning, at least one Grey parrot exhibited an understanding of physical exclusion. My students and I decided to repeat that study with Griffin, to test whether some level of experience with linguistic exclusion would affect the results; for comparison we also tested the aforementioned Arthur, who lacked such experience, and two Greys who were "pet" birds (Pepper and Franco), but who lived with humans who had formerly been trainers in our laboratory and who could be relied upon to carry out trials the same way as in the laboratory (Pepperberg et al., 2013). Given that these four birds were reared in, and lived in, completely different conditions from Mikolasch et al.'s (2011) zoo-based Greys, that two were pets and two lived in a laboratory, our findings would examine the extent to which some degree of reasoning by exclusion is widespread among Greys, whatever their history and level of enculturation with humans. We also performed a subsequent experiment (see the section "Creative Thinking: Transfer Across Dimensionality and Optical Illusions"), designed to test exactly what Grey parrots remembered about the hidings and understood about the conditions of the task.

Based on the original Premack and Premack study (1994) with apes, we hid two different but equally desirable items under two differently colored opaque cups; subsequently the parrots were allowed to witness an experimenter eat one of the items, which had been removed secretly (invisible condition) or in full view (visible condition), or watched an experimenter simply handle an object identical to one that was hidden (association control), and were then allowed to choose one cup. To succeed, the birds would not only have to track the actions of the experimenter in order to avoid the empty cup (the visible condition), but also infer, based on what had been eaten in their presence and what they remembered about the initial hiding (which treat was under which cup), where to search for the remaining treat (invisible condition). The control condition ensured that the birds were not simply avoiding the position that corresponded to the object that the experimenter had handled. Controls for olfactory and possible human cuing were put in place, and interobserver reliability was checked via multiple observers and videotaping (Pepperberg et al., 2013).

All four birds succeeded at equal levels of accuracy (Pepperberg et al., 2013). Not only did all four birds succeed on the visible task, performing at a statistically significant level on 30 trials, but the birds all were also correct on their very first trials, demonstrating that learning was not involved. More importantly, all four birds also succeeded at a statistically significant level on the 30 invisible trials, and all but Griffin were successful on their very first trials; Griffin, however, made only that one mistake on his first five trials. Again, learning was not involved; the birds had deduced the appropriate responses immediately, unlike very young children (Hill et al., 2012). Birds were random on association trials, demonstrating that they were not distracted by the handling of the hidden treats, and no bird was distracted when we tested for the effects of unequal handling of the cups (local enhancement). Thus, the birds seemed capable of true inferential reasoning, and—unlike the effects in the string-pulling task described above—experience in other laboratory tasks did not affect the results.

We next wanted to determine exactly what the birds knew about what was under each cup (Pepperberg et al., 2013). We wanted to see if both sets of birds conceivably might have focused on the lack of removal of *an* object in the previous study, not on a specific object. Thus, we wanted to determine if the bird would still be willing to return to a cup from which a favored reward was partially removed in contrast to a cup that held something to which they were indifferent. Success in these so-called "Favorite" trials would show that the birds were truly focusing on what was under each cup, and, when intermixed with additional Experiment 1 trials, provide another control that their visible and invisible trials were indeed based on inference. Another goal, however, was to test whether the birds knew when to use and when not to use inference by exclusion to solve the problem, a further test of creativity.

In all Favorite trials, birds saw treats such as two jelly beans (very desirable items) placed under one cup and two pieces of parrot chow (which they would eat, but did not favor) under another. Here the experimenter removed either one piece of chow or a bean, invisibly or visibly, and then the birds were allowed to choose. If they were reasoning appropriately, they should always choose the beans, as even one bean is a much better reward than two pieces of chow. To ensure that the birds were not being trained on this task, we also included trials from the first experiment, so that birds really had to attend to the hidings (both the identity and the number of treats and their placement) in order to succeed (Pepperberg et al., 2013). These trials were an additional control for the possibility that the birds had learned to avoid the cup containing what the experimenter had recently eaten, but only when they remembered that multiple items had been stored and inferred what was left. Again, no training preceded this task, so the birds had to demonstrate that they could solve a novel problem.

All four birds succeeded at a statistically significant level on the Favorite trials, 30 visible and 30 invisible, demonstrating that they did not simply avoid a cup from which something had been removed. Three birds (Griffin, Arthur, and Pepper) also appeared to be able to switch between experimental conditions, that is, to track the number of treats, track where removal had occurred, and, on that basis, understand something about when to use or to ignore exclusion. Notably, they inferred the most advantageous choice based on the specific context of the trial.

Creative Thinking: Transfer Across Dimensionality and Optical Illusions

Additional studies with Griffin, this time on optical illusions, demonstrate other ways in which he used his vocal abilities to transfer across domains and demonstrate some level of creative labeling. In these studies, our goal was to determine if Griffin could use his ability to label five various colors and six different shapes with respect to three-dimensional (3D) objects in order to recognize two-dimensional (2D) representations of various optical illusions. Could he still recognize a 2D polygon even if a part was occluded (amodal completion, Figure 1.1A), and could he identify 2D shapes that really did not exist, but were defined by arrangements of pac-men (Kanizsa figures, or modal completion, Figure 1.1B)? To succeed, he would, without any training, first have to infer that 2D line drawings were representations of 3D objects, and then infer the existence of entire figures based on observable parts. All these tasks would be strong tests of his creative approaches to problem solving.

As discussed above, Alex had shown he could transfer his color and shape labels to novel instances of materials after being trained on a small subsample of physically manipulable items, but had never been required to transfer to pictorial representations. The extent to which nonhumans can actually do such a task is unclear. Most studies involve testing the ability to transfer between photographs and real objects (e.g., Bovet & Vauclair, 2000; Dittrich, Adam, & Güntürkün, 2010; Gardner & Gardner, 1984);

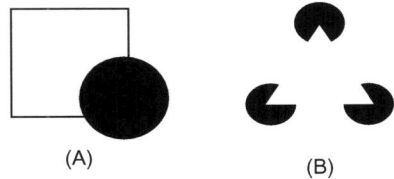

(A)   (B)

FIGURE 1.1   (A) Occluded and (B) illusory objects.

rarely has a subject been required to transfer from a 3D figure to a simple line drawing. Nevertheless, Griffin demonstrated that he could transfer without any explicit training (Pepperberg & Nakayama, 2012). After many trials in which he had correctly identified the shape of numerous 3D items, he began to exhibit perseverance errors, possibly out of boredom (a somewhat common occurrence in Grey parrots during repetitive questioning; see Pepperberg & Carey, 2012). He began calling all items "6-corner," whatever their shape. To try to break the boredom factor, we introduced 2D items. Griffin initially transferred his perseverance to 2D figures, but after only 13 trials (involving questioning without trainers providing the correct responses), he began to label 2D items of any color and shape appropriately.

For amodal completion, portions of variously colored regular poly-gons were occluded by black circles (which Griffin was not trained to label) or other black polygons (Figure 1.2C and L); the questions were "What shape (is the item of) color-X?" Controls were colored polygons missing appropriate pieces and black occluders appropriately displaced (Figure 1.2D and H). An oval or circular shape was cut around each figure so the shape of the paper ("4-corner") would not be a factor.

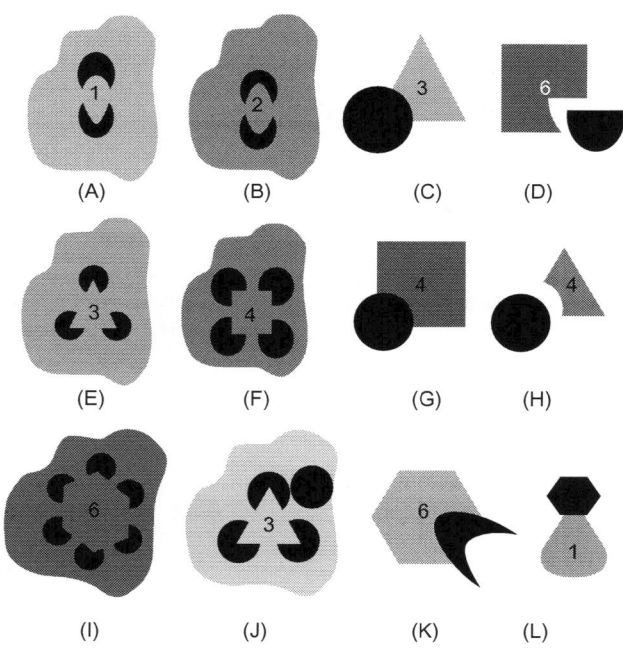

FIGURE 1.2 Samples of Kaniza figures, occluded figures, and controls presented to Griffin.

For modal completion, Kanizsa figures were constructed using black "pac-men" to form regular polygons on colored paper (Figure 1.2A). The same question was used as for amodal completion. Controls involved placing additional circles or "pac-men" near the Kanizsa figure so Griffin could not simply quantify black objects (Figure 1.2J). Stimuli were produced on a color laser printer, and here a random shape was cut out around the target stimuli so that the shape of the paper ("4-corner") would not be a factor (Figure 1.2). The colors used were those tested in a previous study (Pepperberg, Vicinay, & Cavanagh, 2008) to ensure that Griffin would label these printed colors correctly. Although Griffin could label octagons, these figures were not tested because we could not devise appropriately-sized stimuli comparable to the other possible shapes. Stimuli presentations were randomized with respect to modal/amodal task, shape, and color, such that no more than two of these three variants could be presented successively.

Griffin's accuracy was high, at 77% (27/35 trials) for modal completion and 74% (28/38 trials) for amodal completion. Chance could technically be 1/6 because Griffin could chose from six possible shape labels from which to respond (he did not know that we would never question him on "8-corner"), but we chose to use 1/5 because he was actually tested on five different possible shapes. In either case his data would be statistically significant (on a binomial test, $p \ll 0.01$ for both sets of completion questions). Even if we chose to use chance of $1/2$ on the amodal task, proposing that he should respond either with the label of the occluded shape or give the number of visible corners of the actual object, his data are still statistically significant ($p < 0.01$).

Notably, because Griffin had previously been trained to state "x-corner" (the actual number of corners) when asked about a shape, one might have expected him, in the amodal completion trials, to have been more likely to respond to the number of visible corners than to the occluded shape; however, such was not the case, except as expected in controls (Figure 1.2D and H), where he responded to figures never before seen, with the numbers of corners. Thus he inferred that, to be correct, he was to respond to the visible number of corners *only* on a non-occluded trial, that is, when faced with these novel, non-regular objects. Of note is that he was correct on the very first trial with amodal stimuli, which meant he was able to transfer his response from the non-occluded 3D items with which he was familiar, to the novel occluded 2D stimuli used in the task.

For modal completion (Kanizsa figures, illusory contours), Griffin again made few errors; he was correct on 27 of 35 trials. He also was correct on his very first trial, showing yet again that he could recognize these 2D stimuli without any training. Controls ensured that he was not labeling the total number of corners or the number of pac-men, but rather that he understood the nature of the task. Thus he had to infer when the pac-men and circles were distractors and when they were relevant to the task.

CONCLUSIONS

In sum, Grey parrots are capable of using their cognitive abilities in innovative and creative ways in order to solve tasks that differ considerably from the training trials they receive. Although some of these tasks have, as discussed above, been interpreted by other researchers as requiring low-level capacities for their performance, not all aspects of these tasks can be described in that manner. Furthermore, when taken as a whole, the entire corpus of material presented here goes far beyond simple interpretations (e.g., associative behavior), and requires considerable transfer across tasks and even domains.

Acknowledgments

Writing of this chapter was supported by donors to *The Alex Foundation*, particularly the Sterner Family, the Marc Haas Foundation, and Anita Keefe.

References

Bialystok, E., & Codd, J. (2000). Representing quantity beyond whole numbers: Some, none and part. *Canadian Journal of Experimental Psychology, 54*, 117–128.
Bovet, D., & Vauclair, J. (2000). Picture recognition in animals and humans. *Behavioral & Brain Research, 109*, 143–165.
Boysen, S. T., & Berntson, G. G. (1989). Numerical competence in a chimpanzee (*Pan troglodytes*). *Journal of Comparative Psychology, 103*, 23–31.
Brown, R. (1973). *A first language: The early stages*. Cambridge, MA: Harvard University Press.
Carey, S. (1978). The child as word learner. In M. Halle, J. Bresnan, & G. A. Miller (Eds.), *Linguistic theory and psychological reality* (pp. 264–293). Cambridge, MA: MIT Press.
Carey, S. (2009). *The origin of concepts*. New York, NY: Oxford.
de Boysson-Bardies, B. (1999). *How language comes to children*. Cambridge, MA: MIT Press.
Dittrich, L., Adam, R., & Güntürkün, E. (2010). Pigeons identify individual humans but show no sign of recognizing them in photographs. *Behavioural Processes, 83*, 82–89.
Funk, M. S. (2002). Problem solving skills in young yellow-crowned parakeets (*Cyanoramphus auriceps*). *Animal Cognition, 5*, 167–176.
Gardner, R. A., & Gardner, B. T. (1984). A vocabulary test for chimpanzees (*Pan troglodytes*). *Journal of Comparative Psychology, 98*, 381–404.
Gillen, J. (2007). Derwent's doors: Creative acts. *Mind, Culture, and Activity, 14*, 150–159.
Greenfield, P. (1991). Language, tools and brain: The ontogeny and phylogeny of hierarchically organized sequential behavior. *Behavioral & Brain Sciences, 14*, 531–595.
Heinrich, B. (1995). An experimental investigation of insight in Common Ravens (*Corvus corax*). *Auk, 112*, 994–1003.

Hill, A., Collier-Baker, E., & Suddendorf, T. (2012). Inferential reasoning by exclusion in children. *Journal of Comparative Psychology, 126*, 243−254.

Koegel, R. L., Dyer, K., & Bell, L. K. (1987). The influence of child-preferred activities on autistic children's social behavior. *Journal of Applied Behavior Analysis, 20*, 243−252.

Krashen, S. D. (1976). Formal and informal linguistic environments in language learning and language acquisition. *TESOL Quarterly, 10*, 157−168.

Leonard, L. B. (2001). Fillers across languages and language abilities. *Journal of Child Language, 28*, 257−261.

Markman, E. M. (1990). Constraints children place on word meaning. *Cognitive Science, 14*, 57−77.

Marschark, M., Everhart, V. S., Martin, J., & West, S. A. (1987). Identifying linguistic creativity in deaf and hearing children. *Metaphor & Symbolic Activity, 2*, 281−306.

Merriman, W. E. (1991). The mutual exclusivity bias in children's word learning: A reply to Woodward and Markman. *Developmental Review, 11*, 164−191.

Mikolasch, S., Kotrschal, K., & Schloegl, C. (2011). African grey parrots (*Psittacus erithacus*) use inference by exclusion to find hidden food. *Biology Letters, 7*, 875−877.

Patterson, D. K., & Pepperberg, I. M. (1994). A comparative study of human and parrot phonation: Acoustic and articulatory correlates of vowels. *Journal of the Acoustical Society of America, 96*, 634−648.

Patterson, D. K., & Pepperberg, I. M. (1998). Acoustic and articulatory correlates of stop consonants in a parrot and a human subject. *Journal of the Acoustical Society of America, 103*, 2197−2215.

Peperkamp, S. (2003). Phonological acquisition: Recent attainments and new challenges. *Language & Speech, 46*, 87−113.

Pepperberg, I. M. (1981). Functional vocalizations by an African Grey parrot (*Psittacus erithacus*). *Zeitschrift für Tierpsychologie, 55*, 139−160.

Pepperberg, I. M. (1987). Acquisition of the same/different concept by an African Grey parrot (*Psittacus erithacus*): Learning with respect to categories of color, shape, and material. *Animal Learning & Behavior, 15*, 423−432.

Pepperberg, I. M. (1988a). An interactive modeling technique for acquisition of communication skills: Separation of "labeling" and "requesting" in a psittacine subject. *Applied Psycholinguistics, 9*, 59−76.

Pepperberg, I. M. (1988b). Comprehension of "absence" by an African Grey parrot: Learning with respect to questions of same/different. *Journal of the Experimental Analysis of Behavior, 50*, 553−564.

Pepperberg, I. M. (1990). Referential mapping: Attaching functional significance to the innovative utterances of an African Grey parrot (*Psittacus erithacus*). *Applied Psycholinguistics, 11*, 23−44.

Pepperberg, I. M. (1994). Numerical competence in an African Grey parrot. *Journal of Comparative Psychology, 108*, 36−44.

Pepperberg, I. M. (1999). *The Alex studies*. Cambridge, MA: Harvard University Press.

Pepperberg, I. M. (2002). Allospecific referential speech acquisition in Grey parrots: Evidence for multiple levels of avian vocal imitation. In K. Dautenhahn & C. Nehaniv (Eds.), *Imitation in animals and artifacts* (pp. 109−131). Cambridge, MA: MIT Press.

Pepperberg, I. M. (2004a). Human speech: Its learning and use by Grey parrots. In P. Marler & H. Slabbekoorn (Eds.), *Nature's music* (pp. 363−373). London: Elsevier.

Pepperberg, I. M. (2004b). "Insightful" string-pulling in Grey parrots (*Psittacus erithacus*) is affected by vocal competence. *Animal Cognition, 7*, 263−266.

Pepperberg, I. M. (2006a). Grey parrot (*Psittacus erithacus*) numerical abilities: Addition and further experiments on a zero-like concept. *Journal of Comparative Psychology, 120*, 1−11.

Pepperberg, I. M. (2006b). Ordinality and inferential abilities of a Grey parrot (*Psittacus erithacus*). *Journal of Comparative Psychology, 120*, 205−216.

I. EVIDENCES OF CREATIVITY

Pepperberg, I. M. (2007). Grey parrots do not always "parrot": Roles of imitation and phonological awareness in the creation of new labels from existing vocalizations. *Language
Sciences, 29,* 1−13.

Pepperberg, I. M. (2009). Grey parrot vocal learning: Creation of new labels from existing
vocalizations and issues of imitation. *LACUS Forum, 34,* 21−30.

Pepperberg, I. M. (2012). Further evidence for addition and numerical competence by a
Grey parrot (*Psittacus erithacus*). *Animal Cognition, 15,* 711−717.

Pepperberg, I. M., & Brezinsky, M. V. (1991). Relational learning by an African Grey parrot (*Psittacus erithacus*): Discriminations based on relative size. *Journal of Comparative
Psychology, 105,* 286−294.

Pepperberg, I. M., & Carey, S. (2012). Grey parrot number acquisition: The inference of
cardinal value from ordinal position on the numeral list. *Cognition, 125,* 219−232.

Pepperberg, I. M., & Gordon, J. D. (2005). Numerical comprehension by a Grey parrot
(*Psittacus erithacus*), including a zero-like concept. *Journal of Comparative Psychology,
119,* 197−209.

Pepperberg, I. M., & Nakayama, K. (2012). *Recognition of amodal and modally completed
shapes by a Grey parrot (Psittacus erithacus)*. Poster presented at the annual meeting of
the Vision Science Society, Naples, FL.

Pepperberg, I. M., & Shive, H. A. (2001). Simultaneous development of vocal and physical
object combinations by a Grey parrot (*Psittacus erithacus*): Bottle caps, lids, and labels.
Journal of Comparative Psychology, 115, 376−384.

Pepperberg, I. M., & Wilcox, S. E. (2000). Evidence for a form of mutual exclusivity during
label acquisition by Grey parrots (*Psittacus erithacus*)? *Journal of Comparative Psychology,
114,* 219−231.

Pepperberg, I. M., Brese, K. J., & Harris, B. J. (1991). Solitary sound play during acquisition
of English vocalizations by an African Grey parrot (*Psittacus erithacus*): Possible parallels with children's monologue speech. *Applied Psycholinguistics, 12,* 151−177.

Pepperberg, I. M., Gardiner, L. I., & Luttrell, L. J. (1999). Limited contextual vocal learning
in the Grey parrot (*Psittacus erithacus*): The effect of co-viewers on videotaped instruction. *Journal of Comparative Psychology, 113,* 158−172.

Pepperberg, I. M., Koepke, A., Livingston, P., Girard, M., & Hartsfield, L. A. (2013).
Reasoning by inference: Further studies on exclusion in Grey parrots (*Psittacus erithacus*). *Journal of Comparative Psychology, 127,* 272−281.

Pepperberg, I. M., Sandefer, R. M., Noel, D., & Ellsworth, C. P. (2000). Vocal learning in
the Grey Parrot (*Psittacus erithacus*): Effect of species identity and number of trainers.
Journal of Comparative Psychology, 114, 371−380.

Pepperberg, I. M., Vicinay, J., & Cavanagh, P. (2008). The Müller-Lyer illusion is processed
by a Grey parrot (*Psittacus erithacus*). *Perception, 37,* 765−781.

Peters, A. M. (2001). Filler syllables: What is their status in emerging grammar? *Journal of
Child Language, 28,* 229−242.

Peters, A. M., & Boggs, S. T. (1986). Interactional routines as cultural influences upon language acquisition. In B. B. Schieffelin & E. Ochs (Eds.), *Language socialization across cultures* (pp. 80−96). Cambridge, England: Cambridge University Press.

Petkov, C. I., & Jarvis, E. D. (2012). Birds, primates, and spoken language: Behavioral
phenotypes and neurobiological substrates. *Frontiers in Evolutionary Neuroscience, 12,*
1−24.

Premack, D., & Premack, A. J. (1994). Levels of causal understanding in chimpanzees and
children. *Cognition, 50,* 347−362.

Rice, M. L. (1991). Children with specific language impairment: Toward a model of teach-
ability. In N. A. Krasnegor, D. M. Rumbaugh, R. L. Schiefelbusch, & M. Studdert-
Kennedy (Eds.), *Biological and behavioral determinants of language development*
(pp. 447−480). Hillsdale, NJ: Erlbaum.

Rozin, P. (1976). The evolution of intelligence and access to the cognitive unconscious. In J. M. Sprague, & A. N. Epstein (Eds.), *Progress in psychobiology and physiological psychology* (Vol. 6, pp. 245–280). New York, NY: Academic Press.

Salmon, C. M., Rowan, L. E., & Mitchell, P. R. (1998). Facilitating prelinguistic communication: Impact of adult prompting. *Infant-Toddler Intervention, 8*, 11–27.

Schloegl, C., Schmidt, J., Boeckle, M., Weiss, B. M., & Kotrschal, K. (2012). Grey parrots use inferential reasoning based on acoustic cues alone. *Proceedings of the Royal Society B: Biological Sciences, 279*, 4135–4142.

Taylor, A. H., Medina, F. S., Holzhaider, J. C., Hearne, L. J., Hunt, G. R., & Gray, R. D. (2010). An investigation into the cognition behind spontaneous string pulling in New Caledonian crows. *PLoS ONE, 5*, e9345. Available from <http://dx.doi.org/10.1371/journal.pone.0009345>.

Thorndike, E. L. (1943). *Man and his works*. Cambridge, MA: Harvard.

Todt, D. (1975). Social learning of vocal patterns and modes of their applications in Grey parrots. *Zeitschrift für Tierpsychologie, 39*, 178–188.

Tomasello, M. (2003). *Constructing a language*. Cambridge, MA: Harvard.

Treiman, R. (1985). Onsets and rimes as units of spoken syllables: Evidence from children. *Journal of Experimental Child Psychology, 39*, 161–181.

Weir, R. (1962). *Language in the crib*. The Hague: Mouton.

Wellman, H. M., & Miller, K. F. (1986). Thinking about nothing: Development of concepts of zero. *British Journal of Developmental Psychology, 4*, 31–42.

Willatts, P. (1999). Development of means-end behavior in young infants: Pulling a support to retrieve a distant object. *Developmental Psychology, 35*, 651–667.

Winsler, A., Feder, M., Way, E. L., & Manfra, L. (2006). Maternal beliefs concerning young children's private speech. *Infant and Child Development, 15*, 403–420.

Wynne, C. D. L. (1992). *Animal inferences: Complex performances—simple mechanisms*. Paper presented at the meeting of the Midwest Psychological Association, Chicago.

Commentary on Chapter 1: What Can Creativity Researchers Learn from Grey Parrots?

Reading about a topic you know something about (e.g., creativity) in a context in which you know very little (e.g., avian cognition and behavior) is both humbling and provocative. In the case of Pepperberg's excellent chapter, *Creativity and Innovation in the Grey Parrot*, I found myself humbled in the recognition that I, as a creativity researcher, have unnecessarily restricted my conceptions of creativity to humans. Even in my efforts to democratize creativity, my focus has always been on human creativity (Beghetto, 2007, 2013; Beghetto & Kaufman, 2007). I never gave sufficient treatment to how my conceptions of creativity might be extended and demonstrated in nonhumans. This recognition provoked me to ask myself why this might be the case and what creativity researchers might learn from such an effort. Rather than quickly brush away such questions (as I have done in the past), my aim in this brief commentary is to encourage an earnest exploration of these questions.

Why do creativity researchers, such as myself, fail to acknowledge nonhuman creativity? One reason I became aware of when reading Pepperberg's chapter had to do with the way we categorize our beliefs about creativity. The issue for me (and I suspect other creativity researchers) boiled down to viewing nonhuman creativity as a difference *in kind* rather than a difference *in degree*. The reasoning goes something like this: if humans and parrots are different kinds of animals, then they likely exhibit different kinds of creative behavior. Once we endorse this view, it becomes easy to privilege human creative behavior. We can comfortably dismiss ourselves from the question of nonhuman creativity by saying, "Animals may very well demonstrate what might be called creative behavior (i.e., novel and task appropriate), but it is a different kind of creative behavior than what we humans demonstrate. I'm interested in the kind of creativity demonstrated by humans and need not trouble myself with nonhuman creativity." This form of reasoning allows us to insulate ourselves from considering insights from research on nonhuman creativity. We can simply claim that including such insights in our work would result in a logical category error.

Pepperberg's chapter, however, provoked me to uncover my human creativity bias and more closely examine my beliefs about nonhuman creativity. On what basis might one claim that creativity exhibited by nonhumans is a different kind of creativity? When asking such a question, a good place to start is with one's assumptions. Given that assumptions are untested beliefs, they need to be elicited and explored. Fortunately, it did not take long for one of my core assumptions to surface when reading Pepperberg's chapter. I quickly realized that my beliefs about intentional behavior served as the basis for my assumption that nonhuman creativity must be a different kind of creativity. Specifically, I found myself reasoning that human creativity is often (although not always) a form of intentional behavior, whereas nonhuman creativity is likely (if not always) a result of chance behavior.

How did I come to hold this belief? One reason has to do with the way I and other creativity researchers have described creativity as a decision (Kaufman & Beghetto, 2013; Sternberg, 2002; Sternberg & Lubart, 1995). According to this view, creativity results, in part, from people weighing potential costs and benefits and deciding to take creative action. Decisions, including creative decisions, are intentional acts. It's not that creativity researchers who endorse this view fail to recognize that creativity can (and often does) result from unintentional acts. Indeed there are numerous examples of unintentional creativity and innovation. Consider, for example, the invention of the Post-it® note, which resulted from the accidental creation by of a weak adhesive by a researcher in the 3M research lab (Glaveanu & Gillespie, 2014). Still, the central argument made by creativity researchers who hold this view is that creative people are typically aware of the potential risks involved in thinking and acting creatively. Creative people therefore must often choose to be creative in light of sensible risks (Sternberg, 2002).

I further assumed that only humans can intentionally demonstrate creativity and, in turn, believed that nonhuman creative behavior must be a different kind of creativity. Pepperberg's chapter challenged this assumption. She tackled the question of intentionality in nonhuman creative behavior head-on (p. 3):

> The question has been raised, however, as to whether these birds [Grey parrots] can, like young children...intentionally demonstrate creative or innovative use of such speech...several examples exist to demonstrate that at least one Grey parrot was capable...of creativity and innovation.

Pepperberg goes on to discuss several compelling examples from experiments with Grey parrots. In the examples provided, the Grey parrot does, indeed, seem to be intentionally using its knowledge of speech and concepts in creative ways. Examples included the Grey parrot's use of innovative and creative vocalizations to communicate needs (e.g., end a grueling training session) and to achieve goals (e.g., get a nut). Pepperberg also discusses how many of these seemingly intentional creative speech behaviors of Grey parrots are analogous to what human children do when learning language.

One might still protest that the seemingly intentional creative behavior of nonhumans (and perhaps young children) is a different kind of creativity than that of a more accomplished human creator. The accomplished creator, it might be argued, is uniquely capable of being aware of creative risks and, in light of those risks, consciously decides to engage in creative behaviors. Surely, a Grey parrot does not have the ability to be aware of the risks involved in creative behavior. Pepperberg, however, provides compelling evidence that one Grey parrot, Alex, did seem to have some level of awareness about the risks of making novel utterances in the presence of trainers. Consider the following example.

Trainers would scold Alex or take away training materials when he made mistakes in their presence. However, when Alex was alone Pepperberg and her colleagues discovered (through eavesdropping and later through hidden audiotaping) that he would engage in playful and creative speech. Pepperberg explained that Alex seemed to be "practicing in private before attempting the labels in public" (p. 7).

This behavior is similar (albeit on a smaller scale) to the ways that creativity researchers have described the development of creativity and behaviors of more accomplished human creators (Kaufman & Beghetto, 2009; Simonton, 2003). Creators often play with or test out many candidate ideas in the private workshop of their minds before they select the ideas they intend to share and develop into creative contributions.

The persistent naysayer may go on to argue that with humans we can come closer to assessing intent. We have methods (e.g., interviews, talk aloud protocols, etc.), as flawed as they may be, which allow us to draw stronger inferences about awareness and intent than relying on behavioral observations alone. Once we start raising this type of argument, however, the firm stance that there is a difference in kind starts to soften. Indeed, we are now considering different degrees of awareness (i.e., stronger inferences) and different degrees of creativity (i.e., more accomplished creators). Once we start blurring the distinctions between *difference in kind* and *difference in degree*, we become more open to possible similarities. This, in turn, allows us to recognize how examples of nonhuman creativity — such as those presented by Pepperberg — map on to many of the concepts and findings from studies of human creativity.

Here are just a few examples. As I briefly alluded to earlier, Alex playing with creative combinations (outside the evaluative eye of trainers) is similar to what humans do when they generate ideas at the mini-c level of creativity (Beghetto & Kaufman, 2007). It is also similar to work on how the expression of creativity is influenced by the evaluative features of the social context (Amabile, 1996; Beghetto & Kaufman, 2014). Moreover, the role of trainer feedback on the development of the Grey parrot's creative vocalizations and how the Grey parrot was able to demonstrate the ability to use creative vocalizations to "both exert some control over his environment and to demonstrate an expanded understanding" (p. 10) is analogous to descriptions of the developmental trajectory of creativity (i.e., moving from mini-c to little-c creativity, Beghetto & Kaufman, 2014). Pepperberg's chapter also provides examples of nonhuman creative behavior that align well with many longstanding and contemporary issues explored by creativity researchers, including questions of domain specificity, creative polymathy, the roles of context and training, and the development of creative learning trajectories (just to name a few).

So where might human-centric creativity researchers go from here? A good first step is recognizing that when we expand our work to include nonhuman examples of creativity we have a much broader context in which our existing creativity theories and findings can be tested, developed, refuted, and refined. Working alongside researchers who study creativity in nonhuman populations can also help us understand how insights might be shared across fields, identify new possibilities for expanding our theories, as well as discover the limits and differences that really matter.

This chapter and the chapters in this volume make it far more difficult (if not impossible) to as easily dismiss nonhuman creativity as a different kind of phenomenon than what creativity researchers typically study. After reading about Alex the Grey parrot, what is now most remarkable to me is not how innovative and creative Grey parrots can be, but that our creativity research journals are not filled with studies that include nonhuman creativity. Fortunately, Pepperberg's chapter (and all the chapters collected in this volume) may forever change that situation.

References

Amabile, T. M. (1996). *Creativity in context: Update to the social psychology of creativity*. Boulder, CO: Westview.

Beghetto, R. A. (2007). Ideational code-switching: Walking the talk about supporting student creativity in the classroom. *Roeper Review, 29*, 265–270.

Beghetto, R. A. (2013). *Killing ideas softly? The promise and perils of creativity in the classroom*. Charlotte, NC: Information Age Publishing.

Beghetto, R. A., & Kaufman, J. C. (2007). Toward a broader conception of creativity: A case for mini-c creativity. *Psychology of Aesthetics, Creativity, and the Arts, 1*, 73–79.

Beghetto, R. A., & Kaufman, J. C. (2014). Classroom contexts for creativity. *High Ability Studies, 25*, 53–69.

Glaveanu, V. P., & Gillespie, A. (2014). Creativity out of difference: Theorising the semiotic, social and temporal origin of creative acts. In V. P. Glaveanu, A. Gillespie, & J. Valsiner (Eds.), *Rethinking creativity: Contributions from social and cultural psychology*. New York, NY: Taylor & Francis.

Kaufman, J. C., & Beghetto, R. A. (2009). Beyond big and little: The four C model of creativity. *Review of General Psychology, 13*, 1–12.

Kaufman, J. C., & Beghetto, R. A. (2013). In praise of Clark Kent: Creative metacognition and the importance of teaching kids when (not) to be creative. *Roeper Review, 35*, 155–165.

Simonton, D. K. (2003). Creativity as variation and selection: Some critical constraints. In M. A. Runco (Ed.), *Critical creative processes* (pp. 3–18). Cresskill, NJ: Hampton Press.

Sternberg, R. J. (2002). Creativity as a decision. *American Psychologist, 57*, 376.

Sternberg, R. J., & Lubart, T. I. (1995). *Defying the crowd: Cultivating creativity in a culture of conformity*. New York, NY: Free Press.

Creativity in the Interaction: The Case of Dog–Human Play

Robert W. Mitchell

Department of Psychology, Eastern Kentucky University, Richmond, KY, USA

Commentary on Chapter 2: Creativity in the Interaction

Jessica Hoffmann

Yale Center for Emotional Intelligence, Yale University, New Haven, CT, USA

INTRODUCTION

It's a ball. And you have it, or your dog has it. So it's really more than a ball: it's a gateway to fun. For some dogs, your having a ball is tantamount to saying, "Let's play." For some people, a dog with a ball is an invitation to get the ball from the dog while playing with the dog. There are lots of things you and a dog can do with a ball. Most of them are quite commonplace. For example, lots of dogs and people play keep-away with it. Keep-away is fun, because you get to entice your partner to try to get the ball while (mostly) making pretty sure your partner cannot. Most people during keep-away hold on to the ball for a while, and then toss it for the dog to get; most dogs during keep-away hold on to the ball for a while, and then drop it (though they may immediately pick it up again, or chase it). While holding onto the ball, dogs and people entice their partner to try to get the ball by moving it

Animal Creativity and Innovation.
DOI: http://dx.doi.org/10.1016/B978-0-12-800648-1.00002-4

toward the partner, but limiting the likelihood that the partner can actually get the ball. But for the partner, it's not fun if there's no possibility of getting the ball, so whoever has the ball has to calibrate how close to get: enough to entice the partner, but not enough to make it easy for the partner to get the ball. The partner striving to get the ball is essential for the player who has the ball to enjoy using it to play keep-away.

Sometimes the partner gets the ball; clearly, keep-away is a game of (minimal) risk. So is much social play; what is being risked is loss of control. During social play, players attempt to control the other player without being controlled. They do so by engaging in projects, which are repetitive goal-directed action sequences; players vary actions during repetitions to gain practice over these actions (Simpson, 1976). Projects enacted simultaneously by players in relation to each other are called "routines." Most routines are like keep-away—they comprise compatible projects: what each partner does is essential for the other player's satisfaction. Routines based on incompatible projects, where one partner has a project that does not satisfy the other player's project, are less frequent (Mitchell & Thompson, 1991). The same kind of thing happens in sports; players are striving to achieve particular goals in the context of opposing players attempting to thwart these goals in order to achieve their own goals. In both sports and social play (and sports are a type of social play), the players are faced with constraints on their actions, both self-imposed and imposed by the game, within which they strive to achieve their goals. As a consequence of engaging in simultaneously supportive but mutually disruptive projects, unexpected behavioral possibilities arise that allow for innovative action.

My colleague Nick Thompson and I decided to discover if these ideas about projects and routines in play were accurate in dog–human play. So we asked 23 dog owners if they would be willing to play with their own dog and (at a different time) another dog, and if they would allow their own dog to play with someone else (another of the dog owners), and allow us to videotape the interactions. Each person played with his or her own dog and with another unfamiliar dog, and each dog played with his or her owner and an unfamiliar owner. (For the unfamiliar pairings, person A played with person B's dog, and person B played with person A's dog.) We provided them with (new) objects each time—a tennis ball, two strips of sheeting, and a rope—and asked them to begin to play by throwing the ball, and to tell us when they finished (we videotaped a bit after that). We then coded the videotapes using second-by-second analysis to support our ideas about play. For those interested, the research deriving from these videotapes is described in a series of publications (on dog–human play: Mitchell, 2012; Mitchell & Thompson, 1990, 1991, 1993; and on talk and other vocalization to dogs during play: Mitchell, 2001, 2004; Mitchell & Edmonson, 1999; Mitchell & Sinkhorn, 2014).

In this chapter, I examine one pairing of human and dog players, Chris and Hercules, who engaged in a relatively unique routine based on the projects of fakeout and avoid fakeout. (Of the 23 pairs of people playing with both their own dog and an unfamiliar dog, Chris and Hercules were the only ones to play this routine; however, note that fakeout seems to be a variant of the very common project of keep-away.) In this game, the man acted as if about to make the ball move, but often did not, thus trying to get the dog to act toward the ball before it was available: faking out the dog. The dog acted to avoid being faked out: he constrained his actions so as to move as little as possible toward the ball until it was completely free of the man's control, whereupon the dog could gain control over the ball. Once the dog had the ball, he sometimes engaged in fakeout toward the man during the ball's return, taking advantage of opportunities to move or get the ball before the man obtained it. Because the ball's location was to some degree unpredictable at any given moment, players could react to this random process with novel actions to effect their project, using what is called tactical creativity in human sports. Like the creation of a sonnet or haiku, in which novel language allows for creativity within a pre-scribed framework, the pair created novel interactions within the restrictions of their coordinated projects.

Ideas about what happens in sports and what happens in dog—human play are cross-fertilizing (Buytendijk, 1936/1973; compare Mawby & Mitchell, 1986; Mitchell & Thompson, 1986). The literature on creativity in sports, borrowing extensively from the literature on the psychology of human creativity, has supported an extensive array of theories that are usefully applied to dog—human play. (The literature on animal creativity also borrowed extensively from the literature on the psychology of human creativity: see Kaufman, Butt, Kaufman, & Colbert-White, 2011; Kaufman & Kaufman, 2004.) In describing what happened between Chris and Hercules, I view them as part of a collab-orative dyad in which each player strives to gain and retain expertise in his projects within the accepted constraints of the game. The part-ners' striving requires attention, a competitive edge, and a willingness to risk the loss of the ball, much as does the striving in competitive athletic ball games.

The most common definition of creativity used in studying it in sports is that of Sternberg and Lubert (1999, p. 3): creativity is "the abil-ity to produce work that is both novel (i.e., original, unexpected) and appropriate (i.e., useful)." In sports,

> "creative" refers to those varying, rare, and flexible decisions that play an important role in team ball sports like football, basketball, field hockey, and hand-ball. An unexpected no-look pass to a fellow team member (that may not even be

expected by the team member) would be an example of a creative solution in basketball or football. Moreover, creative player A may intend to pass the ball to player B, before perceiving at the last minute that player C is suddenly unmarked, in a better position, and thus player A makes the decision to pass the ball to player C instead. (Memmert, 2011, p. 94)

Evidently, the context in the form of other players' actions may create the possibility for creativity. Bailey, McDaniel, and Thomas (2007) focus on flexibility in behavior as a prerequisite for creativity in animals, and Kaufman and Kaufman (2004; also Kaufman et al., 2011) focus on novelty and appropriateness as signs of creativity, though their sequential model of recognition of novelty, observational learning, and innovative behavior does not easily apply to the creativity I discuss in dog–human play (though their focus does apply to much nonhuman pretense and imagination: see Mitchell, 1990, 2002, 2013). Comparative psychologists reading Memmert's description will likely think of Köhler's (1925/1976, p. 190, italics removed) description of insight: "the appearance of a complete solution with reference to the whole lay-out of the field."

Creative acts (such as deception) in sports, as in animal play, need not indicate that "well-thought-out reason… directs the action, but [rather] an unconscious realization of the possible movements of the adversary" (Buytendijk, 1936/1973, p. 206). Indeed, what are described above as "creative" examples in sports are often well-practiced activities that take into account the current situation (ca. Mitchell, 1999), much like those I will describe for Chris and Hercules: see, for example, the descriptions of Magic Johnson's no-look passes in basketball and Zidane's scoping-out skills in soccer in Memmert, Baker, and Bertsch (2010, p. 3). However much the instances given of creativity in actual sports seem to capture our ideas of creativity, in most of the research about creativity in sports, creativity is measured in a diversity of ways, none of which is relevant to examining creativity in dog–human play (see, e.g., Memmert & Roth, 2007; Memmert et al., 2010). What are relevant are the theories and ideas that lead to the measurements. The most important of these for the present research is work on collaborative circles (Farrell, 2003; elaborated for sports by Corte, 2013) and constraint theory (Elster, 1984, 2000; elaborated for sports by Lewandowski, 2007, Sternberg, 1976); I shall also mention the theory of deliberate practice (Ericsson, Krampe, & Tesch-Römer, 1993; elaborated for sports by Memmert, 2011 and Memmert & Furley, 2007).

Constraint Theory

Elster's (1984, 2000) constraint theory posits that human activity requires the selection of feasible activities from the acknowledged set of possible activities that can achieve a self-set goal; in essence, we choose constraints by our intentions or by the rules of a game, for example, and then can act in several ways within those constraints to fulfill them (see also Mawby & Mitchell, 1986). But because there are several feasible ways one can respond, creativity is possible in the selection of alternatives. Lewandowski (2007) applies Elster's model to sparring by boxers. This particular interaction is compellingly similar to dog—human play, in that, as Lewandowski notes, the actions of each sparring partner set limits on the options for the other partner. Poetry is the ubiquitously used example for creativity under constraint: "Sonneteers must intentionally seek, through reflexively monitored revisions, continued variations, and deliberate experimentations, to maximize their creativity and skills of written expression within their elected constraints" (Lewandowski, 2007, p. 28), which Elster (2000, p. 201) calls "constrained maximization." Similarly, competitive athletes must be engaged in "complex practice aimed at constitutively constrained maximization."

In addition, sparring partners "engage in a form of shared cooperative action and practical improvisation designed to instruct one another in mutually beneficial ways, such as when boxers reflexively 'correct' one another's mistakes with controlled well-placed boxes" (Lewndowski, 2007, p. 34), what Elster (2000, p. 277) labels "mutual self-binding." Lewandowski believes that such mutual self-binding fosters creativity: sparrers learn how to take immediate advantage of minor changes in the other's behavior, and to avoid their own minor changes in behavior that allow the sparring partner to take a shot. Chris and Hercules, like other social players, are mutually self-binding, in that Chris's actions set up the constraints on possible actions of Hercules, and Hercules' actions set up the constraints on possible actions of Chris. This description nicely fits the description of what goes on during play: each player is striving to satisfy his or her project within the confines of the other player striving to satisfy his or her project, and vice versa. For both Chris and Hercules, the other's actions are not only required but offer the possibility of creative responses to these actions.

Expertise and Attention

Memmert (2011) posits that creativity in sports is maximized when players' expertise and attention are maximized. Greater attention allows the attender to be influenced by more stimuli in the environment and to avoid missing stimuli, and attention span is positively related to creativity. Basing his ideas on the theory of deliberate practice, Memmert (p. 94) posits that maximizing expertise requires "extended engagement in high-quality training (i.e., deliberate practice)." This engagement and training must be "performed for the purpose of improving current performance rather than inherent enjoyment." Although Memmert presents these as diametrically opposed options, in fact there could be inherent enjoyment in working to improve one's performance, and we expect that the "love of the game" is essential for quality expertise on the playing field. Certainly play has, for many mammals, an inbuilt enjoyment. But it is clear that there has to be a serious component if one is going to better one's game. Skateboarders enjoy skating, and this drives them to develop expertise through a variety of means, including watching amateur videos from other skaters (Jones, 2011). And both Chris and, especially, Hercules seemed to take their play seriously, so expertise is possible here. Hercules' taut body and directed gaze at the ball and/or Chris's actions indicate that his play was a highly engaging activity for Hercules, not a simple amusement. Hercules took his play seriously. In order to perform his projects, Chris too had to be highly attentive to his partner's actions.

Another sense in one can gain expertise is by having "unusual and unexpected experiences" (Ritter et al., 2012, p. 961). In this view, having diversified experiences leads to creativity in the form of cognitive flexibility. For example, in humans, engaging in schema-violating activities leads to greater imaginative creative finds on unusual uses tasks. Hercules' experiences of engagement with the variability in Chris's fakeouts offers a means by which creative responding might be stimulated in him.

Collaborative Dyads

The idea that innovation can arise through collaboration in humans is well established (John-Steiner, 2005; Paulus & Nijstad, 2003; Sawyer, 2008). Farrell (2003) examined, biographically and historically, specific groups of artists, writers, and scientists to learn how they influenced each other within their group, to develop his ideas about collaborative circles. A collaborative circle is "a primary group consisting of peers who share similar occupational goals and who, through long periods of dialogue and collaboration, negotiate a common vision that guides their work" (p. 13). Although shouts of "anthropomorphism" might be heard in response to my application of Farrell's term to dog–human play, it is not a serious objection. Mead (1934/1974, p. 86) described mutually interactive actions without language as a "conversation of gestures" and hence a dialogue, Mechling (1989) recognized routine activities shared between humans and animals as folk traditions, and a literalist objection that only humans can have occupational goals, while perhaps accurate, denies the gist of Farrell's idea: that the collaborators are working together while developing their own work. That dog and human share a "common" vision is evident in that both want to engage in play activities. Finally, the issue of dog and human being "peers" may seem inherently absurd, in that the dog is not as sophisticated as the man, but in the context of their social play they appear to view each other as equals. This will be evident in the discussion of instrumental intimacy.

Applying Farrell's analysis to the social dynamics and development of BMX riders, Corte (2013) focuses on two concepts: instrumental intimacy and the norm of escalating reciprocity. "Instrumental intimacy is a type of exchange between dyads of the group denoted by trust, mutual support, and free transfer of ideas, resulting from deep knowledge of one another acquired through long and persistent interaction" (Corte, 2013, p. 27). Both Chris and Hercules know what the other is likely to do because they have played with each other a lot, and everything each one does is open to the other. The "norm of escalating reciprocity" is "the dynamic that pushes members to both match and exceed each other's work, ultimately increasing the quality of the work done by the group as a whole" (p. 27). The manner in which members match and exceed each other's work is not imitative; rather, it is matching and exceeding in the quality of the work. In the case of Chris and Hercules, the quality of the work is the continuing satisfaction and success present in the enacted projects.

All of these ideas are essential to understanding the play routines of Chris and Hercules. I filmed their play activities three times (only the last of which was used in the original coded data set), and each time was similar. Their play exhibits an engagement in each other's activities without coercion. Within their play sessions, their routines of fakeout/avoid fakeout are often interrupted by Chris throwing the ball, Hercules obtaining the ball, and then dropping the ball a distance from Chris and rolling about on top of it for a few minutes or more. During this time, Chris often simply waits, watching Hercules who, when he finally moves toward Chris, usually holds the ball in his mouth and chews. Chris at times points to the ground with hand or foot, asking Hercules to "drop the ball." At other times, Hercules retrieves the ball to Chris, who engages in a variety of means to obtain the ball: tug o' war, petting the dog to deflect his attention, massaging his face and pulling out the ball, pushing the dog over to rub him and relax his hold. Their intimacy in the instrumental goal of playing is evident, but becomes dramatized when compared to attempts by Rachel, a woman unfamiliar to Hercules, to play with him. Rachel repeatedly attempted to grab the ball from Hercules, who pulled back every time, resulting in an often unproductive tug o' war. (Note that fakeout/avoid fakeout routines in dog—human play are discussed by Goode (2006).)

The norm of escalating reciprocity is evident in the dynamic of the fakeout/avoid fakeout routine between Chris and Hercules. Hercules was obsessed with not chasing after the ball whenever Chris tried to fake him out, and Chris produced a variety of fakeouts to tempt Hercules to chase when the ball was not available. Each player, working on his own project, strived to make it "better" in relation to the other's project, which required attempting to thwart the other's project: Chris in providing diverse fakeouts trying to entice Hercules to chase an unthrown, unmoving, or otherwise unavailable ball, and Hercules in reacting quickly to avoid chasing after the ball in response to each of these fakeouts. Each player perfected the quality of his project. Chris's fakeouts were highly variable: faked throwing, pausing as if to throw the ball, dropping the ball and retrieving it, fake kicking it, pausing before fake kicking it, leaning over and holding his hand over the ball and moving his hand and/or the ball around, holding the ball with his sneakered foot and moving it around. These fakeouts themselves occurred in variable postures. Chris spiced all this with actual throwing and kicking of the ball so that Hercules could get it. And Hercules showed his extreme excitement about potentially obtaining the ball by jumping about enthusiastically while keeping his eyes on the ball.

Hercules' Fakeouts

Hercules infrequently took advantage of Chris's not-quick-enough grabs at the dropped ball to produce his own minor fakeouts. Hercules' actions likely did not derive from observing Chris's fakeouts and trying to create better fakeouts—Hercules' fakeouts were too opportunistic.

I could only find two types of fakeout from Hercules. In one type, he dropped a ball but made it less available to Chris by standing over or close to the ball and waiting for Chris to attempt to get it; in the other type, he dropped the ball but provided less of a chance for Chris to obtain the ball by immediately rushing after it. I describe instances of each type below.

Standing Over

During their first videotaped encounter, after four bouts of fakeout/ avoid fakeout interrupted by throw/retrieve, Hercules came close to Chris, who was leaning over resting his hands on his knees. Hercules chewed on the ball, and then dropped it, but after a brief pause, jumped to move his body over the ball, so that it became difficult for Chris to obtain it. After a brief pause, as Chris moved one hand slowly toward the ball, Hercules quickly grabbed the ball. Chris moved backward away from Hercules, who moved toward him. The sequence repeated again, though this time Hercules' body was not covering the ball, but his head was close to it, and Chris moved his hand down and let it hang. Hercules had his eyes on Chris's hand. Just as Chris moved his hand to get the ball, Hercules moved his mouth to do so, but Chris got the ball to move toward him and picked it up. A similar encounter occurred during their second videotaped encounter: Hercules dropped the ball near to Chris, who was leaning over, and after a brief pause Hercules jumped over the ball with his head close to the ball. Chris leaned down further, and Hercules grabbed the ball in his mouth, chewing on it. Chris quick reached out and grabbed the ball with both hands and pulled it from Hercules' mouth. Soon after, a similar event occurred, with Hercules dropping the ball near to a leaning over Chris and moving over it, but after 2 s moving back away from it and barking, upon which Chris obtained the ball just as Hercules ran up to get it.

Drop and Get

The other type of fakeout by Hercules, which is perhaps an abbreviation of the first, is to drop the ball and immediately attempt to get it back. During their second episode, Hercules also dropped the ball so that it rolled behind Chris, and then quickly moved to get it without pausing, but Chris got there before Hercules did. The same happened during the third episode (which is the only one we examined in detail in our original studies): Hercules dropped the ball, which bounced off Chris's sneaker, and Hercules tried to grab it as it moved, but once again Chris got it. Once in the second episode, the ball's movement appeared unexpected, and it took the two a few seconds to respond. Hercules dropped the ball on Chris's sneaker, and it bounced away from both of them. They stared at it, and then both moved toward it, Chris getting it before Hercules.

The fakeouts by Hercules occurred amid or between fakeouts by Chris, in which he let the ball sit between them and waited, so that either might get the ball (and either sometimes did), or so that he might kick it or push it with his hands. In some cases in which the ball sat between them, Hercules engaged in avoid fakeout, acting as if waiting for Chris to make the ball move, rather than attempting to get the ball outright.

The fact that Hercules' fakeouts were as not as sophisticated as Chris's led us initially to believe that dogs did not engage in fakeout, as we did not perceive the rather limited fakeout by Hercules in the interaction we coded (Mitchell & Thompson, 1991, p. 217). One might think that the more noticeable and elaborate fakeouts by Chris may have had something to do with his having hands (though Chris also used his feet to fakeout and Hercules did not use his paws). Fakeouts are described in the play of human-reared animals with hands, each of whom had extensive experience playing with people. A bonobo frequently dropped a play object and appeared disinterested, and expressed interest in it again only when his partner did (Savage-Rumbaugh & McDonald, 1988, p. 233). This is a rather common activity by playing dogs that sometimes expresses real rather than feigned disinterest in ape and canids (Mitchell, 2002, p. 308), but the bonobo seemed to vary the distances between himself, the partner, and the object, and remained attentive to the partner's actions. An orangutan once threw a play object toward his partner but in such a way that it returned to the orangutan during an elaborate game of keep-away (Miles, Mitchell, & Harper, 1996, p. 292). Chris and other humans may have an edge over nonhuman animals in fakeout skills through greater cognitive and neural complexity, but until we examine animal play in detail, we cannot be sure what they can accomplish.

Each player's fakeouts derived from wanting to beat the other player while introducing an element of risk that he would not beat the other player. Thus, Chris and Hercules engaged in a collaborative dyad in which intimacy and mutual striving supported the creativity of their responses. They accepted the constraints upon which their play depended, and played within those constraints. And they each strived to perfect their skills at their projects, gaining expertise in their game.

References

Bailey, A. M., McDaniel, W. F., & Thomas, R. K. (2007). Approaches to the study of higher cognitive functions related to creativity in nonhuman animals. *Methods, 42*, 3—11.

Buytendijk, F. J. J. (1936/1973). *The mind of the dog*. New York, NY: Arno Press.

Corte, U. (2013). A refinement of collaborative circles theory: Resource mobilization and innovation in an emerging sport. *Social Psychology Quarterly, 76*, 25—51.

Elster, J. (1984). *Ulysses and the Sirens*. Cambridge, UK: Cambridge University Press.

Elster, J. (2000). *Ulysses unbound: Studies in rationality, precommitment, and constraints*. Cambridge, UK: Cambridge University Press.

Ericsson, K. A., Krampe, R., & Tesch-Römer, C. (1993). The role of deliberate practice in the acquisition of expert performance. *Psychological Review, 100*, 363—406.

Farrell, M. P. (2003). *Collaborative circles: Friendship dynamics and creative work*. Chicago, IL: University of Chicago Press.

Goode, D. (2006). *Playing with my dog Katie: An ethnomethodological study of dog—human interaction*. Lafayette, IN: Purdue University Press.

John-Steiner, J. (2005). *Creative collaboration*. New York, NY: Oxford University Press.

Jones, R. H. (2011). Sport and re/creation: What skateboarders can teach us about learning. *Sport, Education and Society, 16*, 593—611.

Kaufman, A. B., Butt, A. E., Kaufman, J. C., & Colbert-White, E. N. (2011). Towards a neurobiology of creativity in nonhuman animals. *Journal of Comparative Psychology, 125*, 255—272.

Kaufman, J. C., & Kaufman, A. B. (2004). Applying a creativity framework to animal cognition. *New Ideas in Psychology, 22*, 143—155.

Köhler, W. (1925/1976). *The mentality of apes*. New York, NY: Liveright.

Lewandowski, J. (2007). Boxing: The sweet science of constraints. *Journal of the Philosophy of Sport, 34*, 26—38.

Mawby, R., & Mitchell, R. W. (1986). Feints and ruses: An analysis of deception in sports. In R. W. Mitchell, & N. S. Thompson (Eds.), *Deception: Perspectives on human and nonhuman deceit* (pp. 313–322). Albany, NY: SUNY Press.

Mead, G. H. (1934/1974). *Mind, self and society*. Chicago, IL: University of Chicago Press.

Mechling, J. (1989). "Banana cannon" and other folk traditions between humans and non-human animals. *Western Folklore, 48*, 312–323.

Memmert, D. (2011). Creativity, expertise, and attention: Exploring their development and their relationships. *Journal of Sports Sciences, 29*, 93–102.

Memmert, D., Baker, J., & Bertsch, C. (2010). Play and practice in the development of sport-specific creativity in team ball sports. *High Ability Studies, 21*, 3–18.

Memmert, D., & Furley, P. (2007). "I spy with my little eye!": Breadth of attention, inattentional blindness, and tactical decision making in team sports. *Journal of Sport and Exercise Psychology, 29*, 365–381.

Memmert, D., & Roth, K. (2007). The effects of non-specific and specific concepts on tactical creativity in team ball sports. *Journal of Sports Sciences, 25*, 1423–1432.

Miles, H. L., Mitchell, R. W., & Harper, S. (1996). Simon says: The development of imitation in an enculturated orangutan. In A. Russon, K. Bard, & S. T. Parker (Eds.), *Reaching into thought: The minds of the great apes* (pp. 278–299). New York, NY: Cambridge University Press.

Mitchell, R. W. (1990). A theory of play. In M. Bekoff, & D. Jamieson (Eds.), *Interpretation and explanation in the study of animal behavior, vol. 1: Interpretation, intentionality, and communication* (pp. 197–227). Boulder, CO: Westview Press.

Mitchell, R. W. (1999). Deception and concealment as strategic script violation in great apes and humans. In S. T. Parker, R. W. Mitchell, & H. L. Miles (Eds.), *The mentalities of gorillas and orangutans* (pp. 295–315). Cambridge, UK: Cambridge University Press.

Mitchell, R. W. (2001). Americans' talk to dogs during play: Similarities and differences with talk to infants. *Research on Language and Social Interaction, 34*, 182–210.

Mitchell, R. W. (Ed.), (2002). *Pretending and imagination in animals and children* Cambridge, UK: Cambridge University Press.

Mitchell, R. W. (2004). Controlling the dog, pretending to have a conversation, or just being friendly? Influences of sex and familiarity on Americans' talk to dogs during play. *Interaction Studies, 5*, 99–129.

Mitchell, R. W. (2012). Doing and saying in play between dogs and people. In M. DeMello (Ed.), *Animals and society: An introduction to human-animal studies* (pp. 374–376). New York, NY: Columbia University Press.

Mitchell, R. W. (2013). Comparative issues in the study of imagination. In M. Taylor (Ed.), *The Oxford handbook of the development of imagination* (pp. 468–488). Oxford: Oxford University Press.

Mitchell, R. W., & Edmonson, E. (1999). Functions of repetitive talk to dogs during play. *Society and Animals, 7*, 55–81.

Mitchell, R. W., & Sinkhorn, K. (2014). Why do humans laugh during dog–human play interactions? *Anthrozoös, 27*, 235–250.

Mitchell, R. W., & Thompson, N. S. (1986). Deception in play between dogs and people. In R. W. Mitchell, & N. S. Thompson (Eds.), *Deception: Perspectives on human and non-human deceit* (pp. 193–204). Albany, NY: SUNY Press.

Mitchell, R. W., & Thompson, N. S. (1990). The effects of familiarity on dog–human play. *Anthrozoös, 4*, 24–43.

Mitchell, R. W., & Thompson, N. S. (1991). Projects, routines, and enticements in dog–human play. In P. P. G. Bateson, & P. H. Klopfer (Eds.), *Perspectives in ethology* (Vol. 9, pp. 189–216). New York, NY: Plenum Press.

Mitchell, R. W., & Thompson, N. S. (1993). Familiarity and the rarity of deception: Two theories and their relevance to play between dogs (*Canis familiaris*) and humans (*Homo sapiens*). *Journal of Comparative Psychology, 107*, 291–300.

Paulus, P. B., & Nijstad, B. A. (Eds.), (2003). *Group creativity: Innovation through collaboration* New York, NY: Oxford University Press.

Ritter, M. S., Damian, R. I., Simonton, D. K., van Baaren, R. B., Strick, M., Derks, J., et al. (2012). Diversifying experiences enhance cognitive flexibility. *Journal of Experimental Social Psychology, 48,* 961–964.

Savage-Rumbaugh, E. S., & McDonald, K. (1988). Deception and social manipulation in symbol-using apes. In R. W. Byrne, & A. Whiten (Eds.), *Machiavellian intelligence* (pp. 224–237). Oxford: Oxford University Press.

Sawyer, K. (2008). *Group genius: The creative power of collaboration.* New York, NY: Perseus Publishing.

Simpson, M. J. A. (1976). The study of animal play. In P. P. G. Bateson, & R. A. Hinde (Eds.), *Growing points in ethology* (pp. 385–400). Cambridge: Cambridge University Press.

Sternberg, R. J (1976). The study of animal play. In P. P. G. Bateson, & R. A. Hinde (Eds.), *Growing points in ethology* (pp. 385–400). Cambridge: Cambridge University Press.

Sternberg, R. J., & Lubart, T. I. (1999). The concept of creativity: Prospects and paradigms. In R. J. Sternberg (Ed.), *Handbook of creativity* (pp. 3–15). New York, NY: Cambridge University Press.

Commentary on Chapter 2: Creativity in the Interaction

Enjoyable play, whether it is human, animal, or both, involves a delicate dance around control. Play is most enjoyable when it involves free choice, is self-directed, has some element of risk, and maintains a balance of power among the players. As this chapter illustrates, when these basics are achieved, the avenues for enjoyment and the options for creative behaviors become limitless.

Play research with young children indicates that for a behavior to be considered play, there must be some element of choice (Johnson, Christie, & Wardle, 2005; King, 1979). Almost any action can be either work or play partially depending on the level of free will, and that consequent intrinsic motivation, that is present. For example, playing catch with a friend in the yard may be fun when chosen freely and self-directed by the child, while formal baseball practice, implemented too strictly, may feel like work. For Chris and his dog Hercules, both choose to engage in the play routine with each other; if Chris did not feel like playing, he might view the same actions as work (having to take the dog for exercise). If Hercules did not feel like playing, he might simply refuse, or retrieve the ball obediently but without the playful actions described in the chapter.

The theme of control during play is not only relevant in the initial choice to engage in play, but also to the fun that comes from risking loss of control. In many instances, the most exhilarating part of play is when the play approaches the boundary with reality. For example, when children engage in rough and tumble play, such as play wrestling, the risk that one player will be accidentally too aggressive or have his feelings hurt is where the fun lives. Players often enjoy being on the borderline between feeling out of control and overcoming fear (Sandseter, 2009; Stephenson, 2003). For Chris and Hercules, letting either one have the ball too easily without a bit of real frustration, fear, or risk of accidental injury (slipping, falling, knocking heads) would also not be very fun. For both Chris and Hercules, the most entertaining parts of the play are when one almost loses control of the ball but manages to keep it, or when one almost gains control of the ball but not quite. To repeatedly create this emotional experience takes creativity, as the same routine without variation would also result in boredom. Variations in how the ball is thrown or not thrown, and ways in which Hercules can fake out Chris require creativity.

There is also a fragile balance of control that goes on between players, a dynamic that players must attend to carefully. One player with too much control does not allow others to have fun and may inevitably lose his playmates, ending any opportunity for creativity. Collaborative play can be challenging as players compete for power and work to achieve consensus and reciprocity (Sheldon, 1996). For example, in children's pretend play, two boys may pretend to be a policeman and a robber. One boy playing the robber refuses to ever get captured, stating that he has super powers and is invincible. While this is a legitimate use of imagination within the pretense of play, it may eventually lead to frustration or boredom of the part of the boy playing the policeman, prematurely ending the play session and any further opportunities for imagination. It is therefore in all the players' best interests to share control and maintain this balance. As described in the chapter, Chris carefully attends to the balance of power, alternating between faking out Hercules and actually throwing the ball for Hercules to obtain. Similarly, Hercules makes it difficult for Chris to obtain the ball, but not impossible. Chris and Hercules are both skilled enough with the ball to just keep it forever and not let the other one get it; however, this would ruin the fun.

To maintain this power balance also requires creativity, and in some cases, an imbalance of power sparks creative ideas to restore balance. In a recently conducted study, when groups of elementary school girls played together for multiple sessions, similar power struggles were often observed (Hoffmann & Russ, 2012). For example, one playgroup followed a routine each session, where each child picked a doll, and the group of friends went on an adventure. However, each play session, one girl would pretend that her character was sick, "hijacking" the play session and having all other characters visit her in the hospital instead. By the fourth session, the other players had become exasperated with this controlling of the plot. Another player generated an inspired idea: she pretended to have magical powers, magically cured the sick character and allowed the adventure to continue. The magical character returned a balance of power to the group, since magic is even more powerful than sickness for directing the plot of a fantastical story.

Play with a friend is an opportunity to try out new ideas. What would happen if Chris were to throw two balls at once? What if Hercules were to drop the ball so that Chris could reach it, but put his paw on top of it? Some of these actions may result in more fun for the pair, while others may be duds, responded to poorly by the playmate, resulting in less fun. This sequence of playful idea generation, willingness to take risk, trial and error, and attention to feedback from others are all fundamental parts of the creative process.

References

Hoffmann, J., & Russ, S. (2012). *A pretend play group intervention for elementary school children* (Unpublished doctoral dissertation). Cleveland, OH: Case Western Reserve University.

Johnson, J., Christie, J., & Wardle, F. (2005). *Play, development and early education.* New York, NY: Allyn & Bacon.

King, N. R. (1979). Play: The kindergartner's perspective. *The Elementary School Journal, 80,* 80–87.

Sandseter, E. B. H. (2009). Children's expressions of exhilaration and fear in risky play. *Contemporary Issues in Early Childhood, 10,* 92–106.

Sheldon, A. (1996). You can be the baby brother, but you aren't born yet: Preschool girls' negotiation for power and access in pretend play. *Research on Language and Social Interaction, 29,* 57–80.

Stephenson, A. (2003). Physical risk-taking: Dangerous or endangered? *Early Years, 23,* 35–43.

Exploration Technique and Technical Innovations in Corvids and Parrots

Alice M.I. Auersperg

Department of Cognitive Biology, University of Vienna, Austria

Commentary on Chapter 3: Innovations in Corvids and Parrots

Beth A. Hennessey[1] and John H. Stathis[2]

[1]Wellesley College, Psychology Department, Wellesley, MA, USA [2]Connecticut College, Biology Department, New London, CT, USA

PARROTS AND CORVIDS AS AVIAN MODELS FOR PHYSICAL COGNITION

Unusual niches as well as unpredictable or seasonal resources foster a broader spectrum of foraging techniques and often promote flexibility and physical innovations (e.g., Sol, Timmermans, & Lefebvre, 2002; Tomasello & Call, 1997; Webster & Lefebvre, 2001). This may lead to an increase in technical abilities, including enhanced cause—effect exploitation, the capacity to distinguish between functional and nonfunctional object properties and/or the use of physical concepts.

While cognitive research, including technical intelligence, originally focused mainly on primates such as the great apes (e.g., Povinelli, 2000; Tomasello & Call, 1997), today some very important findings also

Animal Creativity and Innovation.
DOI: http://dx.doi.org/10.1016/B978-0-12-800648-1.00003-6

45

derive from bird studies (e.g., Emery, 2004; Emery & Clayton, 2004). Within birds, corvids and psittacines appear to be particularly promising candidate models for studies on physical cognition (note that while corvids comprise a single family of oscine passerine birds, psittacines encompass an entire order, the psittaciformes: when parrots are discussed in this chapter, I refer to any members of the three superfamilies, the "true" parrots, the cockatoos, and the New Zealand parrots). Despite the great evolutionary distance between birds and primates (ca. 300 Ma), their cognitive equipment and development show various similar traits (Emery & Clayton, 2004; Iwaniuk, Dean, & Nelson, 2005). Similarly to the great apes, corvids and parrots live in individualized social groups with complex hierarchical structures, often have extended periods of parent–offspring association and longevity, inhabit variable environments, and face ecological problems such as spatially and temporally heterogeneous resources (Emery, 2004; Timmermans, Lefebvre, Boire, & Basu, 2000). Although their brain anatomy is different, their relative forebrain sizes resemble those of great apes' prefrontal cortices (Timmermans et al., 2000) with the avian high vocal center and caudolateral nidopallium sharing similarities in function, neurophysiology and connectivity with the mammalian prefrontal cortex (Diekamp, Kalt, Ruhm, Koch, & Gunturkun, 2000; Reiner, 1986; Timmermans et al., 2000).

Due to their strong propensity to engage in new activities, both corvids and parrots could be considered innovation-prone taxa. As innovation is a critical ingredient of behavioral flexibility and vice versa, such species are believed to have great technical problem-solving abilities and a high tendency for social learning (Reader & Laland, 2004).

Although innovation is phylogenetically widespread, the highest innovation rates are associated with higher organisms. Lefebvre, Whittle, Lascaris, and Finkelstein (1997) studied avian feeding innovations using analytical techniques on publication notes. Their study resulted in a major survey incorporating more than 2000 examples of foraging innovations. The absolute and relative innovation frequency per order positively correlated with forebrain size. Furthermore, innovation rates seem to explain more of the variance in brain size in corvids and parrots than in other avian species (Emery, 2006; Lefebvre & Bolhuis, 2003; Lefebvre, Reader, & Sol, 2004; Lefebvre et al., 1997; Timmermans et al., 2000).

As expected, members of both families perform at levels similar to the great apes in a number of cognitive tasks, including physical problems such as Piagetian object permanence, comprehension of means–end relationships, connectivity, Gestalt perception, causal reasoning (e.g., Auersperg, Gajdon, & Huber, 2009; Auersperg, Kacelnik, & von

Bayern, 2013; Pepperberg, Willner, & Gravitz, 1997; Seed, Tebbich, Emery, & Clayton, 2006; Pepperberg & Kozak, 1986; Taylor, Hunt, Medina, & Gray, 2009; Tebbich, Seed, Emery, & Clayton, 2007) or innovative tool use (e.g., Auersperg, Gajdon, & Huber, 2010; Auersperg, Szabo, von Bayern, & Kacelnik, 2012; Auersperg, von Bayern, Gajdon, Huber, & Kacelnik, 2011; Bird & Emery, 2009a, 2009b; Cheke, Bird, & Clayton 2011; Gajdon, Lichtnegger, & Huber, 2014; Jones & Kamil, 1973; Taylor, Elliffe, Hunt, & Gray, 2010; Taylor, Hunt, Holzhaider, & Gray, 2007; von Bayern, Heathcote, Rutz, & Kacelnik, 2009; Weir, Chappell, & Kacelnik, 2002; Werdenich & Huber, 2006; Wimpenny, Weir, Clayton, Rutz, & Kacelnik, 2009).

Based on the fossil records, parrots are presently estimated to date back at least around 50 Ma (Dyke & Cooper, 2000), with the earliest record of modern parrots starting approximately 20 Ma (Forshaw, 1989). Passeriformes on the other hand are one of the youngest bird orders appearing around 37 Ma, with the first corvids originating about 17 Ma (Goodwin, 1986). However, both corvids and parrots seem to be subject to recent evolutionary change, (e.g., Omland, Tarr, Boarman, Marzluff, & Fleischer, 2000; Smith, 1975). The phylogenetic distance between the taxa, their innovative capacity, learning ability, as well as their brain size and anatomy may provide an arena for the testing of hypotheses about convergent evolution of high-level cognitive abilities and neural structure within birds (Maclean et al., 2011; Tebbich, Sterelny & Teschke, 2010).

As of now, comparisons taking phylogeny, ecology, as well as the resulting physiological and motivational constraints into account are rare and difficult to establish (Auersperg et al., 2011). Comparisons of cognitive capacities between two distantly related taxa frequently appear following the discovery of a hitherto uncharted ability in one species but not in another, or by applying an established test situation to a new species ignoring their different ecological backgrounds. Furthermore, there still is little focus on the mechanisms underlying differences in performance, such as different explorative strategies (Demery, 2012). Different taxa may have similar cognitive skills but employ them differently.

Perhaps the most prominent corvid model for physical cognition is the New Caledonian crow, which is well known for using and manufacturing a variety of tools in the wild and flexibly adjusting to new tool-related problems in the laboratory (e.g., Auersperg et al., 2011; Hunt, 1996; Taylor et al., 2007, 2009, 2010; von Bayern et al., 2009; Weir et al., 2002; Wimpenny et al., 2009). Some other corvids have also gained some attention within the last decade such as the rook (e.g., Bird & Emery, 2009a, 2009b; Seed et al., 2007; Seed, Clayton, & Emery, 2008), the Eurasian jay (e.g., Cheke et al., 2011), the raven (e.g., Heinrich &

Bugnyar, 2005), the carrion crow (e.g., Hoffmann et al., 2011), or the magpie (e.g., Pollok, Prior, & Gunturkün, 2000).

Likewise, New Zealand's mountain parrot, the neophilic and generalist kea has been used as a parrot model for technical intelligence in Vienna (Auersperg et al., 2009, 2010, 2011; Diamond & Bond, 1999; Gajdon, Amann, & Huber, 2011; Gajdon et al., 2014; Huber & Gajdon, 2006; Werdenich & Huber, 2006). Only recently we started investigating the playful Goffin cockatoo (Auersperg, Szabo, & Bugnyar, 2013; Auersperg et al., 2012). Moreover, there have been studies on physical problem solving in African grey parrots (e.g., Pepperberg & Kozak, 1986; Pepperberg et al., 1997; Schloegl, Schmidt, Boeckle, Weiß, & Kotrschal, 2012), New Zealand kakariki parakeets (e.g., Demery, 2012; Funk, 2002), and some on neotropical species (e.g., Borsari & Ottoni, 2005; de Mendonca-Furtado & Ottoni, 2008; Liedtke, Werdenich, Gajdon, Huber, & Wanker, 2011; Schmuck-Paim, Borsari, & Ottoni, 2009). In this chapter, I will specify some of the factors which may affect exploration mode and problem-solving performance and present the Multi-Access Box (Auersperg, Gajdon, & Huber, 2011), as an experimental approach to compare different advances to problem solving of a corvid and a parrot model.

EXPLORATION AND PLAY

One of the most basic mechanisms influencing the innovative capacity of an animal is the way in which it gathers information about its environment. Exploration encompasses perception as well motor interaction with objects (e.g., Demery, 2012). The exploration of a particular affordance ("affordances" are referred to in the sense of the properties of an object that determine its functionality) is usually not repetitive unless it provides a potential benefit to the actor (Bateson & Martin, 2013).

However, exploration can be hard to segregate from object play. Both are heavily influenced by motivation and curiosity. Both can be linked to creativity (exploration less than play) and to innovation, both are useful to keep constructing knowledge and to avoid suboptimal endpoints through probing and sampling the environment (Bateson & Martin, 2013). While the purpose of exploration is largely limited to information gathering, play has other possible benefits such as exercising the neuromuscular system and rehearsing potentially risky situations (Bateson & Martin, 2013; Burghardt & Graham, 2010). Furthermore, in contrast to exploration, play is repetitive, exaggerated, and regularly leads to dead ends (Bateson & Martin, 2013).

In large brained, highly innovative species, object play and exploration are both advanced and can be hard to disentangle; information gathering can be a mixture of both. Thus, playful object exploration is highly developed in both corvids and parrots, either limited to certain developmental phases or throughout adulthood and is most likely a critical factor underwriting their problem-solving abilities. One aspect of object play, which is often linked to physical cognition are seemingly unrewarded object—object combinations (e.g., Connolly & Dagleish, 1989; Hayashi & Matsuzawa, 2003; Hayashi, Takeshita, & Matsuzawa, 2006; Kenward et al., 2011). In primates, the act of combining two different objects in a nonforaging context in a schematized way such as can be observed in human toddlers (Connolly & Dagleish, 1989; Rat-Fischer, O'Regan, & Fagard, 2012) has so far only been discovered in capuchin monkey and in the great apes (e.g., Hayashi & Matsuzawa, 2003; Hayashi et al., 2006; Takeshita, 2001; Torigoe, 1985). Notably, both groups stand out among primates, not only for their advanced abilities in the technical domain but also for habitually using tools in the wild without depending on them (Fragaszy, Izar, Visalberghi, Ottoni, & Gomes de Oliveira, 2004; Hayashi et al., 2006). Playful object combinations have only recently been addressed analytically in a number of corvids and parrots. Kenward et al. (2011) discovered that throughout their study period of six weeks, object combinations increased steadily in fledgling New Caledonian crows but peaked and declined in common ravens of the same age. Their observations support the assumption that such behavior may have promoted the evolution of habitual tool use in the New Caledonian crow. Kea playfully inserted cylinder-shaped objects into vertical tubes (Gajdon et al., 2014). Notably they grabbed the objects at the one end when the tubes were open (for seemingly intentional insertions) but held them in the center during combinations when the tubes were blocked at the top (placing them on top of the tube). Kea, which playfully combined objects with tubes, were more likely to do so in a problem-solving context (Gajdon et al., 2014). Furthermore, in order to gain a greater synopsis over object combination in parrots and corvids, Auersperg, Oswald, Domansegg & Bugnyar, 2015; Auersperg, van Horik, Bugnyar, Kacelnik, Emery & von Bayern, 2015) recently conducted two studies on unrewarded object combinations using the same setup. The studies incorporated a total of three corvids, and nine parrots. Paralleling previous findings in primates (Hayashi & Matsuzawa, 2003; Hayashi et al., 2006; Takeshita, 2001; Torigoe, 1985), combinatory actions prevailed mostly (Figure 3.1) and tended to be more complex in terms of structure, (combining up to three objects, inserting objects into tubes or stacking rings onto poles) in species associated with advanced technical cognition and tool use. They were mainly observed in the New Caledonian crows within corvids and in kea and Goffin cockatoos within parrots; furthermore in the black palm cockatoo, a parrot which could be classified a habitual tool

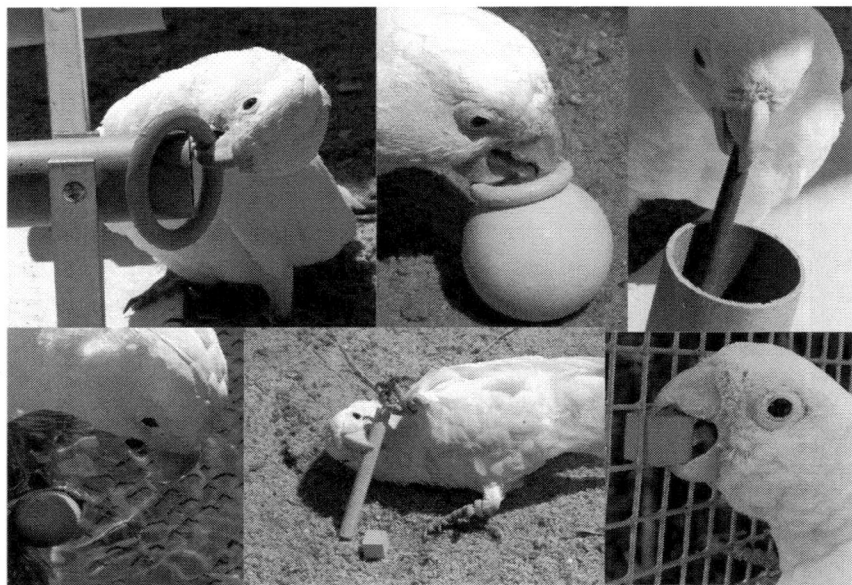

FIGURE 3.1   Examples of combinatory actions during unrewarded play in the Goffin cockatoo, e.g., ring-stacking, put-on-top, insertions, submerging in water. (*Photos: AMI Auersperg.*)

user in a social context as the males beat pieces of wood or hard nutshells with their feet against perches or hollow trees to defend their territory and to attract mates (Lantermann, 1999; Rowley, 1997).

While, in most corvids, object play seems to vanish with or sometime after sexual maturity (Heinrich, 2007; Kenward et al., 2011) it remained at high levels in many of the parrots that were tested (Auersperg et al., 2015) indicating that exploration during adult problem solving is likely to be infused with playful behavior in the respective parrot models.

MORPHOLOGY AND SENSORIMOTOR PLASTICITY

There is some, but limited knowledge about the appendages used during exploration and the subsequent motor plasticity. As an adaption to probing, pecking, and occasionally for turning and tearing during extractive foraging, most corvids have relatively straight, long, sharp, and pointed beaks. Skull measures indicate that in corvids that practice much probing and tearing during foraging, such as the scavenging common raven, the beak tip is slightly more curved than in species that primarily use pecking behavior such as the Eurasian jay (Kuhlmeyer,

Asbar, Gunz, Frahnert, & Barilein, 2009) or the New Caledonian crow, which has a bill morphology perfectly adapted to holding a tool straight between both bill tips (Troscianko, von Bayern, Chappell, Rutz, & Martin, 2012). Wild New Caledonian crows owe a substantial part of their protein intake to their tool use activities (Rutz et al., 2010). In actively probing corvids, the sideward orientation of the eye is marginally increased in contrast to species that mostly peck and turn objects such as the jackdaw, the Eurasian jay, or the magpie (Kuhlmeyer et al., 2009). While corvids seem to explore novel objects initially predominantly through visual inspection, when it comes to direct manipulation they mainly use the bill. As for most birds, corvid feet are anisodactyl with the first digit in the back and the remaining three in front. Thus, they are well adapted for pushing off the substrate and for perching, but not as much for picking up objects (Brooke & Birkhead, 1991). Some corvids use their feet to fixate objects during rostral exploration or to rake away sediments (Heinrich, 2007), but they rarely pick up and hold object in their feet. In contrast, parrots have zygodactylous feet, with relatively strong toes, two in the front (digits 2 and 3) and in two the back (digits 1 and 4). Zygodactily can be found in a few birds such as woodpeckers, certain owls and parrots, as well as in some reptiles such as chameleons, and is most likely an adaption to climbing and clambering (Bock, 1999; Harris, 1989). Parrots regularly use their feet to hold small enough objects up to their bill during rostral manipulation (Luescher, 2006). When using their feet, parrots exhibit motor lateralization, with the majority of individuals within a species being "left footed" (Brown & Magat, 2011; Harris, 1989; Magat & Brown, 2009). Parrot feet are additionally (but less frequently) used for direct manipulative exploration: while five other birds used their bill, one of our Goffin cockatoos used his foot to screw a bolt out of a nut using ca. 25−30 consecutive up and down movements of his claw during problem solving (Auersperg et al., 2013) (Figure 3.2).

A parrot's bill is more flexible than in most other birds, as the upper mandible is not part of the cranium but hinged onto the skull with an additional joint (Forshaw & Cooper, 1989). Consequently it can dexterously be coordinated with the tongue and feet making it an adaptive multipurpose foraging tool (Diamond & Bond, 1999). In comparison to corvids, the upper mandible is considerably larger, pointed and strongly curved, while the lower mandible has a sharp, upward facing edge, which can be used like an anvil against the upper mandible, enabling the animals to extract food from hard-shelled fruits. The strong muscular tongue, which is thick and rounded at its tip (rather than slim and pointed as in most other birds), can be used in a thumb-like manner to turn objects inside the beak or to press objects against the bill tip, compensating for the smaller size of the lower mandible

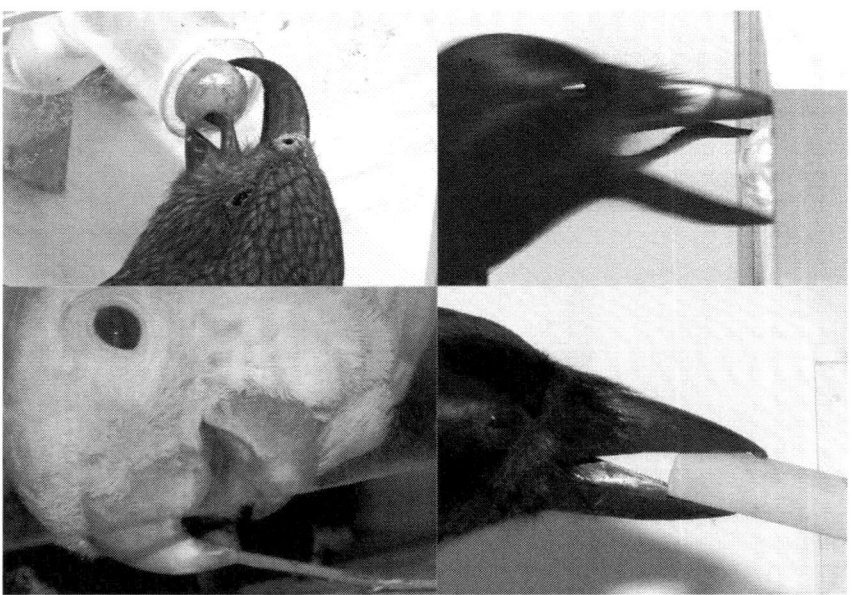

FIGURE 3.2   Top row: kea (left) and New Caledonian crow (right) use their tongue to push a ball into a horizontal tube. Bottom row: Goffin cockatoo (left) holds proximal end of stick tool using his tongue to press it against the tip of its upper mandible. New Caledonian crow (right) holds elongated tool firmly between both mandibles.

(Collar, 1997; Demery, 2012; Zweers, Berkhoudt, & van den Berge, 1994). Interestingly, when inserting a ball-shaped object into a horizontal tube during problem solving, we witnessed both a New Caledonian crow and a kea lifting the ball against the tube entrance between both bill tips and using the tongue to give it a push into the tube (Auersperg et al., 2011). However, while using a slim, stick-type tool, a Goffin cockatoo carefully coordinated the movement of the distal tool end relative to a food reward by pressing its proximal end against the tip of the upper mandible with his tongue, rather than holding it between both bill tips as observed in New Caledonian crows (Auersperg et al., 2012; Troscianko et al., 2012). Bringing objects into contact with the bill tip may further help to collect sensory information, as pits embedded in the keratin along the inner edges of this area harbor highly sensitive mechanoreceptors, responding to small changes in pressure. Other than in parrots, this so-called bill tip organ can be found in several other bird species; for example the New Zealand kiwi, some ibises and shorebirds, as well as waterfowl (Berkhoudt, 1980; Cunningham et al., 2010; Gottschaldt, 1985; Gottschaldt & Lausmann, 1974). Most of the latter species have their mechanoreceptors on the outer surface of the bill and can therefore (in contrast to parrots) use them to perceive objects outside their grip. Interestingly, most of the latter use their bill for exploration and foraging while the manipulated objects are not clearly visible.

Another feature likely to have a profound impact on avian exploration is their visual field (Demery, 2012; Troscianko et al., 2012). Corvids have relatively long binocular regions, which are broader at the front and entail a blind spot above and behind the head. Again, skull data suggests that corvids that primarily use probing actions while foraging have a narrower binocular overlap than species that predominantly use pecking (Kuhlmeyer et al., 2009; Martin, 1985, 2007). Furthermore species that frequently handle objects inside their bill seem to acquire a better horizontal projection of their beak tip (Martin, 2007; Martin & Coetzee, 2004). The most extreme case is the New Caledonian crow: as an adaption to following the movement of the distal end of a stick tool held inside the bill, the binocular overlap of their visual field is larger than in any other corvid examined so far (Troscianko et al., 2012). The eye position of most parrots is relatively high and lateral, suggesting a well-covered overall visual coverage. However, there are, as yet, only limited concrete measurements of the visual field of parrots. Demery, Chappell, and Martin (2011) measured the visual field of Senegal parrots. The respective binocular overlap was relatively broad and, including the two monocular portions, there is an almost complete coverage of the celestial hemisphere, with a blind area starting just below the beak tip. This means that targeting an object with both bill tips would probably be harder for them than for most corvids (Demery et al., 2011; Kuhlmeyer et al., 2009). Objects which are taken under closer inspection are consequently held up to the binocular portion of the visual field using a foot, thereafter the bill tip organ aids by providing additional tactile information, reducing the need for further visual cues (Demery et al., 2011; Whittow & Sturkie, 1999). While the birds are collecting tactile information about an object using the bill tip organ and the tongue, they can simultaneously be vigilant of the region above their heads for predators (Demery et al., 2011).

With the exception of some nocturnal parrots such as the New Zealand kakapo (Gsell, 2012) or nectar feeding lories (Roper, 2003) both corvids (including the food caching species) and parrots have relatively small olfactory bulbs (Cobb, 1960; Macphail, 1982). Although both species are clearly highly visual animals, and judged from the size of the respective brain region their sense of smell is believed to be rather poor, there is some evidence suggesting that it does nevertheless play a role in discrimination during explorative foraging (Buitron & Nuechterlein, 1985; Gsell, Hagelin, & Brunton, 2012; Harriman & Berger, 1986).

The ears are well developed in both corvids and parrots. It is, however, unclear to what extent the sense of hearing is used during exploration of the animal's physical environment. There is some evidence that African grey parrots can use auditory cues (such as an empty and a baited food cup shaken by an experimenter) to identify the location of a hidden food item (Schloegl et al., 2012), whereas results in a similar task were less convincing in Eurasian jays (Shaw, Plotnik, & Clayton, 2013).

In summary, while corvids' visual field allows for better guidance of the beak during direct manipulation, in parrots, appendages such as the feet, tongue, and a flexible bill with sensible mechanoreceptors are used in coordinated action, allowing for a greater repertoire of manipulative behaviors during haptic manipulation.

EMOTIONS AND MOTIVATION

The emotional response to novel environmental situations, closely interlinked with the motivational drive, is similarly contributing to the synopsis of an animal's exploration technique (Greenberg, 2004).

Motivation can be regarded as process that activates, directs, and sustains a particular goal-driven behavior (Franken, 1994). In human psychology, there are several components contributing to motivation: the initiation of the respective behavior (activation), the vigor with which the behavior is conducted (intensity) and the keeping up of the effort (persistence) (Iland, 2013). For example, a species feeding on a large number of unpredictable and difficult to open resources could be expected to be more persistent and vigorous during novel object manipulation. The activation to explore a new environmental situation is further strongly dependent on the aversion and the attraction a subject harbors toward novelty, or its neophobia/neophilia (Greenberg, 2004). It is important to note that the difference between neophobia and neophilia is not believed to be a continuum but that they are two distinct complex and dynamic processes, which are most likely influenced by different selection pressures (Greenberg & Mettke-Hoffmann, 2001) and these processes can interact with one another (Russel, 1973). While both increase the attention to novelty, neophilia is closely linked to curiosity, the desire to create new information (which is a highly important prerequisite for exploration; Berlyne, 1960), and consequently helps to expand the foraging repertoire. Neophobia serves to avoid the dangers of possible injury or death while approaching a new resource (Greenberg & Mettke-Hoffmann, 2001). Neophobia and neophilia are considered instruments for regulating ecological plasticity, linking neophobia to specialism and neophilia to generalism (Greenberg, 2004). For example, using a novel object approach test, Mettke-Hoffman, Winkler, and Leisler (2002) examined neophobia in a number of parrot species.

They found that it was lower in species with complex habitats and in island species.

Interestingly, most parrots and corvids, including generalist species, are highly innovative and do still have high tendencies toward both neophilia and neophobia (with some exceptions, such as the New Zealand kea which seems to lack neophobia; Diamond & Bond, 1999; Greenberg, 2004; Greenberg & Mettke-Hoffmann, 2001). Thus, most parrots and corvids seem to be drawn to novel objects but approach them with fear at the same time. Vernelli (2013) addressed this paradox on a highly neophobic opportunistic generalist, the common magpie. Novelty responses were heavily influenced by environmental context: birds were not afraid of a new object when it was placed in an equally new environment, but were hesitant to approach the same item while being in familiar environment. The author suggests that, in some species, a violation of previous expectations may be more important for neophobia than absolute novelty.

However, neophobia and neophilia do not seem to persist at the same level in corvid and parrot development: while neophobia gradually increases throughout the development of a young corvid, neophilia is largely limited to juvenile and/or subadult life. However, neophilia lasts throughout adulthood in many parrots and cockatoos (Heinrich, 2007; Luescher, 2006).

EXPLORATION TECHNIQUE AND INNOVATIVE PROBLEM SOLVING

Considering the previous and presuming full habituation, a corvid model species such as the New Caledonian crow, the rook or the Eurasian jay (Bird & Emery, 2009a, 2009b; Cheke et al., 2011; Seed et al., 2006; Taylor et al., 2007) could be expected to approach a technical problem in a generally less haptical manner than some parrot models such as the kea, the Goffin cockatoo, or the African grey parrots (Auersperg et al., 2013; Huber & Gajdon, 2006; Schloegl et al., 2012). A species which gathers information predominantly through direct, haptic manipulation rather than through close visual inspection is unlikely to instantly approach a new problem in the most efficient way, but may instead detect its affordances in the process of manipulation. This may be very helpful when the affordances of a task are not directly visible and can only be detected by running through a large repertoire of different motor actions. For example, a food reward inside an "artificial fruit" is locked away by a series of unfamiliar locking devices. In parrots such tasks have been presented to kea and Goffin cockatoos (Auersperg et al., 2013; Huber, Rechberger, & Taborsky, 2001; Miyata,

Gajdon, Huber, & Fujita, 2011). In the most extreme case a male captive Goffin cockatoo Pipin, lacking previous training, dismantled a sequence of five unfamiliar locks in 100 min total trial time (Auersperg et al., 2013). His actions encompassed pulling up a pin, unscrewing a bolt from a nut, pressing a cylinder through a ring, fitting a perforation in a wheel-shaped lock through a t-bar and pushing a bar-lock open. Interestingly, after the initial acquisition of the task he was immediately able to reliably repeat the entire sequence of motor actions. The ability to instantaneously memorize the problem makes his performance innovative, as without such immediate learning, it would be hard to distinguish the occurrence of an innovation from mere exploration (Reader & Laland, 2004).

Pipin solved the problem through vigorous and versatile haptic exploration: he first manipulated the locks close to the reward, moving further away until he detected changes in movability in one of the affordances, which seemed to have a reinforcing effect on the intensity of his manipulation. It is highly unlikely that a similar problem could have been solved by a purely visual explorer (Auersperg et al., 2013). However, as Pipin's performance as well as the performance of the remaining subjects (which needed social and/or stepwise exposure to the problem in order to solve the entire sequence) cannot be segregated from principles of associative learning, it remained unclear whether the animals had learned anything about the physics underlying the problem, namely that one lock was blocking the other. To investigate this, transfer tests were applied after the initial explorative phase was saturated. Interestingly, subjects reacted sensibly and flexibly to disruptions of the sequence in most (not all) conditions, omitting irrelevant locks above the gap, even when the sequence was scrambled (Auersperg et al., 2013). Similarly, when comparing haptic exploration in kakariki and young children, Demery (2012) found that both explore potentially functional changes in an object (weight) more intensely than nonfunctional changes (color). Furthermore, they seem to focus more on object properties that may provide clues about underlying affordances.

A classic benchmark test for causal reasoning in animal cognition is the trap tube problem, which was first used by primatologists to investigate whether the capuchin monkey apprehend surface continuity (Visalberghi & Limongelli, 1994).

The basic idea is a transparent tube bearing an out-of-reach reward, which can be pushed (or pulled in some setups) into both directions by the subject using a tool. The surface of one direction is disrupted by a trap. If the food is pulled inside the trap it is lost to the subject. The setup has been tested integrating numerous controls such as nonfunctional traps and feature changes (e.g., Seed et al., 2006; Taylor et al., 2009; Tebbich et al., 2007; Teschke & Tebbich, 2011; Visalberghi

& Limongelli, 1994). The task is built in a way that the animals only have one single attempt to get it right in each trial. Inhibitory control as well as a cautious approach to the problem would therefore be an important precondition for a good performance. Aberrations of the trap tube problem have so far been tested on two corvids and three parrots: rooks as well as New Caledonian crows mastered the initial tasks following a learning phase (Seed et al., 2006; Taylor et al., 2009; Tebbich et al., 2007). Some subjects of both species performed spontaneously well in a series of control tasks in which fragments of the tasks (functionality of traps, open tube ends or transfer from trap tube to trap table) were altered. In contrary, six New Zealand kea, three red-and-green-winged macaws and a sulfur crested cockatoo failed to complete the original task (Liedtke et al., 2011). In fairness, it should be noted that the kea were offered less trials in total than the minimum that, for example the New Caledonian crows, needed to accomplish the task. However, all parrots were additionally tested in a slot tube task, which was like the original setup, except that instead of the rakes on both sides the tube had a dorsal slot so the birds could move the reward directly inside the tube with their beaks. Most birds quickly detected an alternative solution by lifting the reward within their beak over the trap (the cockatoo used its foot) (Liedtke et al., 2011). Interestingly, two kea initially avoided the trap in the first sessions above chance expectation, but thereafter switched their strategy to lifting the food, as the other birds did, and subsequently ignored the position of the trap (Liedtke et al., 2011).

Liedtke et al. (2011) suggest that the parrots' failure in this task may be caused by lack of inhibition once manipulation starts: as opposed to the parrots, the New Caledonian crows often switched sides at the tube before correctly completing their behavior (Taylor et al., 2008). Alternatively, as parrots cannot see far below their beak, it would be hard for them to observe the food relative to the trap while they are moving it in with their upper mandible along the slot tube setup (Demery et al., 2011; Liedtke et al., 2011). Another study with similar findings was conducted by Schloegl et al. (2009) on common ravens and kea: the birds were confronted with a choice of tasks on bent tubes. Ravens were more efficient in their perceptual exploration of the setup, needing only partial information to form conclusions about the content of bent tubes, while kea examined the tubes more thoroughly, but seemed to ignore the impact of the tube shape on the visibility of its content. The authors suggest that the exploration technique of the kea may prompt them to spend more time in manipulative motor exploration while being a food cacher or having a food caching ancestor may increase pressure to perceptually obtain information about the location of hidden items efficiently.

EXPLORATION TECHNIQUE AND
INNOVATIVE TOOL USE

Despite its benefits, and the many opportunities for tool use to evolve, it remains rare among animals (Hunt, Gray, & Taylor, 2013). Its scarcity does not, however, establish a link to an animals' technical intelligence (Call, 2013). There are numerous examples of stereotyped tool use as an inflexible, innate component of a particular species' behavior (Hunt, 2013; Shumaker, Walkup, & Beck, 2011). Hence, intelligent tool use requires flexibly and the ability to creatively adapt to novel situations (Call, 2013).

Call (2013) suggests three main abilities that influence the emergence of creative tool use in animals: to gather and store information even in the absence of reinforcement, to appropriately link pieces of information that were acquired independently in time and space, as well as a predisposition to manipulate objects during exploration.

Members of both corvids and parrots have been found capable of flexible and creative tool innovation (e.g., Auersperg et al., 2011, 2012; Bird & Emery, 2009a, 2009b; Cheke et al., 2011; Gajdon et al., 2014; Taylor et al., 2007, 2010; von Bayern et al., 2009; Weir et al., 2002; Wimpenny et al., 2009). Within controlled laboratory studies, rooks, Eurasian jays, New Caledonian crows, and kea (Bird & Emery, 2009a, 2009b; Cheke et al., 2011; Gajdon et al., 2014; von Bayern et al., 2009) were recorded to innovate compact objects as tools. Kea, a Goffin cockatoo and rooks, which lack a known predisposition to do so, similarly used stick-shaped objects to access a food reward (Auersperg et al., 2011, 2012; Bird & Emery, 2009a, 2009b). New Caledonian crows and rooks modified objects into known, task-appropriate tools, bending straight pieces of wire into hooks (Bird & Emery, 2009a, 2009b; Weir et al., 2002). New Caledonian crows and rooks used two (and more in New Caledonian crows) similar tools of different sizes sequentially (Bird & Emery, 2009a, 2009b; Taylor et al., 2007, 2010; Wimpenny et al., 2009) Finally blue jays ripped newspaper pieces and a Goffin cockatoo sculpted stick tools out of a wooden beam to rake food toward their own body (Auersperg et al., 2012; Jones & Kamil, 1973). Parrots and corvids are unlikely to share tool-using ancestors. However, other than corvids, most parrots do not construct complex nest cups, they breed in simple burrows and are not likely to have an ecological predisposition for establishing complex objects relationships (Auersperg et al., 2011; Hansell & Ruxton, 2008).

A tool problem applied to two corvids and a parrot yielded rather interesting differences in task acquisition (Bird & Emery, 2009; Gajdon et al., 2014; von Bayern et al., 2009). The basic setup comprised a food reward inside a transparent tube which could only be retrieved by inserting a compact object of the appropriate size into the upper end of the tube and by consequently hitting the food directly with the tool or by collapsing a platform inside the tube. Following a training in which the compact tool was placed on the upper rim of the tube and was accidently nudged inside by the animals, rooks picked up objects, carried them to the tube and dropped them inside (Bird & Emery, 2009a, 2009b).

In order to evaluate the type of experience required to solve this problem, von Bayern et al. (2009) repeated the task with New Caledonian crows but the birds immediately faced the task without prior training. After all subjects failed, they were divided into two groups which received two different types of pre-experiences: one group was trained the same way as the rooks and the other received the shortened tube so the birds could collapse the platform bearing their reward with their bill tip (von Bayern et al., 2009).

Most birds from both groups thereafter mastered the transfer to the original tool task, indicating that learning about certain task affordances was an important precondition for the animals to solve the problem.

When kea were tested on a similar problem facing vertically slanted tubes (one open, one blocked at the top) inside which a food reward was loosely fixed, some birds readily inserted objects into tubes without prior training (Gajdon et al., 2014). This was not surprising, as the same birds had previously also playfully inserted objects into tubes in a non-foraging context. However, the birds initially combined inappropriately sized or objects with the tubes more often than would be expected by chance (combining the object with the lower end or the blocked end of the tube). This strongly contrasts the results in rooks, which always chose the functional stone from a pair (Bird & Emery, 2009a, 2009b). While for a playful explorer such as the kea, the manipulation of the new setup seemed to have become self-rewarding, a species with less adult play drive, such as the rook or the New Caledonian crow, needed to learn about the underlying affordances in order to build up sufficient motivation to attempt the acquisition of the reward. As the kea spontaneously solved the task (most likely accidentally) without requiring controlled pre-experience, it remains unknown what they have learned about the mechanics of the setup.

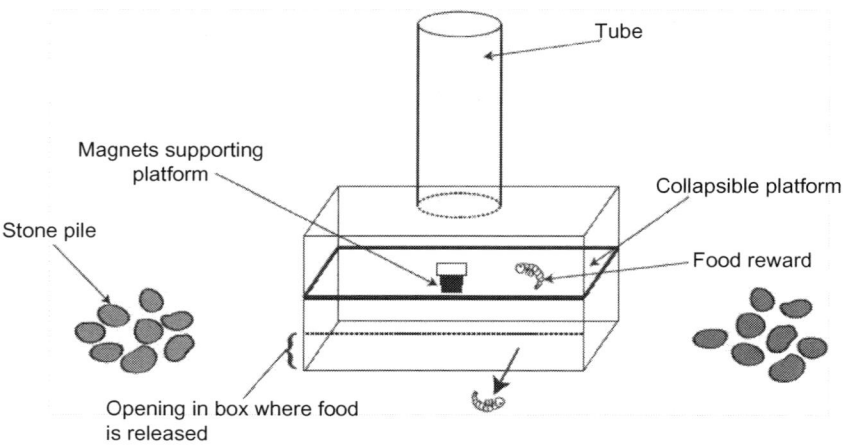

Collapsible platform/tube apparatus. (*As in von Bayern et al., 2009, reprinted with permission.*)

MULTI-ACCESS BOX

Clearly, the way in which different animals gather information is fundamentally important for understanding the way they approach technical problems. It is however extremely hard to find paradigms which are ecologically valid for comparison (Auersperg et al., 2012).

As physical cognition is an umbrella term embracing a diverse variety of different cognitive processes, it is plausible that different species with similar overall technical intelligence will either outperform each other in different processes and/or use different approaches to similar cognitive problems. If performance is compared on the basis of individual tasks tailored to the constraints of one particular species we gain only limited information about the cognitive mechanisms underpinning the respective behavior. If, on the other hand, another species is presented with a converted methodology adjusted to its all-new constraints, we lose comparability (Auersperg et al., 2011, 2012). Direct comparisons between corvids and parrots should be carefully designed considering the constraints and explorative approaches of both taxa.

Auersperg et al. (2011) used a "Multi-Access Box" approach to compare problem-solving performance in two prominent corvid and parrot models: the New Caledonian crow and the kea. The first basic idea was to confront the animals with a battery of different problems at once that lead to the same central food reward so subjects are able to choose from a set of different solutions. This allows us to investigate inter and intraspecific preferences between solutions. The second idea was to let subjects reach criterion on one solution and closing it thereafter. If the same option was used for a given number of times, it was subsequently blocked, forcing subjects to switch to an alternative solution. This way, we can examine not only whether the animals are generally capable of solving all problems presented, but also the way they approach the different problems and their underlying learning strategies. Do they switch between solutions or do they use "win-stay" and only switch when forced? How fast do they accomplish solutions without using ineffective actions, and is there a specific order in which the solutions are accomplished?

The apparatus used was a cubic box with each lateral side presenting a solution to obtain the same central food reward: a string tied around the reward could be pulled through a hole and a window could be opened using a hook-shaped handle. Two additional solutions presented approaches which required the use of a tool: a ball-shaped object could be inserted into a pipe leading toward the food or a stick tool could be inserted through an opening and used to directly push the reward off a central platform. The lateral walls were removable and could be replaced for blocked versions of the original solutions.

One subject of each species both discovered and reached criterion in all four of the solutions indicating that the tasks at hand were within the cognitive capacity of both species. The remaining results turned out to be telling in terms of explorative mode and cognitive competences between species: most likely as a product of high neophilia and a vigorous and haptic exploration technique, the kea seemed to "experiment" with the apparatus, discovering more solutions much faster than the crows, using up to three solutions in the first session alone. Kea quickly abandoned a former option once it was blocked and switched to new solutions faster. Furthermore, they showed individual variation in the order in which solutions were accomplished. However, it consequently took them longer than the crows to reach criterion for one particular solution and for an entrance to be blocked. The crows essentially used a win-stay strategy, sticking to a solution that had worked in the previous trial. Interestingly, all used the same order of solutions, the string, the stick, the ball, and one crow additionally used the window entrance. The fast discovery of the stick solution reflects their adaptive specialization to tool use. As they seemed to explore the box more in a visually guided manner with occasional brief pecking actions with the stick tool against the box, they seemed to have most difficulties solving with the window entrance which required pulling the handle. The kea, in contrast, used a larger repertoire of motor actions on the box, but had problems inserting the stick. After most solutions were blocked they seemed to keep searching for alternative solutions, such as breaking the walls of the box, trying to flip it over (added weights were pushed out of the way), extract the reward platform from the food opening or trying to remove the bolts from the top of the box (personal observations). One kea did succeed in the stick solution using a complex multistep technique for overcoming the morphological constraint of inserting a stick tool with his strongly curved beak.

In summary, both species revealed the cognitive capacity to solve the Multi-Access Box task, coming up with novel- and task-appropriate solutions. However, the way they approached the problems highlighted a number of factors such as differences in neophilia/neophobia, persistence, sensorimotor plasticity, and playfulness reflected though their respective exploration techniques. In order to gain a better overview of explorative predispositions and cognitive skills within and between corvid and parrot models it would be necessary to apply this and other task batteries offering a number of different innovative solutions at the same time to a larger set of species. For example it would be crucial to test some more neophobic parrots. Moreover, testing other corvids on the Multi-Access Box could indicate whether the lack of variability within New Caledonian crows, or their problems with solutions that require direct affordance manipulation such as the window, is caused by their adaptive specialization to tool use.

Nevertheless, when using corvid and parrot subjects not in batteries of tasks but in individual task setups targeting explicit cognitive processes there are a number of complications that require attention. From the previous discussion it is to be expected that a typical parrot model may address a technical problem playfully and with a more haptic exploration technique. Therefore, spontaneous performance will often only provide limited information as a subject may stumble across solutions during playful manipulation of the apparatus (e.g., Gajdon et al., 2014). As extensive habituation, during which animals are allowed to playfully interact with an apparatus, is not always possible without cuing, it may be sensible to use a transfer task, targeting the degree to which the animals consider the physics underlying the task after the initial acquisition is accomplished and playful exploration is saturated (e.g., Auersperg et al., 2013). The most common way to do so is to add or subtract functional and/or nonfunctional features from the original tasks (Shettleworth, 2009). It may moreover be problematic to test certain parrot models, such as the kea, for an extensive number of trials on the same task without alterations as subjects may drop in performance due to their urge to try new strategies (e.g., kea in Liedtke et al., 2011).

Adult corvids in contrast seem to be less playful and more perceptually guided explorers: while they may address a visible technical problem more directly and efficiently, they may have difficulties with a task in which the affordances are not directly visible. Therefore, it can help to use control groups to provide them with distinctive pre-experiences with the affordances of the apparatus (e.g., von Bayern et al., 2009).

When comparing corvids and parrots directly the explorative constraints of both should be considered accordingly.

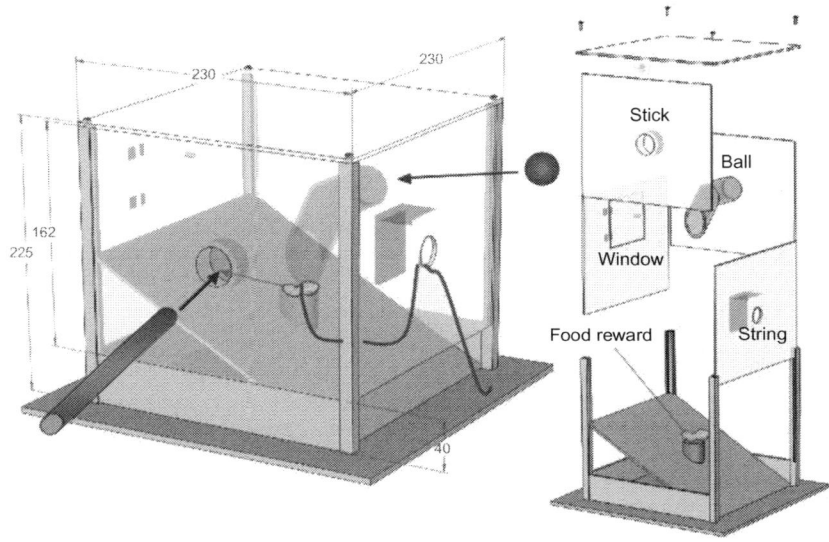

Multi-Access Box (MAB). *(As in Auersperg et al., 2011. Copyright 2011 by the Public Library of Science. Reprinted with permission.)*

References

Auersperg, A. M. I., Gajdon, G. K., & Huber, L. (2009). Kea (*Nestor notabilis*) consider spatial relationships between objects in the support problem. *Biology Letters, 5,* 455–458.

Auersperg, A. M. I., Gajdon, G. K., & Huber, L. (2010). Kea, *Nestor notabilis,* produce dynamic relationships between objects in a second-order tool use task. *Animal Behaviour, 80,* 783–789.

Auersperg, A. M. I., Gajdon, G. K., & Huber, L. (2011). Navigating a tool end in a specific direction: Stick-tool use in kea (*Nestor notabilis*). *Biology Letters.* Available from: http://dx.doi.org/10.1098/rsbl.2011.0388.

Auersperg, A. M. I., Gajdon, G. K., & von Bayern, A. M. P. (2012). A new approach to comparing problem solving, flexibility and innovation. *Journal of Communicative and Integrative Biology, 5*(2), 140–145.

Auersperg, A. M. I., Kacelnik, A., & von Bayern, A. M. P. (2013). Explorative learning and functional inferences on a five-step mechanical problem in juvenile Goffin's cockatoos (*Cacatua goffini*). *Plos One, 8*(7), e68979.

Auersperg, A. M. I., Oswald, N., Domansegg, M., & Bugnyar, T. (2015). Combinatory actions during object play in parrots. *Animal Behavior and Cognition, 1*(4), 470–488.

Auersperg, A. M. I., Szabo, B., & Bugnyar, T. A. (2013). Object permanence in the Goffin's cockatoo (*Cacatua goffini*). *Journal of Comparative Psychology.* Available from: http://dx.doi.org/10.1037/a0033272.

Auersperg, A. M. I., Szabo, B., von Bayern, A. M. P., & Kacelnik, A. (2012). Spontaneous innovation of tool manufacture and use in a Goffin's cockatoo. *Current Biology*, 22(21), 903−R904.

Auersperg, A. M. I., van Horik, J., Bugnyar, T., Kacelnik, K., Emery, N., & von Bayern, A. M. P. (2015). Combinatory actions during object play in psittaciformes (Diopsittaca nobilis, Pionites melanocephala, Cacatua goffini) and corvids (Corvus corax, C. monedula, C. moneduloides). *Journal of Comparative Psychology.*, 129(1), 62−71.

Auersperg, A. M. I., von Bayern, A. M. P., Gajdon, G. K., Huber, L., & Kacelnik, A. (2011). Flexibility in problem solving and tool use of kea and New Caledonian crows in a Multi Access Box paradigm. *PLoS One*, 6(6), e20231.

Bateson, P., & Martin, P. (2013). *Play, playfulness, creativity and innovation.* Cambridge: Cambridge University Press.

Berkhoudt, H. (1980). The morphology and distribution of cutaneous mechanoreceptors (Herbst and Grandry corpuscles) in bill and tongue of the mallard (*Anas platyrhynchos. p. L.*). *Netherlands Journal of Zoology*, 50, 1−34.

Berlyne, D. E. (1960). *Conflict, arousal, and curiosity.* New York, NY: McGraw Hill.

Bird, C. D., & Emery, N. J. (2009a). Insightful problem solving and creative tool modification by captive rooks. *Proceedings of the National Academy of Sciences USA*, 106, 10370−11037.

Bird, C. D., & Emery, N. J. (2009b). Rooks use stones to raise the water level to reach a floating worm. *Current Biology*, 19, 1410−1414.

Bock, W. J. (1999). Functional and evolutionary morphology of woodpeckers. *Journal of African Ornithology*, 70(1). Available from: http://dx.doi.org/10.1080/00306525.1999.9639746.

Borsari, E., & Ottoni, E. B. (2005). Preliminary observations of tool use in captive hyacinth macaws (*Anodorhynchus hyacinthinus*). *Animal Cognition*, 8(1), 48−52.

Brooke, M., & Birkhead, T. (1991). *The Cambridge encyclopedia of ornithology.* New York, USA: Cambridge University Press, ISBN 0521362059.

Brown, C., & Magat, M. (2011). Cerebral lateralization determines hand preferences in Australian parrots. *Biology Letters*, 7(4), 496−498. Available from: http://dx.doi.org/10.1098/rsbl.2010.1121.

Buitron, D., & Nuechterlein, G. L. (1985). Experiments on olfactory detection of food caches by black-billed magpies. *The Condor*, 87(1), 92−95.

Burghardt, G. M., & Graham, K. L. (2010). Current perspectives on the biological study of play: Signs of progress. *The Quarterly Review of Biology*, 84, 4.

Call, J. (2013). Three ingredients for becoming a creative tool-user. In C. M. Sanz, J. Call, & C. Boesch (Eds.), *Tool use in animals: Cognition and ecology* (pp. 89−119). Cambridge, UK: Cambridge University Press.

Cheke, L. G., Bird, C. D., & Clayton, N. S. (2011). Tool-use and instrumental learning in the Eurasian jay (*Garrulus glandarius*). *Animal Cognition*, 14(3), 441−455. Available from: http://dx.doi.org/10.1007/s10071-011-0379-4.

Cobb, S. (1960). A note on the size of the avian olfactory bulb. *Epilepsia*, 1, 394−402.

Collar, N. (1997). Family psittacidae (parrots). In J. del Hoyo, A. Elliott, & J. Sargatal (Eds.), *Handbook of the birds of the world, sandgrouse to cuckoos* (vol. 4, pp. 286−324). Barcelona, Spain: Lynx Editions.

Connolly, K., & Dagleish, M. (1989). The emergence of a tool-using skill in infancy. *Developmental Psychology*, 25, 894−912.

Cunningham, S. J., Alley, M. R., Castro, I., Potter, M. A., Cunningham, M., & Pyne, M. J. (2010). Bill morphology or Ibises suggests a remote-tactile sensory system for prey detection. *The Auk*, 127, 308−316. Available from: http://dx.doi.org/10.1525/auk.2009.09117.

de Mendonça-Furtado, O., & Ottoni, E. B. (2008). Learning generalization in problem solving by a blue-fronted parrot (*Amazona aestiva*). *Animal Cognition*, 11, 719−725.

Demery, Z. (2012). The kakariki model: comparing exploratory learning strategies in parrots and children, *Spoken presentation at 19th international conference on comparative cognition*. Available from: http://f1000.com/posters/browse/summary/1090007.

Demery, Z. P., Chappell, J., & Martin, G. R. (2011). Vision, touch and object manipulation in Senegal parrots *Poicephalus senegalus*. *Proceedings of the Royal Society B, 278*(1725), 3687−3693. Available from: http://dx.doi.org/10.1098/rspb.2011.0374.

Diamond, J., & Bond, A. B. (1999). *Kea, bird of paradox: The evolution and behavior of a New Zealand parrot*. Berkeley, CA: University of California Press.

Diekamp, B., Kalt, T., Ruhm, A., Koch, M., & Gunturkun, O. (2000). Impairment in a discrimination reversal task after D1 receptor blockade in the pigeon 'prefrontal cortex'. *Behavioural Neuroscience, 114*, 1145−1155.

Dyke, G. J., & Cooper, J. H. (2000). A new psittaciform bird from the London clay (Lower Eocene) of England. *Palaeontology, 43*(2), 271−285. Available from: http://dx.doi.org/10.1111/1475-4983.00126.

Emery, N. J. (2004). Are corvids 'feathered apes'? Cognitive evolution in crows, jays, rooks and jackdaws. In S. Watanabe (Ed.), *Comparative analysis of minds* (pp. 181−213). Tokyo, Japan: Keio University Press.

Emery, N. J. (2006). Cognitive ornithology: The evolution of avian intelligence. *Philosophical Transactions of the Royal Society B, 361*, 23−43.

Emery, N. J., & Clayton, N. S. (2004). The mentality of crows: Convergent evolution of intelligence in corvids and apes. *Science, 306*, 1903−1907.

Forshaw, J. M. (1989). *Parrots of the world* (3rd ed.). Australia: Lansdowner.

Forshaw, J. M., & Cooper, W. T. (1989). *Parrots of the world* (2nd ed.). London, UK: David & Charles, Newton Abbot.

Fragaszy, D., Izar, P., Visalberghi, E., Ottoni, E. B., & Gomes de Oliveira, M. (2004). Wild capuchin monkeys (*Cebus libidinosus*) use anvils and stone pounding tools. *American Journal of Primatology, 64*(4), 359−366.

Franken, R. E. (1994). *Human motivation* (3rd ed.). Pacific Grove, CA: Brooks/Cole.

Funk, M. (2002). Problem solving skills in young yellow crowned parakeets (*Cyanoramphus auriceps*). *Animal Cognition, 5*(3), 167−176.

Gajdon, G. K., Amann, L., & Huber, L. (2011). Keas rely on social information in a tool use task but abandon it in favour of overt exploration. *Interaction Studies, 12*, 303−322.

Gajdon, G. K., Lichtnegger, M., & Huber, L. (2014). What a parrot's mind adds to play: The urge to produce novelty fosters tool use acquisition in kea. *Open Journal of Animal Sciences, 4*, 51−58.

Goodwin, D. (1986). *Crows of the world*. London, UK: British Museum (Natural History) Press.

Gottschaldt, K. M. (1985). Structure and function of avian somatosensory receptors. In A. S. King, & J. McLelland (Eds.), *Form and function in birds* (Vol. 3, pp. 375−461). London, UK: Academic Press.

Gottschaldt, K. M., & Lausmann, S. (1974). The peripheral morphological basis of tactile sensibility in the beak of geese. *Cell and Tissue Research, 153*, 477−496.

Greenberg, R. (2004). The role of neophobia and neophilia in the development of innovative behaviour of birds. In S. Reader, & K. Laland (Eds.), *Animal Innovation* (pp. 175−196). Oxford, UK: Oxford University Press.

Greenberg, R., & Mettke-Hoffmann, C. (2001). Ecological aspects of neophobia and neophilia in birds. *Current Ornithology, 16*, 119−169.

Gsell, A. C. (2012). The ecology and anatomy of scent in the critically endangered kakapo (Strigrops habrobtilus) *(Doctoral thesis)*. Auckland, New Zealand: University of Aukland.

Gsell, A. C., Hagelin, J. C., & Brunton, D. H. (2012). Olfactory sensitivity in Kea and Kaka. *EMU, 112*, 60−66.

Hansell, M., & Ruxton, G. (2008). Setting tool use within the context of animal construction behaviour. *Trends in Ecology and Evolution, 23*, 73−78.

Harriman, A. E., & Berger, R. H. (1986). Olfactory acuity in the common raven (*Corvus corax*). *Physiology & Behavior, 36*(2), 257−262.

Harris, L. J. (1989). Footedness in parrots: Three centuries of research, theory, and mere surmise. *Canadian Journal of Psychology, 43*, 369−396. Available from: http://dx.doi.org/10.1037/h0084228.

Hayashi, M., & Matsuzawa, T. (2003). Cognitive development in object manipulation by infant chimpanzees. *Animal Cognition, 6*, 225−233.

Hayashi, M., Takeshita, H., & Matsuzawa, T. (2006). Cognitive development in apes and humans assessed by object manipulation. In T. Matsuzawa, M. Tomonaga, & M. Tanaka (Eds.), *Cognitive development in chimpanzees* (pp. 395−410). Tokyo, Japan: Springer.

Heinrich, B. (2007). *Mind of the raven: Investigations and adventures with wolf-birds*. New York, NY: Harper Collins Publisher.

Heinrich, B., & Bugnyar, T. (2005). Testing problem solving in ravens: String-pulling to reach food. *Ethology, 111*, 962−976.

Hoffmann, A., Ruttler, V., & Nieder, A. (2011). Ontogeny of object permanence and object tracking in the carrion crow, *Corvus corone*. *Animal Behaviour, 82*, 359−367. Available from: http://dx.doi.org/10.1016/j.anbehav.2011.05.012.

Huber, L., & Gajdon, G. (2006). Technical intelligence in animals: The kea model. *Animal Cognition, 9*, 295−305.

Huber, L., Rechberger, S., & Taborsky, M. (2001). Social learning affects object exploration and manipulation in keas, *Nestor notabilis*. *Animal Behaviour, 62*, 945−954. Available from: http://dx.doi.org/10.1006/anbe.2001.1822.

Hunt, G. R. (1996). Manufacture and use of hook-tools by New Caledonian crows. *Nature, 379*, 249−251.

Hunt, G. R., Gray, R. D., & Taylor, A. (2013). Why is tool use rare in animals? In C. M. Sanz, J. Call, & C. Boesch (Eds.), *Tool use in animals: Cognition and ecology* (pp. 89−119). Cambridge, UK: Cambridge University Press.

Iland, A. (2013). *Motivation: Unlock human potential*. North Charleston, USA: Iland Business Pages, CreateSpace Independent Publishing Platform.

Iwaniuk, A. N., Dean, K. M., & Nelson, J. E. (2005). Interspecific allometry of the brain and brain regions in parrots (*Psittaciformes*): Comparisons with other birds and primates. *Brain, Behaviour and Evolution, 65*, 40−59.

Jones, T. B., & Kamil, A. C. (1973). Tool-making and tool-using in the Northern blue jay. *Science, 180*, 1076−1078.

Kenward, B., Schloegl, C., Rutz, C., Weir, A. A. S., Bugnyar, T., & Kacelnik, A. (2011). On the evolutionary and ontogenetic origins of tool-oriented behaviour in New Caledonian crows (*Corvus moneduloides*). *Biological Journal of the Linnean Society, 102*(4), 870−877.

Kuhlmeyer, C., Asbar, K., Gunz, P., Frahnert, S., & Barilein, F. (2009). Functional morphology and integration of corvid skulls—a 3D geometric morphometric approach. *Frontiers in Zoology, 6*(2). Available from: http://dx.doi.org/10.1186/1742-9994-6-2.

Lefebvre, L., & Bolhuis, J. J. (2003). Positive and negative correlations of feeding innovations in birds: Evidence for limited modularity. In S. M. Reader, & K. N. Laland (Eds.), *Animal innovation* (pp. 39−61). Oxford, UK: Oxford University Press.

Lefebvre, L., Reader, S. M., & Sol, D. (2004). Brains, innovations and evolution in birds and primates. *Brain and Behavior Evolution, 63*, 233−246.

Lefebvre, L., Whittle, P., Lascaris, E., & Finkelstein, A. (1997). Feeding innovations and forebrain size in birds. *Animal Behaviour, 53,* 549—560.

Liedtke, J., Werdenich, D., Gajdon, G. K., Huber, L., & Wanker, R. (2011). Big brains are not enough: Parrots fail to solve the trap-tube paradigm. *Animal Cognition, 14,* 143—149.

Lueschr, A. U. (2006). *Manual of parrot behavior.* San Francisco, CA: Wiley-Blackwell.

Maclean, E. L., Matthews, L. J., Hare, B. A., Nunn, C. L., Anderson, R. C., Aureli, F., et al. (2011). How does cognition evolve? Phylogenetic comparative psychology. *Animal Cognition.* Available from: http://dx.doi.org/10.1007/s10071-011-0448-8.

Macphail, E. (1982). *Brain and intelligence in vertebrates.* Oxford, UK: Oxford University Press.

Magat, M., & Brown, C. (2009). Laterality enhances cognition in Australian parrots. *Proceedings of the Royal Society B, 276,* 4155—4162. Available from: http://dx.doi.org/10.1098/rspb.2009.1397.

Martin, G. R. (1985). Eye. In A. S. King, & J. McLalland (Eds.), *Form and function in birds* (Vol. 3, pp. 311—373). London, UK: Academic Press.

Martin, G. R. (2007). Visual fields and their functions in birds. *Journal of Ornithology, 148* (Suppl. 2), 547—562.

Martin, G. R., & Coetzee, H. C. (2004). Visual fields in hornbills: Precision-grasping and sunshades. *Ibis, 146,* 18—26.

Martin, G. R., McNeil, R., & Rojas, L. M. (2007). Vision and the foraging technique of skimmers (*Rhynchopidae*). *Ibis, 149*(4), 750—757.

Mettke-Hoffman, C., Winkler, H., & Leisler, B. (2002). The significance of ecological factors on exploration and neophobia in parrots. *Ethology, 108,* 249—272.

Miyata, H., Gajdon, G. K., Huber, L., & Fujita, K. (2011). How do keas (*Nestor notabilis*) solve artificial-fruit problems with multiple locks? *Animal Cognition, 14*(1), 45—58.

Omland, K. E., Tarr, C. L., Boarman, W. I., Marzluff, J. M., & Fleischer, R. C. (2000). Cryptic genetic variation and paraphyly in ravens. *Proceedings of the Royal Society B*(267), 2475—2482. Available from: http://dx.doi.org/10.1098/rspb.2000.1308. PMC 1690844. <11197122>

Pepperberg, I. (1999). *The Alex studies.* Cambridge, MA: Harvard University Press.

Pepperberg, I. M., & Kozak, F. A. (1986). Object permanence in the African Grey parrot (*Psittacus erithacus*). *Animal Learning & Behavior, 14,* 322—330. Available from: http://dx.doi.org/10.3758/BF03200074.

Pepperberg, I. M., Willner, M. R., & Gravitz, L. B. (1997). Development of piagetian object permanence in a grey parrot. *Journal of Comparative Psychology, 111,* 63—75. Available from: http://dx.doi.org/10.1037/0735-7036.111.1.63.

Pollok, B., Prior, H., & Gunturkün, O. (2000). Development of object permanence in food-storing Magpies (*Pica pica*). *Journal of Comparative Psychology, 114,* 148—157. Available from: http://dx.doi.org/10.1037/0735-7036.114.2.148.

Povinelli, D. J. (2000). *Folk physics for apes: A chimpanzee's theory of how the world works.* Oxford, UK: Oxford University Press.

Rat-Fischer, L., O'Regan, J. K., & Fagard, J. (2012). The Emergence of tool use during the second year of life. *Journal of Experimental Child Psychology, 113*(3), 440—446.

Reiner, A. (1986). Is the prefrontal cortex found only in mammals? *Trends in Neuroscience, 9,* 298—300.

Reader, S., & Laland, K. (2004). Animal innovation, an introduction. In S. Reader, & K. Laland (Eds.), *Animal innovation* (pp. 3—39). Oxford, UK: Oxford University Press.

Roper, T. J. (2003). Olfactory discrimination in yellow-backed chattering lories *Lorius garrulous flavopalliatus*: First demonstration of olfaction in psittaciformes. *Ibis, 145,* 689—691. Available from: http://dx.doi.org/10.1046/j.1474-919X.2003.00195.x.

Russel, R. A. (1973). Relationship between exploratory behaviour and fear: A review. *British Journal of Experimental Psychology, 64,* 417—433.

I. EVIDENCES OF CREATIVITY

Rutz, C., Bluff, L. A., Reed, N., Troscianko, J., Newton, J., Inger, R., et al. (2010). The ecological significance of tool use in New Caledonian crows. *Science, 329*(5998), 1523−1526. Available from: http://dx.doi.org/10.1126/science.1192053.

Schloegl, C., Dierks, A., Gajdon, G. K., Huber, L., Kotrschal, K., & Bugnyar, T. (2009). What you see is what you get? Exclusion performances in ravens and keas. *PLoS One, 4*, e6368.

Schloegl, C., Schmidt, J., Boeckle, M., Weiß, B. M., & Kotrschal, K. (2012). Grey parrots use inferential reasoning based on acoustic cues alone. *Proceedings of the Royal Society of London, Series B, 479*, 4135−4142.

Schmuck-Paim, C., Borsari, A., & Ottoni, E. B. (2009). Means to an end: Neotropical parrots manage to pull strings to meet their goals. *Animal Cognition, 12*(2), 287−301.

Seed, A. M., Clayton, N. S., & Emery, N. J. (2008). Cooperative problem solving in rooks (*Corvus frugilegus*). *Proceedings of the Biological Sciences/The Royal Society, 275*(1641), 1421−1429. Available from: http://dx.doi.org/10.1098/rspb.2008.0111.

Seed, A. M., Tebbich, S., Emery, N. J., & Clayton, N. S. (2006). Investigating physical cognition in rooks. *Current Biology, 16*, 697−701.

Shaw, R. C., Plotnik, J. M., & Clayton, N. S. (2013). Exclusion in corvids: The performance of food-caching Eurasian jays (*Garrulus glandarius*). *Journal of Comparative Psychology, 127*(4), 428−435. Available from: http://dx.doi.org/10.1037/a0032010.

Shettleworth, S. J. (2009). *Cognition, evolution, and behavior* (2nd ed.). New York, NY: Oxford University Press.

Shumaker, R. W., Walkup, K. R., & Beck, B. (2011). *Animal tool behavior: The use and manufacture of tools by animals, Revised and updated edition.* The John's Hopkins University Press.

Smith, G. A. (1975). Systematics of parrots. *Ibis, 117*, 18−68.

Sol, D., Timmermans, S., & Lefebvre, L. (2002). Behavioural flexibility and invasion success in birds. *Animal Behaviour, 63*, 495−502.

Takeshita, H. (2001). Development of combinatory manipulation in infant chimpanzees (*Pan troglodytes*). *Animal Cognition, 4*, 335−345.

Taylor, A. H., Elliffe, D., Hunt, G., & Gray, R. D. (2010). Complex cognition and behavioural innovation in New Caledonian crows. *Proceedings of the Royal Society B, 277*, 2637−2643.

Taylor, A. H., Hunt, G. R., Holzhaider, J. C., & Gray, R. D. (2007). Spontaneous metatool use in New Caledonian Crows. *Current Biology, 17*, 1504−1507.

Taylor, A. H., Hunt, G. R., Medina, F. S., & Gray, R. D. (2009). Do New Caledonian crows solve physical problems through causal reasoning? *Proceedings of the Royal Society B, 276*, 247−254.

Tebbich, S., Seed, A., Emery, N., & Clayton, N. (2007). Non-tool-using rooks (*Corvus frugilegus*) solve the trap-tube task. *Animal Cognition, 10*(2), 225−231.

Tebbich, S., Sterelny, K., & Teschke, I. (2010). The tale of the finch: Adaptive radiation and behavioural flexibility. *Philosophical Transactions of the Royal Society of London Series B, Biological Sciences, 365*, 1099−1109. Available from: http://dx.doi.org/10.1098/rstb.2009.029.

Teschke, I., & Tebbich, S. (2011). Physical cognition and tool-use: Performance of Darwin's finches in the two-trap tube task. *Animal Cognition.* Available from: http://dx.doi.org/10.1007/s10071-011-0390-9.

Timmermans, S., Lefebvre, L., Boire, D., & Basu, P. (2000). Relative size of the hyperstriatum ventrale is the best predictor of feeding innovation rate in birds. *Brain and Behavior Evolution, 56*, 196−203.

Tomasello, M., & Call, J. (1997). *Primate cognition.* Oxford, UK: Oxford University Press.

Torigoe, T. (1985). Comparison of object manipulation among 74 species of non-human primates. *Primates, 26*(2), 182−194.

Troscianko, J., von Bayern, A. M. P., Chappell, J., Rutz, C., & Martin, G. R. (2012). Extreme binocular vision and a straight bill facilitate tool use in New Caledonian crows. *Nature Communications*, 3, 1110.

Vernelli, T. (2013). *The complexity of neophobia in a generalist foraging corvid: The common magpie (pica pica)* (Phd thesis). University of Exeter.

Visalberghi, E., & Limongelli, L. (1994). Lack of comprehension of cause–effect relations in tool-using capuchin monkeys (*Cebus apella*). *Journal of Comparative Psychology (Washington, DC: 1983)*, 108, 15–22. Available from: http://dx.doi.org/10.1037/ 0735-7036.108.1.15.

von Bayern, A. M. F., Heathcote, R. J. P., Rutz, C., & Kacelnik, A. (2009). The role of experience in problem solving and innovative tool use in crows. *Current Biology*, 19(22), 1965–1968.

Webster, S. J., & Lefebvre, L. (2001). Problem solving and neophobia in a columbi-form–passeriform assemblage in Barbados. *Animal Behaviour*, 62, 23–32.

Weir, A. A. S., Chappell, J., & Kacelnik, A. (2002). Shaping of hooks in New Caledonian crows. *Science*, 297, 981.

Werdenich, D., & Huber, L. (2006). A case of quick problem solving in birds: String-pulling in keas (*Nestor notabilis*). *Animal Behaviour*, 71, 855–863.

Whittow, G. C., & Sturkie, P. D. (1999). *Sturkie's avian physiology*. New York, NY: Academic Press.

Wimpenny, J. H., Weir, A. A. S., Clayton, L., Rutz, C., & Kacelnik, A. (2009). Cognitive processes associated with sequential tool use in New Caledonian crows. *PLoS One*, 4 (8), e6471.

Zweers, G. A., Berkhoudt, H., & van den Berge, J. C. (1994). Behavioral mechanisms of avian feeding. In V. L. Bels, M. Chardon, & P. van de Walle (Eds.), *Biomechanics of feeding in vertebrates, advances in comparative environmental physiology* (pp. 241–279). Berlin, Germany: Springer.

Commentary on Chapter 3: Innovations in Corvids and Parrots

Parrots and corvids display fantastically complex behaviors. But do they demonstrate creativity in the human sense of that term? Not long ago, it was thought that only humans make use of and modify tools. Then, definitive evidence from both the field and the laboratory debunked those assertions, and the separator between humans and other species became language use. In the opinion of some scholars, human language shows significant and fundamental differences from animal communication. Yet others have proposed that animal communication systems and human language both fall on the same evolutionary continuum. Recently, studies have demonstrated that bottlenose dolphins develop distinctive, individualized signature whistles that can be used to direct information to specific companions (King, Harley, & Janik, 2014; King & Janik, 2013). This signaling system goes well beyond the "group alert" vocalizations utilized by many social species and offers compelling evidence of the use of "true language" in nonhuman animals. The boundaries are still fuzzy, but at the very least, language does not appear to be the definitive divide that it was once thought to be. Searching further, researchers and theorists wonder what other abilities might separate humans from animals; and in recent years, some researchers and theorists have nominated creative behavior as being a distinctly human capacity. For example, Sawyer (2012) writes that there is a consensus among scientists that "creativity is a uniquely human trait" (p. 30). Importantly however, as evidenced by the many high-level contributions to this volume, a growing number of scholars are coming to believe and provide empirical evidence for the argument that a variety of animal species *do* demonstrate "true creativity."

What would it take to convince the naysayers? What would it take to definitively demonstrate creative behavior in one or more animal species? This is no easy question, as evidenced by the difficulties inherent in the empirical study of human creativity. This research area remains fraught with a number of philosophical and methodological challenges, and to add a consideration of animals to this mix feels a bit overwhelming. In the work on human creativity, the problem of fuzzy boundaries also looms large. Not all exploratory play leads to creativity, but some does. Not all tool use signals creative behavior, but it can. Not all flashes of insight, those AHA! moments, lead to creative responses, but some can. Not all problem solving involves creativity, but some does. In truth, scholars in the area of human creativity have a long way to go before they reach consensus regarding exactly what it is that they are talking about and investigating. However, we would argue that if we are to explore empirically the question of whether animals demonstrate creativity, we must employ the same definitions and operationalizations of variables being employed in the human literature, and, when it comes to problems of definition, researchers generally agree that definitions of creativity must incorporate two distinct hallmarks: newness and usefulness. Building on these ideas, in the research literature, creative behavior is typically defined as a novel and appropriate response to an open-ended task (see Amabile & Mueller, 2008).

How might these criteria be applied to animal behavior? Perhaps the most expedient answer to this question can be found in the literature on young children's creative behavior. In this area, too, not everyone agrees. Some scholars maintain that children's responses can never be deemed creative until they have reached at least the Piagetian formal operational stage of thinking (see Dudek, 1974). And from a psychodynamic perspective, the argument is made that, unlike adults who must work hard to break boundaries, challenge values, and face unconscious fears, children have not yet found the need to regress in the service of the ego (see Sawyer, 2012, p. 19). In other words, they cannot be given credit for thinking outside the box before they even understand that there is a box. On the other side of the argument, those who believe that young children can and do demonstrate creativity typically employ a criterion whereby the product or response generated must be intentional, rather than the result of an accident. Also required is that the response be novel, at least for the individual child in question.

In recent years, there has been developed an especially useful rubric that distinguishes between four types, or levels, of creativity: big-C, pro-c, little-c, and mini-c (Kaufman & Beghetto, 2009). Big-C creativity is associated with celebrated artists, musicians, and scientists. Pro-c levels of creativity are exhibited by professionals who make important strides in their particular areas of expertise. Little-c creativity is everyday creativity. Creative accomplishments at these first three levels rely on the judgment of others. Mini-c creativity, on the other hand, is personally meaningful and is not subject to external assessment. Instead, mini-c creativity comes about when the learner filters new information through existing frameworks of understanding to achieve a new level of awareness or knowledge. Clearly, this notion of mini-c creativity is especially relevant to children's creativity, and we would argue that many of the bird behaviors described by Auersperg also meet mini-c criteria. Like a young child who manipulates, explores, and eventually masters an unfamiliar toy or puzzle, these birds are filtering new information to achieve a new understanding of the tube apparatus or Multi-Access Box.

Moving from the level of mini-c to the more general operationalization of creativity based on the criteria of novelty and appropriateness, here too the birds make an impressive showing. Certainly, many of the behaviors described by Auersperg are novel for these birds. In fact, were they not in captivity and working with experimenters, they would never encounter contraptions like those described in this paper. The birds are definitely engaging in the solution of complex problems. They persist at exploratory behavior and in some cases combine objects in novel ways. At times, risk-taking, another hallmark of human creative behavior, is even involved. Moreover, many of these combinations and behaviors are no accident. Instead, they are the result of intentional actions on the birds' part, actions that are learned, repeated, and sometimes even adapted in transfer tasks.

We propose that the amazing abilities being shown by these birds are demonstrations of their successful completion of what are termed in the human literature as "insight problems." Well-known examples of problems in this category include the "Nine Dot" and Duncker's Candle tasks. These tasks are tests of cognitive performance measuring the influence of functional fixedness on problem-solving abilities.

There is no doubt that at least some of the birds studied moved beyond automatic behaviors and showed insight as they applied a series of novel responses to master the Multi-Access Box. So too, humans faced with Duncker's challenge must break away from the notion that a cardboard box is only a convenient way to hold paperclips and gain the insight that it, too, might play a valuable role in the solution of a problem.

At present, a great many empirical studies of human creativity equate the successful solution of insight problems with creativity. Based upon these criteria, Auersperg's birds have demonstrated creativity. Clearly, moments of insight can often play an important part in the creative process; but in a series of ingenious recent experiments, Beaty, Nusbaum, and Silvia (2014) demonstrate that, contrary to popular opinion, there is no systematic relation between insight problem-solving ability and real-world creative behavior and achievement. Can parrots, corvids, and perhaps other species be shown to move beyond insight problem solving toward a demonstration of other types of mini-c creativity? Might a bird one day surprise researchers by completing an unfamiliar task in a wholly unexpected manner? Amazing strides have already been made by both birds and investigators and we anxiously await further developments.

References

Amabile, T. M., & Mueller, J. S. (2008). Studying creativity, its processes, and its antecedents: An exploration of the componential theory of creativity. In J. Zhou, & C. E. Shalley (Eds.), *Handbook of organizational creativity* (pp. 33−64). New York, NY: Lawrence Erlbaum.

Beaty, R. E., Nusbaum, E. C., & Silvia, P. J. (2014). Does insight problem solving predict real-world creativity? *Psychology of Aesthetics, Creativity, and the Arts, 8*, 287−292.

Dudek, S. Z. (1974). Creativity in young children: Attitude or ability? *The Journal of Creative Behavior, 8*, 282−292.

Kaufman, J. C., & Beghetto, R. A. (2009). Beyond big and little: The four C model of creativity. *Review of General Psychology, 13*, 1−12.

King, S. L., Harley, H. E., & Janik, V. M. (2014). The role of signature whistle matching in bottlenose dolphins, tursiops truncatus. *Animal Behavior, 96*, 79−86.

King, S. L., & Janik, V. M. (2013). Bottlenose dolphins can use learned vocal labels to address each other. *Proceedings of the National Academy of Sciences, 110*(32), 13216−13221.

Sawyer, R. K. (2012). Explaining creativity: The science of human innovation. New York, NY: Oxford.

Cetacean Innovation

Eric M. Patterson[1] and Janet Mann[1,2]

[1]Department of Biology, Georgetown University, Washington DC, USA
[2]Department of Psychology, Georgetown University, Washington DC, USA

A cloud of cigarette smoke was once deliberately released against the glass as Dolly was looking in through the viewing port. The observer was astonished when the animal immediately swam off to its mother, returned and released a mouthful of milk which engulfed her head, giving much the same effect as had the cigarette smoke. (Tayler & Saayman, 1973)

Commentary on Chapter 4: Proto-c Creativity?

Vlad Petre Glăveanu

Aalborg University, Department of Communication and Psychology, Aalborg, Denmark

INTRODUCTION

Dolly's "milk smoking" is just one of many fascinating behaviors Tayler and Saayman observed in Indian Ocean bottlenose dolphins (*Tursiops aduncus*) at the Port Elizabeth Oceanarium in South Africa in the 1970s. Not only did Dolly parade her impressive imitation abilities and potential analogous reasoning, she gave us a peek into her creative mind. But of what use would such creative behavior be in the wild? Milk smoking would certainly be maladaptive for a young, hungry calf, but could similar creative abilities be beneficial, say, for exploiting new

73

resources or obtaining mates? Only recently have such questions been the subject of empirical work in ethology.

To date, most data on animal innovation come from primates and birds. After scouring the literature for anecdotal observations of novel behavior and carefully controlling for confounding factors (e.g., phylogeny, research effort, observation biases, etc.), Laland, Lefebvre, Reader, Sol, and colleagues have provided some of the most intriguing data on animal creativity and innovation (Lefebvre, Reader, & Sol, 2013; Lefebvre, Whittle, Lascaris, & Finkelstein, 1997; Reader & Laland, 2001). Although these studies are impressive, they are taxonomically limited. As such, we need to expand our efforts to include additional taxa, but given the infancy of the study of animal innovation, where does one start? To address this question, we briefly consider the conditions thought to promote innovative and creative behavior across taxa, which can be divided into intrinsic factors (characteristics of the individual itself), and extrinsic factors (external social and ecological conditions).

First, for an individual to be creative and capable of innovation it must be behaviorally plastic so that new or modified behaviors can be introduced into its repertoire (Reader & Laland, 2003). Closely related is that an individual must have the ability to learn (Reader & Laland, 2003). Yet, learning is likely ubiquitous among animals and even argued for plants (Gagliano, Renton, Depczynski, & Mancuso, 2014), so learning itself adds little to our understanding of creative and innovative behavior. More important is an individual's capacity for social learning, which is required for behaviors to spread and persist in a population, and therefore be of evolutionary interest. Additionally, given the central role of novel behavior in both innovation and creativity, an individual with neophilic tendencies may be predisposed to innovate and be creative (Greenberg, 2003; Reader, 2003).

Second, an individual's proclivity to innovate and be creative will depend on both its social and ecological circumstances. For example, social conditions that provide free time are thought to favor innovation and creativity (Kummer & Goodall, 1985). In particular, a period of extended maternal care in which young are relatively free from energetic constraints may be important for allowing individuals to learn and explore their environment with minimal costs. Social or group living may also foster innovation and creativity. Not only do social interactions create opportunities for social learning, they expose individuals to behavioral variation that can be learned and modified to produce novel behavior (Lee, 2003; Reader & Laland, 2002). Furthermore, group living generally reduces predation risk and vigilance (Alexander, 1974) and thus, allows for otherwise costly behaviors and tendencies like exploration, play, boldness, and openness (De Oliveira, Ruiz-Miranda,

Kleiman, & Beck, 2003; Huang, Sieving, & Mary, 2011; Moller & Garamszegi, 2012; Panksepp, 1998). Certain ecological conditions, such as stable environments with abundant resources, may also afford individuals discretionary time and reduce the costs of exploratory behavior (Reader & MacDonald, 2003). On the other hand, highly variable environments present individuals with new challenges that may require innovative and creative solutions (i.e., "necessity is the mother of invention," Sol, 2003), while learning, an intrinsic requirement noted above, is favored in moderately changing environments (Stephens, 1991). Thus, at least some between generation ecological variability in a species' evolutionary history is vital, but beyond this, both ecological stability and variability within generation contribute to the expression of innovative and creative behavior.

Many of these intrinsic and extrinsic factors commonly associated with innovation and creativity are characteristic of cetacea, especially odontocetes. First, cetaceans are extraordinarily plastic in their behavior as exhibited by the great diversity of foraging tactics within and between populations (e.g., *Tursiops* spp., Connor, Wells, Mann, & Read, 2000; *Megaptera novaeangliae*, Clapham, 2000). They are also generally considered curious, inquisitive animals (Birtles & Mangott, 2013). In captivity, odontocetes demonstrate a high degree of neophilia (Defran & Pryor, 1980; Nakahara & Takemura, 1997; Terry, 1986), and in the wild, both odontocetes and mysticetes manipulate almost any new object they come across (Paulos, Trone, & Kuczaj, 2010; Würsig, 2008). Odontocetes also display impressive learning abilities, in particular social learning (e.g., Sargeant & Mann, 2009), with bottlenose dolphins demonstrating some of the most impressive imitation abilities of any nonhuman animal (Herman, 2002). Cetaceans also have an extended period of maternal dependency (Whitehead & Mann, 2000), with calves engaging in extensive social play and exploration (e.g., Gibson & Mann, 2008; Mann & Watson-Capps, 2005). Female cetacea give birth to a single (except in rare cases of twins, Olesiuk, Bigg, & Ellis, 1990), large, precocial calf, often born into a highly social environment (Chivers, 2009). Odontocete calves are particularly slow growing and generally nurse for over a year (Oftedal, 1997), with some delphinids nursing for 9 years or more (Kasuya & Marsh, 1984; Mann, Connor, Barre, & Heithaus, 2000). Furthermore, because cetaceans can store energy in their blubber, they can capitalize on ephemeral abundant resources, which may reduce foraging demands and allow for exploratory behavior. Finally, as top predators often living in groups, cetaceans may be less vigilant or "fearful" than species with high predation risk, a state which may otherwise inhibit exploration. Thus, these social, behaviorally plastic, neophilic, mammalian predators not only exhibit all the conditions thought to promote innovative and creative behavior, if studied in greater detail, they may prove to be one of the more innovative and creative taxa, a subject this chapter attempts to examine in detail.

Although some authors note the apparent innovative abilities of cetaceans (e.g., Kaufman, Butt, Kaufman, & Colbert-White, 2011; Lee, 2003; Sol, 2003), direct literature on the subject is sparse. This is in part due to the logistical difficulties of studying at sea, but even existing accounts of innovative and creative cetacean behavior offer only basic descriptions and rarely consider such observations in a comparative context. In this chapter, we hope to address this gap by reviewing possible cases of innovative and creative behaviors in both captive and wild cetaceans. However, before we do, we must operationalize the terms creativity and innovation.

Defining creativity and innovation is no simple task even though both concepts have intuitive meanings. The subject has been covered in great depth by others (e.g., Kaufman & Kaufman, 2004; Plucker, Beghetto, & Dow, 2004; Ramsey, Bastian, & van Schaik, 2007; Reader & Laland, 2003; Simonton, 2003), so we will only briefly touch on it here. While some treat creativity and innovation as if they are synonymous, we agree with Bateson and Martin (2013) that the two overlap but are not interchangeable. For us, creativity refers more to the cognitive processes that *directly* generate novelty, whereas innovation involves implementing novelty in a useful way. Not all innovations arise from creativity, nor does all creativity result in innovation. To illustrate the first point, consider the classic animal innovation of milk bottle opening by European blue tits (*Cyanistes caeruleus*) (Fisher & Hinde, 1949). Perhaps the first bird to open a milk bottle did so accidentally, by pecking at an insect crawling on the bottle's foil cap. After fishing out the insect, the bird would immediately be rewarded with milk and, consequently, might learn to open bottles solely for milk. In this scenario, learning would be involved, but the initial milk reward would just be a fortuitous *by-product* of cognition related to insect foraging, and so this innovation would not involve creativity. Our second point, that creative acts do not always lead to innovations, is more obvious since probably all human artworks involve creativity, but most are not considered innovations.

In this chapter we apply the term innovation to all novel or modified, functional behaviors. However, we only consider the modification of a behavior innovative if the modified component has some functional outcome unique from that of the original behavior. We use the term creative when it seems likely that a novel behavior is the *direct* product, rather than the by-product, of some underlying cognitive process, regardless of what that process is and whether or not the behavior has some function. Unfortunately, in animal studies usually only the product, novel behavior, rather than the process that creates it is known. Because of this gap, we describe most behaviors simply as innovations and only discuss a few in the context of creativity. In our coverage of cetaceans, we rely on researcher reports of species-atypical novel, unique, or unusual behaviors that might be innovative or creative, which is similar to the approach that has been used for birds and primates (Lefebvre et al., 1997; Reader & Laland, 2001).

In the following sections, we will (i) provide examples of innovative and creative behavior from captivity, (ii) examine possible cases of such behavior in wild cetaceans, (iii) discuss what both of these datasets tell us about cetaceans' creative and innovative abilities, and finally (iv) contrast cetacean data with other taxa to provide comparative insight on the evolution of animal creativity and innovation.

TALES FROM THE TANK

Observations from captivity provide some of the best examples of innovative and creative behavior in cetaceans. With no predators, foraging demands, nor obvious costs, captive animals have ample time to spare and their captive environments present new, albeit artificial, social, and ecological challenges (Kummer & Goodall, 1985). Furthermore, most aquaria have trainers that encourage new behaviors. While many novel behaviors from aquaria are unlikely to be of use in the wild (e.g., milk smoking), examining a species' tendency to innovate and be creative in captivity can be informative, especially for difficult to study taxa such as cetaceans.

The first published record of a captive cetacean dates back to the early 1860s (Wyman, 1863), although some reports go as far back as the first century AD (Corkeron, 2009). To date, some 35 species have been held in captivity, just over a third of the 92 currently recognized (Defran & Pryor, 1980; Perrin, 2014). This captive sample is extremely odontocete-biased, with only two species of mysticetes, gray (*Eschrichtius robustus*) and minke whales (*Balaenoptera acutorostrata*), represented. Moreover, most are delphinids, particularly bottlenose dolphins (*Tursiops* spp. [*truncatus* and *aduncus* species], which here we will discuss as one taxonomic group). In fact, of the current 576 reported captive cetaceans in the United States, around 94% are delphinids (83% from genus *Tursiops*), with only a few representatives of the phocoenidae and monodontidae families (National Marine Fisheries Service Office of Protected Resources, 2014). Nonetheless, in the taxa represented, examples of novel or modified behavior are plentiful and widespread, the most impressive of which we review below. To help illustrate the breadth of these behaviors, we group them by their apparent function (play, vocal communication, foraging).

Play

Play, or play-like behavior, is notoriously difficult to study, but has been of great interest to animal behaviorists for well over a century (Bekoff & Byers, 1998; Groos, Baldwin, & Baldwin, 1898). Generally speaking, play behavior consists of actions that appear to have no immediate purpose other than their own enjoyment (Bekoff & Byers, 1981). However, play is often thought to be a form of practice and important for normal development (Bekoff & Byers, 1981), so here we consider it functional. In what is almost certainly the first published report on the behavior of a captive cetacean, Wyman (1863) describes the "playful disposition" of a captive male beluga (*Delphinapterus leucas*) who used to amuse himself by tossing stones, or, on occasion, capturing fishes in his tank only to later release them unharmed. Observations similar to this are common, with captive odontocetes voluntarily manipulating and playing with a variety of natural and unnatural objects (Paulos et al., 2010), in addition to any items they are deliberately trained to manipulate. In aquaria, odontocetes clearly regularly interact and play with novel objects, but are these behaviors innovative and do they tell us anything about cetacean creativity?

We can attempt to answer this by examining a few play behaviors in more detail. In noting the behavior of bottlenose dolphins at Marineland of Florida, Tavolga (1966) describes two calves playing a rather unique game of fetch. One calf would bring a pelican feather (or sometimes a handkerchief) to an intake jet in the facility and allow the strong current to carry the feather up toward the surface. The other calf, stationed along the feathers path, would then catch the feather, after which the two would switch positions and repeat the process all over. At the same facility, a short-finned pilot whale (*Globicephala macrorhynchus*) invented a game almost the exact opposite in nature. Kritzler (1952) describes how the whale would remove a piece of flotsam from the drain near the center of the tank, swim a short distance away, release the item, and then follow it slowly as it was sucked back in by the vortex. Meanwhile, at Marineland of the Pacific, a male bottlenose dolphin named Frankie was also observed playing the feather jet game, although only by himself. Here the behavior eventually spread to other individuals and sometimes dolphins even resorted to plucking feathers from pelicans if none were available (Brown & Norris, 1956). While such catch and release behavior is common among wild delphinids during prey handling and play (e.g., Mann & Smuts, 1999; Mann, Sargeant, & Minor, 2007; Whitehead & Mann, 2000), these captive individuals employed water jets and drains in a novel way. Unfortunately, nothing is known about the cognitive processes that lead in these behaviors so we cannot say for certain whether or not they are creative, but they do appear to be clever.

Some odontocetes also manipulate objects to create effects and/or create objects of their own to manipulate and play with. For example, Paulos et al. (2010) describe how some bottlenose dolphins tow inflatable pool toys around their tank, effectively creating a pressure wave which the dolphins can then surf. Probably the best captive example of creative and innovative object manipulation in play comes in the form of bubbles. Amazon River dolphins (*Inia geoffrensis*), belugas, and bottlenose dolphins have all been seen creating and playing with bubbles in captivity in diverse ways (Paulos et al., 2010). Amazon River dolphins use sticks and brushes to strike the water's surface to produce bubble curtains (possibly a modification of their wild behavior described later), which they then swim through or lie inside. They sometimes create curtains for each other and even rub one another with their bubble-creating brushes (Gewalt, 1989; Renjun, Gewalt, Neurohr, & Winkler, 1994). In the absence of brush or stick tools, they create bubble rings by releasing air from their blowhole (Gewalt, 1989). Belugas create bubble rings with their mouths (Delfour & Aulagnier, 1997), while bottlenose dolphins do so with both their blowholes and mouths, often in combination with other elaborations such as injecting air into vortices they created (Marten, Shariff, Psarakos, & White, 1996). Individuals also interact with bubbles in a diversity of ways from simple bubble biting and moving, to more complex bubble splitting and joining (Kuczaj, Makecha, Trone, Paulos, & Ramos, 2006; Marten et al., 1996; McCowan, Marino, Vance, Walke, & Reiss, 2000; Paulos et al., 2010). Much of this behavior appears to be planned as bottlenose dolphins monitor bubble quality and anticipate bubble movement (McCowan et al., 2000). Some authors have suggested that bubble techniques might be passed down through social learning and represent traditions or perhaps a "[bubble] ring culture" (Marten et al., 1996). While creating and playing with bubbles in and of itself may be common and not represent an innovation, the great diversity of ways in which odontocetes create, modify, and manipulate their bubbles suggests their behavior is the direct result of some underlying cognitive process and thus, in addition to representing their innovative abilities, speaks to their creativity.

As noted earlier, delphinids are well known for the imitative abilities, and they often employ these during play. Dolly's milk smoking (Tayler & Saayman, 1973) is particularly striking because Dolly was only around 6 months old at the time and inhibited her normal swallowing behavior to instead release milk into the water. It seems plausible to infer that Dolly's behavior was the direct result of creative cognition, but Dolly was not the only imitator Tayler and Saayman observed in their study. Haig, one of Dolly's tank mates, regularly imitated the resident cape fur seal's (*Arctocephalus pusillus*) swimming, sleeping, and even grooming behavior, all of which were quite awkward and unnatural for Haig. She also appeared to imitate the behavior of skates, turtles, and penguins. Meanwhile, the male in the group, Daan, regularly imitated the behavior of human divers. On one occasion, after observing divers clean algae off of the underwater viewing port in his tank, Daan was seen using a seagull feather to scrape at the glass while mimicking the sounds and slow stream of bubbles produced by divers. Over the course of 54 days, Daan used food-fish, sea slugs, stones, and paper to keep the window clean and aggressively defended his new found territory. He even mimicked divers' hand positions by placing his pectoral fin on the window when cleaning. Later, Haig also attempted to clean the window and was actually given a brush to do the job but failed to hold the tool correctly. Bottlenose dolphins are not alone in their ability to imitate other species. Kritzler (1952) observed the same pilot whale that invented the flotsam-drain game attempt to mimic bottlenose dolphins who used their rostra to spear and throw inflatable inner tubes. Limited by his morphology (a short rostrum), after several failed attempts the whale began to use his pectoral fin to spear and play with the inner tube instead. It is true that intraspecific imitation may require very little, if any, innovation or creativity, but interspecies imitation is different. Such behavior requires individuals to not only first map their bodies onto the anatomy of a sometimes very different model, they must then create analogous behaviors within their own anatomical limits. Thus, these types of behaviors represent innovations and since many likely involve analogous cognitive mapping, also indicate some level of creativity.

Vocal Communication

Cetaceans have extraordinarily diverse communication systems (Tyack, 2000), aspects of which speak to their innovative and creative abilities. The strongest example is the signature whistle. During the 1960s, Melba and David Caldwell recorded the vocalizations of captive bottlenose and common dolphins (*Delphinus delphis*) and noticed that each individual appeared to have its own distinct whistle contour, which they appeared to develop within the first months of life (Caldwell & Caldwell, 1965, 1968, 1979). They called these vocalizations signature whistles and hypothesized that they might be used for individual identification. Later, evidence of signature whistles was found in captive Pacific white-sided dolphins (*Lagenorhynchus obliquidens*) (Caldwell & Caldwell, 1971), Atlantic spotted dolphins (*Stenella plagiodon*) (Caldwell, Caldwell, & Miller, 1973), and most recently, Pacific humpback dolphins (*Sousa chinensis*) (van Parijs & Corkeron, 2001). In these studies, signature whistles were recorded when individuals were isolated, usually for medical attention, a period in which an individual's signature whistle constitutes about 94% of its vocalizations. This led the Caldwells to further hypothesize that signature whistles might be used for maintaining group cohesion (Caldwell, Caldwell, & Tyack, 1990). In free swimming captive bottlenose dolphins, Janik and Slater (1998) found that individuals tended to produce signature whistles when one group member voluntarily isolated itself, lending further support to the group cohesion hypothesis. More recent evidence also supports the individual identification and vocal labeling hypothesis. For example, Bruck (2013) found that bottlenose dolphins in aquaria can remember each other's signature whistles even when separated for over 20 years. Some investigators have contested the presence of signature whistles (e.g., McCowan & Reiss, 2001), but more recent evidence from wild bottlenose dolphins provides fairly unequivocal support for both the presence of signature whistles and their importance in communication (reviewed in Janik & Sayigh, 2013). In fact, as we will discuss, field studies not only provide a wealth of information on signature whistles, they also indicate that every individual engages in true innovation during the first year of life, although the underlying cognitive processes are not known. What is somewhat unique here among animal innovations is that in order for a signature whistle to function as an individual identifier, it *cannot* spread through the population, although signature whistle copying occurs, possibly as a way to address conspecifics (Janik & Slater, 1998; Tyack, 1986). Thus, with every animal a signature whistle innovation is born and dies.

Foraging

While captive animals are usually well fed, some individuals seem to have invented novel ways in which to supplement their diet. For example, in addition to imitating other animals in her tank, Haig also imitated human divers in order to obtain food (Tayler & Saayman, 1973). After several days of closely observing divers scrape sea lettuce (*Ulva* sp.) off the tank floor, Haig scraped the tank floor with the divers' tool herself and ate the sea lettuce that came loose. After the scraper was taken away, she found a piece of broken tile to do the job. Later, her tank mate Lady Dimple learned the behavior and both dolphins continued to forage in this way until all tiles were removed from the tank. A few decades earlier at Marineland of the Pacific, bottlenose dolphins Frankie and Floyd were observed attempting to capture a moray eel (*Gymnothorax mordax*), which they had trapped by positioning themselves on either side of a crevice the eel was hiding in (Brown & Norris, 1956). At some point, one of the dolphins left to capture a scorpion fish (*Scorpaena guttata*), and upon its return, poked the eel with the fish's sharp, venomous spines. The eel vacated the crevice and was quickly caught, although not consumed. Meanwhile, at Marineland of Florida, a young male bottlenose dolphin was observed attempting to coerce a large red grouper (*Epinephelus morio*) out of a rock crevice using dead squid and fish food as bait. Captive killer whales (*Orca orcinus*) have also been seen using their food to bait live prey especially gulls (Kuczaj, Lacinak, Garver, & Scarpuzzi, 1998; Noonan, 2005). While all these behaviors likely constitute innovations, Frankie and Floyd's eel poking is perhaps the most intriguing and seems likely to involve creative cognition.

Testing for Innovation

While these anecdotes provide some fairly convincing evidence that odontocetes are creative and capable innovators, most of the systematic research on cetacean creativity and innovation comes from Louis Herman's Kewalo Basin Marine Mammal Laboratory in Honolulu, Hawaii, which in its 30 plus years, provided the most comprehensive view of dolphin cognition to date (Herman, 2012a). By training their dolphins to respond to auditory and gestural cues, Herman and colleagues tested dolphins on a variety of tasks carefully constructed to probe their cognitive abilities. Animals were trained to perform various actions such as *fetch, leap over, swim under*, etc. often in conjunction with various items such as *ball, hoop*, and *frisbee* (Herman, 2010). In most of Herman's studies, objects were positioned appropriately to allow the dolphins to perform the requested action. However, on at least one occasion, a female dolphin named Akeakamai (Ake) was asked to perform behaviors that were impossible given the current orientation of the objects in her tank. She spontaneously solved this problem by rearranging the items in a way that would allow her to complete the requested action (Herman, 2006). For example, if asked to swim through a hoop lying flat on the bottom of the tank, Ake would lift the hoop into an upright position and swim through it. She also spontaneously rearranged objects in her tank to help in cleaning up after an experimental session. At the end of a session, trainers gave Ake the *fetch* gesture, which asks her to retrieve all the objects in her tank. Rather than fetching the objects one by one as she was trained, Ake "round up" several objects at once, effectively inventing a novel strategy for efficiently cleaning up her tank (Herman, 2006).

Ake again showed her innovative and creative skills when she was asked to perform a behavior with an object that was absent from her tank. Ake was taught to press one of two paddles corresponding to either yes or no depending on whether or not an object was present in her tank (Herman, 2006). During one experiment, she was asked to bring one object over to another that was missing, and she responded by retrieving the first object and bringing it to the no paddle (Herman, 2006), essentially coming up with a way to communicate to the experimenter that the task could not be completed. This was not the only time she came up with a novel way to communicate with trainers. Occasionally wind blew debris into Ake's tank, after which she would produce a distinct, loud whistle. Eventually staff learned that Ake seemed to be calling to them to come remove the debris, which she had brought to the side of the tank (Herman, 2006). Importantly, all of Ake's responses above were spontaneous, untrained, unexpected, and novel, and it seems likely that at least some involved creativity.

We have already noted that delphinids are impressive imitators, and Herman's dolphins are no exception. In trying to teach two young dolphins Hiapo and Elele to associate specific behaviors with gestures, trainers gave a gesture and then immediately modeled the behavior for the dolphins. While neither dolphin learned the gestural cues, both imitated their human trainers' actions, sometimes with remarkable accuracy on the first try (Herman, 2002). This somewhat accidental result was examined in more detail later by teaching Ake and Phoenix a gesture for *mimic* and testing their ability to imitate human models. Like Haig and Daan's behavior, these cross-species imitations require abnormal body mapping meaning they are both creative and innovative. From these studies and others, we know that bottlenose dolphins are exceptional imitators and even appear to have a conceptual level understanding of imitation (Herman, 2002).

While the above studies corroborate some of the anecdotes presented earlier, no study does so better than the famous "Creative Porpoise" experiment conducted at Sea Life Park at the Makapuu Oceanic Center in Hawaii in the 1960s.[1] In their study, Karen Pryor (also see Chapter 17) and colleagues used a paradigm in which "only those actions will be reinforced, which have not been reinforced previously" (Pryor, Haag, & O'Reilly, 1969). The technique was initially trialed on

---

[1]See video clip at http://reachingtheanimalmind.com/chapter_05.html.

Malia, a female rough-toothed dolphin (*Steno bredanensis*), during regular dolphin shows. After several days of reinforcing one unique behavior each show, Malia exhausted her normal repertoire and began showing signs of frustration. However, on the morning of day four, Malia seemed to spontaneously "get it" and upon entering the main tank, quickly circled around to build up speed, rolled onto her back and stuck her tail fluke in the air and coasted as if she were sailing. She had created a new behavior, one she could not have practiced in her small holding tank, and one Pryor and colleagues had never seen before (Pryor, 2009). After this breakthrough, Malia continued to come up with novel behaviors. Perhaps one of the more interesting examples is what Pryor calls Malia's "art project" (Pryor, 2009). Showing off her sketching abilities, Malia once used her dorsal fin to draw "beautiful looping lines" in the silt on the bottom of her tank (Pryor, 2009). Pryor et al. (1969) followed up on their initial trials with a more detailed study using a new female rough-toothed dolphin, Hou. The same general procedure was used in that only one novel behavior was reinforced per training session, and after 16 sessions Hou too seem to grasp the concept. In fact, by session 33 Hou's behaviors became so complex and novel that the experimenters had trouble discriminating and describing them, so the study had to be stopped. To examine just how unique some of these behaviors were, Pryor et al. (1969) asked facility staff to rank Hou's behaviors according to how often they spontaneously occurred in other captive cetaceans. While some of Hou's behaviors had been commonly observed in a variety of species, 11 of Hou's 16 behaviors had either rarely been observed in any species, only observed in *Stenella* spp., or had never been observed by staff before in any cetacean. After the experiment was over, Hou was introduced into the dolphin show and trained, like Malia, to perform certain behaviors in response to specific cues. However, if trainers took up the position used during the creative experiments, both Malia and Hou would readily produce novel behavior, demonstrating their remarkable flexibility in being able to switch from producing behavior under specific stimulus control, to that which requires their own creative input.

While some have questioned the validity of Pryor's et al. (1969) original findings (e.g., Holth, 2012), similar methods have since been used to elicit novel behavior in a variety of other species (dogs, *Canis lupus familiaris*; cats, *Felis catus*; horses, *Equus ferus caballus*; parrots, Psittaciformes; gorillas, *Gorilla gorilla gorilla*; budgerigars, *Melopsittacus undulates*; and walruses, *Odobenus rosmarus*) (Manabe, 1997; Manabe & Dooling, 1997; Pryor, 2004a, 2004b, 2006, 2009; Schusterman & Reichmuth, 2007), including bottlenose dolphins (Herman, 1991; Kuczaj & Eskelinen, 2014). Herman and colleagues taught bottlenose dolphins a *create* gesture, which asked for a behavior different from the preceding behavior (Braslau-Schneck, 1994; Herman, 1991, 2002, 2006, 2010; Mercado, Murray, Uyeyama, Pack, & Herman, 1998; Mercado, Uyeyama, Pack, & Herman, 1999). This is slightly different than Pryor's et al. (1969) study in that creative behavior is now under control of a gesture, allowing trainers to repeatedly ask for new behaviors within a single trial. Herman's dolphins were also taught a *repeat* gesture, which asked them to repeat the behavior they had just performed. Elele and Hiapo successfully learned both the *repeat* and *create* commands, and could even repeat previously self-selected behavior (Herman, 2002; Mercado et al., 1998, 1999). Furthermore, the two were able to create new behaviors together. After individually mastering the *create* sign, the dolphins where given the *tandem* sign (previously used to elicit synchronous trainer selected behavior) and *create* sign in conjunction (*tandem* + *create*) to see how they would respond. After a brief period of side-by-side swimming, the dolphins executed a new behavior in almost perfect synchrony (Braslau-Schneck, 1994; Herman, 2002, 2006). Later, Ake and Phoenix also successfully learned the *tandem*, *repeat*, and *create* signs and could use them in combination (Braslau-Schneck, 1994). While, the *create* command does not actually require completely novel behavior, the only restriction being that the behavior not be identical to the one immediately preceding it, all dolphins in Herman's studies regularly came up with never before seen behaviors, both individually and in tandem (Braslau-Schneck, 1994; Herman, 2002, 2006). More recently, Kuczaj and Eskelinen (2014) examined individual variation in response to a *create* cue, which they prefer to call *vary*, using bottlenose dolphins at Dolphin Cove in Key Largo, Florida. Leo, the youngest of the three males in their study, exhibited more novel, more complex, and higher energy behaviors than the older adults Alfonz and Kimbit. Kuczaj and Eskelinen (2014) suggest this could be due to differences in cognitive style and strategy, but at around 10 years old, Leo was much younger than the two 20-year-old adults so his higher performance could also be the result of differences in motivation, trainability, and fatigue that may come with age. Regardless, these *create/vary* studies not only demonstrate that dolphins are remarkably innovative and creative and can be so on command, they seem to grasp the concept of novelty itself (Herman, 2006).

Given this brief review in captivity, what can we conclude about cetacean innovation and creativity? Since so few have been kept in captivity, nothing about mysticetes, but odontocete appear to innovate in a variety of contexts. Most often these skills are used in play, rather than in foraging or communication. This is not surprising since captive animals are encouraged to play, but have little forage demands and experience an unnatural social environment with few incentives for interspecific communication. While this situation poorly represents the conditions faced by wild odontocetes, the anecdotes and experiments discussed *do* suggest that many odontocetes are highly innovative and sometimes creative, skills that prove useful in the wild as discussed next.

INNOVATIONS AT SEA

The study of marine organisms, even at the most basic level, lags behind that of terrestrial species for obvious logistical reasons. Cetologists at least have the advantage that their study subjects are large and breathe air, but this provides quite literally only a surface level understanding of cetacean behavior. The end result is that compared to those on terrestrial fauna, field studies on marine fauna are fewer, suffer greater observational biases, and tend to be more descriptive in nature. Recognizing these limitations is critical when examining innovative and creative behavior. For example, we know very little about deep, underwater behavior, particularly that which occurs far offshore, in the oceanic zone, and even near shore, at the surface we usually can only observe behavior at a coarse grain (Russon, 2003). In all, these limitations almost certainly lead us to underestimate the prevalence of innovative and creative behaviors in wild cetaceans. Nevertheless, there are many examples of rare, unusual, novel, and/or atypical species behavior scattered throughout the literature that provide insight into how cetacean innovative and creative abilities might have evolved and are used in the wild. As before, we describe these within their greater functional contexts.

Play

Like those in captivity, free ranging cetaceans demonstrate a great diversity of play behaviors, many of which can be considered innovative and some, creative. As before, some include intraspecific imitation, and thus likely involve creativity. For example, Würsig (2008) witnessed a bottlenose dolphin off the coast of the Bahamans mimic the awkward swimming of a snorkeling tourist. However, most examples of innovative play behavior in the wild involve cetaceans interacting and playing with novel objects (Paulos et al., 2010). As before, simply manipulating and playing with such objects may not be all that innovative, but the details concerning a few of these behaviors inform us on innovation and creativity. Examples exist for both odontocetes and mysticetes, and include some mysticete—odontocete interactions.

Many wild delphinids regularly interact with birds in a playful manner (Heubeck, 2001; Hewitt, 1986; Mann & Smuts, 1999; Würsig, 2008). For example, after foraging on anchovy schools off the coast of Argentina, some dusky dolphins (*Lagenorhynchus obscurus*) have been seen carefully grabbing the dangling legs of unsuspecting gulls, quickly surging underwater, and then releasing the gulls', effectively dunking the birds. The dolphins are very gentle in their grasp, never causing any harm, and simply appear to be having a little fun with their feathered counterparts (Würsig, 2008). As yet unreported elsewhere, this gull-dunking game resembles some of the innovative play behavior of captive delphinids described earlier.

Another possible example of play innovation among wild odontocetes comes from the "southern resident," fish-eating community of killer whales in Puget Sound, Washington. In 1987, a female from a pod known as K-pod was observed carrying a dead salmon around on top of her head (Baird, 2002; Whitehead, Rendell, Osborne, & Würsig, 2004). Within that same year, the behavior quickly spread to two additional pods in the southern resident community, but disappeared from the community shortly after. While the function of dead salmon carrying is unknown, some suggest it may have been a cultural fad (Baird, 2002; Whitehead et al., 2004), although here we consider it in the context of play since it appears to have no immediate purpose. This example not only demonstrates potential innovative play behavior, but also its horizontal transmission and eventual disappearance.

Free ranging mysticetes also play with objects in ways that suggest innovative behavior, with bowhead whales (*Balaena mysticetus*) providing probably the strongest example. During the boreal summer and fall, bowhead whales in the Beaufort Sea can be seen interacting with floating logs in rather "artistic" ways (Würsig, 2008, p. 887). This includes nudging, propelling, and dunking logs using their flippers and tails, but more impressively, balancing logs on their backs or bellies and even adjusting their body position to account for ocean swell (Würsig & Dorsey, 1989; Würsig, 2008). Similar log play might also occur in sperm (*Physeter macrocephalus*) (Nishiwaki, 1962) and perhaps humpbacks whales (*Megaptera novaeangliae*) (Couch, 1930). Playing with logs itself is likely not all that innovative, but the diversity of ways in which whales manipulate their log toys may indicate some creative processes is at work.

Occasionally odontocetes and mysticetes interact with each other in ways that demonstrate their innovative abilities. For example, recently off the coast of Hawaii Deakos, Branstetter, Mazzuca, Fertl, and Mobley (2010) observed humpback whales using their rostra to lift bottlenose dolphins entirely out of the water, after which the dolphins slid down the whales' backs and entered the water. This behavior was observed on two occasions and in both instances neither the dolphin nor the humpback behaved aggressively. Given this, and that the behavior has no obvious function, it is presumed to be a form of interspecific play and may constitute a cross-species innovation. Other odontocete-mysticete interactions seem to be more one sided. Bottlenose and dusky dolphins have been known to "coerce" large whales (balaenids and sperm whales) into helping them surf by swimming on either side of the whales' eyes perhaps to agitate them (Würsig, 2008). After a while, the whales appear to become irritated and surge forward, creating large waves that the dolphins eagerly surf. Perhaps dolphins' use of pool toys to create waves as noted earlier is a modification of this behavior adapted to a captive lifestyle. While intent is not known, the dolphins are probably the innovators, and the whales, just the object of their harassment.

Cetaceans also interact with humans in playful ways that suggest innovation and creativity. Many cetologists have firsthand experience with this, as wild delphinids often invent and play games with their scientific audience (e.g., passing seagrass or leaves back and forth) (Johnson & Norris, 1994; Mann & Smuts, 1999). However, one of the most well-known human–cetacean interactions involves gray whales. In 1975, off the coast of Baja California, Mexico in Laguna San Ignacio, Gilmore (1976) observed several gray whales that behaved unusually curious and "friendly" toward boats. Several whales deliberately approached whale-watching vessels and allowed tourists to stroke them. At the time, such behavior was quite surprising since less than a century ago these long-lived mammals were referred to as "devil fish," a name attesting to their violent interactions with whalers. A later study confirmed that while the so called "friendly" phenomena may have been around slightly earlier (1960s), the behavior seemed to be relatively new (Jones & Swartz, 1984). This study also found that by 1982, the friendly behavior had rapidly spread throughout the Baja California region, and even to the northern end of the whales' migration route, off Vancouver Island and in the Bering Sea (Jones & Swartz, 1984). Friendly whales of both sexes and all ages, individually and in groups, approach tour vessels, usually from the stern and mainly when the engine is in neutral perhaps to investigate the source of the engine noise (Jones & Swartz, 1984). In addition to allowing tourist to pet them, the whales often blow bubbles under the boats, produce a variety of vocalizations, and probe and lift the boats, occasionally even knocking passengers overboard (Jones & Swartz, 1984). Some whales seem particularly attached to boats and will follow them for an entire day, even at speed, and repeatedly do so year after year (Jones & Swartz, 1984). Although this behavior does not appear all that creative, like dead salmon carrying in killer whales, it illustrates how innovations that likely have little impact on fitness can rapidly spread throughout a population, even across some 5,000 km or more of ocean.

Vocal Communication

Research on wild cetacean vocal communication may suffer the greatest from the logistical difficulties of studying at sea, with the most notable obstacle being one of the first steps in studying vocal communication: identifying the vocalizer. Most terrestrial species produce visual cues that indicate they are vocalizing and, even in the absence of such cues, researchers can often localize the sound with their own ears. However, cetaceans produce very few, if any, visual cues and it is impossible to localize sound underwater without sophisticated hydrophone arrays. That said, in the last 25 years or so great advances in acoustic technologies have provided a wealth of information on cetacean vocal communication, and much of this work points to communication as a fruitful domain in which to examine cetacean innovation and creativity.

In the two largest odontocetes, killer and sperm whales, neighboring groups of individuals exhibit distinct vocal repertoires called dialects (Ford, 2008). Among killer whales, dialects seem to be present at two hierarchical social levels, both based on matrilineal relatedness. Smaller stable groups known as pods have distinct acoustic repertoires but share a portion of this repertoire with several other related pods to form what is known as an acoustic clan (Ford, 1991; Nousek, Slater, Wang, & Miller, 2006). In contrast, sperm whale dialects (termed coda dialects) also differ among matrilines, but are only distinct at a higher clan level (Whitehead, 2003). In both taxa, dialects are thought to be socially learned within matrilines and function in communication and maintaining group cohesion (Ford, 1991; Miller, 2002; Weilgart & Whitehead, 1993). In killer whales, changes in dialects have been documented and appear to mostly be the result of cultural drift (Deecke, Ford, & Spong, 2000). However, some evidence suggests that cultural selection, possibly through inbreeding avoidance, plays a role in the evolution of more complex vocal calls (Barrett-Lennard, 2000; Yurk, Barrett-Lennard, Ford, & Matkin, 2002). Although no long-term changes in sperm whale coda dialects have been recorded (Rendell & Whitehead, 2005), high dialect differentiation and low genetic differentiation across ocean basins suggests coda dialects have changed through time and are subject to cultural evolution (Lyrholm, Leimar, Johanneson, & Gyllensten, 1999; Rendell & Whitehead, 2003). In both sperm and killer whales, a single change in a group's dialect, if functional, could constitute an innovation, but whether or not this is the case is unclear. Regardless, most evidence indicates that these vocal changes are not the result of some creative process, but rather errors in vocal copying (Deecke et al., 2000; Yurk et al., 2002).

While sperm and killer whale dialects may hint at vocal innovation, signature whistles provide direct evidence of innovative behavior in wild cetaceans. Since the time of the Caldwells' initial work in captivity, evidence of signature whistles has been documented in wild bottlenose dolphins (Gridley et al., 2013; Smolker, Mann, & Smuts, 1993), Guiana dolphins (*Sotalia guianensis*) (Duarte de Figueiredo & Simão, 2009), narwhals (*Monodon monoceros*) (Shapiro, 2006) and possibly pilot whales (Sayigh, Quick, Hastie, & Tyack, 2013). In bottlenose dolphins, field data continue to support the hypothesis that signature whistles aid in group cohesion and communicate individual identification. For example, in Sarasota Bay, Florida bottlenose dolphins increased signature whistle production in larger groups and during social behavior (Cook, Sayigh, Blum, & Wells, 2004), and in Shark Bay, Australia bottlenose dolphin calves produced signature whistles primarily when separated from their mother, particularly at far distances and near the end of separations (Smolker et al., 1993). In addition, wild bottlenose dolphins seem to recognize their own signature whistles (King & Janik, 2013), and copy the whistles of close associates (mothers and calves, alliance partners) possibly as a way to label and address conspecifics (King, Sayigh, Wells, Fellner, & Janik, 2013). Vocal copying also seems to be important in the development of signature whistles. In Sarasota Bay, Florida calves appear to model their signature whistles from those of their conspecifics, but modify these to invent their own distinct, new whistles (Fripp et al., 2005). However, males and females seem to differ in terms which model they use. Male calves' signature whistles more closely resemble those of their mothers' compared to females', possibly to help avoid inbreeding (Sayigh, Tyack, & Wells, 1995; Sayigh, Tyack, Wells, & Scott, 1990). Male signature whistles are also less stable than females, often converging with those of their male alliance partners in adulthood (Smolker & Pepper, 1999; Watwood, Tyack, & Wells, 2004). However, for both male and female calves there is low similarity between mother and offspring signature whistles, further suggesting that signature whistles are not genetically determined, but rather individually and socially learned (Janik & Sayigh, 2013). In all, signature whistles clearly serve an important social function, and are somewhat unique among animal communication signals in just how individually distinct they are.

Foraging

Cetaceans exhibit a great diversity of foraging behaviors in the wild, many of which speak to their innovative and creative skills. There are many examples of cetaceans interacting with fishers or tourists in novel ways to obtain food. For example, dolphins and whales in numerous locations depredate long-lines (Ashford, Rubilar, & Martin, 1996;

Hamer, Childerhouse, & Gales, 2012; Nolan, Liddle, & Elliot, 2000; Purves, Agnew, & Balguerias, 2004; Visser, 2000; Yano & Dahlheim, 1995), steal prey from traps or fish farms (Kemper et al., 2003; Noke & Odell, 2002), and follow trawlers or trammel nets in hopes of consuming stray prey or discards (Chilvers & Corkeron, 2001; Gonzalvo, Valls, Cardona, & Aguilar, 2008; Jefferson, 2000; Leatherwood, 1975; Pennino, Mendoza, Pira, Floris, & Rotta, 2013). In many cases these behaviors start out at low frequency, but rapidly spread to the rest of the population, possibly indicating an initial innovation event and its subsequent social transmission among sympatric cultural units or clusters (e.g., Chilvers & Corkeron 2001; Donaldson, Finn, Bejder, Lusseau, & Calver, 2012; Fearnbach et al., 2013; Whitehead et al., 2004). Despite the risks of becoming bycatch themselves, these cetaceans benefit by avoiding the energetic costs of long prey chases and deep dives.

Some cetacean—fisher interactions have become elaborated into what can be called cooperative foraging. Famously, for 80 years the "killer whales of Eden" assisted whalers during hunts of humpback and southern right whales (*Eubalaena australis*) in Twofold bay, Australia (Dakin, 1934; Jefferson, Stacey, & Baird, 1991). In exchange for the help, whalers allowed the killer whales to feed on the favored tongues and "lips" in an agreement known as "the law of the tongue" (Brady, 1909). In coastal fisheries off Laguna, Brazil, bottlenose dolphins and fishers regularly cooperate to catch schools of mullet (*Mugil* spp.) (Daura-Jorge, Cantor, Ingram, Lusseau, & Simões-Lopes, 2012; Pryor, Lindbergh, Lindbergh, & Milano, 1990; Simões-Lopes, Fabián, & Menegheti, 1998). Here, dolphins drive mullet toward shore and after a series of more subtle movements, "signal" with tail and head slaps where the fishers should throw their nets (Simões-Lopes et al., 1998). This foraging tactic appears to be socially facilitated as dolphins who perform the behavior associate more than those who do not, leading some to suggest that the behavior has been passed down for over 160 years (Daura-Jorge et al., 2012; Pryor et al., 1990). In Myanmar, a similar cooperation between dolphins and fishers exist. For over 130 years fishers have used a variety of signals to advertise their interest in cooperating with Irrawaddy dolphins (*Orcaella brevirostris*), who then herd fishes into tight schools for easy netting (Smith, Tun, Chit, Win, & Moe, 2009). In all well-documented cases, both fishers and dolphin appear to benefit with higher catch rates and/or volumes of fishes (Simões-Lopes et al., 1998; Smith et al., 2009). Although the underlying cognition and transmission of these behaviors is not known, they illustrate how innovations between humans and wild animals can develop and persist for over a century.

Probably the most common form of human—cetacean interaction involves recreational fishers directly feeding dolphins, which has resulted in extensive begging behaviors and harm to wildlife in south-eastern United States (e.g., Samuels & Bejder, 2004) and several parts of Australia (Donaldson et al., 2012; Foroughirad & Mann, 2013). Although these animals are making use of a novel food resource, such feeding is not particularly unusual (Orams, 2002). However, in some locations dolphin provisioning has taken other forms and involves more elaborated behaviors that are not typical. In Monkey Mia, Shark Bay, Australia, a small number of bottlenose dolphins (mostly female and from 3 matrilines) have been provisioned by tourists since the 1960s (Foroughirad & Mann, 2013), a program that is now supervised and strictly regulated by Western Australia's Department of Parks and Wildlife. In the 1980s, several unusual behaviors emerged among the provisioned dolphins (Smolker, 2001); one, started by a young male named Snubnose, involved a posture in which a dolphin arches its back, holds its tail and head out of the water, and plants its pectoral fins and belly on the seafloor (Figure 4.1). This gesture soon spread to other provisioned dolphins and is still seen today, even by young calves that are not fed (JM, personal observation). Another unusual behavior involves "gift-giving." Even though dolphins do not share prey with each other (Mann et al., 2007), in Monkey Mia they occasionally attempt to give fishes to humans, a behavior that has been seen ~30 times over the course of 27 years. In each case, a dolphin brings a large fish to the beach and drops or holds it gently next to a person, sometimes nudging him or her repeatedly with the fish. In the most striking example, after

FIGURE 4.1 A male dolphin named Snubnose performing a begging gesture near Monkey Mia, Australia in the 1980s, a behavior that spreads and persists to this day among other provisioned dolphins. *Photo taken by Janet Mann.*

months of a ranger trying to initiate a 6-year-old female named Piccolo into the provisioning program, she finally accepted a fish and swam off. The next day, she caught a large whiting (*Sillago* sp.), brought it into the beach, and spent 20 min trying to get that same ranger to take it, even rising up on the ranger's chest with the fish in her mouth (JM, personal observation). It is tempting to speculate that because humans were sharing fishes with her, Piccolo was trying to reciprocate, a behavior that could be considered creative. In the end, the ranger did not take the fish and Piccolo swam off. After that, she refused fish offers for the next 5 years until she had her own offspring and was perhaps hungry enough to give it a try. Food gift-giving has occurred at human dolphin provisioning sites elsewhere (Holmes & Neil, 2012), but to our knowledge, has otherwise not been reported between nondomesticated species and humans.

Some human—cetacean interactions may indirectly result in innovative foraging behavior. For example, whaling practices (Williams, Estes, Doak, & Springer, 2004), oil spills (Loughlin, 1994), and other human activities have depleted preferred prey of killer whales. From large whales to Steller sea lions (*Eumetopias jabatus*), harbor seals (*Phoca vitulina*) (Loughlin, 1994), and eventually sea otters (*Enhydra lutris*) (Estes, Tinker, Williams, & Doak, 1998; Williams et al., 2004), killer whales turned to smaller and smaller prey to meet their energetic demands (but see DeMaster et al., 2006). Although the causes have been debated, these killer whales are utilizing both new resources and new foraging tactics to obtain them. Such behavior however, is not surprising since killer whales are known to consume a great diversity of prey (Jefferson et al., 1991) (although see later for discussion of foraging tactics). Here, prey switching may simply constitute a change in expression of an existing behavior and involve little creativity or innovation *per se*. Humpback whales have also responded to prey collapses over the decades by switching diets from herring (*Clupea harengus*) to sand lance (*Ammodytes americanus*), and in the process, foraging tactics from lunge feeding to lobtail feeding and bottom-side rolling (Allen, Weinrich, Hoppitt, & Rendell, 2013; Read, 2001; Ware et al., 2014; Weinrich, Schilling, & Belt, 1992). Lobtail feeding is characterized by whales striking their tail on the water surface several times, followed by creating bubble-streams, but the whales also scrape along the bottom where sand lance congregate (Ware et al., 2014). This behavior contrasts with the more typical lunge-feeding pattern, and is argued to have been culturally transmitted given the pattern of diffusion (Allen et al., 2013). Although this striking shift in foraging tactics suggests flexibility and social transmission among whales, sand lance feeding clearly occurred at least at low frequencies well before spreading (Hain, Carter, Kraus, Mayo, & Winn, 1982; Weinrich et al., 1992).

Clearly direct (e.g., tourism, fishing, feeding) and indirect (e.g., prey collapse) human impacts create new challenges for which cetaceans have innovative responses, but cetaceans also naturally exhibit diverse and innovative foraging tactics. Both humpback and minke whales have been intensively observed at some locations so their foraging tactics are relatively well known. Humpback whales engage in bubblenet and bubblecloud feeding (Hain et al., 1982; Ingebrigtsen, 1929; Jurasz & Jurasz, 1979; Wiley, Ware, & Bocconcelli, 2011) where they cooperatively or singly create circular walls of bubbles or bursts of bubbles that form a cloud to aid in prey capture (Hain et al., 1982; Wiley et al., 2011). Given the apparent cooperation between individuals that engage in bubblenet feeding, some have suggested that the behavior is socially learned (Clapham, 2000; Weinrich, 1991; Weinrich, Rosenbaum, Scott Baker, Blackmer, & Whitehead, 2006) and possibly a case of tool use depending on one's definition (Mann & Patterson, 2013). Similar to the creative use of bubbles described earlier, the cooperation and planning that appears to be involved indicates that bubblenet feeding may involve some creative cognition and qualify as an innovation.

Humpback whales are not alone in their use of bubbles during foraging as such behavior has been previously reported in killer whales (Sigurjónsson, 1988; Similä & Ugarte, 1993) and Atlantic spotted dolphins (*Stenella attenuate*) (Fertl & Würsig, 1995). For example, in a behavior that may also involve creative cognition and constitute innovative tool use, killer whales have been observed releasing bubbles "downward" toward stingray prey possibly to startle them from their burrowed location (Visser, 1999). In fact, killer whales exhibit a great diversity of feeding tactics within and across sites (Baird, 2000; Jefferson et al., 1991). Many of these tactics are thought to be socially learned traditions passed on primarily through matrilines (Riesch, Barrett-Lennard, Ellis, Ford, & Deecke, 2012). Most famous of these occur with the two ecotypes of killer whales in the Pacific Northwest: the "resident" or fish-eating killers whales (which eat predominantly salmon, *Oncorhyncus* spp.) and the "transient" or mammal-eating killer whales (which eat predominantly harbor seals) (Baird, 2000). Over 40 years of study has revealed that these ecotypes differ genetically and in diet, ranging patterns, vocal and social behavior, and pod structure (Baird, 2000; Dahlheim & White, 2010; Deecke, Nykänen, Foote, & Janik, 2011; Ford, Ellis, Barrett-Lennard, Morton, & Balcomb, 1998; Herman et al., 2005; Saulitis, Matkin, Barrett-Lennard, Heise, & Ellis, 2000). Although occasionally near each other, they do not interact or breed, and it has been suggested that they are currently undergoing speciation (Riesch et al., 2012). More recently, studies have revealed what is thought to be yet another Pacific Northwest ecotype, called the "offshores," that specialize in other fishes such as pacific halibut (*Hippoglossus stenolepis*) and pacific sleeper sharks (*Somniosus pacificus*), but currently little is known about their social structure and behavior (Dahlheim et al., 2008; Ford & Ellis, 2014; Ford et al., 2011; Jones, 2006; Krahn et al., 2007).

The killer whales of the Pacific Northwest are not unique among killer whales in terms of their innovative and creative foraging behavior. At several disparate sites from Patagonia, the Faroes, to the Crozet Islands, killer whales beach or temporarily strand themselves to catch pinnipeds and occasionally seabirds (Bloch & Lockyer, 1988; Guinet, 1991; Guinet & Bouvier, 1995; Hoelzel, 1991; Lopez & Lopez, 1985). Within a population, members of some pods engage in the behavior while others do not. Sometimes the killer whales create pressure waves to help wash unsuspecting seals off the beach, or lunge suddenly out of the water and even synchronize their beaching to reduce the chances of prey escaping (Bloch & Lockyer, 1988; Guinet, 1991; Guinet & Bouvier, 1995; Hoelzel, 1991; Lopez & Lopez, 1985). Wave washing, singly and in coordinated groups, also occurs on ice floes in the Southern Ocean off Antarctica (Pitman & Durban, 2012; Smith & Siniff, 1981; Visser et al., 2008). These dramatic and planned attacks involve inspecting ice floes for preferred prey, Weddell seals (*Leptonychotes weddellii*), and recruiting fellow pod-members to aid in washing seals off the ice and into the water where they can be pursued. For larger floes, whales sometimes first use a wave to break the floes up and have even been observed lifting ice with their heads (Pitman & Durban, 2012). The variation in these attacks suggests that the killer whales' behavior is planned and adaptable to their changing circumstances. Like in the Pacific Northwest, killer whales in the Southern Ocean appear to represent several different ecotypes, one that specializes on seals, a second on penguins (Pitman & Durban, 2010), a third on minke whales, and a fourth on fishes (Pitman & Ensor, 2003). In the North Atlantic, this pattern again repeats itself with some pods specializing on marine mammals (Deecke et al., 2011; Foote, Newton, Piertney, Willerslev, & Gilbert, 2009), and others on fishes (predominantly Atlantic herring, *Clupea harengus*) (Similä & Ugarte, 1993; Similä, Holst, & Christensen, 1996). The striking feature within and across killer whale populations is that genetic and behavioral data suggest that many of these foraging innovations have become specializations, which are then matrilineally transmitted and can result in reproductive isolation and eventually speciation (Riesch et al., 2012).

Like killer whales, bottlenose dolphins also exhibit great diversity in their natural foraging behaviors, many of which can be considered innovative. Strand-feeding, similar to that described for killer whales above, occurs at several sites including South Carolina and Georgia (Duffy-Echevarria, Connor, & St. Aubin, 2008; Hoese, 1971; Rigley, VanDyke, Cram, & Rigley, 1981), Mexico (Silber & Fertl, 1995), Portugal (dos Santos & Lacerda, 1987), Mozambique (for *Sousa plumbea*, Peddemors & Thompson, 1994) and Australia (Sargeant, Mann, Berggren, & Krützen, 2005). In all of these cases, dolphins either individually or collectively trap fishes (almost always mullet) by chasing them onto the shoreline, and then fully or partially lunge out of the water and onto the shore to catch their prey. Only certain individuals engage in this risky behavior, even though in some cases (e.g., Sargeant et al., 2005) other dolphins repeatedly witness these beaching dolphins successful catch prey. At each location, distinct variations of strand-feeding are evident. In Bull Creek, South Carolina, typically 3–4 dolphins cooperatively surge onto mudbanks to catch fishes (Duffy-Echevarria et al., 2008). In Shark Bay, only 6 dolphins currently engage in the behavior out of hundreds near shore (Sargeant et al., 2005; JM & EMP, personal observation), and usually do so individually even when they use the same beach simultaneously.

Other unusual foraging behaviors have been reported from most sites where bottlenose dolphins have been studied. For example, off the coast of Florida several forms of mud plume feeding have been seen, where dolphins either cooperatively (Torres & Read, 2009) or singly (Lewis & Schroeder, 2003) use their tails to create mud like "nets" around schools of mullet, which attempt to escape by leaping over the mud-ring, but end up in the dolphins' waiting jaws. However, most of what we know about foraging innovations among wild bottlenose dolphins comes from one population, the Shark Bay Indian Ocean bottlenose dolphins, where we have documented over 20 distinct foraging tactics. In what is perhaps one of the most striking innovations among cetaceans, a subset of the Shark Bay bottlenose dolphin population perform a tool-use behavior known as sponging. This behavior was first discovered in 1984, when Rachel Smolker observed a dolphin named "Halfluke" with a marine sponge on her rostrum (Smolker, Richards, Connor, Mann, & Berggren, 1997). The cone-shaped sponge (*Echinodictyum mesenterinum*) was seen to fit over the dolphin's beak (Figure 4.2) (Smolker et al., 1997) and hypothesized to act as protection when foraging. Since the 1980s, we have now identified dolphins using several other species of conical sponges (Mann & Patterson, 2013) and have further confirmed that sponges are used as tools during foraging to allow dolphins to exploit a unique niche (Krützen et al., 2014; Patterson & Mann, 2011). Sponging dolphins appear to detach basket sponges from the seafloor of deep (8–14 m) channels, and then wear these over their rostra for protection while probing rock, shell and other debris in search of prey, primarily barred sandperch (*Parapercis nebulosa*) (Patterson, 2012; Patterson & Mann, 2011). Five female spongers were recognized in the 1980s, but today over 100 have been documented (Mann & Patterson, 2013). While this increase is not due to the

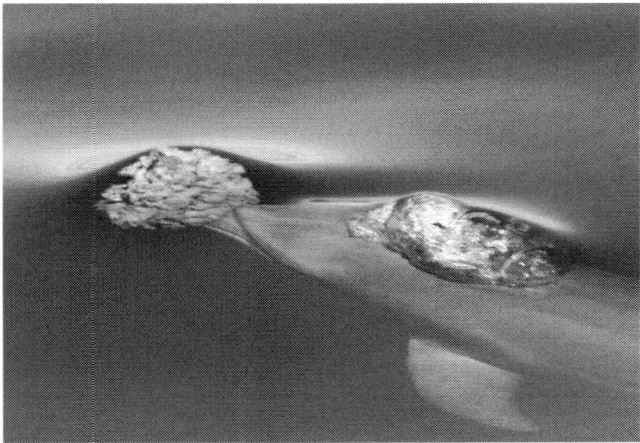

FIGURE 4.2   A male dolphin named Dali foraging with a marine basket sponge (*Ircinia* sp.) in Shark Bay, Australia. *Photo taken by Eric M. Patterson.*

behavior's spread, just an expansion in research effort, we have documented the spread of sponging vertically, within matrilines. Whereas daughters of spongers have over a 95% chance of becoming spongers, sons have only a 50% chance of doing so (Mann et al., 2008). Although spongers preferentially associate, the majority of their associates are nonspongers, but the behavior does not seem to spread horizontally (Mann, Stanton, Patterson, Bienenstock, & Singh, 2012). In fact, only about 4% of the dolphin population uses this foraging method, which is restricted almost entirely to deep-water channels. Data indicate that sponging occurs in at least three sites in Shark Bay, the eastern gulf (Mann et al., 2008; Smolker et al., 1997), the western gulf (Kopps, Krützen, Allen, Bacher, & Sherwin, 2013), and a point 50 km north of both sites, near the point of the Peron Peninsula (Mann & Patterson, 2013). This could indicate three independent innovation events, given the strict matrilineal transmission and genetic haplotypes observed at the sites (Kopps et al., 2014; Krützen et al., 2005; Patterson et al., in preparation), or historical horizontal transmission of an initial sponging innovation. That said, spongers have only been sighted a maximum of 6 km from the channels where they "sponge," and since the three sponging locations are separated by more than this length of nonchannel habitat, mixing between the subpopulations would be rare (Mann & Patterson, 2013).

For the last 20 years, we have also periodically observed another potential foraging innovation called "shelling." In this behavior, dolphins retrieve the shells of large dead molluscs (*Syrinx aruanus* and *Melo amphora*) from the seafloor and then, at the water's surface, balance and wave them around in what appears to be an effort to drain the shells and extract the prey hiding inside (Allen, Bejder, & Krützen, 2011; Mann & Patterson, 2013). In the eastern gulf, the same individuals, from only a few families, have been seen engaging in the behavior on more than one occasion. Currently, the nature of this innovation is not completely understood. We do not know whether dolphins chase fishes into the shells, happen upon shells with fishes inside them, and/or revisit or reuse the same "fish traps" repeatedly. Nevertheless, the fact that we have observed a few matrilines repeatedly "shell" may suggest that this foraging tactic is socially learned and not an incidental event.

Another equally dramatic behavior, called "golden trevally hunting," thus far has primarily been observed in one female dolphin named Wedges. Golden trevally (*Gnathanodon speciosus*) are a large schooling fish with a wide distribution in tropical and subtropical waters of the Indian and Pacific Oceans, including Shark Bay. Although feeding on juvenile golden trevally is common in Shark Bay, Wedges hunts the largest adults (reaching up to 1.2 m). Only Wedges has been directly observed catching large golden trevally, which entails prolonged chasing in deep water (>7 m) with dozens of high leaps to catch the fish, taking the fish down to the substrate, possibly to snap its neck, and finally, travelling to shallow water (<2 m) to break the fish into smaller piece using the substrate, all of which often takes over an hour (Mann & Sargeant, 2003). Although dolphins regularly break up large shallow water fishes on the seafloor, to date no dolphin other than Wedges has been observed catching deep-water fishes and bringing them to shallow water for processing. We have observed dolphins catching large snapper (Sparidae) in deep water on numerous occasions, but here dolphins spend considerable time and effort repeatedly diving to the deep bottom (>7 m) to break up the fish. Some of Wedges' associates from the Puck family appear to have adopted her method as they have also been observed breaking up golden trevally in shallow water. However, only Wedges appears to specialize in the behavior, catching golden trevally every 2.6 h (Mann & Sargeant, 2003), whereas members of the Puck family, who have been followed intensively, have only been seen with trevally a few times.

Another intriguing behavior we call "shag robbing." In Shark Bay, pied cormorants (*Phalacrocorax varius*), often called shags, regularly interact with dolphins, typically by stealing prey that dolphins ferret from the seafloor. Sometimes, dolphins try to get their prey back, but a young female named Sequel once reversed the roles of this "hunter−thief" relationship by instead, robbing the shags of their prey. After riding the bow wave of our boat, Sequel darted off in the direction of a shag more than 40 m away that was swallowing a fish. Almost as soon as Sequel was out of sight she goosed the shags' underbelly, causing it startle, fly off, and drop the fish that she then quickly consumed. During the next 20 min of observation, she "robbed" a total of three different shags separated by hundreds of meters (JM & EMP, personal observation). What is so striking about this behavior is that Shark Bay dolphins virtually never steal prey from each other (Mann et al., 2007), making this technique novel in more ways than one.

Social and Sexual Behavior

While social living is thought to promote the occurrence and spread of all innovative behavior, it presents specific challenges that may elicit innovations with a social function. Earlier we described one social challenge solved by signature whistles and dialects: identifying preferred associates. However, nonvocal social innovations can also be used for this purpose. When two pods of southern resident killer whales meet, they perform a type of "greeting ceremony" in which the two pods first line up across from each other, then slowly begin their approach, and finally halt and pause before making physical contact (Osborne, 1986). After a brief moment, the whales submerge and form tight mixed groups and often engage in social and sexual behavior (Baird, 2000; Osborne, 1986). While the exact function of these ceremonies is unknown, it may be a form of cultural greeting, innovated within the southern resident clan. The northern residents too have their own unique tradition termed "beach rubbing" (Ford, 1989). Near Johnstone Strait, off Vancouver Island northern resident killer whales frequently visit one of several shallow beaches to rub on smooth pebbles. Although this was initially suggested to be a scratching exercise and may help to remove ectoparasites (Thomas, 1970), given that it usually occurs in groups and in conjunction with resting and social behavior, some have proposed a social function (Ford, 1989; Ford, Ellis, & Balcomb, 2000).

Among other delphinids, social innovation is less group specific and more widespread. For example, male bottlenose dolphins in Shark Bay, Australia perform elaborate physical displays, individually and synchronously, when in the presence of adult females (Connor, Smolker, & Richards, 1992a, 1992b). A variety of displays have been observed, including "rooster struts" (Figure 4.3) and "butterfly" displays, some of which are more common than others (Connor et al., 2000). Males vary considerably in their display repertoire and many displays have only ever been observed once (Connor et al., 2000; JM & EMP, personal observation) suggesting that males are flexible in their display behavior and regularly create novel displays. While the exact function of these

FIGURE 4.3   A male dolphin named Enchilada Starlet performing a display called a rooster strut in Shark Bay, Australia. *Photo taken by Janet Mann.*

displays is still unknown, the complexity, novelty, and synchrony of displays may advertise sexually selected characteristics (Trivers, 1972). In fact, such a hypothesis has been put forth regarding similar innovative display behavior performed by Amazon River dolphins, where males appear to manipulate sticks, grass, and clay to attract females (Martin, da Silva, & Rothery, 2008). From Herman and colleagues' work using the *tandem + create* command we know that bottlenose dolphins are certainly capable of inventing synchronous displays, and Kuczaj and colleagues' have demonstrated that males may vary in their ability to do so. Thus, if females exercise some mate choice based on the novelty of a male's display, innovative behavior in this species, as has been suggest for some spiders (Elias, 2006), could potentially be under sexual selection. Alternatively, or in addition, displays may have an agonistic or affiliative function that could be directed at females, males, or both (Connor et al., 2000). Irrespective of the function of male displays, they provide an impressive example of bottlenose dolphin innovative and creative behavior.

Another example of potential innovative social and sexual behavior comes from the vocalizations of some mysticetes, humpback whales being the best studied. Payne and McVay (1971) were perhaps the first to formally describe this behavior, which they call "singing." From recordings taken off the coast of Bermuda in the 1950s and 1960s, they noticed that humpbacks "produce a series of beautiful and varied sounds for a period of 7−30 min and then repeat the same series with considerable precision" (p. 597). Similar but simpler songs have now also been reported in bowhead whales, blue whales (*Balaenoptera musculus*), fin whales (*Balaenoptera physalus*), and minke whales (Darling, 2009). At the time of Payne and McVay's recordings, very little was known concerning the function of these "sonorous moans and screams" (Schevill, 1964), and while this remains largely the case, we have learned a great deal about the context of singing. To date, all singers are male, and while song has been recorded during summer and winter near the poles (Garland et al., 2013; Magnúsdóttir, Rasmussen, Lammers, & Svavarsson, 2014), most singing occurs in warm, low latitude waters during the breeding season (Darling, 2009). As such, many suggest that song functions in reproduction. For example, it may act as a display or secondary sexual characteristic (Tyack, 1981), and given that males seem to congregate when singing, some suggest a lekking function (Clapham, 1996). It could serve to attract females (Tyack, 1981) and/or to signal status to other males (Darling & Bérubé, 2001). Others have proposed singing could help synchronize estrus between females, provide a way for males to appropriately space themselves on the breeding grounds, or even assist males in finding females by acting as a type of sonar (Darling, 2009). At the very least, singing seems to signal where males are and that breeding is underway.

While generally speaking all males in the same assemblage sing the same song at any one time (Payne & Guinee, 1983), over time this song changes in what might be considered innovation. Long-term studies have documented a gradual change in the songs of several populations of humpbacks (Garland et al., 2011; Noad, Cato, Bryden, Jenner, & Jenner, 2000; Payne & Payne, 1985), with some songs being completely replaced within only a few years (Noad et al., 2000). These changes, which include new sounds and/or the loss of old sounds, appear to occur during the winter breeding season and rapidly spread throughout an assemblage, presumably through social learning (Darling, 2009). Through time, song changes in one assemblage may also be transmitted to another. For example, Garland et al. (2011) recently documented the rapid spread of song types from eastern to western breeding populations in the South Pacific, in what they suggest was large scale horizontal social transmission. One song called the black song, which was first recorded in Western Australia in 1995, spread to Polynesia by 2001. In the northern hemisphere, similar transmission of songs may occur with recent findings suggesting a negative correlation between geographic proximity and assemblage song similarity (Darling, Acebes, & Yamaguchi, 2014). Exactly how these song variants get from one population to another is unclear. Low-level interchange between populations has been recorded suggesting that some males may physical carry songs from one group of whales to another. However, recent recordings of singing from the Antarctic, suggest that males from different populations could exchange songs during the summer while at their common feeding grounds (Garland et al., 2013). As with killer whale and sperm whale dialects, humpback song may have some function, but whether a modification to a song alters its function, and thus is an innovation, is unclear. Given the current evidence, we cannot rule out that changes in both songs and dialects are nonfunctional and simply the result of errors in vocal copying and learning. In fact, this somewhat simpler explanation seems to be the case for many changes in the songs of passerine bird (Slater & Lachlan, 2003).

CETACEANS' INNOVATION ABILITIES

Given the intriguing novel, unique, and atypical behaviors described for both captive and wild cetaceans thus far, what can we conclude about their innovative and creative abilities? While most of the innovative and creative behavior performed by wild cetacea probably goes unobserved, odontocetes appear to innovate across a great breadth of functional domains including play, vocal communication, foraging, and social and sexual behavior. This may indicate that their ability to innovate, as has been suggested for humans, is domain general and underpinned by a single cognitive capacity, adaptable to a variety of contexts (Hauser, 2003). Such a view is supported by data indicating that at least in bottlenose dolphins, other related cognitive processes like imitation, are also domain general (Herman, 2012b). A similar argument has been recently made regarding primate cognition and intelligence (Reader, Hager, & Laland, 2011). Primate and odontocete societies share many characteristics (Yamagiwa & Karczmarski, 2014), which some believe helps explain their convergent cognitive abilities (Marino, 2002). In some primates, cognitive abilities appear to have coevolved across functional domains leading Reader et al. (2011) to suggest that some form of flexible, general intelligence was selected for in these species. Given that some odontocetes excel in a variety of cognition abilities (Herman, 2010), their innovative and creative skills could be an extension of a similar general intelligence. However, much greater detail on the cognitive abilities of a wider range of odontocete taxa are needed to better test this hypothesis.

Much less can be concluded regarding innovation and creativity in mysticetes. The understandable lack of captive studies is partly to blame, but with their expansive ranges and long dives, free-ranging mysticetes are often more difficult to study than odontocetes. Nonetheless, mysticetes show innovation in both play and foraging, and perhaps social and sexual behavior depending on how one interprets the data on humpback whale song. However, beyond simply documenting their ability to be innovative and perhaps creative, there is little we can say given the paucity of data.

What is somewhat clear is that examples for mysticetes are fewer and less diverse compared to odontocetes, for which there are several explanations. This discrepancy could reflect merely differences in research effort and ease of observation, or real differences in the taxa's innovative and creative abilities. For instance, it could be that being smaller, more agile, and having more maneuverable appendages and beaks, odontocetes simply have more manipulative ability than mysticetes. Greater manipulative ability increases the diversity of ways in which an organism can interact with its environment (both ecological and social) and thus we might expect the most manipulative of animals to show the greatest diversity in object manipulation. Whether or not manipulative ability and innovation/creativity are related remains to be seen, but such a hypothesis may be testable using data from primates and birds. Odontocetes and mysticetes also differ in their life history and ecology. First, while both suborders experience maternal care, odontocetes have a relatively longer period of maternal dependency (Whitehead & Mann, 2000), a feature thought to promote innovate and creative behavior. Second, many odontocetes live in relatively shallow water, near shore in a dynamic, diverse, and complex habitat, very different to that of most mysticetes. Of particular importance here may be the increase in ecological complexity (Sol, 2003) and availability of objects for manipulation (Mann & Patterson, 2013). In contrast, mysticetes, living primarily in deep-open water, may experience greater environmental stability and less complexity. Like many pelagic animals, mysticetes likely interact with few objects other than prey and conspecifics, and thus, may simply have little opportunity to be innovative or creative.

Combined, these examples from mysticetes and odontocetes provide much needed comparative data on animal innovation and creativity. Yet, many of these data are anecdotal and rarely discussed within the context of innovation and creativity. As such, we suggest that future cetacean researchers, particularly those studying wild populations, should (i) document in detail potential innovative and creative behavior, (ii) monitor the spread of such behaviors when possible, and (iii) place their data within a comparative framework to inform our understanding of the evolution of animal innovation and creativity.

COMPARATIVE INSIGHTS

While rigorous studies on animal innovation and creativity are still relatively recent, comparative data have provided information on both *who* innovates (among and within species) and *why* (proximate and ultimate causes). In regards to who, within species the literature is currently mixed. Some suggest that younger individuals are more likely to innovate given that they are actively learning about their environment, often under the protection of kin (Kummer & Goodall, 1985; Lee, 2003). Yet others suggest that having fully developed their physical and cognitive skills, adults are the most likely to innovate (Reader & Laland, 2001). Evidence is also mixed regarding the sexes proclivity to innovate, but this discrepancy may be largely due to differences in reproductive interests and the context in which innovation occurs (i.e., females would be expected to be more innovative in foraging, and males in social behavior that provides access to mates) (Box, 2003; Laland & Bergen, 2003; Lee, 2003; Reader & Laland, 2001; Trivers, 1972). For social species, some evidence indicates social status may also affect innovative behavior with lower ranking chimpanzees (*Pan troglodytes*) innovating more than their higher-ranking conspecifics (Reader & Laland, 2001).

In captive cetaceans, many of the innovation examples, particularly those that occur in play, occur in younger individuals. For example, Kuczaj et al. (2006) found that approximately 80% of novel play behaviors observed in captivity were performed by calves and suggested that younger individuals may be an important source of behavioral variation and innovation (Kuczaj & Walker, 2006; Kuczaj et al., 2006). McBride and Hebb (1948) also note that younger individuals appear more playful and engage in more novel behavior. Yet these observations are not surprising since captive animals have plenty of time for play, and in younger individuals such behavior likely promotes both cognitive and physical development. What matters is if these behaviors are functional, transmit throughout the population, and are important for a species ecology and evolution, and for this we must turn to wild cetaceans. Unfortunately, here we rarely know the age of the innovator or even its species social system well enough to know if social status might matter. That said, data from wild cetaceans do suggest there may be sex differences in cetacean innovation, perhaps the most informative of which come from the Shark Bay. In Shark Bay, female bottlenose dolphins engage in a wide diversity of foraging tactics (e.g., Figure 4.2), many of which are likely innovations, while males perform extravagant novel social displays, some of which have only ever been observed once (e.g., Figure 4.3). Here, both sexes may be equally innovative and creative but the context in which they use these skills seems to reflect differences in fitness limiting factors: for female, access to food, and for males, access to mates. A similar situation may occur with male humpback whale song and male Amazon River dolphin displays, although in these species it is unclear if one sex is more innovative than the other in foraging behavior.

To address *why* animals might innovate or be creative it is helpful to first examine some of the possible consequences or benefits of innovation. In other words, what does being innovative and creative allow a species or individual to do? Based on data primarily from birds and primates, Sol (2003) proposed several potential consequences of innovative and creative behavior. First, the ability to innovate is predicted to increase a species niche width (Sol, 2003). Within cetaceans, there may be some support for this predication as both killer whales and bottlenose dolphins, two of perhaps the most innovative and creative species, consume a wide variety of prey using a diverse set of innovative foraging tactics (Riesch et al., 2012; Sargeant & Mann, 2009). As such, these two species occupy a wide niche, but individual bottlenose dolphins and killer whales often utilize a much narrow niche, sometime associated with a particular innovation (Foote et al., 2013; Krützen et al., 2014; Patterson & Mann, 2011). Innovative species are also expected to have broad distributions (Sol, 2003). Again, as a species bottlenose dolphins and killer whales occupy a wide variety of habitats all over the world, but individually, dolphins and killer whales may have restricted ranges and even show habitat specialization (Riesch et al., 2012; Patterson, 2012). Such data are indicative of the taxa's great behavioral flexibility, which some suggest may be the result of their cognitive abilities. Indeed Herman (2006) suggests that it is bottlenose dolphins' cognitive abilities that may have lead to their widespread distribution and colonization of a great diversity of habitats. Such a hypothesis is consistent with data from birds and other mammals suggesting that increased cognitive ability positively correlate with a species ability to invade and colonize new habitats (Lefebvre & Bolhuis, 2003; Sol, Bacher, Reader & Lefebvre, 2008), although this may not universally be the case (Jønsson, Fabre, & Irestedt, 2012; Reader & MacDonald, 2003). Innovative species are also predicted to exhibit less migratory behavior (Sol, 2003), the idea being that species that are inflexible in their foraging must migrate to meet their energetic demands. The discrepancy between odontocete and mysticete innovation could provide preliminary support for this idea as nearly all mysticetes migrate, and most odontocetes do not (Lockyer & Brown, 1981), but we emphasize caution in this interpretation due to the sampling biases and difficulties of studying mysticetes mentioned above. The ability to innovate is also predicted to reduce susceptibility to extinction and lead to faster rates of evolution (Lefebvre & Bolhuis, 2003; Nicolakakis, Sol, & Lefebvre, 2003; Sol, 2003; Wyles, Kunkel, & Wilson, 1983). Many of the foraging innovations described above involve individuals adapting to increased anthropogenic threats, which could provide support for first of these claims, but if this will ultimately increase species' resistance to extinction is unknown. Data from killer whales may support the second claim given that maternally transmitted foraging innovations may be leading to speciation (Riesch et al., 2012). Yet, the relationship between innovation and evolution could go either way depending on ecological stability. As with learning (Stephens, 1991), if an innovation is consistently adaptive between generations, we might expect it to become fixed and encoded genetically, and if this somehow leads to reproductive isolation, this could result in speciation. However, if there is some ecological instability between generations, behavioral flexibility, and thus the *ability* to be innovative, might be favored which would not increase the rate of evolution. After all, innovative species are flexible by definition and it may be this behavioral plasticity, rather than some specific, innovation that is most adaptive.

Having reviewed some of the benefits, we can hypothesize as to possible driving forces behind the selection of innovation and creativity, or whether or not it is selected for at all. It is important to note that simply because an innovation may appear to provide some benefit, say allowing for an increase in niche width, it does not necessarily follow that this benefit was driving force behind ones innovative abilities (Gould & Lewontin, 1979). Being innovative and creative likely has many benefits across a wide range of social, ecological, and evolutionary contexts so it is likely that multiple forces are at work. Two key functional domains that may favor innovation and creativity are resource acquisition (foraging) and courtship or mating (sexual). For example, innovations in foraging behavior and diet may reduce intraspecific competition, which could ultimately allow for increased population density. Support for this comes from the Shark Bay dolphins who have one of the highest population densities reported for bottlenose dolphins (Watson-Capps, 2005) and may also exhibit the greatest diversity in foraging behavior. While we know of no clear link between foraging innovations and population density in killer whales, some of their innovative behavior may actually be in response to interspecific competition with humans (Estes et al., 1998). Since foraging pressures are perhaps greatest on females due to the increased reproductive demands, it would not be surprising if natural selection favored foraging innovation in females. Sexual selection on the other hand may select for innovative and creative behavior in males, particularly in the context of displays used to attract mates. Among cetaceans, male odontocete displays and mysticete song may be an example of such sexually selected innovative and creative behavior. Outside of cetaceans, one of the best examples of sexually selected innovation comes from male satin bowerbirds (*Ptilonorhynchus violaceus*) that selectively decorate their bowers to increase its uniqueness and novelty (Borgia, Kaatz, & Condit, 1987). Thus, as has been argued for in humans (Simonton, 2003), both sexual and natural selection seems to favor innovative and creative behavior in some animals species, the extent to which we are only beginning to understand.

References

Alexander, R. (1974). The evolution of social behavior. *Annual Review of Ecology and Systematics, 5*, 325–383.

Allen, J., Weinrich, M. T., Hoppitt, W., & Rendell, L. E. (2013). Network-based diffusion analysis reveals cultural transmission of lobtail feeding in humpback whales. *Science, 340*, 485–488.

Allen, S. J., Bejder, L., & Krützen, M. (2011). Why do Indo-Pacific bottlenose dolphins (*Tursiops* sp.) carry conch shells (*Turbinella* sp.) in Shark Bay, Western Australia? *Marine Mammal Science, 27*, 449–454.

Ashford, J., Rubilar, P., & Martin, A. R. (1996). Interactions between cetaceans and longline fishery operations around South Georgia. *Marine Mammal Science, 12*, 452–457.

Baird, R. W. (2000). The killer whale. In J. Mann, R. C. Connor, P. L. Tyack, & H. Whitehead (Eds.), *Cetacean societies, field studies of dolphins and whales* (pp. 127–154). Chicago, IL: The University of Chicago Press.

Baird, R. W. (2002). *Killer whales of the world: Natural history and conservation*. Stillwater, MN: Voyageur Press.

Barrett-Lennard, L.G. (2000). Population structure and mating patterns of killer whales (*Orcinus orca*) as revealed by DNA analysis (Ph.D. thesis) (p. 97). Vancouver, Canada: Department of Zoology, University of British Columbia.

Bateson, P., & Martin, P. (2013). *Play, playfulness, creativity and innovation*. Cambridge, UK: Cambridge University Press.

Bekoff, M., & Byers, J. A. (1981). A critical reanalysis of the ontogeny and phylogeny of mammalian social and locomotor play: An ethological hornet's nest. In M. Main, K. Immelmann, G. W. Barlow, & L. Petrinovich (Eds.), *Behavioral development: The bielefeld interdisciplinary project* (pp. 296–337). Cambridge, UK: Cambridge University Press.

Bekoff, M., & Byers, J. A. (1998). *Animal play: Evolutionary, comparative and ecological perspectives*. Cambridge, UK: Cambridge University Press.

Birtles, A., & Mangott, A. (2013). Highly interactive behaviour of inquisitive dwarf minke whales. In *Whales and dolphins: cognition, culture, conservation and human perceptions* (pp. 140–148). New York, NY: Routledge.

Bloch, D., & Lockyer, C. H. (1988). Killer whales (*Orcinus orca*) in Faroese waters. *Rit Fiskideildar, 11*, 55–64.

Borgia, G., Kaatz, I., & Condit, R. (1987). Flower choice and bower decoration in the satin bowerbird *Ptilonorhynchus violaceus*: A test of hypotheses for the evolution of male display. *Animal Behaviour, 35*, 1129–1139.

Box, H. O. (2003). Characteristics and propensities of marmosets and tamarins: Implications for studies of innovation. In S. M. Reader, & K. N. Laland (Eds.), *Animal innovation* (pp. 197–222). Oxford, UK: Oxford University Press.

Brady, E. J. (1909). The law of the tongue: Whaling by compact at Twofold Bay. *Australia To-Day, 37*–39.

Braslau-Schneck, S. (1994). Innovative behaviors and synchronization in bottlenose dolphins (MA thesis). Honolulu, HI: Department of Psychology, University of Hawaii.

Brown, D., & Norris, K. (1956). Observations of captive and wild cetaceans. *Journal of Mammalogy, 37*, 311–326.

Bruck, J. N. (2013). Decades-long social memory in bottlenose dolphins. *Proceedings of the Royal Society B: Biological Sciences, 280*, 20131726.

Caldwell, M. C., & Caldwell, D. K. (1965). Individualized whistle contours in bottlenosed dolphins (*Tursiops truncatus*). *Nature, 207*, 434–435.

Caldwell, M. C., & Caldwell, D. K. (1968). Vocalization of naive captive dolphins in small groups. *Science, 159*, 1121–1123.

Caldwell, M. C., & Caldwell, D. K. (1971). Statistical evidence for individual signature whistles in Pacific white-sided dolphins, *Lagenorhynchus obliquidens*. *Cetology*, *3*, 1–9.

Caldwell, M. C., & Caldwell, D. K. (1979). The whistle of the Atlantic bottlenosed dolphin (*Tursiops truncatus*)—ontogeny. In H. E. Winn, & B. L. Olla (Eds.), *Behavior of marine animals: Current perspectives in research. Volume 3: Cetaceans* (pp. 369–401). New York, NY: Plenum Press.

Caldwell, M. C., Caldwell, D. K., & Miller, J. F. (1973). Statistical evidence for individual signature whistles in the spotted dolphin, *Stenella plagiodon*. *Cetology*, *16*, 1–21.

Caldwell, M. C., Caldwell, D. K., & Tyack, P. L. (1990). Review of the signature-whistle hypothesis for the Atlantic bottlenose dolphin. In S. Leatherwood, & R. R. Reeves (Eds.), *The bottlenose dolphin* (pp. 199–234). San Diego, CA: Academic Press.

Chilvers, B. L., & Corkeron, P. J. (2001). Trawling and bottlenose dolphins' social structure. *Proceedings of the Royal Society B: Biological Sciences*, *268*, 1901–1905.

Chivers, S. J. (2009). Cetacean life history. In W. F. Perrin, B. Würsig, & J. Thewissen (Eds.), *Encyclopedia of marine mammals* (pp. 215–220). Burlington, MA: Academic Press.

Clapham, P. J. (1996). The social and reproductive biology of humpback whales: An ecological perspective. *Mammal Review*, *26*, 27–49.

Clapham, P. J. (2000). The humpback whale. In J. Mann, R. C. Connor, P. L. Tyack, & H. Whitehead (Eds.), *Cetacean societies, field studies of dolphins and whales* (pp. 173–196). Chicago, IL: The University of Chicago Press.

Connor, R. C., Smolker, R. A., & Richards, A. F. (1992a). Two levels of alliance formation among male bottlenose dolphins (*Tursiops* sp.). *Proceedings of the National Academy of Sciences of the United States of America*, *89*, 987–990.

Connor, R. C., Smolker, R. A., & Richards, A. F. (1992b). Dolphin alliances and coalitions. In A. H. Harcourt, & F. B. M. de Waal (Eds.), *Coalitions and alliances in humans and other animals* (pp. 415–443). Oxford, UK: Oxford University Press.

Connor, R. C., Wells, R. S., Mann, J., & Read, A. J. (2000). The bottlenose dolphin: Social relationships in a fission–fusion society. In J. Mann, R. C. Connor, P. L. Tyack, & H. Whitehead (Eds.), *Cetacean societies, field studies of dolphins and whales* (pp. 91–125). Chicago, IL: University of Chicago Press.

Cook, M. L. H., Sayigh, L. S., Blum, J. E., & Wells, R. S. (2004). Signature-whistle production in undisturbed free-ranging bottlenose dolphins (*Tursiops truncatus*). *Proceedings of the Royal Society B: Biological Sciences*, *271*, 1043–1049.

Corkeron, P. J. (2009). Captivity. In W. F. Perrin, B. Würsig, & J. Thewissen (Eds.), *Encyclopedia of marine mammals* (pp. 183–188). Burlington, MA: Academic Press.

Couch, L. (1930). Humpback whale killed in Puget Sound, Washington. *The Murrelet*, *11*, 75.

Dahlheim, M. E., & White, P. A. (2010). Ecological aspects of transient killer whales *Orcinus orca* as predators in southeastern Alaska. *Wildlife Biology*, *16*, 308–322.

Dahlheim, M. E., Schulman-Janiger, A., Black, N., Ternullo, R., Ellifrit, D., & Balcomb, K. C., III (2008). Eastern temperate North Pacific offshore killer whales (*Orcinus orca*): Occurrence, movements, and insights into feeding ecology. *Marine Mammal Science*, *24*, 719–729.

Dakin, W. J. (1934). *Whalemen adventures: the story of whaling in Australian waters and other southern seas related thereto, from the days of sails to modern times* (1st ed.). Sydney, Australia: Angus and Robertson.

Darling, J. D., & Bérubé, M. (2001). Interactions of singing humpback whales with other males. *Marine Mammal Science*, *17*, 570–584.

Darling, J. D. (2009). Song. In W. F. Perrin, B. Würsig, & J. Thewissen (Eds.), *Encyclopedia of marine mammals* (2nd ed., pp. 1053–1056). Burlington, MA: Academic Press.

Darling, J. D., Acebes, J., & Yamaguchi, M. (2014). Similarity yet a range of differences between humpback whale songs recorded in the Philippines, Japan and Hawaii in 2006. *Aquatic Biology*, *21*, 93–107.

Daura-Jorge, F. G., Cantor, M., Ingram, S. N., Lusseau, D., & Simões-Lopes, P. C. (2012). The structure of a bottlenose dolphin society is coupled to a unique foraging cooperation with artisanal fishermen. *Biology Letters, 8*, 702–705.

De Oliveira, C. R., Ruiz-Miranda, C. R., Kleiman, D. G., & Beck, B. B. (2003). Play behavior in juvenile golden lion tamarins (Callitrichidae: Primates): Organization in relation to costs. *Ethology, 109*, 593–612.

Deakos, M. H., Branstetter, B. K., Mazzuca, L., Fertl, D., & Mobley, J. R. (2010). Two unusual interactions between a bottlenose dolphin (*Tursiops truncatus*) and a humpback whale (*Megaptera novaeangliae*) in Hawaiian waters. *Aquatic Mammals, 36*, 121–128.

Deecke, V. B., Ford, J. K. B., & Spong, P. (2000). Dialect change in resident killer whales: Implications for vocal learning and cultural transmission. *Animal Behaviour, 60*, 629–638.

Deecke, V. B., Nykänen, M., Foote, A., & Janik, V. M. (2011). Vocal behaviour and feeding ecology of killer whales *Orcinus orca* around Shetland, UK. *Aquatic Biology, 13*, 79–88.

Defran, R. H., & Pryor, K. W. (1980). The behavior and training of cetaceans in captivity. In L. M. Herman (Ed.), *Cetacean behavior: Mechanisms and functions* (pp. 319–362). New York, NY: John Wiley & Sons.

Delfour, F., & Aulagnier, S. (1997). Bubbleblow in beluga whales (*Delphinapterus leucas*): A play activity?. *Behavioural Processes, 40*, 183–186.

DeMaster, D. P., Trites, A. W., Clapham, P. J., Mizroch, S., Wade, P., Small, R. J., et al. (2006). The sequential megafaunal collapse hypothesis: Testing with existing data. *Progress in Oceanography, 68*, 329–342.

Donaldson, R., Finn, H., Bejder, L., Lusseau, D., & Calver, M. (2012). The social side of human–wildlife interaction: Wildlife can learn harmful behaviours from each other. *Animal Conservation, 15*, 427–435.

Dos Santos, M. E., & Lacerda, M. (1987). Preliminary observations of the bottlenose dolphin (*Tursiops truncatus*) in the Sado Estuary (Portugal). *Aquatic Mammals, 13*, 65–80.

Duarte de Figueiredo, L., & Simão, S. M. (2009). Possible occurrence of signature whistles in a population of *Sotalia guianensis* (Cetacea, Delphinidae) living in Sepetiba Bay, Brazil. *The Journal of the Acoustical Society of America, 126*, 1563–1569.

Duffy-Echevarria, E. E., Connor, R. C., & St. Aubin, D. J. (2008). Observations of strand-feeding behavior by bottlenose dolphins (*Tursiops truncatus*) in Bull Creek, South Carolina. *Marine Mammal Science, 24*, 202–206.

Elias, D. O. (2006). Female preference for complex/novel signals in a spider. *Behavioral Ecology, 17*, 765–771.

Estes, J. A., Tinker, M. T., Williams, T. M., & Doak, D. F. (1998). Killer whale predation on sea otters linking oceanic and nearshore ecosystems. *Science, 282*, 473–476.

Fearnbach, H., Durban, J. W., Ellifrit, D. K., Waite, J. M., Matkin, C. O., Lunsford, C. R., et al. (2013). Spatial and social connectivity of fish-eating "resident" killer whales (*Orcinus orca*) in the northern North Pacific. *Marine Biology, 161*, 459–472.

Fertl, D., & Würsig, B. (1995). Coordinated feeding by Atlantic spotted dolphins (*Stenella frontalis*) in the Gulf of Mexico. *Aquatic Mammals, 21.1*, 3–5.

Fisher, J., & Hinde, R. A. R. (1949). The opening of milk bottles by birds. *British Birds, 42*, 347–357.

Foote, A. D., Newton, J., Ávila-Arcos, M. C., Kampmann, M.-L., Samaniego, J. A., Post, K., et al. (2013). Tracking niche variation over millennial timescales in sympatric killer whale lineages. *Proceedings of the Royal Society B: Biological Sciences, 280*, 20131481.

Foote, A. D., Newton, J., Piertney, S. B., Willerslev, E., & Gilbert, M. T. P. (2009). Ecological, morphological and genetic divergence of sympatric North Atlantic killer whale populations. *Molecular Ecology, 18*, 5207–5217.

Ford, J. K. B. (1989). Acoustic behaviour of resident killer whales (*Orcinus orca*) off Vancouver Island, British Columbia. *Canadian Journal of Zoology, 67*, 727–745.

Ford, J. K. B. (1991). Vocal traditions among resident killer whales (*Orcinus orca*) in coastal waters of British Columbia. *Canadian Journal of Zoology, 69*, 1454–1483.

Ford, J. K. B. (2008). Dialects. In W. F. Perrin, B. Würsig, & J. Thewissen (Eds.), *Encyclopedia of marine mammals* (2nd ed., pp. 310–311). Burlington, MA: Academic Press.

Ford, J. K. B., & Ellis, G. M. (2014). You are what you eat: foraging specializations and their influence on the social organization and behaviour of killer whales. In J. Yamagiwa, & L. Karczmarski (Eds.), *Primates and fetaceans: Field research and conservation of complex mammalian societies* (pp. 75–98). New York, NY: Springer.

Ford, J. K. B., Ellis, G. M., & Balcomb, K. C. (2000). *Killer whales: The natural history and genealogy of Orca orcinus in British Columbia and Washington* (2nd ed.). Vancouver, BC: University of British Columbia Press.

Ford, J. K. B., Ellis, G. M., Barrett-Lennard, L. G., Morton, A. B., & Balcomb, K. C. (1998). Dietary specialization in two sympatric populations of killer whales (*Orcinus orca*) in coastal British Columbia and adjacent waters. *Canadian Journal of Zoology, 1471*, 1456–1471.

Ford, J. K. B., Ellis, G. M., Matkin, C. O., Wetklo, M. H., Barrett-Lennard, L. G., & Withler, R. E. (2011). Shark predation and tooth wear in a population of northeastern Pacific killer whales. *Aquatic Biology, 11*, 213–224.

Foroughirad, V., & Mann, J. (2013). Long-term impacts of fish provisioning on the behavior and survival of wild bottlenose dolphins. *Biological Conservation, 160*, 242–249.

Fripp, D., Owen, C., Quintana-Rizzo, E., Shapiro, A. D., Buckstaff, K., Jankowski, K., et al. (2005). Bottlenose dolphin (*Tursiops truncatus*) calves appear to model their signature whistles on the signature whistles of community members. *Animal Cognition, 8*, 17–26.

Gagliano, M., Renton, M., Depczynski, M., & Mancuso, S. (2014). Experience teaches plants to learn faster and forget slower in environments where it matters. *Oecologia, 175*, 63–72.

Garland, E. C., Gedamke, J., Rekdahl, M. L., Noad, M. J., Garrigue, C., & Gales, N. (2013). Humpback whale song on the Southern Ocean feeding grounds: Implications for cultural transmission. *PLoS One, 8*, e79422.

Garland, E. C., Goldizen, A. W., Rekdahl, M. L., Constantine, R., Garrigue, C., Hauser, N. D., et al. (2011). Dynamic horizontal cultural transmission of humpback whale song at the ocean basin scale. *Current Biology, 21*, 687–691.

Gewalt, W. (1989). Orinoco-freshwater-dolphins (*Inia geoffrensis*) using self-produced air bubble "rings" as toys. *Aquatic Mammals, 15*, 73–79.

Gibson, Q. A., & Mann, J. (2008). Early social development in wild bottlenose dolphins: Sex differences, individual variation and maternal influence. *Animal Behaviour, 76*, 375–387.

Gilmore, R. M. (1976). The friendly whales of Laguna San Ignacio. *Terra, 15*, 24–28.

Gonzalvo, J., Valls, M., Cardona, L., & Aguilar, A. (2008). Factors determining the interaction between common bottlenose dolphins and bottom trawlers off the Balearic Archipelago (western Mediterranean Sea). *Journal of Experimental Marine Biology and Ecology, 367*, 47–52.

Gould, S., & Lewontin, R. (1979). The spandrels of San Marco and the Panglossian paradigm: A critique of the adaptationist programme. *Proceedings of the Royal Society B: Biological Sciences, 204*, 581–598.

Greenberg, R. (2003). The role of neophobia and neophilia in the development of innovative behaviour of birds. In S. M. Reader, & K. N. Laland (Eds.), *Animal innovation*. Oxford, UK: Oxford University Press.

Gridley, T., Cockcroft, V. G., Hawkins, E. R., Blewitt, M. L., Morisaka, T., & Janik, V. M. (2013). Signature whistles in free-ranging populations of Indo-Pacific bottlenose dolphins, *Tursiops aduncus*. *Marine Mammal Science, 30*, 512–527.

Groos, K., Baldwin, E. L., & Baldwin, J. M. (1898). *The play of animals.* New York, NY: Appleton.

Guinet, C. (1991). Intentional stranding apprenticeship and social play in killer whales (*Orcinus orca*). *Canadian Journal of Zoology, 69*, 2712–2716.

Guinet, C., & Bouvier, J. (1995). Development of intentional stranding hunting techniques in killer whale (*Orcinus orca*) calves at Crozet Archipelago. *Canadian Journal of Zoology, 73*, 27–33.

Hain, J. H. W., Carter, G. R., Kraus, S. D., Mayo, C. A., & Winn, H. E. (1982). Feeding behavior of the humpback whale, *Megaptera novaeangliae*, in the western North Atlantic. *Fishery Bulletin, 80*, 259–268.

Hamer, D. J., Childerhouse, S. J., & Gales, N. J. (2012). Odontocete bycatch and depredation in longline fisheries: A review of available literature and of potential solutions. *Marine Mammal Science, 28*, E345–E374.

Hauser, M. D. (2003). To innovate or not to innovate? That is the question. In S. M. Reader, & K. N. Laland (Eds.), *Animal innovation* (pp. 329–338). Oxford, UK: Oxford University Press.

Herman, D. P., Burrows, D. G., Wade, P. R., Durban, J. W., Matkin, C. O., Leduc, R. G., et al. (2005). Feeding ecology of eastern Northern Pacific killer whales *Orcinus orca* from fatty acid, stable isotope and organochlorine analyses of blubber biopsies. *Marine Ecology Progress Series, 302*, 275–291.

Herman, L. M. (1991). What the dolphin knows, or might know, in its natural world. In K. Pryor, & K. S. Norris (Eds.), *Dolphin societies: Discoveries and puzzles* (pp. 349–364). Berkeley, CA: University of California Press.

Herman, L. M. (2002). Vocal, social, and self imitation by bottlenosed dolphins. In K. Dautenhahn, & C. Nehaniv (Eds.), *Imitation in animals and artifacts* (pp. 63–108). Cambridge, MA: MIT Press.

Herman, L. M. (2006). Intelligence and rational behaviour in the bottlenosed dolphin. In S. Hurley, & M. Nudds (Eds.), *Rational animals?* (pp. 437–468). Oxford: Oxford University Press.

Herman, L. M. (2010). What laboratory research has told us about dolphin cognition. *International Journal of Comparative Psychology, 23*, 310–330.

Herman, L. M. (2012a). Historical perspectives: Birthing a dolphin research laboratory: The early history of the Kewalo Basin Marine Mammal Laboratory. *Aquatic Mammals, 38*, 102–125.

Herman, L. M. (2012b). Body and self in dolphins. *Consciousness and Cognition, 21*, 526–545.

Heubeck, M. (2001). Pilot whale apparently playing with moulting common eiders. *Scottish Birds, 22*, 62.

Hewitt, O. (1986). Dolphin interferes with loon. *Florida Field Naturalist, 14*, 100.

Hoelzel, A. R. (1991). Killer whale predation on marine mammals at Punta Norte, Argentina; Food sharing, provisioning and foraging strategy. *Behavioral Ecology and Sociobiology, 29*, 197–204.

Hoese, H. (1971). Dolphin feeding out of water in a salt marsh. *Journal of Mammalogy, 52*, 222–223.

Holmes, B., & Neil, D. (2012). "Gift giving" by wild bottlenose dolphins (*Tursiops* sp.) to humans at a wild dolphin provisioning program, Tangalooma, Australia. *Anthrozoos, 25*, 397–413.

Holth, P. (2012). The creative porpoise revisited. *European Journal of Behavior Analysis, 1*, 1–5.

Huang, P., Sieving, K. E., & Mary, C. M. S. (2011). Heterospecific information about predation risk influences exploratory behavior. *Behavioral Ecology, 23*, 463–472.

Ingebrigtsen, A. (1929). Whales caught in the North Atlantic and other seas.. *International Council for the Exploration of the Sea. Rapports et Proces-Verbaux des Reunions, 56*, 1–26.

Janik, V. M., & Sayigh, L. S. (2013). Communication in bottlenose dolphins: 50 years of signature whistle research. *Journal of Comparative Physiology. A, Neuroethology, Sensory, Neural, and Behavioral Physiology, 199*, 479–489.

Janik, V. M., & Slater, P. J. B. (1998). Context-specific use suggests that bottlenose dolphin signature whistles are cohesion calls. *Animal Behaviour, 56*, 829–838.

Jefferson, T. A. (2000). Population biology of the Indo-Pacific hump-backed dolphin in Hong Kong waters. *Wildlife Monographs, 144*, 1–65.

Jefferson, T. A., Stacey, P., & Baird, R. W. (1991). A review of killer whale interactions with other marine mammals: predation to co-existence. *Mammal Review, 21*, 151–180.

Johnson, C. M., & Norris, K. S. (1994). Social behavior. In *The Hawaiian spinner dolphin* (pp. 243–286). Berkeley, CA: The University of California Press.

Jones, I. (2006). A northeast Pacific offshore killer whale (*Orcinus orca*) feeding on a Pacific halibut (*Hippoglossus stenolepis*). *Marine Mammal Science, 22*, 198–200.

Jones, M. L., & Swartz, S. L. (1984). Demography and phenology of gray whales and evaluation of whale-watching activities in Laguna San Ignacio, Baja California Sur, Mexico. In M. L. Jones, & S. L. Swartz (Eds.), *The gray whale:* Eschrichtius robustus (Vol. 1, pp. 309–374). Orlando, FL: Academic Press.

Jønsson, K. A., Fabre, P.-H., & Irestedt, M. (2012). Brains, tools, innovation and biogeography in crows and ravens. *BMC Evolutionary Biology, 12*, 1–12.

Jurasz, C. M., & Jurasz, V. P. (1979). Feeding modes of the humpback whale, *Megaptera novaeangliae*, in southeast Alaska. *Scientific Reports of the Whales Research Institute, 31*, 69–83.

Kasuya, T., & Marsh, H. (1984). Life history and reproductive biology of the short-finned pilot whale, Globicephala macrorhynchus, off the Pacific coast of Japan. *Report of the International Whaling Commission, 6*, 259–310.

Kaufman, A. B., Butt, A. E., Kaufman, J. C., & Colbert-White, E. N. (2011). Towards a neurobiology of creativity in nonhuman animals. *Journal of Comparative Psychology, 125*, 255–272.

Kaufman, J. C., & Kaufman, A. B. (2004). Applying a creativity framework to animal cognition. *New Ideas in Psychology, 22*, 143–155.

Kemper, C. M., Pemberton, D., Cawthorn, M., Heinrich, S., Mann, J., Würsig, B., et al. (2003). Aquaculture and marine mammals: Coexistence or conflict? In N. Gales, M. Hindell, & R. Kirkwood (Eds.), *Marine mammals: Fisheries, tourism and management issues* (pp. 208–228). Collingwood, Australia: CSIRO Publishing.

King, S. L., & Janik, V. M. (2013). Bottlenose dolphins can use learned vocal labels to address each other. *Proceedings of the National Academy of Sciences of the United States of America, 110*, 13216–13221.

King, S. L., Sayigh, L. S., Wells, R. S., Fellner, W., & Janik, V. M. (2013). Vocal copying of individually distinctive signature whistles in bottlenose dolphins. *Proceedings of the Royal Society B: Biological Sciences, 280*, 20130053.

Kopps, A. M., Ackermann, C. Y., Sherwin, W. B., Allen, S. J., Bejder, L., & Krützen, M. (2014). Cultural transmission of tool use combined with habitat specializations leads to fine-scale genetic structure in bottlenose dolphins. *Proceedings of the Royal Society B: Biological Sciences, 281*, 2013245.

Kopps, A. M., Krützen, M., Allen, S. J., Bacher, K., & Sherwin, W. B. (2013). Characterizing the socially transmitted foraging tactic "sponging" by bottlenose dolphins (*Tursiops* sp.) in the western gulf of Shark Bay, Western Australia. *Marine Mammal Science, 30*, 847–863.

Krahn, M. M., Herman, D. P., Matkin, C. O., Durban, J. W., Barrett-Lennard, L. G., Burrows, D. G., et al. (2007). Use of chemical tracers in assessing the diet and foraging regions of eastern North Pacific killer whales. *Marine Environmental Research, 63*, 91–114.

Kritzler, H. (1952). Observations on the pilot whale in captivity. *Journal of Mammalogy, 33,* 321–334.

Krützen, M., Kreicker, S., MacLeod, C. D., Learmonth, J., Kopps, A. M., Walsham, P., et al. (2014). Cultural transmission of tool use by Indo-Pacific bottlenose dolphins (*Tursiops* sp.) provides access to a novel foraging niche. *Proceedings of the Royal Society B: Biological Sciences, 281,* 20140374.

Krützen, M., Mann, J., Heithaus, M. R., Connor, R. C., Bejder, L., & Sherwin, W. B. (2005). Cultural transmission of tool use in bottlenose dolphins. *Proceedings of the National Academy of Sciences of the United States of America, 102,* 8939–8943.

Kuczaj, S. A., & Eskelinen, H. C. (2014). The "creative dolphin" revisited: what do dolphins do when asked to vary their behavior? *Animal Behavior and Cognition, 1,* 66–76.

Kuczaj, S. A., Lacinak, C. T., Garver, A., & Scarpuzzi, M. (1998). Can animals enrich their own environment? In V. J. Hare, & K. E. Worley (Eds.), *Proceedings of the third international conference on environmental enrichment* (pp. 168–170). Orlando, FL: The Shape of Enrichment, Inc.

Kuczaj, S. A., Makecha, R., Trone, M., Paulos, R. D., & Ramos, J. A. (2006). Role of peers in cultural innovation and cultural transmission: Evidence from the play of dolphin calves. *International Journal of Comparative Psychology, 19,* 223–240.

Kuczaj, S. A., & Walker, R. T. (2006). How do dolphins solve problems? In E. A. Wasserman, & T. R. Zentall (Eds.), *Comparative Cognition: Experimental Explorations of Animal Intelligence* (pp. 580–601). New York, NY: Oxford University.

Kummer, H., & Goodall, J. (1985). Conditions of innovative behaviour in primates. *Philosophical Transactions of the Royal Society of London. Series B, Biological Sciences, 308,* 203–214.

Laland, K. N., & Bergen, Y. Van (2003). Experimental studies of innovation in the guppy. In S. M. Reader, & K. N. Laland (Eds.), *Animal innovation* (pp. 155–174). Oxford, UK: Oxford University Press.

Leatherwood, S. (1975). Some observations of feeding behavior of bottle-nosed dolphins (*Tursiops truncatus*) in the northern Gulf of Mexico and (*Tursiops* cf. *T. gilli*) off southern California, Baja California, and Nayarit, Mexico. *Marine Fisheries Review, 37,* 10–16.

Lee, P. C. (2003). Innovation as a behavioural response to environmental challenges: A cost and benefit approach. In S. M. Reader, & K. N. Laland (Eds.), *Animal innovation* (pp. 261–268). Oxford, UK: Oxford University Press.

Lefebvre, L., & Bolhuis, J. J. (2003). Positive and negative correlations of feeding innovations in birds: Evidence for limited modularity. In S. M. Reader, & K. N. Laland (Eds.), *Animal innovation* (pp. 39–62). Oxford, UK: Oxford University Press.

Lefebvre, L., Reader, S. M., & Sol, D. (2013). Innovating innovation rate and its relationship with brains, ecology and general intelligence. *Brain, Behavior and Evolution, 81,* 143–145.

Lefebvre, L., Whittle, P., Lascaris, E., & Finkelstein, A. (1997). Feeding innovations and forebrain size in birds. *Animal Behaviour, 53,* 549–560.

Lewis, J. S., & Schroeder, W. W. (2003). Mud plume feeding, a unique foraging behavior of the bottlenose dolphin (*Tursiops truncatus*) in the Florida Keys. *Gulf of Mexico Science, 21,* 92–97.

Lockyer, C. H., & Brown, S. G. (1981). The migration of whales. In D. J. Aidley (Ed.), *Animal migration* (Issue 13., pp. 105–137). Cambridge, UK: Cambridge University Press.

Lopez, J., & Lopez, D. (1985). Killer whales (*Orcinus orca*) of Patagonia, and their behavior of intentional stranding while hunting nearshore. *Journal of Mammalogy, 66,* 181–183.

Loughlin, T. R. (Ed.), (1994). *Marine mammals and the exxon valdez* New York, NY: Academic Press.

Lyrholm, T., Leimar, O., Johanneson, B., & Gyllensten, U. (1999). Sex-biased dispersal in sperm whales: Contrasting mitochondrial and nuclear genetic structure of global populations. *Proceedings of the Royal Society B: Biological Sciences, 266,* 347—354.

Magnúsdóttir, E. E., Rasmussen, M. H., Lammers, M. O., & Svavarsson, J. (2014). Humpback whale songs during winter in subarctic waters. *Polar Biology, 37,* 427—433.

Manabe, K. (1997). Control of vocal repertoire by reward in budgerigars (*Melopsittacus undulatus*). *Journal of Comparative Psychology, 111,* 50—62.

Manabe, K., & Dooling, R. J. (1997). Control of vocal production in budgerigars (*Melopsittacus undulatus*): Selective reinforcement, call differentiation, and stimulus control. *Behavioural Processes, 41,* 117—132.

Mann, J., & Patterson, E. M. (2013). Tool use by aquatic animals. *Philosophical Transactions of the Royal Society of London. Series B, Biological Sciences, 368,* 20120424.

Mann, J., & Sargeant, B. L. (2003). Like mother, like calf: The ontogeny of foraging traditions in wild Indian Ocean bottlenose dolphins (*Tursiops* sp.). In D. Fragaszy, & S. Perry (Eds.), *The biology of traditions: Models and evidence* (pp. 236—266). Cambridge, UK: Cambridge University Press.

Mann, J., & Smuts, B. (1999). Behavioral development in wild bottlenose dolphin newborns (*Tursiops* sp.). *Behaviour, 136,* 529—566.

Mann, J., & Watson-Capps, J. J. (2005). Surviving at sea: ecological and behavioural predictors of calf mortality in Indian Ocean bottlenose dolphins, *Tursiops* sp.. *Animal Behaviour, 69,* 899—909.

Mann, J., Connor, R. C., Barre, L. M., & Heithaus, M. R. (2000). Female reproductive success in bottlenose dolphins (*Tursiops* sp.): Life history, habitat, provisioning, and group-size effects. *Behavioral Ecology, 11,* 210—219.

Mann, J., Sargeant, B. L., & Minor, M. (2007). Calf inspections of fish catches in bottlenose dolphins (*Tursiops* sp.): Opportunities for oblique social learning? *Marine Mammal Science, 23,* 197—202.

Mann, J., Sargeant, B. L., Watson-Capps, J. J., Gibson, Q. A., Heithaus, M. R., Connor, R. C., et al. (2008). Why do dolphins carry sponges? *PLoS One, 3,* e3868.

Mann, J., Stanton, M. A., Patterson, E. M., Bienenstock, E. J., & Singh, L. O. (2012). Social networks reveal cultural behaviour in tool-using dolphins. *Nature Communications, 3,* 980.

Marino, L. (2002). Convergence of complex cognitive abilities in cetaceans and primates. *Brain, Behavior and Evolution, 59,* 21—32.

Marten, K., Shariff, K., Psarakos, S., & White, D. J. (1996). Ring bubbles of dolphins. *Scientific American, 275,* 82—87.

Martin, A. R., da Silva, V. M. F., & Rothery, P. (2008). Object carrying as socio-sexual display in an aquatic mammal. *Biology Letters, 4,* 243—245.

McBride, A., & Hebb, D. (1948). Behavior of the captive bottle-nose dolphin, *Tursiops truncatus*. *Journal of Comparative and Physiological Psychology, 41,* 111—123.

McCowan, B., & Reiss, D. (2001). The fallacy of "signature whistles" in bottlenose dolphins: A comparative perspective of "signature information" in animal vocalizations. *Animal Behaviour, 62,* 1151—1162.

McCowan, B., Marino, L., Vance, E., Walke, L., & Reiss, D. (2000). Bubble ring play of bottlenose dolphins (*Tursiops truncatus*): Implications for cognition. *Journal of Comparative Psychology, 114,* 98—106.

Mercado, E., Murray, S. O., Uyeyama, R. K., Pack, A. A., & Herman, L. M. (1998). Memory for recent actions in the bottlenosed dolphin (*Tursiops truncatus*): Repetition of arbitrary behaviors using an abstract rule. *Animal Learning & Behavior, 26,* 210—218.

Mercado, E., Uyeyama, R. K., Pack, A. A., & Herman, L. M. (1999). Memory for action events in the bottlenosed dolphin. *Animal Cognition, 2,* 17—25.

Miller, P. (2002). Mixed-directionality of killer whale stereotyped calls: a direction of movement cue? *Behavioral Ecology and Sociobiology, 52*, 262−270.

Moller, A. P., & Garamszegi, L. Z. (2012). Between individual variation in risk-taking behavior and its life history consequences. *Behavioral Ecology, 23*, 843−853.

Nakahara, F., & Takemura, A. (1997). A survey on the behavior of captive odontocetes in Japan. *Aquatic Mammals, 23.3*, 135−143.

National Marine Fisheries Service Office of Protected Resources. (2014). *U.S. marine mammal inventory report.* Available at <http://www.nmfs.noaa.gov/pr/permits/inventory.htm>.

Nicolakakis, N., Sol, D., & Lefebvre, L. (2003). Behavioural flexibility predicts species richness in birds, but not extinction risk. *Animal Behaviour, 65*, 445−452.

Nishiwaki, M. (1962). Aerial photographs show sperm whales' interesting habits. *Norsk Hvalfangst-Tidende, 51*, 395−398.

Noad, M. J., Cato, D. H., Bryden, M. M., Jenner, M. N., & Jenner, K. C. (2000). Cultural revolution in whale songs. *Nature, 408*, 537−538.

Noke, W., & Odell, D. (2002). Interactions between the Indian River Lagoon blue crab fishery and the bottlenose dolphin, *Tursiops truncatus. Marine Mammal Science, 18*, 819−832.

Nolan, C., Liddle, G., & Elliot, J. (2000). Interactions between killer whales (*Orcinus orca*) and sperm whales (*Physeter macrocephalus*) with a longline fishing vessel. *Marine Mammal Science, 16*, 658−664.

Noonan, M. (2005). Gull baiting in captive orcas: A possible instance of cultural transmission. In *Animal behavior society meeting, August 6−10*, Snowbird, UT.

Nousek, A. E., Slater, P. J. B., Wang, C., & Miller, P. J. O. (2006). The influence of social affiliation on individual vocal signatures of northern resident killer whales (*Orcinus orca*). *Biology Letters, 2*, 481−484.

Oftedal, O. T. (1997). Lactation in whales and dolphins: Evidence of divergence between baleen- and toothed-species. *Journal of Mammary Gland Biology and Neoplasia, 2*, 205−230.

Olesiuk, P. F., Bigg, M., & Ellis, G. (1990). Life history and population dynamics of resident killer whales (*Orcinus orca*) in the coastal waters off British Columbia and Washington state. *Report of the International Whaling Commission, 12*, 209−243.

Orams, M. B. (2002). Feeding wildlife as a tourism attraction: A review of issues and impacts. *Tourism Management, 23*, 281−293.

Osborne, R. W. (1986). A behavioral budget of Puget Sound killer whales. In B. C. Kirkevold, & J. S. Lockard (Eds.), *Behavioral biology of killer whales* (pp. 211−249). New York, NY: A. Liss.

Panksepp, J. (1998). *Affective neuroscience: The foundations of human and animal emotions.* Oxford, UK: Oxford University Press.

Patterson, E.M. (2012). Ecological and life history factors influence habitat and tool use in wild bottlenose dolphins (*Tursiops* sp.) (PhD thesis) (p. 170). Washington, DC, Georgetown: Department of Biology, Georgetown University.

Patterson, E. M., & Mann, J. (2011). The ecological conditions that favor tool use and innovation in wild bottlenose dolphins (*Tursiops* sp.). *PLoS One, 6*, e22243.

Paulos, R., Trone, M., & Kuczaj, S. A. (2010). Play in wild and captive cetaceans. *International Journal of Comparative Psychology, 23*, 701−722.

Payne, K., & Payne, R. S. (1985). Large scale changes over 19 years in songs of humpback whales in Bermuda. *Zeitschrift für Tierpsychologie, 68*, 89−114.

Payne, R. S., & Guinee, L. N. (1983). Humpback whale (*Megaptera novaeangliae*) songs as an indicator of "stocks.". In R. S. Payne (Ed.), *Communication and behavior of whales* (AAAS Selec ed., pp. 333−358). Boulder, CO: Westview Press.

Payne, R. S., & McVay, S. (1971). Songs of humpback whales. *Science, 173*, 585−597.

Peddemors, V. M., & Thompson, G. (1994). Beaching behaviour during shallow water feeding by humpback dolphins Sousa plumbea. *Aquatic Mammals, 20,* 65–76.

Pennino, M. G., Mendoza, M., Pira, A., Floris, A., & Rotta, A. (2013). Assessing foraging tradition in wild bottlenose dolphins (*Tursiops truncatus*). *Aquatic Mammals, 39,* 282–289.

Perrin, W.F. (2014). *World cetacea database.* <http://www.marinespecies.org/cetacea> Accessed 27.02.15.

Pitman, R. L., & Durban, J. W. (2010). Killer whale predation on penguins in Antarctica. *Polar Biology, 33,* 1589–1594.

Pitman, R. L., & Durban, J. W. J. (2012). Cooperative hunting behavior, prey selectivity and prey handling by pack ice killer whales (*Orcinus orca*), type B, in Antarctic Peninsula waters. *Marine Mammal Science, 28,* 16–36.

Pitman, R. L., & Ensor, P. (2003). Three forms of killer whales (*Orcinus orca*) in Antarctic waters. *Journal of Cetacean Research and Management, 5,* 131–139.

Plucker, J. A., Beghetto, R. A., & Dow, G. T. (2004). Why isn't creativity more important to educational psychologists? Potentials, pitfalls, and future directions in creativity research. *Educational Psychologist, 39,* 83–96.

Pryor, K. W. (2004a). *On behavior: Essays & research.* Waltham, MA: Sunshine Books, Inc..

Pryor, K. W. (2004b). *Lads before the wind: Diary of a dolphin trainer* (4th ed.). Waltham, MA: Sunshine Books, Inc..

Pryor, K. W. (2006). *Don't shoot the dog!: The new art of teaching and training* (3rd ed.). Lydney, UK: Ringpress Books Limited.

Pryor, K. W. (2009). *Reaching the animal mind: Clicker training and what it teaches us about all animals.* New York, NY: Scribner.

Pryor, K. W., Haag, R., & O'Reilly, J. (1969). The creative porpoise: Training for novel behavior. *Journal of the Experimental Analysis of Behavior, 12,* 653–661.

Pryor, K. W., Lindbergh, J., Lindbergh, S., & Milano, R. (1990). A dolphin–human fishing cooperative in Brazil. *Marine Mammal Science, 6,* 77–82.

Purves, M., Agnew, D., & Balguerias, E. (2004). Killer whale (*Orcinus orca*) and sperm whale (*Physeter macrocephalus*) interactions with longline vessels in the Patagonian toothfish fishery at South Georgia, South Atlantic. *CCAMLR Science, 11,* 111–126.

Ramsey, G., Bastian, M. L., & van Schaik, C. (2007). Animal innovation defined and operationalized. *Behavioral and Brain Sciences, 30,* 393–432.

Read, A. J. (2001). Trends in the maternal investment of harbour porpoises are uncoupled from the dynamics of their primary prey. *Proceedings of the Royal Society B: Biological Sciences, 268,* 573–577.

Reader, S. M. (2003). Innovation and social learning: Individual variation and brain evolution. *Animal Biology, 53,* 147–158.

Reader, S. M., & Laland, K. N. (2001). Primate innovation: Sex, age and social rank differences. *International Journal of Primatology, 22,* 787–805.

Reader, S. M., & Laland, K. N. (2002). Social intelligence, innovation, and enhanced brain size in primates. *Proceedings of the National Academy of Sciences of the United States of America, 99,* 4436–4441.

Reader, S. M., & Laland, K. N. (2003). Animal innovation: an introduction. In S. M. Reader, & K. N. Laland (Eds.), *Animal innovation* (pp. 4–35). Oxford, UK: Oxford University Press.

Reader, S. M., & MacDonald, K. (2003). Environmental variability and primate behavioral flexibility. In S. M. Reader, & K. N. Laland (Eds.), *Animal innovation* (p. 4). Oxford, UK: Oxford University Press.

Reader, S. M., Hager, Y., & Laland, K. N. (2011). The evolution of primate general and cultural intelligence. *Philosophical Transactions of the Royal Society of London. Series B, Biological Sciences, 366,* 1017–1027.

Rendell, L. E., & Whitehead, H. (2003). Vocal clans in sperm whales (*Physeter macrocephalus*). *Proceedings of the Royal Society B: Biological Sciences, 270,* 225–231.

Rendell, L. E., & Whitehead, H. (2005). Spatial and temporal variation in sperm whale coda vocalizations: Stable usage and local dialects. *Animal Behaviour, 70,* 191–198.

Renjun, L., Gewalt, W., Neurohr, B., & Winkler, A. (1994). Comparative studies on the behaviour of *Inia geoffrensis* and *Lipotes vexillifer* in artificial environments. *Aquatic Mammals, 20.1,* 39–45.

Riesch, R., Barrett-Lennard, L. G., Ellis, G. M., Ford, J. K. B., & Deecke, V. B. (2012). Cultural traditions and the evolution of reproductive isolation: Ecological speciation in killer whales? *Biological Journal of the Linnean Society, 106,* 1–17.

Rigley, L., VanDyke, V. G., Cram, P., & Rigley, I. (1981). Shallow water behavior of the Atlantic bottlenose dolphin (*Tursiops truncatus*). *Proceedings of the Pennsylvania Academia of Science, 55,* 157–159.

Russon, A. E. (2003). Innovation and creativity in forest-living rehabilitant orang-utans. In S. M. Reader, & K. N. Laland (Eds.), *Animal innovation* (pp. 279–308). Oxford, UK: Oxford University Press.

Samuels, A., & Bejder, L. (2004). Chronic interaction between humans and free-ranging bottlenose dolphins near Panama City Beach, Florida, USA. *Journal of Cetacean Research and Management, 5,* 69–77.

Sargeant, B. L., & Mann, J. (2009). Developmental evidence for foraging traditions in wild bottlenose dolphins. *Animal Behaviour, 78,* 715–721.

Sargeant, B. L., Mann, J., Berggren, P., & Krützen, M. (2005). Specialization and development of beach hunting, a rare foraging behavior, by wild bottlenose dolphins (*Tursiops* sp.). *Canadian Journal of Zoology, 83,* 1400–1410.

Saulitis, E., Matkin, C., Barrett-Lennard, L. G., Heise, K., & Ellis, G. (2000). Foraging strategies of sympatric killer whale (*Orcinus orca*) populations in Prince William Sound, Alaska. *Marine Mammal Science, 16,* 94–109.

Sayigh, L. S., Quick, N., Hastie, G., & Tyack, P. L. (2013). Repeated call types in short-finned pilot whales, *Globicephala macrorhynchus. Marine Mammal Science, 29,* 312–324.

Sayigh, L. S., Tyack, P. L., & Wells, R. (1995). Sex difference in signature whistle production of free-ranging bottlenose dolphins, *Tursiops truncatus. Behavioral Ecology and Sociobiology, 36,* 171–177.

Sayigh, L. S., Tyack, P. L., Wells, R., & Scott, M. (1990). Signature whistles of free-ranging bottlenose dolphins *Tursiops truncatus:* Stability and mother–offspring comparisons. *Behavioral Ecology and Sociobiology, 26,* 247–260.

Schevill, W. E. (1964). Underwater sounds of cetaceans. In Tavo (Ed.), *Marine bioacoustics* (pp. 307–316). New York, NY: Pergamon.

Schusterman, R. J., & Reichmuth, C. (2007). Novel sound production through contingency learning in the Pacific walrus (*Odobenus rosmarus divergens*). *Animal Cognition, 11,* 319–327.

Shapiro, A. D. (2006). Preliminary evidence for signature vocalizations among free-ranging narwhals (*Monodon monoceros*). *The Journal of the Acoustical Society of America, 120,* 1695–1705.

Sigurjónsson, J. (1988). Photoidentification of killer whales, *Orcinus orca,* off Iceland, 1981 through 1986. *Rit Fiskideildar, 11,* 99–114.

Silber, G. K., & Fertl, D. (1995). Intentional beaching by bottlenose dolphins (*Tursiops truncatus*) in the Colorado River Delta, Mexico. *Aquatic Mammals, 21,* 183–186.

Similä, T., & Ugarte, F. (1993). Surface and underwater observations of cooperatively feeding killer whales in northern Norway. *Canadian Journal of Zoology, 71,* 1494–1499.

Similä, T., Holst, J. C., & Christensen, I. (1996). Occurrence and diet of killer whales in northern Norway: Seasonal patterns relative to the distribution and abundance of Norwegian spring-spawning herring. *Canadian Journal of Fisheries and Aquatic Sciences, 53,* 769–779.

Simões-Lopes, P. C., Fabián, M. E., & Menegheti, J. O. (1998). Dolphin interactions with the mullet artisanal fishing on southern Brazil: A qualitative and quantitative approach. *Revista Brasileira de Zoologia, 15*, 709−726.

Simonton, D. K. (2003). Human creativity: Two Darwinian analyses. In S. M. Reader, & K. N. Laland (Eds.), *Animal innovation* (pp. 309−328). Oxford, UK: Oxford University Press.

Slater, P. J. B., & Lachlan, R. F. (2003). Is innovation in bird song adaptive? In S. M. Reader, & K. N. Laland (Eds.), *Animal innovation* (pp. 117−136). Oxford, UK: Oxford University Press.

Smith, B. D., Tun, M. T., Chit, A. M., Win, H., & Moe, T. (2009). Catch composition and conservation management of a human−dolphin cooperative cast-net fishery in the Ayeyarwady River, Myanmar. *Biological Conservation, 142*, 1042−1049.

Smith, T., & Siniff, D. (1981). Coordinated behavior of killer whales, *Orcinus orca*, hunting a crabeater seal, Lobodon carcinophagus. *Canadian Journal of Zoology, 59*, 1185−1189.

Smolker, R. A. (2001). *To touch a wild dolphin.* New York, NY: Anchor Books.

Smolker, R. A., & Pepper, J. (1999). Whistle convergence among allied male bottlenose dolphins (Delphinidae, *Tursiops* sp.). *Ethology, 105*, 595−617.

Smolker, R. A., Mann, J., & Smuts, B. (1993). Use of signature whistles during separations and reunions by wild bottlenose dolphin mothers and infants. *Behavioral Ecology and Sociobiology, 33*, 393−402.

Smolker, R. A., Richards, A. F., Connor, R. C., Mann, J., & Berggren, P. (1997). Sponge carrying by dolphins (Delphinidae, *Tursiops* sp.): A foraging specialization involving tool use? *Ethology, 103*, 454−465.

Sol, D. (2003). Behavioural flexibility: A neglected issue in the ecological and evolutionary literature. In S. M. Reader, & K. N. Laland (Eds.), *Animal innovation* (pp. 63−82). Oxford, UK: Oxford University Press.

Sol, D., Bacher, S., Reader, S. M., & Lefebvre, L. (2008). Brain size predicts the success of mammal species introduced into novel environments. *The American Naturalist, 172*, S63−S71.

Stephens, D. W. (1991). Change, regularity, and value in the evolution of animal learning. *Behavioral Ecology, 2*, 77−89.

Tavolga, M. C. (1966). Behavior of the bottlenose dolphin (*Tursiops truncatus*): Social interactions in a captive colony. In K. S. Norris (Ed.), *Whales, dolphins and porpoises* (pp. 718−730). Oakland, CA: University of California Press.

Tayler, C., & Saayman, G. (1973). Imitative behaviour by Indian Ocean bottlenose dolphins (*Tursiops aduncus*) in captivity. *Behaviour, 44*, 286−298.

Terry, R. (1986). The behaviour and trainability of *Sotalia fluviatilis guianensis* in captivity: A survey. *Aquatic Mammals, 12.3*, 71−79.

Thomas, B. (1970). Notes on the behavior of the killer whale *Orcinus orca* (Linneaus). *The Murrelet, 51*, 10−11.

Torres, L. G., & Read, A. J. (2009). Where to catch a fish? The influence of foraging tactics on the ecology of bottlenose dolphins (*Tursiops truncatus*) in Florida Bay, Florida. *Marine Mammal Science, 25*, 797−815.

Trivers, R. (1972). Parental investment and sexual selection. In B. Campbell (Ed.), *Sexual selection and the descent of man, 1871−1971* (pp. 136−179). Chicago, IL: Aldine.

Tyack, P. L. (1981). Interactions between singing Hawaiian humpback whales and conspecifics nearby. *Behavioral Ecology and Sociobiology, 8*, 105−116.

Tyack, P. L. (1986). Whistle repertoires of two bottlenosed dolphins, *Tursiops truncatus*: Mimicry of signature whistles? *Behavioral Ecology and Sociobiology, 18*, 251−257.

Tyack, P. L. (2000). Functional aspects of cetacean communication. In J. Mann, R. C. Connor, P. L. Tyack, & H. Whitehead (Eds.), *Cetacean Societies, field studies of dolphins and whales* (pp. 270−307). Chicago, IL: The University of Chicago Press.

Van Parijs, S. M., & Corkeron, P. J. (2001). Evidence for signature whistle production by a Pacific humpback dolphin, *Sousa chinensis*. *Marine Mammal Science, 17*, 944−949.

Visser, I. N. (1999). Benthic foraging on stingrays by killer whales (*Orcinus orca*) in New Zealand waters. *Marine Mammal Science, 1*, 220−227.

Visser, I. N. (2000). Killer whale (*Orcinus orca*) interactions with longline fisheries in New Zealand waters. *Aquatic Mammals,* 241−252.

Visser, I. N., Smith, T. G., Bullock, I. D., Green, G. D., Carlsson, O. G. L., & Imberti, S. (2008). Antarctic peninsula killer whales (*Orcinus orca*) hunt seals and a penguin on floating ice. *Marine Mammal Science, 24*, 225−234.

Ware, C. R., Wiley, D. N., Friedlaender, A. S., Weinrich, M. T., Hazen, E. L., Bocconcelli, A., et al. (2014). Bottom side-roll feeding by humpback whales (*Megaptera novaeangliae*) in the southern Gulf of Maine, U.S.A.. *Marine Mammal Science, 30*, 494−511.

Watson-Capps, J.J. (2005). Female mating behavior in the context of sexual coercion and female ranging behavior of bottlenose dolphins (*Tursiops* sp.) in Shark Bay, Western Australia (PhD thesis) (p. 195). Washington, DC, Georgetown: Department of Biology, Georgetown University.

Watwood, S. L., Tyack, P. L., & Wells, R. S. (2004). Whistle sharing in paired male bottlenose dolphins, *Tursiops truncatus*. *Behavioral Ecology and Sociobiology, 55*, 531−543.

Weilgart, L. S., & Whitehead, H. (1993). Coda communication by sperm whales (*Physeter macrocephalus*) off the Galapagos Islands. *Canadian Journal of Zoology, 71*, 744−752.

Weinrich, M. T. (1991). Stable social associations among humpback whales (*Megaptera novaeangliae*) in the southern Gulf of Maine. *Canadian Journal of Zoology, 69*, 3012−3019.

Weinrich, M. T., Rosenbaum, H., Scott Baker, C., Blackmer, A. L., & Whitehead, H. (2006). The influence of maternal lineages on social affiliations among humpback whales (*Megaptera novaeangliae*) on their feeding grounds in the southern gulf of Maine. *The Journal of Heredity, 97*, 226−234.

Weinrich, M. T., Schilling, M., & Belt, C. (1992). Evidence for acquisition of a novel feeding behaviour: Lobtail feeding in humpback whales, *Megaptera novaeangliae*. *Animal Behaviour, 44*, 1059−1072.

Whitehead, H. (2003). *Sperm whales: Social evolution in the Ocean*. Chicago, IL: University of Chicago Press.

Whitehead, H., & Mann, J. (2000). Female reproductive strategies of cetaceans. In J. Mann, R. C. Connor, P. L. Tyack, & H. Whitehead (Eds.), *Cetacean societies, field studies of dolphins and whales* (pp. 219−246). Chicago, IL: The University of Chicago Press.

Whitehead, H., Rendell, L. E., Osborne, R. W., & Würsig, B. (2004). Culture and conservation of non-humans with reference to whales and dolphins: Review and new directions. *Biological Conservation, 120*, 427−437.

Wiley, D., Ware, C. R., & Bocconcelli, A. (2011). Underwater components of humpback whale bubble-net feeding behaviour. *Behaviour, 148*, 575−602.

Williams, T. M., Estes, J. A., Doak, D. F., & Springer, A. M. (2004). Killer appetites: Assessing the role of predators in ecological communities. *Ecology, 85*, 3373−3384.

Würsig, B. (2008). Playful behavior. In W. F. Perrin, B. Würsig, & J. Thewissen (Eds.), *Encyclopedia of marine mammals* (2nd ed., pp. 885−888). Burlington, MA: Academic Press.

Würsig, B., & Dorsey, E. (1989). Feeding, aerial and play behaviour of the bowhead whale, *Balaena mysticetus*, wummering in the Beaufort Sea. *Aquatic Mammals, 15.1*, 27−37.

Wyles, J. S., Kunkel, J. G., & Wilson, A. C. (1983). Birds, behavior, and anatomical evolution. *Proceedings of the National Academy of Sciences of the United States of America, 80*, 4394−4397.

Wyman, J. (1863). Description of a "white fish," or "white whale," (*Beluga Borealis*, Lesson). *Boston Journal of Natural History, 7*, 603−612.

Yamagiwa, J., & Karczmarski, L. (2014). *Primates and cetaceans: Field research and conservation of complex mammalian societies*. New York, NY: Springer.

I. EVIDENCES OF CREATIVITY

Yano, K., & Dahlheim, M. E. (1995). Killer whale, *Orcinus orca*, depredation on longline catches of bottomfish in the southeastern Bering Sea and adjacent waters. *Fishery Bulletin, 93*, 355–372.

Yurk, H., Barrett-Lennard, L. G., Ford, J. K. B., & Matkin, C. (2002). Cultural transmission within maternal lineages: Vocal clans in resident killer whales in southern Alaska. *Animal Behaviour, 63*, 1103–1119.

Commentary on Chapter 4: Proto-c Creativity?

Learning from nature is almost absent from discussions of creativity in psychology. Building on the more or less implicit assumption that creativity is an eminently human capacity, researchers focus first and foremost on understanding the high-end cultural manifestations of creativity as embodied in the work of geniuses. It is in the last decades that everyday life creative acts came to be studied intensely (see Glăveanu, 2011; Richards, 2007). However, the boundaries of creativity are still strictly guarded by constant references to intentionality, consciousness, knowledge, etc., to the marginalization and almost exclusion of children, the "mentally ill," and nonhuman animals. When they are inspired by nature, creativity researchers tend to keep this inspiration at the level of metaphor or analogy (see for example Simonton's (1999) account of how ideas are generated and selected through what resemble Darwinian processes). Real efforts to bridge human and nonhuman creativity are still rare (for an exception, see Kaufman & Kaufman, 2014) and yet we can't help but feel intrigued by behaviors like the milk smoking in the case of Dolly, a bottlenose dolphin calf, the starting point for Patterson and Mann's (this volume) discussion of cetacean innovation. "Is this really creativity?" one might ask, persuaded by the current status quo. If nothing else, it is a behavior that generates surprise and, at least for Bruner (1962), this is a key creativity marker.

The central question a creativity scholar has when approaching a chapter like the one referred to above is: what exactly is the "core" of creativity in cetaceans and, by extension, in nonhuman animals? The two authors offer us valuable insights into this by referring to behavioral plasticity, to the capacity of dolphins and whales to act flexibly in an ever-changing environment and to respond in novel ways to both new and old stimuli. The essence of creativity in human and nonhuman animals alike relates in the end to *difference*, acting in an open and nonpredetermined manner in ways that generate novelty. When our behavior is not the direct consequence of environmental conditions, not a reflex reaction but a new response to what the environment has to offer, then I believe we are facing the roots of all creative expression. In order to act differently (and, thus, creatively to some extent), organisms need to find mechanisms to distance themselves from the here and now of perception and of their immediate situation. Such distanciation is ensured in the case of humans by the development of the symbolic function (Vygotsky, 1997). Not many nonhuman animals are credited with the capacity to form and use signs however so, consequently, they are denied "human" (symbolic) creativity. But the theoretical question remains of whether action can or should be catalogued simply as either symbolic or non/pre-symbolic. In their chapter, Patterson and Mann offer us a wealth of illustrations of flexible and atypical behavior in wild and captive whales and dolphins that come to challenge strict divisions between the two.

The starting point in their analysis of cetacean innovation can be found in the relation between organism and its environment. The two authors list preconditions for creativity that include the close relation between features of the organism (like behavioral plasticity, ability to learn, neophilic tendencies), and properties of its context and surroundings (e.g., the existence of free time and extended maternal care, social or group living, and reduced predation risk). In an effort to summarize what seem to be a set of "minimal" conditions for creativity to develop, we are being introduced in the chapter to a detailed discussion of how variations in these conditions can lead to differences in creative expression, in both its content and level, within and most of all between species. At the end it is hard to say for the two authors if a stable environment with abundant resources will always increase the chances of innovation or if, on the contrary, challenging circumstances (marked by tourism, fishing, food shortages, etc.) are not actually more beneficial as they constrain the animal to create or perish. The idea of "optimal conditions" for creativity, where challenges stimulate but don't block new behaviors, would perhaps be useful here.

The complexity of cetacean innovation is reflected in their often intricate cooperative behavior and interaction with objects. Just as in the case of human creativity (see Glăveanu, 2011, 2014), social relations are crucial in making behaviors flexible and thus potentially creative. Interacting with members of one's own species not only "socializes" the individual but also offers examples of plastic and adaptive actions that can be, in turn, copied and modified. Moreover, contact with members of other, distant groups or individuals can increase one's repertoire of actions; e.g., in the case of whales, the transmission and variations of "songs" during the breeding season. This is reminiscent of Bartlett's (1923) old assumption that creativity emerges when members of two different communities or cultures come into contact. Perhaps more intriguingly, Patterson and Mann also give examples of cooperation with members of other species, including fishermen, and even episodes of cetaceans trying to reciprocate in the relationship. On the other hand, we have numerous illustrations of whales and especially dolphins creating and using tools. In ways that were reminiscent of functional creativity (Cropley & Cropley, 2010), dolphins in captivity were noticed scraping the tank floor with tools used by divers (and later, when these were removed, with pieces of broken tile) and eating the sea lettuce that came loose or, in the wild, detaching sponges from the seafloor to use them later as protection when probing rocks or shells in search of prey. If the examples above foreground the utility of such potentially creative acts, it is play activities, primarily among calves, that offer an even wider range of opportunities for creativity serving no apparent practical reason (but only apparently, see Kuczaj & Eskelinen, 2014a, 2014b). Patterson and Mann describe the playful interaction of two bottlenose dolphin calves at Marineland in Florida, in which they engaged in a game of fetch. Most interestingly, the two partners often switched roles and repeated the whole process in ways that remind of position-exchange exercises, a precondition for the development of agency and symbolic representation (Gillespie, 2012). In fact, the roots of an early capacity to take the perspective of another in the situation is demonstrated also by the surprising finding that dolphins can produce novelty on demand (when instructed to "create"), both individually and in tandem. I would argue that this is not only a matter of remembering what was done in that session (Kuczaj & Eskelinen, 2014a, 2014b), but a sign of grasping what is intended and perceived by the trainer.

The capacity to recognize novelty is considered by Kaufman and Kaufman (2004) as the first level of animal creativity. The two authors devised an interesting creativity framework for animal cognition that includes, alongside the recognition of novelty, observational learning and innovative behavior. Although one might be tempted to see this as a hierarchy, the authors made clear the fact that each creativity marker should be studied in its own right and that animals, for instance, don't necessarily have to master observational learning before showing signs of innovative behavior (which in this case is represented by the ability to create a tool or behavior that is new and different and perceive it as such). In order to avoid hierarchical readings of their model, the framework was elaborated further as a creativity spectrum (see Kaufman, Butt, Kaufman, & Colbert-White, 2011). This conception resembles to some extent an understanding of creative action as discussed previously, including habitual, improvisational, and innovative creativity (Glăveanu, 2012). In both cases, innovative behavior is considered the one that results in more or less "deliberate" novelty (although the intentionality of this process in the case of animals can be debated). Patterson and Mann actually chose to focus in their chapter on cetacean innovations because they consider them to be the implementation of novelty in a useful way. In contrast, creativity for the two authors relates more to the underlying cognitive processes that generate novelty. The difficulty of capturing such processes in the case of nonhuman animals in fact led many animal cognition scholars to study innovative outcomes (see also Boogert, Reader, Hoppitt, & Laland, 2008) and consider innovations at an individual level instead of entire populations (Ramsey, Bastian, & van Schaik, 2007). This separation between mental processes and behavior, reflected also in the broader distinction made today between idea generation (creativity) and idea implementation (innovation), is problematic as it portrays creativity as a purely cognitive affair (for a critique, see Glăveanu, 2014). Such a reductionist reading is not productive for either human or animal creativity; in the case of the former it ignores the materiality of creative action, in that of the latter it makes creativity virtually impossible to study as such without direct access to cognition.

Instead of using this problematic criterion I consider it more useful to think about what type of creativity we are talking about in the case of cetaceans and nonhuman animals more widely. A comprehensive typology for human creativity, taking into account person, product, and process, has been proposed by Kaufman and Beghetto (2009) and it differentiates between mini, little-, Pro-, and Big-c creative acts. Kaufman and Kaufman (2014) applied this four C model to animal studies and concluded that mini-c is characterized in this case by the situation-specific innovations of a single individual, little-c emerges when a second animal joins the first in performing the novel behavior, and Pro-c is reserved for expert innovators (like Imo, the Japanese macaque who discovered potato and wheat washing). Of course, Big-c creativity, which in humans relates to achievements celebrated by an entire society and therefore requires the existence of accumulated culture, translates poorly to the animal kingdom. While the aim of making distinctions between forms or levels of animal creativity is certainly worthwhile, a discussion of the four C model in this area remains at the level of analogy. Between the mini-, little-, or Pro-c creativity of humans and that of nonhuman animals there is not only a difference of degree but, in most cases, a qualitative leap enforced by the fact that humans structure their (creative) action symbolically (Glăveanu, 2014; Vygotsky, 1997). This reflection, as well as the rich illustrations of whale and dolphin innovations offered by Patterson and Mann, make me consider whether it wouldn't be more appropriate to actually expand the four C model to a five C one and include what I would call, in lack of a better term, *proto-c creativity*. This for me is the creativity intrinsic to organisms exist as dynamic, open systems, acting within environments marked by variability and change. Behavioral flexibility and adaptability are the essence of proto-c creativity and they constitute the basis for other, more elaborate creations or innovations in both humans and animals. Proto-c is the creativity of action that precedes and accompanies symbolic forms of creativity such as little-c, built on reinterpretations of experience. Postulating the existence of proto-c therefore solves a complicated theoretical dilemma: is there creativity outside and beyond symbolically mediated activity, before the use of language? The playful activities of cetacean calves, just like those of a human infant, compel us not only to at least consider this conceptual possibility, but also engage in a much closer study of habits, acts of imitation and copying (see also Glăveanu, 2012). Patterson and Mann report, for the latter, cases of dolphins imitating the behavior of turtles, penguins, and even humans, and of whales copying the whistle sounds of close associates in the process of developing their signature vocalizations.

In the end, proto-c is not outside or before "real" creativity but it is the basis for the creative action of humans and animals alike. It is also not separate from the creativity involved in mini-, little-, Pro-, and Big-c creative acts. In the case of cetaceans in fact, most of the examples offered by Patterson and Mann illustrate simultaneously proto- and mini-c forms of creativity in which cetaceans experiment with new forms of action and interaction and become then capable of reproducing and varying them, arguably while recognizing their novelty. The domain generality of creativity claimed by the two authors is in fact the generality of a proto-c form of creative expression. As we move toward the more elaborate little-, Pro-, and especially Big-c types there is much more room for variation depending on the characteristics of the individual, the domain, and the field or audience. Kaufman and Kaufman (2004, p. 144) were right to say that conceptualizing creativity in animals can not only shed light on animal cognition but also expands our current understanding of creativity. Big-c creative achievements in human societies might stand out like mountaintops in our current perception, but we should not forget they emerge out of a sea of proto- and mini-c expressions and are constantly nurtured by these. Looking toward and inside this sea (pun intended) is greatly facilitated by ongoing research on cetacean innovation aptly summarized in this book.

References

Bartlett, F. C. (1923). *Psychology and primitive culture*. Cambridge: Cambridge University Press.

Boogert, N. J., Reader, S. M., Hoppitt, W., & Laland, K. (2008). The origin and spread of innovations in starlings. *Animal Behaviour, 75*, 1509–1518.

Bruner, J. (1962). *On knowing: Essays for the left hand*. Cambridge: Belknap Press.

Cropley, D., & Cropley, A. (2010). Functional creativity: "Products" and the generation of effective novelty. In J. C. Kaufman, & R. J. Sternberg (Eds.), *The Cambridge handbook of creativity* (pp. 301–317). Cambridge: Cambridge University Press.

Gillespie, A. (2012). Position exchange: The social development of agency. *New Ideas in Psychology, 30*(1), 32–46.

Glăveanu, V. P. (2011). Creativity as cultural participation. *Journal for the Theory of Social Behaviour, 41*(1), 48–67.

Glăveanu, V. P. (2012). Habitual creativity: Revising habit, reconceptualizing creativity. *Review of General Psychology, 16*(1), 78–92.

Glăveanu, V. P. (2014). *Distributed creativity: Thinking outside the box of the creative individual.* Cham: Springer.

Kaufman, A. B., & Kaufman, J. C. (2014). Applying theoretical models on human creativity to animal studies. *Animal Behavior and Cognition, 1*(1), 77–89.

Kaufman, A. B., Butt. A. E., Kaufman, J. C., & Colbert-White, E. N. (2011). Towards a neurobiology of creativity in nonhuman animals. *Journal of Comparative Psychology, 125*(3), 255–272.

Kaufman, J. C., & Beghetto, R. A. (2009). Beyond big and little: The four C model of creativity. *Review of General Psychology, 13*(1), 1–12.

Kaufman, J. C., & Kaufman, A. B. (2004). Applying a creativity framework to animal cognition. *New Ideas in Psychology, 22*, 143–155.

Kuczaj, S. A., & Eskelinen, H. C. (2014a). Why do dolphins play? *Animal Behavior and Cognition, 1*(2), 113–127.

Kuczaj, S. A., & Eskelinen, H. C. (2014b). The "creative dolphin" revisited: What do dolphins do when asked to vary their behavior? *Animal Behavior and Cognition, 1*(1), 66–77.

Ramsey, G., Bastian, M. L., & van Schaik, C. (2007). Animal innovation defined and operationalized. *Behavioral and Brain Sciences, 30*, 393–437.

Richards, R. (2007). *Everyday creativity and new views of human nature: Psychological, social, and spiritual perspectives.* Washington, DC: American Psychological Association.

Simonton, D. K. (1999). *Origins of genius: Darwinian perspectives on creativity.* New York, NY: Oxford University Press.

Tayler, C., & Saayman, G. (1973). Imitative behaviour by Indian Ocean bottlenose dolphins (*Tursiops aduncus*) in captivity. *Behaviour, 44*, 286–298.

Vygotsky, L. S. (1997). The history of the development of higher mental functions. In R. W. Rieber (Ed.), *The collected works of L.S. Vygotsky* (Vol. IV, pp. 1–251). New York, NY: Plenum Press.

PART II

REQUIREMENTS FOR CREATIVITY

Creativity, Play, and the Pace of Evolution

Gordon M. Burghardt

Departments of Psychology and Ecology & Evolutionary Biology, University of Tennessee, Knoxville, TN, USA

Play serves two ends—for experimenting: as such it is an introduction to knowledge, gives certain vague notions concerning the nature of things; for creating: this is its principle function. **Ribot (1906: 114)**

Commentary on Chapter 5: Play—A Multipurpose Vehicle

Sandra W. Russ

Case Western Reserve University, Department of Psychological Sciences, Cleveland, OH, USA

INTRODUCTION

Male Mientian tree frogs living in urban areas in Taiwan have begun using artificial human-made storm drains to enhance the loudness and duration of their mating calls, presumably giving them a reproductive advantage over those frogs not using perches in such drains (Tan, Tsai, Lin, & Lin, 2014). Popular accounts in *Nature* and *Scientific American* refer to these city frogs using drains as a "mating megaphone." Is this creative and innovative behavior? Is it at all different from the spread of grain washing by Japanese monkeys (Itani & Nishimura, 1973)? If not, why? Regardless, how did this behavior arise and spread? And if we could say anything about such matters, could play be a factor underlying the behavior in either frogs or monkeys? Does such flexibility or plasticity in behavior have evolutionary consequences beyond the specific population studied? Can anything useful, in fact, be said?

Here is another example. Chimpanzees in three semi-wild populations in large 25–77 hectare naturalistic enclosures in Zambia developed six different ways of opening hard-shelled fruits that varied across groups (Rawlings, Davila-Ross, & Boysen, 2014). The ecological settings were similar. While social learning was involved in that multiple animals used methods in a population that were not used in others, how the animals initially came upon the strategies used is unknown. However, the authors "suggest that chimpanzees were displaying hierarchical mental construction—the capacity to hold and integrate several cognitive, motoric or perceptual components to achieve the goal" (p. 897). Could play have been a facilitating aspect in the development of unique methods by a population?

In the final sentences of an extensive treatment on the neural basis of play some years ago, I wrote the following in relation to play (Burghardt, 2001):

> The role of the neocortex may be to provide the animal with a greater ability to utilize the information gathered during interactions with the world, provide neural resources for rapid synapse formation, and derive novel ways of dealing with environmental challenges. From here it is not too far to thought processes, planning, and especially creativity as consequences of playful activities. But much play may still be nonfunctional, evolutionary detritus, or byproducts of developmental retardation. Thus, the paradoxical nature of play endures (p. 350).

Play, unlike most learning or even instinctive, innate, or reflexive action, is not necessarily adaptive in most situations and may occur for many reasons (Burghardt, 2005). Play is a process that evolution has co-opted or used, and one of the questions addressed here is whether play is an important process underlying creative or innovative behavior.

A GUIDING RATIONALE

It is difficult to be creative about creativity. It may not be so difficult to be creative. This seems to be the impression gained from the prodigious output of acknowledged creative geniuses such as Newton, da Vinci, and Picasso. But how accurate is this impression? To the extent that play and a playful attitude are involved in creativity, it may seem that creativity is not that difficult in the right circumstances for some people. Others have pointed out that creativity is not so simple; it is based on hard work, determination, and grit. As I have argued in trying to understand play, a bottom up, not top down, approach may be most illuminating (Burghardt, 2005) and will be explored in this chapter in which I review some studies on nonhuman animals that seem to link creativity and play, discuss some of the terminological problems involved with creativity, novelty, and innovation, critique some of the preceding work (playfully!) and offer an argument that play was and is an important, if not essential, factor in evolutionary change and progress in the long history of evolution (Burghardt, 2014). If there is any merit to this approach, then the fact that evolutionary biologists have largely ignored play, and those studying animal play behavior have largely restricted it to mammals and some birds, is a major handicap in understanding evolutionary processes. On the other hand, this volume and others on the role of key innovations in evolution: morphological, physiological, behavioral, and cultural, show that interest is growing in understanding such changes that have far-reaching and enduring effects.

Novel phenotypic expressions are the hallmark of evolution and occur in many guises such as in anatomy, external appearance, physiology, life history, diet, behavior, nervous system, motivation and emotion processes, parental care, and sociality. If such novelties facilitate new modes of living, they are often placed as nodes on phylogenetic trees. Evolutionary biologists are very interested in morphological innovations such as cell types, feathers, and flowers, and some creative ways of looking at such novel traits are at the forefront of research on evo−devo, the relationships between developmental processes at the individual level and natural selection at the population level (Wagner, 2014). Typically, and in this book, we are primarily interested in behavior variously termed creative, novel, innovative, inventive, and original. Sometimes such behavior also is referred to as ingenious, insightful, resourceful, and artistic. The question is, how do we determine if behavior in animals, human and nonhuman, has these qualities, and, if they do, how do they arise and come about? What is their role in subsequent evolution, both biological and cultural, and what role has evolution and prior behavioral experience played in their initial appearance?

In this chapter, the focus is on exploring play as an important aspect of animal creativity. I addressed this first, and briefly, some years ago (Burghardt, 1999), but in the interim my ideas on play have altered and much interesting new research has appeared. Here I present an expanded perspective. This appears timely. For example, a recent article on "The role of behavior in the establishment of novel traits" (Zuk, Bastiaans, Langkilde, & Swanger, 2014) decried how behavior and individual level processes were ignored in recent treatments. However, they ignored virtually the entire literature on animal innovation (Reader & Laland, 2003b) focusing almost entirely on phenotypic plasticity, the role of plasticity in buffering organisms from selection, and the role of novel and changing environments from a genetics perspective. It is not surprising, then, that neither creativity nor play are mentioned in their otherwise informative paper, but it is surprising that such a narrow approach was so recently published.

SOME IMPORTANT ISSUES

We are immediately confronted with some questions, also certainly addressed in other chapters, such as what is meant by creativity and innovation, as well as the attributes and possible role of play. The precise meaning of these terms outside of specific examples may not seem essential, but previous theoretical and conceptual work does need attention along with empirical studies and evolutionary speculation. In preparing this chapter I gladly acknowledge the timely and excellent book by Bateson & Martin (2013) and a summarizing article (Bateson, 2014) that together provide a current review and background of play behavior as well as creativity and innovation. I have used some sources they identified along with those found elsewhere in the literature (Graham & Burghardt, 2010; Kaufman, Butt, Kaufman, & Colbert-White, 2011; Kaufman & Kaufman, 2004) as well as more recent examples such as the megaphone using frogs cited above.

As *Creativity and Innovation in Animals* is the title of this book, these two terms are the key along with play. Interestingly, these are also the terms used by Bateson and Martin in their book title. But what is meant by each term? While detailed elsewhere in this volume, the common usage of the terms by Bateson and Martin is straightforward. Creativity is having a new idea; innovation is applying that idea in a productive and novel manner so that a better outcome is developed. This is close to the distinction between the roles of basic research or brainstorming and applied science or engineering. I view this distinction, though clear enough at the ends of the continuum and applicable in many aspects of human culture, not so useful in the nonhuman animal literature, child development literature, and some other areas where creativity is discussed and thus I downplay, even conflate, the distinction in this chapter. Although the following may seem to be tangentially related to the topic, I think that these points are deeply relevant in considering play in the creative process.

First of all, having an original idea is not that obvious when observing animals as compared to hearing or reading a verbal exemplar of a purported new idea by a person. In art, perhaps, it can be claimed that a new painting style, use of a new medium, applying different tools (spray paint rather than a brush) could be counted as creative, as well as a new useful construction. Musical composition is similar; indeed

novel sounds and combinations are much discussed in the avian literature as well in humpback whale songs (Payne & McVay, 1971). Such
behavioral products can be discerned in non-literate animals and an
underlying creative process inferred. But what if such productions are
not really new? Was Roger Bacon creative when he discovered gun-
powder or Guttenberg the printing press, as actually both were already
known in Asia? Or is something creative only if it is new or novel in
the culture or population of a specific group? But then, even if gunpow-
der or the printing press were new in that culture, did the creators
"just" apply or extend in an incremental way knowledge or skills
already present, which only in retrospect, led to truly momentous
change? This is not, we shall see, a trivial issue in creativity research.

Second, consider science itself. When is a new finding creative?
We often put little stock in replication as compared to a new result.
But what about applying a finding or method to a new species—
demonstrating that not only rhesus monkeys develop learning sets but
also Japanese macaques? A replication with a similar species doesn't
seem very creative. But what about showing the process occurs in a
bird, for which rather different methods needed to be developed than
those used by Harlow (1949) to demonstrate the effect? Was this more
creative? Perhaps the true creativity was only to be seen in the initial
Harlow studies. But then, learning sets were an extension of habit
reversal studies that similarly looked at a type of "learning to learn."
This all seems a bit beside the point. Still, we do not consider all scientists equally creative. But are not all scientists, qua scientists (e.g., not
technicians) necessarily creative, more or less, whenever they discover
something new in nature. Are only the recognized stars, the Nobel
Laureates, the creative elite truly creative? On the other hand, perhaps
true creativity resides only in scientists who fostered the paradigm
shifts that so intrigued Kuhn (1996). The names of Copernicus, Galileo,
Newton, Faraday, Einstein, Watson and Crick come to mind. But then
these also were often dependent on predecessors as in Newton's
famous, and now trite, line that he saw further since he stood on the
shoulders of giants. You can see why discussions of the distinction
between what is creative, novel, and innovative begin to lose traction
and what and who is creative seems to fade before our tired eyes. But
does any of this relate to animals?

Third, the appellation "creative" is one that we do not typically
apply to behavioral development, where animals often go through
stages of personal discovery, as growth and maturation allow new
behavioral and cognitive milestones to be traversed. Thus, every child
develops "ideas" and new behavioral abilities that are, for them, truly
innovative when they begin to grasp objects, speak, construct sentences,
crawl, walk, run, open doors, feed themselves, etc. And it is not only

these behaviors. Consider the statement of one of the earliest scientists of child development (Preyer, 1893) that also introduces play:

> A satisfactory theory of play is still wanting, and yet a man does not learn through any kind of instruction or study in later life anything like so much as the child learns in the first four years of his careless existence, through the perceptions and ideas acquired in his play... as I have previously spoken of the experimenting of little children as play, I may now mention the internal resemblance of their procedure to that of the naturalist (Preyer, 1893)

Here, the distinction between child and scientist collapses and play, curiosity, and exploration enter the room. But do not many scientists and artists often view their accomplishments as based on the play of youth extending into adulthood and even old age! This quotation from Preyer allows us to see how play may be at the heart of creativity and innovation, however defined, and it is this claim that will animate this chapter. It has, in various disguises, been at the heart of much discussion of childhood education and play. In any event, certainly non-human animals go through developmental milestones just as do children (Parker & McKinney, 1999).

INTRODUCING PLAY

What do we mean by play? How do we recognize it? Who plays? Why do animals play? What are the origins of play? I will only answers these questions briefly, and in turn, referring interested readers to more thorough discussions, before exploring how play may inform our understanding of creativity and innovation.

Traditionally three types of play in animals are recognized. These are locomotor/rotational play, object play, and social play (Fagen, 1981). While the first two are typically solitary, all three can occur together, as when animals use objects in a chase game with another individual. Types of play such as pretense, construction, vocal, and covert (cognitive, mental) may also occur.

Although the three types of play behavior often seem obvious and easily recognized, they actually are not. The difficulties in formulating a satisfactory definition of play has led to anthropomorphic attributes (e.g., having fun) and biases (play is limited to smart animals), warm-blooded endotherms (birds and mammals) that prevented even looking for play in other taxa. For play to be recognized in species in which we did not already knew it occurred, a more objective definitional approach was needed. This I tried to do (Burghardt, 2005, 2011) by combining and refining the most useful aspect common to all types of play. They have been found useful in the play literature (Bateson &

Martin, 2013; Pellegrini, 2009; Pellis & Pellis, 2009; Smith, 2010). A recent brief statement (Burghardt, 2014) is: *Play is repeated, seemingly non-functional behavior differing from more adaptive versions structurally, contextually, or developmentally, and initiated when the animal is in a relaxed, unstimulating, or low stress setting* (p. 91). While the details of the five play criteria that are the basis for this statement are presented in detail elsewhere (Burghardt, 2005, 2011), the aim here is to find a means of applying a common framework to all of the major types of animal and human play. However, it does not capture many of the special nuances or features of social play. In examining the literature on innovations, it is apparent that most have to do with objects rather than social or loco-motor behavior. Social factors are certainly key to understanding the spread of innovations, but social innovations in, for example, communication, have received a fair share of attention.

Applying the play criteria has allowed the recognition of putative play behavior in many animals thought bereft of the capability including members of all vertebrate classes and invertebrates as well, including cephalopods, insects, crustaceans, and spiders (Burghardt, 2005; Graham & Burghardt, 2010; Pruitt, Burghardt, & Riechert, 2012). Certainly there are major reasons why play occurs in some groups more, and more complexly, than in others, but what the taxonomic diversity indicates is that play may have a long evolutionary history, and thus may have a role in evolution far beyond any importance to the limited taxa it has been heretofore restricted (Bekoff & Byers, 1981).

This broader phylogenetic perspective also broadens our search for the benefits of play and why it occurs in the first place. Many claims have been made that play facilitates learning, socialization, behavioral flexibility, mental agility, and creativity (Bateson & Martin, 2013; Burghardt, 1999, 2005; Fagen, 1981). Evolutionary arguments based on comparative data often accompany these assertions. However, strong empirical findings are sparse (Burghardt, 2005; Martin & Caro, 1985). A few recent reviews revisiting this situation are Lillard et al. (2013), Pellis and Pellis (2009), and Smith (2010). However, progress is being made on the role of specific types of play in limited species (van den Berg et al., 1999; Pellis, Pellis, Barrett, & Henzi, 2014). In other words, rather than grand theories of play function, mini theories may prove most needed currently (Burghardt, 2014). As discussed below, this seems to be the case in creativity research as well. Nonetheless, play, with its costs in time, energy, and risk will surely be shown to be valuable in many contexts. Creativity and novel behavior production are included in these values/benefits (Bateson, 2014).

Knowledge about the origins of play may help provide answers to its function and particularly its role in evolution, as elaborated earlier. What may be valuable is to ground the phenomena of play in a phylogenetic context by analyzing the processes leading to playfulness and not just the putative and most likely derived consequences. The similarity with innovation should be noted. Just as the study of play was hindered by using definitions based on putative functions, so too the focus on "useful" innovations and novel products has, I will argue below, held back the study of creativity or novelty generation.

Play is not a unitary phenomenon and the role of play may differ for different kinds of play, differ throughout ontogeny, and differ across species. I will present a way of looking at play as a potential source of the rapid evolution of behavioral complexity in endothermic vertebrates and as a source of developmental changes in behavior that can be both conservative and novel.

While theories of play abound, by focusing on function they are less useful in understanding the reasons why play first arose in evolution. I have developed an approach to play in animals from an evolutionary/developmental perspective called Surplus Resource Theory or SRT (Burghardt, 1984, 1988, 2005). This view has several components including a reliance on studying animals, including ectothermic vertebrates, with less rich play in contrast to the exclusive focus on highly playful taxa (primates, canids, felids, ungulates) usually pursued. SRT incorporates physiology (e.g., activity metabolism, thermoregulation), life history (e.g., parental care, altriciality, food niche, sociality), and psychology (e.g., stress, learning, and the richness and flexibility of the behavioral repertoire). In short, play is facilitated when animals have ample time, energy, and a richness of behavioral resources (based on both innate processes and prior experiences). Supporting the role of surplus resources is the extensive research on the stone tool use to open nutrition-rich nuts by capuchin monkeys in Brazil. Populations where the trait arises are those where food is NOT in short supply and so the animals seem to have the luxury of engaging in, and learning, a rather arduous but rewarding activity (Verderane, Izar, Visalberghi, & Fragaszy, 2013). Juvenile monkeys begin playing with stones while very young and later use them as nut-cracking tools.

Play may be both a product and cause of evolutionary processes; that is, playful activities may be both sources of enhanced behavioral and mental improvements as well as a byproduct or remnant of prior evolutionary and ontogenetic events. It is the conflict between these two views of play that I view as being at the root of the paradoxical nature of play. A way of viewing these differing aspects of play is to distinguish among primary, secondary, and tertiary play processes (Burghardt, 2005, 2014). It is only tertiary process play that seems to be a good candidate for producing novel behavior that can lead to innovations that may have important consequences for the individual and, perhaps subsequently, to conspecifics. The next move is to distinguish

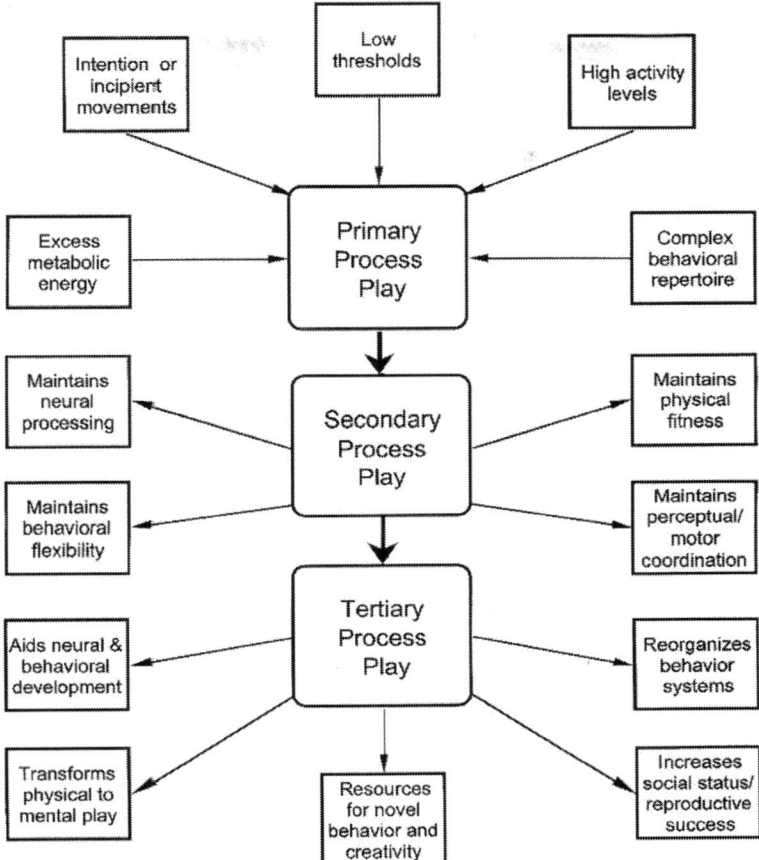

FIGURE 5.1 The three play processes (from Burghardt, 2005).

between primary and secondary processes in play. An example of a secondary process would be that vigorous rough and tumble play of young rats and dogs enhances adult performance, promotes socialization, or increases behavioral flexibility. However, it is precisely because the more primary processes of play have been ignored that predictions made from secondary processes have fared poorly. An example of an important primary process derived from SRT would be the role of metabolic rate or boredom in production of "surplus" behavioral "mutants" that could in turn be selected ontogenetically and phylogenetically. A secondary process derived from a detailed consideration of primary processes would be the correlation, by Byers and Walker (1995), of the timing of playful activity with the timing of permanent long-term changes in the muscular and cerebellar systems of several species of domesticated mammals. Secondary process play may underlie normal behavioral development and maintain behavioral and physiological capabilities, as exercise can maintain cardiovascular fitness. It has recently been found that exercise in and of itself can turn on and reprogram the "epigenome" (Lindholm et al., 2014) with consequent health benefits. Vigorous play is one way such beneficial consequences can be attained, in the absence of having to exercise by "work" or "serious" behavior.

It is in the tertiary play process, where play goes beyond normal developmental processes or routine learning, and aids the animal in making advances in how it deals with the world that play is reputed to have its most useful (innovative?) effects. Again, this does not mean that play is the only way the animal can make such advances, but that it may be nonetheless important. Among the various consequences products of tertiary play is that it provides resources for generating novel and creative behavior (Figure 5.1).

DIVERSE ATTEMPTS TO BE CREATIVE ABOUT CREATIVITY

Whether novel behavior or more interior, cognitive "ideas" are viewed as creative is not that applicable to animal play, nor whether these need to be distinguished from "mere" scaffolded innovation based on prior reinforcement histories and cultural familiarity (Simonton, 2003). In their valuable edited book on animal innovation, Reader and Laland (2003a) summarize many of the ways the term innovation is used by animal researchers. They particularly contrast innovation as a product, as used by Bateson and Martin above, and innovation as a process. While the chapters in Reader and Laland (2003b) struggle with many ways of conceiving innovation, Reader and Laland (2003a) themselves include creativity as one of several processes underlying or facilitating innovations. On the other hand Simonton (2003) views creativity as logically prior to innovation and, importantly, whether the innovation is adaptive or functional is not a criterion, as it is for many views of innovation. If the novel behavior is "crazy" it is not creative! This reflects the same problem as noted in defining play in terms of function. Here I use creativity as a process, often facilitated by play, that may leads to unusual or novel results that may or may not be useful to either the individual (as in solving a problem) or group (exploiting a new resource). Note that while behavior must be repeated to be play, in my definition, a creative product may be a sudden insight or behavioral combination resulting from play.

The history of work on creativity cannot be summarized easily, and there are numerous books, journals, and competing theories attempting to deal with creativity and innovation in humans. I think it is instructive to look at some previous ways of looking at creativity generally as they highlight the way various formulations bias or constrain our view of creativity and underlying some of the differences identified by Reader and Laland (2003a). Three approaches spanning the last 60 years highlight some of the complexities and issues, and illuminate problems and perils when we consider animal creativity as well as play as a formative factor in evolution.

Consider first a small book by the renowned British experimental psychologist, Sir Frederic Bartlett (1951), titled *The Mind at Work and Play*. Written before the cognitive revolution, it seems completely alien to the current volume, for words such as play, creativity, innovation, flexibility, imagination, etc. are not in the short abbreviated index, nor, except for play (used primarily for games and puzzles) in the text as well. So what is the mind at work and play? Well, mind is deployed basically by Bartlett to characterize how we measure/estimate length, distance, height, and position and studied with what today are primitive tasks such as moving levers, recognizing perceptual cues in ambiguous figures, and so forth. He is concerned with fatigue and loss of accuracy over time in performing often repetitive acts or watching for very rare events in defense monitoring (as in animals being vigilant for predators?). In discussing a ball game he notes that "it is astonishing how often the good player will make the correct response when it can be proved beyond doubt that he cannot possibly have observed more than a very small part of the things that are prompting him" (p. 65).

Bartlett then moves on to perception and uses now common gestalt figures, where one can only see only one of two versions of the figure at a time, such as in figure—ground reversals, reversible figures and movements, and figures within figures. For many people, and even animals, these are suddenly apprehended—a light bulb moment akin to creative insight. Consider also visual search images used in identifying cryptic prey or objects (Pietrewicz & Kamil, 1979; von Uexküll & Kriszat, 1934) that could, for an individual, be an important cognitive and functionally adaptive breakthrough. When a baby makes a sudden novel (for him or her) connection or association we are often greatly impressed. In discussing thinking, Bartlett, too, distinguishes between suddenly apprehending a solution compared to those problems where solutions are not instantaneous but "when evidence has to be collected from instances not quite complete in themselves and carried further, is in the identification of the critical points of agreement and disagreement between the instances" (p. 135).

Indeed, all thinking involves transfer of training and the filling in of gaps. Thus, while most puzzles and gestalt apprehensions allow of only one solution, it is when there are alternatives that the most useful thinking appears. Bartlett claims that effective thinking is based on "a good deal of freedom in the selection of the first step, and, within the limits of consistency, of any others as well" (p. 141). The initial steps involve apprehending relevant similarities and interpolation across gaps, rather than focusing on differences. He refers to gaps in knowledge and needing new solutions. Building on an early version of the now common dual process theory (thinking fast and slow) he writes: "If the gap can be filled only by a very radical change in the ordinary way of looking at the evidence available, it seems that there must be either laborious trial and error or a sudden flash of insight or intuition" (p. 141). Is this then creativity? Does this apply to other species as noted above? When a visually hunting bird suddenly is able to identify a type of camouflaged or novel appearing insect as edible and henceforth is able to find them much more quickly and systematically, is this akin to insight, a sudden, for that individual, new ability? I present Bartlett because he represents the "establishment" view from experimental psychology prior to the cognitive "revolution" and reflects my historian's hat assertion that good ideas can come from reading old books (Burghardt, 2009).

Another early book explicitly on creativity included chapters by many influential scholars such as Torrence, and presented a rather different approach (Taylor, 1964b). Here the focus is specifically on creativity, particularly scientific creativity in adults where it is viewed as a quite qualitatively different process from the context in which Bartlett embeds it. Creativity (quality) is distinguished from productivity (quantity). The book's editor reviews definitions of creativity and highlights two that he thinks are the best. These are the proposal by Ghiselin "that the measure of a creative product be the extent to which it restructures our universe of understanding" (p. 6), and Lacklen who prefers "the extent of the area of science that the contribution underlies: the more creative the contribution, the wider its effects" (p. 6) (Taylor, 1964a). But the same book also proposes the more general view (Torrence, 1964), that creativity emerges in childhood and is a property found in all children to varying extents. Except for perhaps a Mozart or other prodigy, such creativity does not meet Taylor's conditions above. This introduces an important issue: how relevant is it to identify creativity as the production of a novel or important landmark in an individual as compared to it being one having a broad social, cultural, or phylogenetic impact? This important distinction plagues discussions of creativity down to the present.

The influential book by Csikszentmihalyi (1996) follows the Taylor approach. Consider his definition (p. 28): "Creativity is any act, idea, or product that changes an existing domain, or that transforms an existing domain into a new one." A creative person is someone who does this. By this token, creativity is not a trait but a socially determined product. He is clearer than other writers on where his approach takes us. Csikszentmihalyi states that Mendel was not creative during his lifetime because his accomplishment did not have a social impact in a domain until well after his death. Similarly, the waxing and waning of reputations and influence of artists (e.g., Raphael) and composers (e.g., Bach) means that their creativity also fluctuates through the ages. Thus, creativity is neither a process nor a product; not even an individual developing a novel or seminal, idea. Creativity is limited to those who themselves or through others effectively promote a "product," are successful in gaining its acceptance/influence, and finally having their name attached to said idea or product. Although this view may be disconcerting to scientists, artists, and others who just want to concentrate on their "creative" work rather than self-promotion, this conception may have, arguably, some face validity, though only in the cultural history sphere. Furthermore, this view is more related to identifying creative individuals rather than the creative result. The former may not be accurately identified if ideas are stolen, plagiarized, and promoted by those who then get the credit, at least by the general populace. However, Csikszentmihalyi's view may not be that relevant to nonhuman animal research where, with few exceptions, creative individuals are not identified. In short, I think that ideas of creativity comingled with creators and social impact deter us from a more useful and general approach, especially in regards to animals, of how creative ideas or innovations arise.

Bruce MacLennan (2009) reviews some additional views developed in the creativity literature—also see Kaufman (2009) and other chapters in this book. First is the distinction between little-c and Big-C creativity. While little-c is here viewed as finding run of the mill new and nonobvious solutions to a problem, big-C is the kind of "creative accomplishments that loom large in history books" and scientific advances "that effect conceptual revolutions" (p. 1) and reflects Taylor cited above. This may preclude nonhuman animal creativity *unless* we take a broader view of history, and include major behavioral advances, or discontinuities, in evolution and culture in other species. This is supported by MacLennan's use of Boden's distinction between P(sychological)-creativity and H(istorical)-creativity. The latter is producing "something new and interesting" in the producer's culture, while P-creativity is producing "something new and interesting" to the individual producer (p. 2). Note that the operative word is "interesting," not useful.

By making distinctions such as these it is possible to resolve some of the conundrums attending to creativity recognition in nonhuman animals. For example, a Four-C approach has been suggested for creativity studies (Kaufman & Beghetto, 2009). The Four-C model is meant to represent "a developmental trajectory of creativity in a person's life" (p. 6). Here I translate their model into a way I think helps in looking at animals. The category of mini-c creativity is the ordinary process of adapting behavior using one's "experiences, actions, and events" (p. 3). Thus, individual milestones and leaps of development, continuous or discontinuous, might fall into this category. It is specifically related to child development by Kaufman and Beghetto (2009). An example of the category termed little-c creativity would be an art project produced by an elementary student. It may be innovative and surprising for this class and school, but would probably not be so viewed as creative if produced by a professional or even an advanced art student. This little-c creativity may go nowhere in terms of cultural spread, even in the student's school, but still merit having the hallmarks of being creative and the underlying process may have similarities with Big-C creativity. Finally, and perhaps least appropriate for animals, is Pro-c creativity. This is where persons with advanced training and expertise produce ideas or artifacts. These would include "individuals who are professional creators, but have not reached eminent status" (p. 4). Most scientists, artists, composers, and chefs would be in this category—doing successful, useful work but not worthy of Big-C accolades. So, based on their descriptions, mini-c could be considered tinkering and trying out variations; little-c goes beyond this by involving some reflection and reasoning, Pro-c involves formal and informal training and apprenticeship (training apes and elephants to paint?) without producing breakout ideas or products; Big-C is the kind of legendary greatness thrust upon a few who, usually, but not always, have population-wide effects on their culture (Japanese monkey who started the grain washing tradition). Of course, even with humans many originators of seminal innovations, such as the hand axe or pottery making, are unknown, and this is certainly the case with behavioral innovations in the animal world.

Most writers on creativity discuss four common typical stages of the creative process: preparation, incubation, inspiration and elaboration or testing of the viability/utility of the purported solution. These stage areas have been elaborated and refined in several ways, but it suffices to point out that in animals, preparation (e.g., prior experience) and inspiration (measured in the novelty of the behavior or problem solving attempt) are possible to measure. Incubation could be the time between when an animal is confronted with a problem or situation and producing a solution; this can typically best be assessed in an experimental setting. The incubation period most of us can appreciate from our own experience: the advantage of letting a problem rusticate without determined effort for varying times until a solution arises. Incubation may be the most neglected aspect of creativity, however (Ritter & Dijksterhuis, 2014), and the most difficult to assess in animals. Indeed, the quickness of finding a solution to a problem is often viewed as a mark of intelligence. Perhaps we need ways to measure the time animals are confronted with a challenge and if and when they return to it after time away, rather than exhibiting persistence that often wastes time and energy.

CREATIVITY AND INNOVATION IN ANIMALS

In our opinion, if we may with any truthfulness attribute a creative power to animals... it is purely motor, and expresses itself through the various kinds of play
Ribot (1906: 97)

Evolutionary innovations, which can include behavior, are prominent topics today (Greig & Webster, 2014; Wagner, 2014; Zuk et al., 2014). However, while behavior may be involved, it is usually not considered that important innovations arose through the kinds of processes discussed above. Consider a recent paper on a species that created a "novel" signal, which was actually transferring a signal from one context to another. Thus, the transfer of a trill vocalization in most fairy-wrens from a response to avian predators with a predator alarm or mobbing function to being used in conspecific displays in others (Greig & Webster, 2014) is considered a process of "generating novel signal phenotypes" (p. 58). In fact, the venerable but now seeming neglected ethological concept of ritualization (Foster, 1995) is being invoked, though without recognition that this process is one with a long theoretical history with many putative examples. Interestingly, the signal shift is later described as one from predator-prompted trilling to "unprompted" trilling that occurs during dawn choruses. Whether these more spontaneous unprompted calls were actually instigated as play, if predator risk was low, and so subsequently became incorporated into the new context was not entertained. While this is an elegant study, the more proximate mechanisms underlying the shifts in putative function and context were slighted. This is just one example of a recent study where play may have been involved. Indeed, rituals may derive from play in both human and nonhuman animals (Burghardt, in press) through both genetic and cultural processes.

By what processes can creative behavior that may involve play, arise? Several approaches will be briefly described before moving on to some examples. An early treatment of innovative behavior proposed several "Candidate processes underlying or influencing innovation" (Reader & Laland, 2003a). Interestingly, their key table (1.1, p. 16), listed creativity as one of several processes supporting their contention that creativity is a process rather than a product. This would then seem to suggest that the usefulness of such a product is immaterial to whether creativity is involved, which turns out to be an important issue in what follows. The candidate processes they list include novelty responses, exploration, individual learning, insight, creativity, behavioral flexibility, and social processes. But is creativity able to be separated from any or all of these other candidates? The answer must be negative.

Perhaps the most thorough reviews of processes of animal creativity are those by Kaufman et al. (2011) and Kaufman & Kaufman (2004). They cover the basic theories of creativity and apply them to animals more thoroughly than I have here and also review the basic characteristics of creativity (though in terms of ideas): fluency (quantity), flexibility (different types of ideas), originality, and elaboration (development and application). With animals these "ideas" need to be either translated into behavioral measures or inferred from behavior. The model of animal creativity provided by the Kaufman group posits four components/levels on a hierarchy of complexity: (i) novelty recognition, (ii) novelty seeking (neophilia) and risk taking, (iii) observational learning, and (iv) innovation. The last level results in the creative product, while the four stages taken together reflect the creative process. While the authors claim early levels are not necessary for the latter ones, they do claim that innovation rarely occurs in the absence of observational learning The later paper (Kaufman et al., 2011) is particularly useful in that several examples of novel behavior and the processes involved in animals are described, including bower birds and dolphins, along with suggestions for the neural substrates of creativity in animals. This paper also briefly mentions that play may be involved in level 1 and, although not specifically stated, probably also in level 2.

These authors are among those who emphasize that novel behavior may often not be innovative or creative. For example, if a bower bird makes too elaborate or complex a display, nest predators may be attracted and thus the bird will be less successful in reproduction. As the authors write "Such excessive or elaborate behavior, while novel, would not be deemed innovative because it is ultimately not appropriate to the task of finding a mate" (p. 263). But, if only innovative behavior is creative, then creativity is limited to novel productions that actually are beneficial or adaptive in the long run. Outside of laboratory studies this would be difficult to assess. When Mike, the male chimpanzee, discovered that banging kerosene cans, ostensibly through play, intimidated rivals and could be used to traverse up the dominance hierarchy and become the alpha male in the troop, the behavior was clearly adaptive and beneficial in the reproductive sense (Kummer & Goodall, 2003). When a novel behavior spreads throughout a population it certainly is not detrimental and may, like grain washing, be useful and spread for that reason (Itani & Nishimura, 1973).

On the other hand, we know from human experience that fads and fashions, created novelty, can spread rapidly but also peter out quickly as they are functionally useless. Consider the recent documentation of a chimpanzee (Julie) who, probably through play, began to put grass in her ear and so adorned herself repeatedly. This behavior spread through the colony, living in large enclosures in native habitat, and was acquired by others, undoubtedly by observational learning (van Leeuwen, Cronin, & Haun, 2014). I find it hard not to consider this behavior both innovative and a creative use of grass. It should not be argued that just because the behavior spread it was necessarily functional or adaptive. This is the kind of circular reasoning that plagues too many discussions, as noted above with play. Indeed, given some views of play as nonfunctional behavior, as soon as we substantiate a putative adaptive function the behavior should no longer be considered play! We should not engage in the reverse style of argument either. There are many examples of persons who developed products or had ideas that were ahead of their time and did not fit into their society and its current culture, economics, or technology. Were these still not creative concepts even if unsuccessful? It is also frequently observed that modern economies are based on *creating* (manufacturing) a cultural need for a product in order to sell products people did not even realize they needed. Here the creativity is in manipulating the cultural transmission process itself. This is probably not generally applicable to animals, although animals can be lured into liking non-typical foods, games, and addictions.

Bateson and Martin (2013) (also see Bateson, 2014) independently developed a treatment of play as an important aspect of creativity, and in terms more attuned to this author's perspective. However, they discuss the process with examples drawn largely from the human literature, although they do refer to some examples of innovation in animals. For example, many cetaceans produce bubbles used in play, but humpback whales use these bubbles in group foraging for herring, a truly remarkable behavior with many of the hallmarks of planning and group coordination (Wiley et al., 2011). These authors emphasize that a playful mood or attitude is crucial for creativity and point to several features of play, related to the five play criteria, such as operating in protected environments with intrinsic motivation and rewards. They posit two processes. Spontaneous play producing novel productions and generalizing from experiences such as those gained through playful manipulation of objects. This leads directly to a third recent and independent approach to play and creativity.

Josep Call explored the concept of creativity in animals specifically in regards to tool use (Call, 2013). He outlines three "ingredients" for becoming a creative tool user. He claims that these three are sufficient, perhaps even necessary, in devising innovative tool-using solutions. What are these three key ingredients? They fall under the headings of information hoarding, recombining or restructuring information, and a propensity to manipulate objects. Note this is not a sequential process. These ideas reflect Call's call for a renewed distinction between learning and reasoning, a feature, he notes, of early comparative psychology (Maier & Schneirla, 1935). So for Call, one way to become a tool user is via learning, contiguity, and reinforcement. For reasoning, animals need to combine separate experiences, including exploration. Call makes a distinction between *reproductive* thinking: "applying familiar procedures to solve problems that have been encountered before, or slight variations on those problems" (p. 7) and *productive* thinking, which is "inventing new procedures for solving a problem, either familiar or unfamiliar" (p. 7). While not limiting reasoning or productive thinking to apes, Call is focused on tool use. Note that this distinction is reminiscent of Bartlett (1951).

Call specifically mentions play and exploration as integral to both information hoarding and object manipulation as precursors to effective tool use solutions. This recalls the statement in Ben Beck's classic book on tool use (Beck, 1980) that learned tool use may derive from associations developed through play with objects in settings not related to problem solving. As Call is focused on tool use, he calls for systematic study of the relationship between the frequency and complexity of object manipulation with tool use across species, particularly birds and nonhuman primates.

Extending these ideas, I suggest that when an animal has the requisite prior experience with objects, along with good vision, he or she may be able to forgo trial and error physical manipulation of objects and mentally manipulate them to come up with the innovative creative solution. Some animals are able to solve, for example, mental rotation problems (Köhler, Hoffmann, Dehnhardt, & Mauck, 2005), although there are few data available on this possibility and the results are mixed.

Chimps choose rigid over flexible tools to solve a task, but visual inspection of the tools was not as effective as physical manipulation or observing the experimenter do so (Manrique, Gross, & Call, 2011). However, orangutans, chimpanzees, and bonobos, when confronted with a task of modifying a tool to use as a straw, showed a difference in that only the orangutans were able to select the proper tool on the basis of visual information alone without physical manipulation. Here "playing" with the object was not beneficial. But the authors point out that the superior ability of the orangutans may be related to their more extensive use of the mouth in foraging. Thus, perhaps they are in some way mentally playing with the object (Manrique & Call, 2011).

Many of the chapters in Reader and Laland (2003b) also mention the role of exploration, and neophilia as compared to neophobia, in attracting animals to objects or situations so that opportunities for novel behavior can result. This subject is treated rather systematically in relation to birds (Greenberg, 2003) as well as the related concept of behavioral flexibility (Lefebre, Whittle, Lascaris, & Finkelstein, 1997). Here we may have an entree into testing ideas on play and creativity. Interestingly, while correlates of innovation have been looked for in relation to ecology, neurology, and life history, the actual process of creativity or novel behavior production is rarely documented, primarily in a few examples of problem solving. For example, if one animal learns a new way of gaining food or some other novel behavior and is introduced to naïve animals, will the behavior spread throughout the population (Hauser, 2003). Such studies could model if and how novel behavior is disseminated, but it does not tell us anything about what went into the novelty generation. Perhaps, even in experimenter designed problems, play or play-like processes may be key.

The role of play may differ for different kinds of play, differ throughout ontogeny, and differ across species. Problem solving or novel behavior is sometime often claimed to be more frequent or effective in younger animals rather than older ones (Huffman, 1984). Could this be partially explained by younger animals being more prone to play, both socially, physically, and with objects (Burghardt, 2005)? But, just as all young animals are not the same in terms of playfulness, learning ability, and social learning neither are adults. Personality factors such as openness seem related to play. In an intriguing study, adult chimpanzees of both sexes were given two novel foraging puzzles to solve. While age and sex did not correlate with success, personality factors did show interesting relationships in males, but not females (Hopper et al., 2014). Specifically, scoring high on dominance, openness, and methodical correlated with high levels of manipulating the puzzles and success in solving them in males. Although problem solving often rewards being persistent and systematic in eliminating alternatives that do not work, no sex differences were found in this study, suggesting that alternative explanations are important.

EXAMPLES OF NOVEL BEHAVIOR IMPLICATING A ROLE FOR PLAY

Rather than trying to provide an exhaustive list of examples where play may be involved in the development of novel behavior and thus be creative in the broad sense, I will briefly discuss several diverse examples that suggest play is a factor in the production of novel, often adaptive, behavioral phenotypes. While the authors involved in these examples recognize the role of play, it is surprising that those studying innovation rarely acknowledge its importance. For example, in the pioneering book on animal innovation (Reader & Laland, 2003b), the only entry for play in the index refers to a photo of a Japanese monkey making a snowball, but this is not even discussed. To be fair, I should point out that in the chapter on neophilia in birds, Greenberg (2003) does mention play in discussing comparative aspects of curiosity and it may have been mentioned in passing elsewhere in this pioneering book.

Cetacean Play

Bottlenose dolphins and other cetaceans are well-known for their rich repertoire of object and social play, as reviewed in several reports (Kuczaj & Eskelinen, 2014; Paulos, Trone, & Kuczaj, 2010). Play has been implicated in the generation and subsequent transmission of novel behavior, including communication signals. Cetaceans are also known for manufacturing their own ephemeral toys, air bubbles or varying shapes they interact with and which are copied by others or used in social foraging as noted above. A particularly fine example is found in a Belugas (*Delphinapterus leucas*) in a captive setting, who form long, smooth, helix-like tubes which were interacted with by others and then transmitted throughout the social group (Jones & Kuczaj II, 2014). Furthermore, novel forms of communication that arise in bottlenose dolphins and other animals are convincingly described as arising from play (Kuczaj II & Makecha, 2008). This group of researchers hypothesize that the mechanism underlying the production of novelty through play is the production of "moderately discrepant events" as dolphins challenge themselves and avoid boredom by repeating the same games over and over again (Kuczaj II & Makecha, 2008; Kuczaj & Eskelinen, 2014). Bubble play has also recently been recorded in a young female California sea lion (Ibler & Kuhne, 2014).

Gestural Communication in Apes

Communication by gestures, postures, and movements has long been an area of study in animals, but recent decades have seen a great emphasis on gestural communication in apes, which is increasingly being viewed as a forerunner of language in humans. Much is available on the topic (Call & Tomasello, 2007; Hobaiter & Byrne, 2011, 2014).

It has been reported that through play in chimpanzees novel gestures are developed, formalized into signals, and disseminated in population, with different captive groups having different repertoires. Indeed, a literature has developed on what has been termed "ontogenetic ritualization." Studies on nonhuman primates document the development of novel communicative signals, both gestures and sounds, which are then incorporated into a population's behavioral repertoire. Too large to review here, some of the key papers are those on gestural communication in chimpanzees (Liebel & Call, 2012; Tomasello et al., 1997) and on the development of an attention getting sound in chimpanzees (Taglialatela et al., 2012). Tomasello et al. (1997) mention play as a key source for the novel gestural signals, a point also made in the introduction to a volume devoted to gestural communication (Tomasello & Call, 2007).

Such studies are not without critics. It has been pointed out that these studies on captive animals did not consider the natural gestural repertoire of wild chimpanzees. Furthermore, the hypothesis of ontogenetic ritualization of novel gestures is claimed to be not necessary, although theoretically an attractive idea (Hobaiter & Byrne, 2011). Yet these same authors, in a comprehensive analysis of the meaning of gestures in chimpanzees, find that many are used in play, some exclusively so, and thus had to be eliminated from an analysis of the meaning (function, purpose) of such gestures (Hobaiter & Byrne, 2014). This does not, then, rule out the possibility that in the past, if not ongoing, play was the source for the gestures that then later became incorporated into signals with specific meaning and intent, thus no longer appearing as play, or not exclusively so.

One way play was considered a viable source for novel signals was that it was not "evolutionarily urgent" and thus encompassed the more relaxed setting in which play typically occurs (Tomasello & Call, 2007). The opening word in a recent critique (Hobaiter & Byrne, 2012) of this idea is "play." This critique points out that sex and consortship are evolutionary urgent and critical aspects of a chimpanzee's life, more so than the grooming and more gentle social interactions mentioned by Tomasello and Call (2007). The data presented by Hobaiter and Byrne certainly suggest that males are particularly intent on successful outcomes. But, we also know that adult play, including sexual play, is an important part of the lives of many primates, including apes (Pellis & Iwaniuk, 1999, 2000). Thus, I do not think that play can be ruled out here either.

Play is, thus, a controversial, but important topic in the origins of intentional communication and language, as also noted by (Kuczaj II & Makecha, 2008), who also discuss the role of babbling in vocal behavior in many species (see also Burghardt, 2005; Pepperberg, Brese, & Harris, 1991). The role of play in communicative novelty is thus still open.

Indeed, do not poets, humorists, and others play with words, yet no one demands that the words be completely novel or made up. Context is everything.

Orangutan Tool Use and Water Innovation

Although extensively discussed in this book by Russon, Kuncoro, and Ferisha (see Chapter 15), mention needs to be made of the remarkable findings based on long-term field studies of orangutans (*Pongo pygmaeus*) in Indonesia. Russon et al. make the useful distinction between an innovation, not based on social learning, and improvisations, which are "on the spot, spontaneous solutions to a task facing the actor" (p. 419); these may be locomotor decisions adjusted to a particular arboreal locomotor problem faced in negotiating highly variable terrain. These may never be used again, and reflect, perhaps, what in other contexts may result from play, though not with an end "in mind." Their chapter goes into these issues in depth with much detailed observations, but here an earlier paper that directly relates to play is highlighted (Russon, Kuncoro, Ferisa, & Handayani, 2010).

Orangutans typically do not engage with water as it can harbor predators, little typical food, and drowning is a risk as orangutans cannot swim. Innovations in water use were studied in a population of 43 ex-captive orangutans being rehabilitated on an over 100 ha island in native habitat, though supplemental food was provided. Animals were observed with focal sample techniques and water-related behavior recorded. Behaviors seen were divided into variants, which were minor differences without an obvious function, and innovations, which were rare. By recording individual behavior they were able to identify the presumed innovator and his/her age, status, and conspecific associates. Dozens of variants and innovations were found such as retrieving floating items, using a rake to obtain them, groping around with hands underwater, scooping water with a shell, absorbing water sponge-like, traveling on a floating log to cross water, splashing water as a display, and many more. They argue that most innovations are not gestalt like creative leaps but incremental steps.

They consider four contributing events to innovation: 1) applying pre-existing knowledge to a new function, 2) accidental co-occurrence of behaviors that typically are segregated, 3) generating novel behavioral components independently (where they view play as being involved), and 4) social cross-fertilization, where more and different variants are produced than by single individuals (brainstorming?). The authors, from their observations over time, were able to reconstruct possible pathways to innovation. Animals using precursors were more likely to innovate, supporting the first event. They also found that there was transfer of a pre-existing task to a new function, and some evidence that play was a source of at least one independent innovation. More studies such as these are needed.

Stone Play in Japanese Monkeys

Perhaps the best documented novel behavior emerging through play and then disseminated through a population, based on long-term studies spanning over 30 years, is stone play in Japanese macaques (*Macaca fuscata*) as described in the many publications by Michael Huffman and colleagues (Huffman, 1984; Huffman, Leca, & Nahallage, 2010; Leca, Gunst, & Huffman, 2012). This behavior was first observed in 1979 in Arashiyama B troop as a nonadaptive behavior begun by juveniles and spread throughout half the large troop in 5 years. Eight behaviors were recorded with the stone toys including gathering, carrying, scattering, clacking them together, rolling, rubbing, and cuddling them. Many subsequent studies have uncovered other features of this creative use of a common object. For example after some years a "faddish" shift occurred whereby monkeys moved the stones to foraging sites and mingled the use of stones with foraging behavior. The stones then were sometimes used as tools (Huffman & Quiatt, 1986). By investigating the use of stones in other troops, which indeed differed substantially even though stones were available to all, the role of play as a formative factor in culture was proposed (Huffman, 1996; Leca, Gunst, & Huffman, 2007). Indeed, based on their studies, six species level predispositions for innovations were proposed (Huffman & Hirata, 2003), but their provocative suggestions have yet to be systematically tested.

The data collected and many published papers by Huffman and colleagues over decades up to the present, many in leading journals in comparative psychology, animal behavior, and primatology, and recently reviewed (Leca et al., 2012), provide a rich trove of findings that support how behavior originating in play can become not only disseminated, ritualized, and even functional (Nahallage & Huffman, 2008), but also details the processes underlying the role of play in creativity perhaps better than any other. The absence of reference to these studies from recent treatments (Bateson, 2014; Bateson & Martin, 2013; Call, 2013; Kaufman et al., 2011; Kaufman & Kaufman, 2004) is unfortunate.

A SCENARIO FOR CREATIVITY EVOLUTION

Fagen (1981), Burghardt (2005) and others have noted that in play animals often vary and recombine behavioral elements and signals. In this way, play can generate new behavior or combinations that can then be selected. This can aid the individual in, for example, winning a play fight, negotiating rough terrain, escaping from a predator, or exploiting a novel food source. Similarly, in the operant literature, spontaneous and variable behavior, under the rubric of volition, can occur (Neuringer, 2014). This is exemplified in the now classical studies by Karen Pryor on generating and reinforcing novel responses in dolphins (Pryor & Chase, 2014; Pryor, Haag, & O'Reilly, 1969). In so far as the behavior or plasticity has some heritable component, such behavior can become established through processes such as genetic assimilation or the Baldwin effect (Burghardt, 2005). But, even if no heritable aspect is present, such novel behavior can be spread through social learning and cultural processes. These can also occur in tandem (Henrich & McElreath, 2007).

If play is a factor in innovation and creativity, then the amount and complexity of play in a species should be related to such behavior. Obviously, such a link does not prove a causal or formative process, but does suggest looking more closely at the behavior of such species. However, since play consists of heterogeneous phenomena, the type of play may be important. Thus, extractive foragers and scavengers that engage in considerable object play may also be species where novel and eventually useful skills emerge. This might also apply to escape tactics in prey species and social deception in highly social groups. Novel loco-motor innovations may occur in those species, such as arboreal ones, that have to navigate highly variable and risky environments (Spinka, Newberry, & Bekoff, 2001). That is, mini-c creativity may be involved here. Surplus Resource Theory posits that play is more likely to occur when animals have a rich behavioral repertoire that provides more opportunity for novel variants to arise.

Although all organisms are continually subjected to the processes of evolution and selection, it is also true that the pace of evolutionary change, in terms of speciation and higher level processes, can differ remarkably among lineages. Many animals, such as salamanders and horseshoe crabs have not changed in basic body plan for millions of years, whereas other groups have. Mammals and birds evolved dramat-ically and quite quickly over the 65,000,000 years since the Cretaceous extinctions, whereas other phyla and classes evolved more slowly. Apes and humans experienced much change over a few million years and *Homo sapiens* had much of their biological evolution and speciation in less than a million years, with major cultural evolution over less than 10,000 years, Many animals, however, have not changed much in hun-dreds of millions of years (e.g., cockroaches, millipedes) while endo-thermic vertebrates, birds and mammals, have evolved more rapidly and in extreme forms, generating many qualitative behavioral and cognitive innovations, as compared to less extreme adaptations of much older successful Baupläne.

Does play have any role in this? Endothermic animals generally have high metabolic rates and can engage in sustained and vigorous behavior for long periods of time. If they have a complex behavioral repertoire and appendages and faces capable of highly variable behavior, then they are more likely to produce variants that are useful. Furthermore, advanced cognitive and social learning abilities enhance the chances that innovations that do occur can spread. A genius solitary animal will most likely be a dead end. Note that in this view, characterizing creativity as only the production of useful innovations misses a critical aspect of Darwin's radical message. It is only in the aftermath, measured perhaps in generations, that we can assess the importance of a novel production, but in many ways this is a separate issue from the creative process itself, just as mutation, recombination, and other genetic processes can be viewed as critical sources for evolutionary change.

Play, then, can be seen as a way for animals, with the appropriate underlying motivational, physiological, and behavioral attributes, to facilitate the production of variants upon which evolutionary processes, biological and cultural, can act. In this way, viewing play as a means of generating novelties that MAY prove adaptive can help us understand the pace of evolution and the remarkable changes in brain size and cognitive attainments in endothermic animals, in spite of the enormous energetic costs of maintaining high levels of aerobic metabolism and large brains. Endothermy was a catalyst for cognitive evolution and play and play-like processes may have been critical (Burghardt, 1988). It should be possible to test these ideas by comparing the rate of production, and type, of behavioral "mutants" in animals with different metabolic rates, behavioral repertoires, foraging and social tactics and strategies, individual and social learning capacities, and so forth, in a phylogenetic context.

WHAT HAPPENS AFTER THE CREATIVE LEAP?

A creative innovation, however viewed, that has an impact on a population may spread. Then what? If the innovative change becomes fixed in the population it can become so via an innate or instinctive mechanism or through a learned or cultural change via imitation, social learning, parental shaping or guidance, etc. But changes generated via play may become not only widespread but "fixed," to the extent that they are viewed as species typical behavior that no longer fit the criteria for play. This can happen through population level "traditions," with or without a genetic component. They may also evolve into behavioral displays and rituals that having been successful may become essential in social and other contexts. Cultural rituals that we may not view as adaptive or functional may also arise from play such as artistic, musical, dance, and theatrical productions, all of which have precursors in other species. That is pretense, socio-dramatic play, and other forms of play may, like stone clacking, spread and be incorporated in other behavioral systems (Hogan, 1988). Recently it has been experimentally documented that different novel foraging techniques can be introduced into sub populations of wild birds (great tits, *Parus major*) by trained demonstrators and differentially spread throughout each subpopulation socially, become stable, and be maintained over generations (Aplin et al., 2014). Thus, a single bird who playfully innovates a new technique can be an effective seed in animal as well as in human societies!

Indeed, rituals, including those in religion (Huizinga, 1955) have long been linked with play, though with inadequate support that current work might begin to rectify. In fact, play criteria overlap those of ritual in surprising ways. (Burghardt, in press), and this may belie a connection to their playful origins. Certainly such behavior may not be what most of us have in mind when we think of creativity and innovations. But we must overcome the idea that creativity is something magical, mysterious, and beyond being rooted, ultimately, in the behavioral processes of evolution that, just as with play, need a bottom up approach.

Acknowledgments

I thank the editor and Peter Smith for useful comments on early drafts of this chapter and the working group on Play, Evolution, and Sociality at the National Institute of Mathematical and Biological Synthesis at the University of Tennessee supported by the National Science Foundation through NSF Award #DBI-1300476 for creative conversations.

References

Aplin, L. M., Farine, D. R., Morand-Ferron, J., Cockburn, A., Thornton, A., & Sheldon, B. C. (2015). Experimentally induced innovations lead to persistent culture via conformity in wild birds. *Nature, 518*, 538—541.
Bartlett, F. (1951). *The mind at work and play.* London: George Allen and Unwin, Ltd..
Bateson, P. (2014). Play, playfulness, creativity, and innovation. *Animal Behavior and Cognition, 1*, 99—112. Available from: http://dx.doi.org/10.12966/abc.05.02.2014.
Bateson, P., & Martin, P. (2013). *Play, playfulness, creativity, and innovation.* Cambridge, UK: Cambridge University Press.
Beck, B. B. (1980). *Animal tool behavior.* New York, NY: Garland.
Bekoff, M., & Byers, J. A. (1981). A critical reanalysis of the ontogeny and phylogeny of mammalian social and locomotor play: An ethological hornet's nest. In K. Immelmann, G. W. Barlow, L. Petrinovich, & M. Main (Eds.), *Behavioral development: The Bielefeld interdisciplinary project* (pp. 296—337). Cambridge: Cambridge University Press.
Burghardt, G. M. (1984). On the origins of play. In P. K. Smith (Ed.), *Play in animals and humans* (pp. 5—41). Oxford: Basil Blackwell.

Burghardt, G. M. (1988). Precocity, play, and the ectotherm-endotherm transition: Superficial adaptation or profound reorganization? In E. M. Blass (Ed.), *Handbook of behavioral neurobiology* (Vol. 9, pp. 107–148). New York, NY: Plenum.

Burghardt, G. M. (1999). Conceptions of play and the evolution of animal minds. *Evolution and Cognition, 5*, 115–123.

Burghardt, G. M. (2001). Play: Attributes and neural substrates. In E. M. Blass (Ed.), Handbook of behavioral neurobiology *(Vol. 13, Developmental psychobiology* (pp. 327–366). New York, NY: Plenum.

Burghardt, G. M. (2005). *The genesis of animal play: Testing the limits.* Cambridge, MA: MIT Press.

Burghardt, G. M. (2009). Darwin's legacy to comparative psychology and ethology. *American Psychologist, 64*, 102–110.

Burghardt, G. M. (2011). Defining and recognizing play. In A. D. Pellegrini (Ed.), *The Oxford handbook of the development of play* (pp. 9–18). New York, NY: Oxford University Press.

Burghardt, G. M. (2014). A brief glimpse at the long evolutionary history of play. *Animal Behavior and Cognition, 1*, 90–98. Available from: http://dx.doi.org/10.12966/abc.05.01.2014.

Burghardt, G.M. (in press). The origins, evolution, and interconnections of play and ritual: Setting the stage. In C. Renfrew, I. Morley, & M. Boyd (eds.), *From play to faith: Ritual and play in animals, and early human societies.* Cambridge, UK: Cambridge University Press.

Byers, J. A., & Walker, C. (1995). Refining the motor training hypothesis for the evolution of play. *American Naturalist, 146*, 25–40.

Call, J. (2013). Three ingredients for becoming a creative tool user. In C. Sanz, J. Call, & C. Boesch (Eds.), *Tool use in animals: Cognition and ecology* (pp. 3–20). Cambridge, UK: Cambridge University Press.

Call, J., & Tomasello, M. (Eds.), (2007). *The gestural communication of apes and monkeys* Mahwah, NJ: Lawrence Erlbaum Associates.

Csikszentmihalyi, M. (1996). *Creativity: Flow and the psychology of discovery and invention.* New York, NY: Harper Collins.

Fagen, R. (1981). *Animal play behavior.* New York, NY: Oxford University Press.

Foster, S. A. (1995). Constraint, adaptation, and opportunism in the design of behavioral phenotypes. In N. S. Thompson (Ed.), *Perspectives in ethology. Vol. 11: Behavioral design* (pp. 61–81). NY: Plenum Press.

Graham, K. L., & Burghardt, G. M. (2010). Current perspecitves on the biological study of play: Signs of progress. *Quarterly Review of Biology, 85*, 393–418.

Greenberg, R. (2003). The role of neophobia and neophilia in the development of innovative behaviour of birds. In R. M. Reader, & K. N. Laland (Eds.), *Animal innovation* (pp. 176–196). Oxford, UK: Oxford University Press.

Greig, E. I., & Webster, M. S. (2014). How do novel signals originate? The evolution of fairy-wren songs from predator to display contexts. *Animal Behaviour, 88*, 57–65.

Harlow, H. F. (1949). The formation of learning sets. *Psychological Review, 56*, 51–65.

Hauser, M. D. (2003). To innovate or not to innovate? That is the question. In R. M. Reader, & K. N. Laland (Eds.), *Animal innovation* (pp. 329–338). Oxford, UK: Oxford University press.

Henrich, J., & McElreath, R. (2007). Dual-inheritance theory: The evolution of human cultural capacities ard cultural evoluition. In R. I. M. Dunbar, & L. Barrett (Eds.), *The Oxford handbook of evolutionary psychology* (pp. 555–570). Oxford, UK: Oxford University Press.

Hobaiter, C., & Byrne, R. W. (2011). The gestural repertoire of the wild chimpanzee. *Animal Cognition, 14*, 745–767.

Hobaiter, C., & Byrne, R. W. (2012). Gesture use in consortship. Wild chimpanzees' use of gesture for an "evolutionarily urgent" purpose. In S. Pika, & K. Liebel (Eds.), *Developments in primate gesture research* (pp. 129–146). Amsterdam: John Benjamins Publishing Co.

Hobaiter, C., & Byrne, R. W. (2014). The meaning of chimpanzee gestures. *Current Biology*, *24*, 1–5.

Hogan, J. A. (1988). Cause and function in the development of behavior systems. In E. M. Blass (Ed.), *Handbook of behavioral neurobiology* (Vol. 9, pp. 63–106). New York, NY: Plenum.

Hopper, L. M., Price, S. A., Freeman, H. D., Lambeth, S. P., Schapiro, S. J., & Kendal, R. L. (2014). Influence of personality, age, sex, and estrous state on chimpanzee problem-solving success. *Animal Cognition*, *17*, 835–847.

Huffman, M. A. (1984). Stone-play of *Macaca fuscata* in Arashiyama B troop: Transmission of a non-adaptive behaviour. *Journal of Human Evolution*, *13*, 725–735.

Huffman, M. A. (1996). Acquisition of innovative cultural behaviors in non-human primates: A case study of stone handling, a socially transmitted behavior in Japanese macaques. In B. J. Galef, & C. Hayes (Eds.), *Social learning in animals: The roots of culture* (pp. 267–289). Orlando, FL: Academic Press.

Huffman, M. A., & Hirata, S. (2003). Biological and ecological foundations of primate behavioral tradition. In D. M. Fragaszy, & S. Perry (Eds.), *The biology of tradition: Models and evidence* (pp. 267–296). Cambridge, UK: Cambridge University Press.

Huffman, M. A., & Quiatt, D. (1986). Stone handling by Japanese Macaques (*Macaca fuscata*): Implications for tool use of stone. *Primates*, *27*, 413–423.

Huffman, M. H., Leca, J.-B., & Nahallage, C. A. D. (2010). Cultured Japanese maqaques: A multidisciplinary approach to stone handling behavior and its implications for the evolution of behavioral tradition in non-human primates. In N. Nakagawa, et al. (Eds.), *The Japanese macaques* (pp. 191–219). New York, NY: Springer.

Huizinga, J. (1955). *Homo ludens: A study of the play element in culture* (R. F. C. Hull, Trans.). Boston, MA: Beacon.

Ibler, B., & Kuhne, R. (2014). Spiel eine Kalifornischen Seelowen, *Zalophus californianus* (Lesson, 1828), mit selbst produzierten Luftblasenringen im Zoologischen Garten Berlin. *Der Zoologische Garten*, *83*(1–3), 28–32.

Itani, J., & Nishimura, A. (1973). The study of infrahuman culture in Japan: A review. In E. W. Menze (Ed.), *Precultural primate behavior* (pp. 26–50). Basel: Karger.

Jones, B. L., & Kuczaj, S. A., II (2014). Beluga (*Delphinapterus leucas*) novel bubble helix play behavior. *Animal Behavior and Cognition*, *1*, 206–214. Available from: http://dx.doi.org/10.12966/abc.05.10.2014.

Kaufman, A. B., Butt, A. E., Kaufman, J. C., & Colbert-White, E. N. (2011). Towards a neurobiology of creativity in nonhuman animals. *Journal of Comparative Psychology*, *125*, 255–272. Available from: http://dx.doi.org/10.1037/a0023147.

Kaufman, J. C. (2009). *Creativity 101*. New York, NY: Springer Publishing Co.

Kaufman, J. C., & Beghetto, R. A. (2009). Beyond big and little: The four C model of creativity. *Review of General Psychology*, *13*, 1–12.

Kaufman, J. C., & Kaufman, A. B. (2004). Applying a creativity framework to animal cognition. *New Ideas in Psychology*, *22*, 143–155.

Köhler, C., Hoffmann, K. P., Dehnhardt, G., & Mauck, B. (2005). Mental rotation and rotational invariance in the Rhesus monkey (*Macaca mulatta*). *Brain, Behavior, and Evolution*, *66*(3), 158–166.

Kuczaj, S. A., & Eskelinen, H. C. (2014). Why do dolphins play. *Animal Behavior and Cognition*, *1*, 113–127.

Kuczaj, S. A., II, & Makecha, R. (2008). The role of play in the evoluiton and ontogeny of contextully flexible communication. In D. K. Oller, & U. Griebel (Eds.), *Evolution of communicative flexibility: Complexity, creativity, and adaptability in human and animal communication* (pp. 253–277). Cambridge, MA: MIT Press.

Kuhn, T. S. (1996). *The structure of scientific revolutions* (3rd ed.). Chicago, IL: University of Chicago Press.

Kummer, H., & Goodall, J. (2003). Conditions of innovative behaviour in primates. In R. M. Reader, & K. N. Laland (Eds.), *Animal innovation* (pp. 223–235). Oxford, UK: Oxford University Press.

Leca, J. B., Gunst, N., & Huffman, M. A. (2007). Japanese macaque cultures: Inter- and intra-troop behavioural variability of stone handling patterns across 10 troops. *Behaviour, 144*, 251–281.

Leca, J.-B., Gunst, N., & Huffman, M. A. (2012). Thirty years of stone handling tradition in Arashiyama-Kyoto macaques: Implications for cumulative culture and tool use in non-human primates. In J.-B. Leca, M. A. Huffman, & P. L. Vasey (Eds.), *The monkeys of stormy mountain: 60 years of primatological research on the Japanese macaques of arashiyama* (pp. 223–257). Cambridge, UK: Cambridge University Press.

Lefebre, L., Whittle, P., Lascaris, E., & Finkelstein, A. (1997). Feeding innovations and forebrain size in birds. *Animal Behaviour, 53*, 549–560.

Liebel, K., & Call, J. (2012). The origins of non-human primates' manual gestures. *Philosophical Transactions of the Royal Society B, 367*, 118–128.

Lillard, A. S., Lerner, M. D., Hopkins, E. J., Dore, R. A., Smith, E. D., & Palmquist, C. M. (2013). The impact of pretend play on children's development: A review of the evidence. *Psychological Bulletin, 139*, 1–34.

Lindholm, M. E., Marabita, F., Gomez-Cabrero, D., Rundqvist, H., Ekström, T. J., & Sundberg, C. J. (2014). An integrative analysis reveals coordinated reprogramming of the epigenome and the transcriptome in human skeletal muscle after training. *Epigenetics.* Available from: http://dx.doi.org/10.4161/15592294.2014.982445.

MacLennan, B. J. (2009) "Computer-Enhanced Scientific Creativity," University of Tennessee Dept. of Electrical Engineering and Computer Science Technical Report UT-CS-09-639.

Maier, N. R. F., & Schneirla, T. C. (1935). *Principles of animal psychology.* New York, NY: McGraw-Hill.

Manrique, H. M., & Call, J. (2011). Spontaneous use of tools as straws in great apes. *Animal Cognition, 14*, 213–226.

Manrique, H. M., Gross, A. M.-M., & Call, J. (2011). Great apes select tools on the basis of their rigidity. *Journal of Experimental Psychology: Animal Behavior Processes, 36*, 409–422.

Martin, P., & Caro, T. M. (1985). On the function of play and its role in behavioral development. *Advances in the Study of Behavior, 15*, 59–103.

Nahallage, C. A. D., & Huffman, M. A. (2008). Environmental and social factors associated with the occurrence of stone-handling behavior in a captive troop of *Macaca fuscata*. *International Journal of Primatology, 29*, 795–806.

Neuringer, A. (2014). Operant variability and the evolution of volition. *International Journal of Comparative Psychology, 27*, 62–81.

Parker, S. T., & McKinney, M. L. (1999). *Origins of intelligence: The evolution of cognitive development in monkeys, apes, and humans.* Baltimore, MA: Johns Hopkins University Press.

Paulos, R. D., Trone, M., & Kuczaj, S. A. (2010). Play in wild and captive cetceans. *International Journal of Comparative Psychology, 23*, 701–722.

Payne, R. S., & McVay, S. (1971). Songs of humpback whales. *Science, 173*, 585–597.

Pellegrini, A. D. (2009). *The role of play in human development.* Oxford, UK: Oxford University Press.

Pellis, S. M., & Iwaniuk, A. N. (1999). The problem of adult play fighting: A comparative analysis of play and courtship in primates. *Ethology, 105*, 783–806.

II. REQUIREMENTS FOR CREATIVITY

Pellis, S. M., & Iwaniuk, A. N. (2000). Adult–adult play in primates: Comparative analyses of its origin, distribution, and evolution. *Ethology, 106*, 1083–1104.

Pellis, S. M., & Pellis, V. C. (2009). *The playful brain, venturing to the limits of neuroscience.* Oxford, UK: Oneworld Press.

Pellis, S. M., Pellis, V. C., Barrett, L., & Henzi, S. P. (2014). One good turn deserves another: Combat versus other functions of acrobatic maneuvers in the play fighting of vervet monkeys (*Chlorocebus aethiops*). *Animal Behavior and Cognition, 1*, 128–143.

Pepperberg, I. M., Brese, K. J., & Harris, B. J. (1991). Solitary sound play during acquisition of English vocalizations by an African Grey parrot (*Psittacus erithacus*): Possible parallels with children's monologue speech. *Applied Psycholinguistics, 12*, 151–178.

Pietrewicz, A. T., & Kamil, A. C. (1979). Search image formation in the blue jay (*Cyanocitta cristata*). *Science, 204*, 1332–1333.

Preyer, W.. (1893). *Mental development in the child* (H. W. Brown, Trans.). New York, NY: D. Appleton.

Pruitt, J. N., Burghardt, G. M., & Riechert, S. E. (2012). Non-conceptive sexual behavior in spiders: A form of play associated with body condition, personality type, and male intrasexual selection. *Ethology, 118*, 33–40.

Pryor, K., & Chase, S. (2014). Training for variable and innovative behavior. *International Journal of Comparative Psychology, 27*, 218–225.

Pryor, K. W., Haag, R., & O'Reilly, J. (1969). The creative porpoise: Training for novel behavior. *Journal of the Experimental Analysis of Behavior, 12*, 653–661.

Rawlings, B., Davila-Ross, M., & Boysen, S. T. (2014). Semi-wild chimpanzees open hard-shelled fruits differently across communities. *Animal Cognition, 17*, 891–899.

Reader, R. M., & Laland, K. N. (2003a). Animal innovation: An introduction. In R. M. Reader, & K. N. Laland (Eds.), *Animal innovation* (pp. 3–35). Oxford, UK: Oxford University Press.

Reader, R. M., & Laland, K. N. (Eds.), (2003b). *Animal innovation* Oxford, UK: Oxford University Press.

Ribot, T. (1906). *Essay on the creative imagination.* Chicago, IL: Open Court Publishing Co..

Ritter, S. M., & Dijksterhuis, A. (2014). Creativity—The unconscious foundations of the incubation period. *Frontiers in Human Neuroscience, 8*(215), 1–10.

Russon, A. E., Kuncoro, P., Ferisa, A., & Handayani, D. P. (2010). How orangutans (*Pongo pygmaeus*) innovate for water. *Journal of Comparative Psychology, 134*, 14–28.

Simonton, D. K. (2003). Human creativity: Two Darwinian analyses. In R. M. Reader, & K. N. Laland (Eds.), *Animal innovation* (pp. 309–325). Oxford, UK: Oxford University Press.

Smith, P. K. (2010). *Children and play.* Oxford, UK: Wiley-Blackwell.

Spinka, M., Newberry, R. C., & Bekoff, M. (2001). Mammalian play: Training for the unexpected. *Quarterly Review of Biology, 76*, 141–168.

Taglialatela, J. P., Reamer, L., Schapiro, S. J., & Hopkins, W. D. (2012). Social learning of a communicative signal in captive chimpanzees. *Biology Letters, 8*, 498–501.

Tan, W.-H., Tsai, C.-G., Lin, C., & Lin, Y. K. (2014). Urban canyon effect: Storm drains enhance call characteristics of the Mientian tree frog. *Journal of Zoology, 294*, 77–84. Available from: http://dx.doi.org/10.1111/jzo.12154.

Taylor, C. W. (1964a). Introduction. In C. W. Taylor (Ed.), *Creativity: Progress and potential* (pp. 1–14). New York, NY: McGraw-Hill.

Taylor, C. W. (Ed.), (1964b). *Creativity: Progress and potential* New York, NY: McGraw-Hill.

Tomasello, M., & Call, J. (2007). Introduction: Intentional communication in nonhuman primates. In J. Call, & M. Tomasello (Eds.), *The gestural communication of apes and monkeys* (pp. 1–15). Mahwah, NJ: Lawrence Erlbaum Associates.

Tomasello, M., Call, J., Warren, J., Frost, G. Y., Carpenter, M., & Nagell, K. (1997). The ontogeny of chimpanzee gestural signals: A comparison across groups and generations. *Evolution of Communication, 1*, 223–253.

Torrence, E. P. (1964). Education and creativity. In C. W. Taylor (Ed.), *Creativity: Progress and potential* (pp. 49–128). New York, NY: McGraw-HIll.

van den Berg, C. L., Hol, T., van Ree, J. M., Spruijt, B. M., Everts, H., & Koolhaas, J. M. (1999). Play is indispensable for an adequate development of coping with social challenges in the rat. *Developmental Psychobiology, 34*, 129–138.

van Leeuwen, E. J. C., Cronin, K. A., & Haun, D. B. M. (2014). A group-specific arbitrary tradition in chimpanzees (*Pan troglodytes*). *Animal Cognition, 17*, 1421–1425.

von Uexküll, J., & Kriszat, G. (1934). *Streifzüge durch die Umwelten von Tieren und Menschen.* Berlin: Julius Springer.

Verderane, M. P., Izar, P., Visalberghi, E., & Fragaszy, D. M. (2013). Socioecology of wild bearded capuchin monkeys (*Sapajus libidinosus*): An analysis of social relationships among female primates that use tools in feeding. *Behaviour, 150*, 659–689.

Wagner, G. P. (2014). *Homology, genes, and evolutionary innovation.* Princeton, NJ: Princeton University Press.

Wiley, D., Ware, C., Bocconocelli, A., Cholewiak, D., Friedlaender, A., Thompson, M., et al. (2011). Underwater components of humpback whale bubble-net feeding behaviour. *Behaviour, 148*, 575–602.

Zuk, M., Bastiaans, E., Langkilde, T., & Swanger, E. (2014). The role of behaviour in the establishment of novel traits. *Animal Behaviour, 92*, 333–344.

Commentary on Chapter 5: Play—A Multipurpose Vehicle

Burghardt's key thesis in this chapter, that play is an important activity in vertebrate evolution, is consistent with theories and research about the role of pretend play in child development. Pretend play in particular is important in creativity in children. There are a number of different processes that occur in pretend play, which we can observe and measure. A number of these processes are important in creativity. Divergent thinking (the ability to generate a variety of ideas) is an important component of pretend play and of creativity. Symbolism and metaphor occur in both pretend play and creativity. Affect is involved in the pretend play narrative (fear of monsters; love for a baby doll) and affect is crucial in many forms of artistic creativity. The expression of joy and deep engagement occurs in both play and creative activities. There is a substantial body of research that supports the association of these processes in play with creativity in children, even when possible bias of researchers is controlled for (see Russ, 2014, for a review).

Play is unique in that it is a vehicle through which many processes can be expressed and practiced. Different processes can be important for different species and for different children at different points in their development. In that sense, play can function for many purposes depending upon the circumstances and developmental level of the individual. It can be simple or complex. Play is a multipurpose vehicle. And the fact that play is fun and joyful provides built-in reinforcement that encourages one to return to this engaging activity. This is certainly true in children and in many higher order species. For example, a recent study by Ross, Bard, and Matsuzawa (2014) concluded that in chimpanzees, playful expressions in social and solitary play during the second year of life are emotional expressions of joy.

Many animal play theorists have proposed that play is rehearsal for adult activities (Burghardt, 2005). What are the necessary activities for humans that pretend play can prepare children for? I (Russ, 2014, p. 33) proposed two broad categories of necessary activities for humans:

1. *Problem solving*. Pretend play can be rehearsal for a variety of problems that children face daily. Practice with divergent thinking can be useful for creative problem solving and for coping with problems. Practice with manipulating ideas and images should develop flexibility in thinking.
2. *Processing of emotion*. The use of pretend play to express, process, and integrate emotions should help the child feel comfortable with a variety of emotions. In the safe arena of pretend play, children can learn to modulate emotions, manipulate affect-laden cognition, and manage negative emotions such as fear and anger and memories of stressful or traumatic events. Child psychotherapists have used these mechanisms in play to help children overcome fears and resolve conflicts.

An exciting question for future research is "what are the neurological processes that account for the relationship between play and creativity?". There are probably a number of underlying mechanisms. A broad associative network has long been thought to be one mechanism important in helping individuals to generate original ideas (Mednick, 1962). For children, pretend play could help a broad associative network develop. Affective experiences, mental representations, and ideas are expressed, manipulated, and encoded (Fein, 1989; Russ, 1993). This associative process is consistent with Burghardt's (2005) notion that animal play involves recombining behavioral elements and signals. He also expands on the concept and posits that some animals could mentally manipulate objects to come up with creative solutions.

One of the current debates in the literature is whether or not engagement in play actually does facilitate creativity, emotion regulation, and other aspects of child development (Lillard et al., 2013; Russ & Wallace, 2013). Research in this area is difficult and labor intensive. I have concluded, as has Dansky (1999), that there are rigorous studies that demonstrate that pretend play intervention can facilitate creativity. It is essential that research in this area continue. Placing pretend play within an evolutionary framework is useful in understanding the functions of play in children.

References

Burghardt, G. (2005). *The genesis of animal play: Testing the limits.* Cambridge, MA: MIT Press.

Dansky, J. (1999). Play. In M. Runco, & S. Pritzker (Eds.), *Encyclopedia of creativity* (pp. 393–408). San Diego, CA: Academic Press.

Fein, G. (1987). Pretend play: Creativity and consciousness. In P. Gorlitz, & J. Wohlwill (Eds.), *Curiosity, imagination and play* (pp. 281–304). Hillsdale, NJ: Lawrence Erlbaum Associates.

Lillard, A., Lerner, M., Hopkins, E., Dore, R., Smith, E., & Palmquist, C. (2013). The impact of pretend play on children's development: A review of empirical evidence. *Psychological Bulletin, 139*, 1–34.

Mednick, S. (1962). The associative bases of the creative process. *Psychology Review, 69*, 220–232.

Ross, K., Bard, K., & Matsuzawa, T. (2014). Playful expressions of one-year-old chimpanzee infants in social and solitary play contexts. *Frontiers in Psychology, Cognitive Science*, 1–22.

Russ, S. (1993). *Affect and creativity: The role of affect and play in the creative process.* Hillsdale, NJ: Lawrence Erlbaum Associates.

Russ, S. (2014). *Pretend play in childhood: Foundation of adult creativity.* Washington, DC: American Psychological Association Books.

Russ, S., & Wallace, C. (2013). Pretend play and creative processes. *American Journal of Play*, 136–148.

The Evolution of Innovativeness: Exaptation or Specialized Adaptation?

Daniel Sol

CREAF (Centre for Ecological Research and Applied Forestries), CSIC (Spanish National Research Council), Bellaterra, Catalonia, Spain

Commentary on Chapter 6: Can Sol's Explanation for the Evolution of Animal Innovation Account for Human Innovation?

Liane Gabora and Apara Ranjan

Department of Psychology, University of British Columbia, Vancouver, BC, Canada

Innovation is the process that generates novel learned behaviors in the population (Lefebvre, Whittle, Lascaris, & Finkelstein, 1997; Reader & Laland, 2003; Ramsey, Bastian, & van Schaik, 2007). In the last decades, innovations have attracted much attention because of their presumed relationship with intelligence and culture (Lefebvre et al., 1997; Ramsey et al., 2007; Reader, Hager, & Laland, 2011). Evidence has accumulated showing that animals vary considerably in innovation propensity and that this in part reflects differences in their neural structures (Lefebvre et al., 1997; Webster & Lefebvre, 2001; Reader & Laland, 2002; Overington, Morand-Ferron, Boogert & Lefebvre, 2009; Reader et al., 2011). Innovation propensity has also been associated with individual

Animal Creativity and Innovation.
DOI: http://dx.doi.org/10.1016/E978-0-12-800648-1.00006-1

163

and social learning, yielding support to the existence of general intelligence (Reader & Laland, 2002; Lefebvre, Reader, & Sol, 2004; Reader et al., 2011). More recently, the innovation process has attracted the attention of evolutionary ecologists. Innovations allow animals to solve ecological problems and hence change their relationship with the environment (Figure 6.1). This may have important ecological and evolutionary implications, like allowing animals to adopt novel ecological opportunities (Lefebvre et al., 1997), to shift to alternative resources when the traditional ones become scarce (Sol, Lefebvre, & Rodríguez-Teijeiro, 2005), and to improve survival in novel environments (Sol, Duncan, Blackburn, Cassey, & Lefebvre, 2005; Sol, Lapiedra, & Gonzalez-Lagos, 2013). Behavioral innovations even open the possibility that animals can shape the course of their own evolution (Wyles, Kunkel, & Wilson, 1983; Sol & Price, 2008; Lapiedra, Sol, Carranza, & Beaulieu, 2013).

Despite the progress in our understanding of the causes and consequence of the innovation process, the evolutionary origin of innovative capacities remains a mystery. For some authors, innovative ability is an adaptive specialization to particular environmental demands (Healy & Rowe, 2007). The problem with this hypothesis, however, is that it lacks a plausible evolutionary mechanism. Natural selection can finely adapt organisms to common situations but not to situations they have never or rarely experienced before. In fact, there is currently no evidence that innovation propensity is an adaptive specialization. Although two

FIGURE 6.1   Innovations allow animals to device solutions to important problems. African elephants (photo), for example, require over 190 L of water per day, and to secure it during the dry season they have learned to dig water holes. In Asia, elephants have even been observed filling in the holes with chewing bark ripped from a tree and covering over it with sand to avoid evaporation.

recent studies provide evidence for a link between innovation and some fitness components (Cauchard, Boogert, Lefebvre, Dubois, & Doligez, 2012; Cole, Morand-Ferron, Hinks, & Quinn, 2012), such evidence is insufficient in the absence of information on whether innovative capacity is heritable and can be selected directly rather than as a by-product of other traits.

An alternative is considering innovation propensity not as the direct target of selection, at least in the early stages of evolution, but as a by-product of a combination of traits that have evolved for other functions yet predispose individuals to solve problems by adopting novel behaviors (Figure 6.2). If so, innovativeness would not be an adaptation but an exaptation (sensu Gould & Vrba, 1982). As Darwin (1872) already noted over 140 years ago, the evolution of many complex features can result from the combination of traits that have originally served for other functions. Applied to the evolution of innovativeness, such an interpretation is in my view more consistent with evolutionary theory and fits better with current empirical evidence. Although similar arguments have previously been suggested by others (Reader, 2007; Chiappe & Gardner, 2011), these have never been formally developed.

In this chapter, I develop the argument that innovativeness initially evolved as an exaptation by showing that the innovation process is the outcome of a combination of adaptations that can be useful by themselves, and that these adaptations can have evolved together in animals in contexts where innovations yield higher benefits than costs. While the ontogenetic and causal mechanisms (proximate causes) behind the innovation

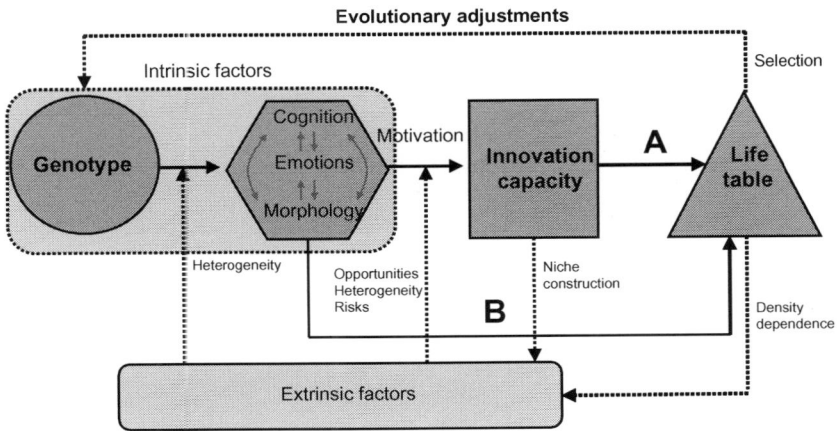

FIGURE 6.2 Under appropriate environmental conditions, selection can enhance innovation capacity not only directly (A) but also indirectly (B) by favoring correlations between mechanisms that favor innovations.

process have received a great deal of recent attention, the ultimate factors that explains its evolutionary origin have largely been ignored. I consequently use the former to speculate over the evolutionary processes that may have favored or constrained the evolution of creative lifestyles. My ultimate goal is to encourage a discussion about the evolutionary processes that have led to the evolution of creative animals and identify the key elements that remain insufficiently understood.

THE INNOVATION PROCESS DEPENDS ON A COMBINATION OF FACTORS

Behavioral innovations are frequent in nature, particularly in the foraging domain (Lefebvre et al., 1997; Reader & Laland, 2002). Over the past years, for example, Lefebvre and coworkers have compiled more than 2000 observations of feeding innovations in over 800 avian species, including jackdaws (*Corvus monedula*) opening horse-chestnuts by dropping them on pavement, green jays (*Cynaocorax yncas*) using twigs as probes and levers, and herring gulls (*Larus argentatus*) catching small rabbits and killing them by dropping them on rocks (Lefebvre et al., 1997; Overington et al., 2009). In the foraging domain, the innovation process requires four steps: encountering the new opportunity (sampling), assessing that this is a new opportunity (exploring), solving the problem of acquiring the new opportunity (problem solving), and incorporating the new behavior into the individual's repertoire (learning) (Figure 6.3). While the definition of innovations is often restricted to problem solving, considering all the steps is important to understand

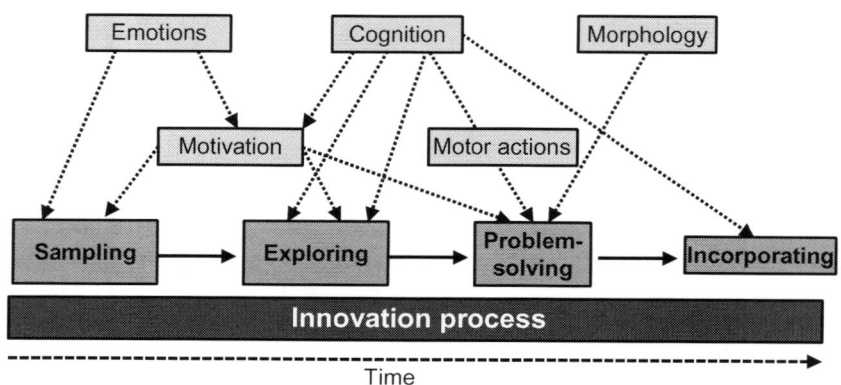

FIGURE 6.3 Proximate factors affecting the different steps of the innovation process. See main text for details.

why some organisms are more innovative than others in the wild. In practice, separating the steps is difficult and hence I will discuss all them together.

The mechanisms underlying innovations have generally been investigated via laboratory experiments in which individuals have to solve a task, like opening a lid on a box hiding food (e.g., Laland & Reader, 1999; Webster & Lefebvre, 2001; Sol, Griffin, Bartomeus, & Boyce, 2011). Performance in innovation is then measured as whether or not the individual solves the task and/or the time it takes to solve. Although the relevance of the proposed tasks and the extrapolation of the results to other species are open to criticism (Lefebvre et al., 2004), these experiments have allowed us to unambiguously establish that variation within species in the propensity to innovate reflects differences in motivation, emotional responses, cognitive abilities, and morphology. Next I will discuss how each of these factors can influence the likelihood to innovate.

Motivation

Confirming the proverb "necessity is the mother of invention," motivation has been identified as a major predictor of the likelihood of innovation (Kacelnik, 2009). Differences in motivation can reflect the physiological state of the animal, its social status and/or the stage of its life cycle. Common mynahs *Sturnus tristis*, for example, largely vary in motivation to forage even under controlled environmental conditions, and this variation significantly correlates with the latency to solve problems, such as adopting a novel food or removing a lid from a wooden well that hides food (Sol et al., 2011). Although the mechanisms are not yet fully understood, motivation seems in this case to be driven by hunger and affects innovation latency by favoring exploration and persistence in trying to solve the problem.

Emotional Responses

Emotional responses, also called animal personality or temperament traits when consistent over time and across situations (Sih, Bell, & Johnson, 2004; Réale et al., 2007), encompass two conflicting forces: the need to approach and to explore new resource opportunities, on the one hand, and the need to avoid unnecessary risks, on the other (Greenberg, 2003). Many animals exhibit an aversion to approaching novel objects, which can reduce their performance in experiments of technical innovations involving new objects. However, even highly neophobic animals can be good innovators if they exhibit an attraction to novel objects (neophilia). Thus, as stated by Greenberg (2003), it is the balance between neophobia and neophilia which ultimately determines that an animal explores novel opportunities conducive to innovation. Other emotional responses might also be important for innovation, but here evidence is less clear. Activity levels, for example, can reflect anxiety and can reduce attention to solve a task. However, activity has received far less attention than neophobia and neophilia, and there is currently no evidence that it can influence the innovation process (Sol, Griffin, & Bartomeus, 2012).

Cognition

Cognition is considered central in innovation because information gathering, problem solving, and learning are essential components of the process. In addition, cognition can indirectly influence innovation through its effect on motivation and emotional responses by altering the balance between costs and benefits of problem solving. The most general evidence that cognition is involved in the innovation process comes from comparative analyses in birds and primates showing that innovation propensity, in particular innovations associated with the emergence of new foraging techniques, is higher in lineages that have larger brains relative to their body size (Lefebvre et al., 1997; Reader & Laland, 2002; Overington et al., 2009; Reader et al., 2011).

While innovation propensity has often been associated with superior cognitive abilities, in fact many innovations in nonhuman animals seem to require simple cognitive abilities like trial-and-error learning (e.g., Sol et al., 2011). Even many instances of tool use do not seem to be particularly demanding in cognitive terms (Hansell & Ruxton, 2008). Still, cognition is important in terms of sampling style, attention, choosiness, discrimination, physical cognition, episodic learning, flexible decision making, and learning (Sih & Del Giudice, 2012). For example, because of the speed−accuracy trade-off, individuals that explore quickly do so more superficially and hence make faster, often less accurate decisions (Marchetti & Drent, 2000; Sih & Del Giudice, 2012). This should make these individuals less prone to innovating. For most sophisticated innovations, causal reasoning and planning have also been suggested to play a role, although evidence for these features in nonhuman animals remains controversial (Kacelnik, 2009).

Morphology

Morphology can limit innovation by affecting the type and diversity of motor patterns exhibited by the animal. The use of the appropriate motor actions is important in innovation, making some species fail when either they do not use them at all or they do but are unable to shift to using the most appropriate ones. A broad diversity of motor patterns is not itself necessary to innovation if individuals exhibit the appropriate behavior to solve a given problem, but it likely increases the likelihood of solving a broader variety of tasks (e.g., Benson-Amram & Holekamp, 2012). In addition, the most sophisticated innovations probably require some flexible alterations of motor actions. In animals that make tools, for instance, the challenge is not only to make them but to flexibly change the sequence of actions needed to solve a new problem (Güntürkün, 2014). This involves cognition and it is highly facilitated by appendages for manipulating objects, like hands in primates, slender bills in birds, feet in parrots, or the trunk in elephants. The importance of morphology in innovativeness is evident in New Caledonian crows (*Corvus moneduloides*), one of the most exceptional tool users of all animals. Compared with other Corvids, their unusually straight bill enables a stable grip on tools, and raises the tip of the tool into their visual field's binocular sector (Troscianko, von Bayern, Chappell, Rutz, & Martin, 2012). In addition, the eyes are more forward-positioned, rather than sideways-positioned, which should provide better binocular overlap and hence enhance visual perception. This later possibility is in line with the finding that relative eye size and relative brain size covary positively in birds (Garamszegi, Møller, Erritzøe, & Erritzoe, 2002).

Adaptive Nature of the Traits Influencing Innovativeness

For a trait to be considered adaptive, the trait has to be heritable and it has to provide some kind of fitness benefit. Admittedly, some of the presumed components that make a species innovative are either state-dependent, like motivation, or can be plastically altered during the life of the individual. In vervet monkeys (*Chlorocebus pygerythrus*), for instance, mothers have considerable impact on how their offspring react to novel situations. Nevertheless, some other components of innovativeness, like life history and morphology, are highly heritable. Studies of human twins, for example, have revealed high heritability (up to 95%) in overall brain volume and in areas involved in learning (Peper, Brouwer, Boomsma, Kahn, & Hulshoff Pol, 2007). Even personality traits like neophobia can have a heritable component (Réale et al., 2007), which limits the extent to which they can be plastically altered.

A growing literature has also reported evidence for fitness benefits of traits underlying innovativeness. Evidence is ample for the fitness benefits of morphological and life history traits (e.g., Endler, 1986; Reznick & Ghalambor, 2005), aspects that have traditionally interested evolutionary biologists, but it is also growing for cognitive and person-ality traits (Dukas & Bernays, 2000; Dingemanse & Réale, 2005). In big-horn sheep (*Ovis canadiensis*), in which individuals are known to consistently differ in boldness, predation by cougars (*Puma concolor*) can select for this trait (Réale & Festa-Bianchet, 2003). With low preda-tion, survival is high and selection does not affect personality. When predation by cougars is intense, however, survival rates drop and selection acts on boldness.

Finally, there is some evidence, albeit scarce, that some of the compo-nents underlying innovative capacity are genetically correlated. In gup-pies (*Poecilia reticulata*), artificial selection experiments for large- and small-brained individuals demonstrate that brain size can evolve rap-idly in response to divergent selection (Kotrschal et al., 2013). Importantly, guppies selected for larger brains improve, as a by-product, in certain cognitive abilities, like numerical essays or spatial learning (Kotrschal et al., 2013), and are faster to habituate to, and more exploratory in, open field tests (Kotrschal, Lievens, & Dahlbom, 2013).

Admittedly, we do not know whether the traits underlying innova-tive capacity have been selected for the function they currently have, a requisite to distinguish adaptations from exaptations (Gould & Vrba, 1982). However, note that for my argument it does not matter whether these traits are adaptations or exaptations, but only that innovativeness is the by-product of traits that have evolved for other functions. In fact, strictly speaking, many complex features that we usually classify as adaptations should be more formally called exaptations because they are the result of the combination of traits that have been initially selected for other functions.

WHY CAN WE EXPECT ADAPTATIONS UNDERLYING INNOVATIVENESS TO EVOLVE TOGETHER?

In the previous section, I have argued that innovativeness depends on a combination of traits, some of which can themselves be adapta-tions or exaptations. The question is now how these features ended up together in highly innovative animals. There are a number of mechan-isms that can drive the combination of these traits and hence lead to innovativeness syndromes. The most important ones are shared mechanisms and functional links.

Correlation Due to Shared Mechanisms

Correlations between cognitive, behavioral, and physiological traits conducive to innovativeness can evolve as a result of shared mechanisms related to pleiotropic genetic effects or common physiological pathways resulting from past selection pressures on the developmental stability and homeostasis of an organism (Réale et al., 2010; Sih & Del Giudice, 2012). The life history of the species, which subsumes a variety of physiological mechanisms controlling how individuals allocate the time and energy into reproduction, development, and survival, creates an obvious link between the traits underlying innovations (Sol, 2009a, 2009b). The fast—slow continuum, in particular, reflects a differential need to collect, assess, and retain information and hence it can affect the speed over accuracy trade-off when the need to store information slows decision making (Sih & Del Giudice, 2012). Species at the "slow" extreme of the continuum prioritize survival over reproduction, and hence should be more risk-averse, explore slowly but more accurately, and have higher problem-solving abilities (Sol, 2009a, 2009b; Sih & Del Giudice, 2012). The latter should be facilitated by the fact that slow-lived species are more likely to have larger brains relative to body mass than fast-lived species, as they are less constrained by development (larger brains take longer to grow) and obtain greater benefits by investing in adaptations that enhance survival (Sol, 2009a; Barton & Capellini, 2011). It should also be noted that time does not have the same value for all animals. For a short-lived animal, spending too much time trying to solve a problem can be more costly in terms of fitness than for a long-lived animal. Taken together, the above arguments suggest that, under appropriate environmental conditions (see next section), the combination of traits needed to innovate is likely to evolve together in slow-lived animals.

Endocrine control mechanisms, often related to life history strategies, can also produce incompatible physiological states that create trade-offs and restrict the combinations of traits that are possible (Ricklefs & Wikelski, 2002). For example, glucocorticoids influence the response of individuals to stress and have been shown to vary in a consistent way among individuals within species and across species (see Lendvai et al., 2013). This indicates that animals differ in the way they respond to stressful conditions, like novel objects. In house sparrows (*Passer domesticus*), for example, individuals with lower levels of corticosterone, the main avian stress hormone, solved a difficult task faster and were more efficient learners than birds with higher corticosterone levels (Bókony et al., 2013). Corticosterone has also been related to brain size. In birds, species with larger brains (relative to body size) tend to show lower baseline and peak corticosterone levels than species with smaller brains (Lendvai et al., 2013). Because these endocrine control mechanisms can produce incompatible physiological states that restrict life histories to a single dominant axis of variation (Ricklefs & Wikelski, 2002), they represent a mechanistic nexus between life history and key components of innovativeness.

Finally, processes related to domain-general cognition, where the animal acquires certain rules to discriminate and respond to information that can be applied in different contexts, provides a shared mechanism that links a variety of cognitive components of innovative capacities. This issue relates to the controversy over the modularity of cognition, the idea that the mind consists of independent, domain-specific processing modules (Lefebvre et al., 2004; Lefebvre, 2011; Sih & Del Giudice, 2012). Although the idea was popular in the 1990s, there is no evidence that innovativeness requires any cognitive adaptive specialization. Tebbich and coworkers, for instance, have been unable to identify any cognitive specialization that can differentiate tool-using Galapagos finches from non-tool-using close relatives (Teschke, Cartmill, Stankewitz, & Tebbich, 2011, 2013). Thus, tool users, like the woodpecker finch *Cactospiza pallida*, do not outperform non-tool users (e.g., *Camarrhynchus parvulus*) in cognitive skills required for innovation. Bird and Emery (2009) came to a similar conclusion with their experiments in rooks *Corvus frugilegus*. This species does not use tools in the wild but readily uses them in captivity, implying that they are not using specialized domain-specific intelligence to innovate. In contrast, there is growing evidence that some important cognitive components covary both within and across species. In comparisons across primate species, different cognitive measures like behavioral innovation, social learning, tool use, extractive foraging, tactical deception, and captive learning performance tests tend to be associated, suggesting general intelligence, and covary with brain volume (Deaner, Burkart, Schaik, Isler, & van Schaik, 2007; Reader et al., 2011). Within species, exploration and flexible decision making have been shown to positively covary in great tits (*Parus major*; Marchetti & Drent, 2000), whereas in pigeons (*Columba livia*) a positive relationship exists between innovative problem solving and social learning (Bouchard, Goodyer, & Lefebvre, 2007).

Correlation due to Selection for Trait Combinations Under Certain Environments

Even when the traits underlying innovativeness are governed by independent mechanisms, correlations can also evolve through correlated selection resulting from some combinations of phenotypes attaining higher fitness in certain environments (Sinervo & Svensson, 2002). Given that the environmental context of animals often changes over time and space, challenging individuals with new problems, environmental induction is likely to be a powerful driver of innovations

(Giraldeau, Lefebvre, & Morand-Ferron, 2007). In a search of primate journals, Reader and Laland (2002) found that ecological challenges such as periods of food shortage, dry seasons, or habitat degradation accounted for approximately half of recorded innovations where data were available. Although it remains unclear what are the environments that would favor correlated evolution of traits favoring innovation, theory suggests that the correlations should be more frequent in environments where individuals are often exposed to new problems for which they have no specialized adaptive solutions. These include poor environments, environments where there are many ecological opportunities, heterogeneous environments where individuals constantly have to update environmental information and stochastic environments where there are good and bad years that make optimum adaptation difficult (Sol, 2003, 2009b).

Although still limited, there is evidence that the environment can lead to correlated evolution of traits associated with innovativeness. In a multi-population study of black-capped chickadees, for instance, Roth, LaDage, and Pravosudov (2010) found that populations from the latitudinal extremes of the species' range not only exhibited significant differences in learning ability but also in neophobia, memory, and brain morphology. Harsh environments, where there is uncertainty in resource availability, are thought to promote innovative behaviors because by being able to exploit alternative foods, individuals can alleviate the adverse conditions (Sol, 2003; Sol, Lefebvre et al., 2005).

When traits are governed by independent mechanisms, the resulting correlations should be less difficult to decouple than correlations resulting from shared mechanisms (Sih et al., 2004). This opens the possibility that there exist different ways to be innovative. For example, living in a low risk environment can favor some components of innovation, like reduced neophobia and increased neophilia (Greenberg, 2003). However, because neophobia and neophilia are not necessarily associated, animals living in a risky environment can still be innovative if they are highly neophilic.

THE EVOLUTION OF CREATIVE LIFESTYLES

Almost all animals have the capacity to innovate, but a few really excel in their innovative capacities. These include some parrots, crows, primates, and cetaceans (Lefebvre et al., 2004; Lefebvre & Sol, 2008). Interestingly, these animals possess disproportionally large brains, enhanced cognitive abilities, are very exploratory and capable of manipulating objects, and tend to be at the slow extreme of the fast–slow continuum of life history variation (Emery, 2006; Lefebvre & Sol, 2008). Thus, in these animals the key features favoring innovativeness are particularly well developed and genetically integrated. I propose here that this more advanced stage in the evolution of innovativeness, which I will call a "creative lifestyle," results from correlated selection for combinations of the traits underlying innovativeness that enhance fitness in animals (including humans) that actively expose themselves to novel problems. Two possible, nonmutually exclusive pathways to develop such a creative lifestyle are sociality and ecological generalism.

Sociality

Living in stable, cohesive social groups is often considered to favor the evolution of enhanced cognitive abilities, which are needed to recognize group members, track their social status, and infer relationships among them (Bond, Kamil, & Balda, 2004; Shultz & Dunbar, 2007). Groups are also more likely to contain individuals competent at solving problems (Liker & Bókony, 2009; Morand-Ferron & Quinn, 2011), and provide opportunities for social learning, expanding the benefits of the innovation to more individuals. All this predicts that some key cognitive components of innovativeness vary with the social context. In Zenaida doves (*Zenaida aurita*) from Barbados, social individuals learn more readily from a conspecific tutor, while territorial individuals learn from the heterospecific they most often interact with, the Carib grackle, *Quiscalus lugubris* (Carlier & Lefebvre, 1997). Social doves also learn individually faster than territorial doves (Carlier & Lefebvre, 1996). If in addition we consider that sociality facilitates encounters with new opportunities while reducing the risk for individuals, and that more complex animal societies are found in long-lived animals (Covas & Griesser, 2007), it seems plausible that the evolution of creative lifestyles can be influenced by the social context. This would explain why the most innovatory animals are all social.

Ecological Opportunist-Generalist

Opportunist-generalist lifestyles, where individuals are frequently confronted with new problems, might also lead to correlated evolution of different innovation components. In spatially and temporally heterogeneous environments, for instance, many animals rely on a variety of food resources that are too scarce or unstable over time for specialization. Food resources might not only differ in quality, but some can even contain toxins. Such a scenario should select for individuals that are more persistent in sampling and discriminating between different options and that are less impulsive and risk-prone in taking decisions (Sih & Del Giudice, 2012). Although there is a cost in delaying decisions, this can be compensated for if the rewards are high and the risks of taking a bad decision decrease. This should particularly be true in long-lived species, which prioritize survival over reproduction and hence tend to have risk-averse lifestyles (Sol, 2009a, 2009b; Réale et al., 2010). Thus, an opportunistic-generalist lifestyle should select for correlated evolution of key traits needed to innovate. Indeed, there is evidence that innovation rate and habitat generalism covary in birds (Overington, Griffin, Sol, & Lefebvre, 2011). In some cases, the opportunist can adopt a resource that is very abundant and/or of high quality but underutilized due to the difficulty of its exploitation, which can then derive to a specialized innovation. This seems to be the case of new Caledonian crows, which primarily forage on the ground yet the preys collected there are considerably smaller than those that individuals obtain by using stick tools (Rutz et al., 2010).

SYNTHESIS

In this review, I have argued that innovation propensity should not been seen as a specialized adaptation but an exaptation that is a by-product of the evolution of other traits whose benefits to the organism were initially unrelated to innovation. Innovativeness can result from a combination of life history, morphological, cognitive, and physiological features that have evolved together because they share common mechanisms and/or because together increase fitness under a variety of environmental circumstances. All these processes can be reinforced by learning, and can be further developed and integrated in animals that actively expose themselves to novel problems, like highly social animals and animals with opportunist-generalist lifestyles.

The view that innovation capacity is not a specialized adaptation but a combination of adaptations can imply that this is not a unitary process but that there may be different ways to be innovative. This is exemplified in the finding that technical (e.g., using novel tools) and consumer (e.g., eating a novel food) innovations are driven by different mechanisms (Overington et al., 2009; Sol et al., 2012). In experiments in common mynas, for instance, the influence of neophobia was restricted to performance on the technical innovation task, and did not extend to the consumer innovation test (Sol et al., 2012). Likewise, comparative evidence suggest that the diversity of technical innovations displayed by birds is correlated with brain size, while the diversity of consumer innovations is not, again implying differences in the underlying mechanisms (Overington et al., 2009). However, this does not deny the possibility of convergent evolution in animals with highly developed creative lifestyles (Lefebvre et al., 2004).

The proposed framework allows us to understand a number of puzzling observations:

1. Why do species like rooks readily show tool use innovations in the laboratory despite never using tools in the wild? This can be understood if they already possess all the machinery needed to use tools (Kacelnik, 2009).
2. Why do some presumed innovative behaviors appear to be innate? New Caledonian crows and woodpecker finches exhibit tool-oriented behavior even if they are raised in the absence of tool-using models (Kacelnik, 2009; Teschke et al., 2013). A process that can generate this is genetic assimilation, that is, the process by which a phenotypic feature induced as a response to an environmental condition later becomes genetically encoded via natural selection (Waddington, 1961; Crispo, 2007). Consider for instance the classical example of blue tits (*Cyanistes caeruleus*) opening milk bottles in England during the 1950s. Imagine also that the efficiency in exploiting the new food depends on a variety of heritable traits (e.g., bill morphology, motor ability, enzymatic system to digest milk) and that individuals that are better at exploiting the new food opportunity attain higher fitness (e.g., have higher probability of surviving the winter). Over time, the frequency of individuals possessing the combination of traits will increase in the population and, after many generations subject to natural selection, individuals will end up looking like if they have an innate ability to open milk bottles. Although the new behavior will become increasingly efficient through genetic assimilation, it will still be learned behavior (see Hunt & Gray, 2007).
3. Why are innovative animals characterized by a large brain relative to body size (Lefebvre et al., 1997; Laland & Reader, 1999)? If innovation depends on the combination of a variety of perceptual, motor, and cognitive abilities, then we should not expect it to be associated with particular regions of the brain but with the size of large brain areas or even with the whole brain (Lefebvre & Sol, 2008; Lefebvre, 2011). This should be true even if the brain is organized in independent domain-specific processing modules.
4. Why is assembling evidence for the fitness benefits of innovation so difficult? When examining long-term calving records of bottlenose dolphins in Shark Bay, Australia, Mann et al. (2008) found no difference in the reproductive output of individuals using sponges as tools compared to nonspongers, suggesting that the use of tools did

not offer significant fitness benefits. Leaving apart the fact that demonstrating fitness effects is always difficult, under the proposed framework this observation would not be problematic: As by-product of other processes more directly associated with fitness, innovations *per se* are not required to enhance fitness. Exceptions would be cases where the species has developed a creative lifestyle or when there are episodes of environmental stress (e.g., a drought which reduces food availability) that can be alleviated with innovative behaviors.

Arguing that innovation propensity is not a unitary process is not to say that it cannot be selected. When innovations become essential for the survival of an animal, that is, the animal adopts a creative lifestyle, then it is possible that the key traits required to innovate becomes increasingly integrated and further developed through correlated selection. This process can for instance have contributed to the evolution of large brains and advanced cognitive abilities in highly innovative animals (Sol, 2009a, 2009b). Innovativeness itself can also be directly selected through mechanisms other than natural selection, like mating competition. As noted by Madden (2007), sexual displays provide an unusual situation where selection favors innovation per se. Females may prefer innovative males because unusual males are less likely to be relatives, or because innovative males possess a general cognitive ability correlated with a sexually selected trait (Madden, 2001, 2007).

The thesis that innovation propensity is, in its early evolutionary stages, mainly a by-product of a combination of traits that predispose some animals to solve problems by adopting new behaviors is based on theoretical arguments and backed for some empirical evidence. However, important aspects of the theory remain poorly understood. To further develop the theory, we need to better understand how the phenotype, the environment, and the interaction between phenotype and environment facilitate or inhibit innovations. We need to use evolutionary approaches to assess whether innovation propensity affects fitness and, if so, determine whether the effect is direct or indirectly caused by any of its components. We need to better know what common mechanisms and/or environmental pressures lead to the covariation of traits favoring innovativeness. Finally, we need retrospective approaches, like phylogenetic reconstructions, to unravel how past environmental and phenotypic changes have shaped current variation in innovation propensity and in their functional components. There is thus a long way to go to understand the evolution of innovative capacities, but past progress allows at least thinking in a theoretical framework from which to derive future empirical tests.

Acknowledgments

I am grateful to Allyson and James Kaufman for the invitation to contribute to this volume, to Louis Lefebvre and Allyson Kaufman for revising the chapter, and to Louis Lefebvre, Simon Reader, Andrea Griffin, Ferran Sayol, Joan Maspons, Cesar González-Lagos and Oriol Lapiedra for fruitful discussions over the past years. This work was supported by a Proyecto de Investigación (ref. CGL2013-47448-P) from the Spanish Government.

References

Barton, R. A., & Capellini, I. (2011). Maternal investment, life histories, and the costs of brain growth in mammals. *Proceedings of the National Academy of Sciences of the United States of America, 108*, 6169–6174. Available from: http://dx.doi.org/10.1073/pnas.1019140108.

Benson-Amram, S., & Holekamp, K. E. (2012). Innovative problem solving by wild spotted hyenas. *Proceedings of the Royal Society of London, Series B, 279*, 4087–4095. Available from: http://dx.doi.org/10.1098/rspb.2012.1450.

Bird, C. D., & Emery, N. J. (2009). Insightful problem solving and creative tool modification by captive nontool-using rooks. *Proceedings of the National Academy of Sciences of the United States of America, 106*, 10370–10375. Available from: http://dx.doi.org/10.1073/pnas.0901008106.

Bókony, V., Lendvai, A. Z., Vagasi, C. I., Patras, L., Pap, P. L., Nemeth, J., et al. (2013). Necessity or capacity? Physiological state predicts problem-solving performance in house sparrows. *Behavioral Ecology, 25*, 124–135. Available from: http://dx.doi.org/10.1093/beheco/art094.

Bond, A., Kamil, A., & Balda, R. (2004). Pinyon jays use transitive inference to predict social dominance. *Nature, 430*, 5–8. Available from: http://dx.doi.org/10.1038/nature02720.1.

Bouchard, J., Goodyer, W., & Lefebvre, L. (2007). Social learning and innovation are positively correlated in pigeons (*Columba livia*). *Animal Cognition, 10*, 259–266. Available from: http://dx.doi.org/10.1007/s10071-006-0064-1.

Carlier, P., & Lefebvre, L. (1996). Differences in individual learning between group-foraging and territorial. *Behaviour, 133*, 1197–1207. Available from: http://dx.doi.org/10.1163/156853996X00369.

Carlier, P., & Lefebvre, L. (1997). Ecological differences in social learning between adjacent, mixing, populations of zenaida doves. *Ethology, 103*, 772–784. Available from: http://dx.doi.org/10.1111/j.1439-0310.1997.tb00185.x.

Cauchard, L., Boogert, N. J., Lefebvre, L., Dubois, F., & Doligez, B. (2012). Problem-solving performance is correlated with reproductive success in a wild bird population. *Animal Behaviour, 85*, 1–8. Available from: http://dx.doi.org/10.1016/j.anbehav.2012.10.005.

Chiappe, D., & Gardner, R. (2011). The modularity debate in evolutionary psychology. *Theory & Psychology, 22*, 669–682. Available from: http://dx.doi.org/10.1177/0959354311398703.

Cole, E. F., Morand-Ferron, J., Hinks, A. E., & Quinn, J. L. (2012). Cognitive ability influences reproductive life history variation in the wild. *Current Biology, 22*, 1808–1812. Available from: http://dx.doi.org/10.1016/j.cub.2012.07.051.

Covas, R., & Griesser, M. (2007). Life history and the evolution of family living in birds. *Proceedings of the Royal Society of London, Series B, 274*, 1349–1357. Available from: http://dx.doi.org/10.1098/rspb.2007.0117.

Crispo, E. (2007). The Baldwin effect and genetic assimilation: Revisiting two mechanisms of evolutionary change mediated by phenotypic plasticity. *Evolution, 61*, 2469−2479. Available from: http://dx.doi.org/10.1111/j.1558-5646.2007.00203.x.

Darwin, C. (1872). *The origin of species by means of natural selection; or, the preservation of favored races in the struggle for life.* London: John Murray. Available from: http://dx.doi. org/10.5962/bhl.title.2109.

Deaner, R. O., Burkart, J., Schaik, V., Isler, K., & van Schaik, C. P. (2007). Overall brain size, and not encephalization quotient, best predicts cognitive ability across non-human primates. *Brain Behavior and Evolution, 70*(2), 115−124. Available from: http://dx.doi.org/10.1159/000102973.

Dingemanse, N., & Réale, D. (2005). Natural selection and animal personality. *Behaviour, 142*(9), 1159−1184. Available from: http://dx.doi.org/10.1163/156853905774539445.

Dukas, R., & Bernays, E. A. (2000). Learning improves growth rate in grasshoppers. *Proceedings of the National Academy of Sciences of the United States of America, 97*, 2637−2640.

Emery, N. J. (2006). Cognitive ornithology: The evolution of avian intelligence. *Philosophical Transactions of the Royal Society of London. Series B, Biological Sciences, 361*, 23−43.

Endler, J. A. (1986). *Natural selection in the wild.* Princeton, NJ: Princeton University Press.

Garamszegi, L. Z., Møller, A. P., Erritzøe, J., & Erritzoe, J. (2002). Coevolving avian eye size and brain size in relation to prey capture and nocturnality. *Proceedings of the Royal Society of London. Series B, 269*(1494), 961−967. Available from: http://dx.doi.org/10.1098/rspb.2002.1967.

Giraldeau, L.-A., Lefebvre, L., & Morand-Ferron, J. (2007). Can a restrictive definition lead to biases and tautologies? *Behavioral and Brain Sciences, 30*, 393−437. Available from: http://dx.doi.org/10.1017/S0140525X07002427.

Gould, S. J., & Vrba, E. S. (1982). Exaptation—A missing term in the science of form. *Paleobiology, 8*, 4−15. Available from: http://dx.doi.org/10.2307/2400563.

Greenberg, R. (2003) The role of neophobia and neophilia in the development of innovative behaviour of birds. In S. M. Reader, & K. N. Laland (Eds.), *Animal innovation* (pp. 176−196). Oxford: Oxford University Press.

Güntürkün, O. (2014). Is Dolphin Cognition Special? Commentary on Manger PR (2013): Questioning the interpretations of behavioral observations of cetaceans: Is there really support for a special intellectual status for this mammalian order? Neuroscience 250:664−696. *Brain, Behavior and Evolution.* Available from: http://dx.doi.org/10.1159/000357551.

Hansell, M., & Ruxton, G. D. (2008). Setting tool use within the context of animal construction behaviour. *Trends in Ecology & Evolution, 23*, 73−78. Available from: http://dx.doi.org/10.1016/j.tree.2007.10.006.

Healy, S. D., & Rowe, C. (2007). A critique of comparative studies of brain size. *Proceedings of the Royal Society of London, Series B, 274*, 453−464. Available from: http://dx.doi.org/10.1098/rspb.2006.3748.

Hunt, G. R., & Gray, R. D. (2007). Genetic assimilation of behaviour does not eliminate learning and innovation. *Behavioral and Brain Sciences.* Available from: http://dx.doi.org/10.1017/S0140525X07002439.

Kacelnik, A. (2009). Tools for thought or thoughts for tools? *Proceedings of the National Academy of Sciences of the United States of America, 106*, 10071−10072. Available from: http://dx.doi.org/10.1073/pnas.0904735106.

Kotrschal, A., Lievens, E., & Dahlbom, J. (2013). Artificial selection on relative brain size reveals a positive genetic correlation between brain size and proactive personality in the guppy. *Evolution, 68*, 1−33. Available from: http://dx.doi.org/10.1111/evo.12341.

Kotrschal, A., Rogell, B., Bundsen, A., Svensson, B., Zajitschek, S., Brännström, I., et al. (2013). Artificial selection on relative brain size in the guppy reveals costs and benefits of evolving a larger brain. *Current Biology*, *23*, 168–171. Available from: http://dx.doi.org/10.1016/j.cub.2012.11.058.

Laland, K. N., & Reader, S. M. (1999). Foraging innovation in the guppy. *Animal Behaviour*, *57*, 331–340. Available from: http://dx.doi.org/anbe.2000.1450.

Lapiedra, O., Sol, D., Carranza, S., & Beaulieu, J. (2013). Behavioural changes and the adaptive diversification of pigeons and doves. *Proceedings of the Royal Society of London, Series B*, *280*, 20122893. Available from: http://dx.doi.org/10.1098/rspb.2012.2893.

Lefebvre, L. (2011). Taxonomic counts of cognition in the wild. *Biology Letters*, *7*, 631–633. Available from: http://dx.doi.org/10.1098/rsbl.2010.0556.

Lefebvre, L., & Sol, D. (2008). Brains, lifestyles and cognition: Are there general trends? *Brain Behavior and Evolution*, *72*(2), 135–144. Available from: http://dx.doi.org/10.1159/000151473.

Lefebvre, L., Reader, S. M., & Sol, D. (2004). Brains, innovations and evolution in birds and primates. *Brain Behavior and Evolution*, *63*(4), 233–246. Available from: http://dx.doi.org/10.1159/000076784.

Lefebvre, L., Whittle, P., Lascaris, E., & Finkelstein, A. (1997). Feeding innovations and forebrain size in birds. *Animal Behaviour*, *53*, 549–560. Available from: http://dx.doi.org/10.1016/anbe.1996.0330.

Lendvai, Á., Bókony, V., Angelier, F., Chastel, O., & Sol, D. (2013). Do smart birds stress less? An interspecific relationship between brain size and corticosterone levels. *Proceedings of the Royal Society of London, Series B*, *280*, 20131734. Available from: http://dx.doi.org/10.1098/rspb.2013.1734.

Liker, A., & Bókony, V. (2009). Larger groups are more successful in innovative problem solving in house sparrows. *Proceedings of the National Academy of Sciences of the United States of America*, *106*, 7893–7898. Available from: http://dx.doi.org/10.1073/pnas.0900042106.

Madden, J. (2001). Sex, bowers and brains. *Proceedings of the Royal Society of London, Series B*, *268*, 833–838. Available from: http://dx.doi.org/10.1098/rspb.2000.1425.

Madden, J. (2007). Innovation in sexual display. *Behavioral and Brain Sciences*, *30*, 417–418. Available from: http://dx.doi.org/10.1017/S0140525X07002488.

Mann, J., Sargeant, B., Watson-Capps, J., Gibson, Q., Heithaus, M., Connor, R., et al. (2008). Why do dolphins carry sponges? *PLoS One*, *3*, e3868. Available from: http://dx.doi.org/10.1371/journal.pone.0003868.

Marchetti, C., & Drent, P. (2000). Individual differences in the use of social information in foraging by captive great tits. *Animal Behaviour*, *60*, 131–140. Available from: http://dx.doi.org/10.1006/anbe.2000.1443.

Morand-Ferron, J., & Quinn, J. L. (2011). Larger groups of passerines are more efficient problem solvers in the wild. *Proceedings of the National Academy of Sciences of the United States of America*, *108*, 15898–15903. Available from: http://dx.doi.org/10.1073/pnas.1111560108.

Overington, S. E., Griffin, A. S., Sol, D., & Lefebvre, L. (2011). Are innovative species ecological generalists? A test in North American birds. *Behavioral Ecology*, *22*, 1286–1293. Available from: http://dx.doi.org/10.1093/beheco/arr130.

Overington, S. E., Morand-Ferron, J., Boogert, N. J., & Lefebvre, L. (2009). Technical innovations drive the relationship between innovativeness and residual brain size in birds. *Animal Behaviour*, *78*, 1001–1010. Available from: http://dx.doi.org/10.1016/j.anbehav.2009.06.033.

Peper, J. S., Brouwer, R. M., Boomsma, D. I., Kahn, R. S., & Hulshoff Pol, H. E. (2007). Genetic influences on human brain structure: A review of brain imaging studies in twins. *Human Brain Mapping, 28,* 464–473. Available from: http://dx.doi.org/10.1002/hbm.20398.

Ramsey, G., Bastian, M. L., & van Schaik, C. (2007). Animal innovation defined and operationalized. *The Behavioral and Brain Sciences, 30,* 393–407, 407–432. Available from: http://dx.doi.org/10.1017/S0140525X07002373.

Reader, S. M. (2007). Environmentally invoked innovation and cognition. *Behavioral and Brain Sciences, 30,* 4. Available from: http://dx.doi.org/10.1017/S0140525X07002518.

Reader, S. M., & Laland, K. N. (2002). Social intelligence, innovation, and enhanced brain size in primates. *Proceedings of the National Academy of Sciences of the United States of America, 99,* 4436–4441. Available from: http://dx.doi.org/10.1073/pnas.062041299.

Reader, S. M., & Laland, K. N. (Eds.), (2003). *Animal Innovation.* Oxford: Oxford University Press. Available from: http://dx.doi.org/10.1093/acprof.

Reader, S. M., Hager, Y., & Laland, K. N. (2011). The evolution of primate general and cultural intelligence. *Philosophical Transactions of the Royal Society of London. Series B, Biological Sciences, 366,* 1017–1027. Available from: http://dx.doi.org/10.1098/rstb.2010.0342.

Réale, D., & Festa-Bianchet, M. (2003). Predator-induced natural selection on temperament in bighorn ewes. *Animal Behaviour, 65,* 463–470. Available from: http://dx.doi.org/10.1006/anbe.2003.2100.

Réale, D., Garant, D., Humphries, M. M., Bergeron, P., Careau, V., & Montiglio, P.-O. (2010). Personality and the emergence of the pace-of-life syndrome concept at the population level. *Philosophical Transactions of the Royal Society of London. Series B, Biological Sciences, 365,* 4051–4063. Available from: http://dx.doi.org/10.1098/rstb.2010.0208.

Réale, D., Reader, S. M., Sol, D., McDougall, P. T., Dingemanse, N. J., & Réale, D. (2007). Integrating animal temperament within ecology and evolution. *Biological Reviews, 82,* 291–318. Available from: http://dx.doi.org/10.1111/j.1469-185X.2007.00010.x.

Reznick, D. N., & Ghalambor, C. K. (2005). Selection in nature: Experimental manipulations in natural populations. *Integrative and Comparative Biology, 45,* 456–462.

Ricklefs, R. E., & Wikelski, M. (2002). The physiology/life history nexus. *Trends in Ecology & Evolution, 17,* 462–468. Available from: http://dx.doi.org/10.1016/s0169-5347(02)02578-8.

Roth, T. C., LaDage, L. D., & Pravosudov, V. V. (2010). Learning capabilities enhanced in harsh environments: A common garden approach. *Proceedings of the Royal Society of London, Series B, 277,* 3187–3193. Available from: http://dx.doi.org/10.1098/rspb.2010.0630.

Rutz, C., Bluff, L. a, Reed, N., Troscianko, J., Newton, J., Inger, R., et al. (2010). The ecological significance of tool use in New Caledonian crows. *Science, 329,* 1523–1526. Available from: http://dx.doi.org/10.1126/science.1192053.

Shultz, S., & Dunbar, R. I. M. (2007). Evolution in the social brain. *Science, 317,* 1344–1347. Available from: http://dx.doi.org/10.1126/science.1145463.

Sih, A., & Del Giudice, M. (2012). Linking behavioural syndromes and cognition: A behavioural ecology perspective. *Philosophical Transactions of the Royal Society of London. Series B, Biological Sciences, 367,* 2762–2772. Available from: http://dx.doi.org/10.1098/rstb.2012.0216.

Sih, A., Bell, A., & Johnson, J. C. (2004). Behavioral syndromes: An ecological and evolutionary overview. *Trends in Ecology & Evolution, 19,* 372–378. Available from: http://dx.doi.org/10.1016/j.tree.2004.04.009.

Sinervo, B., & Svensson, E. (2002). Correlational selection and the evolution of genomic architecture. *Heredity, 89,* 329–338. Available from: http://dx.doi.org/10.1038/sj.hdy.6800148.

Sol, D. (2003). Behavioural flexibility: A neglected issue in the ecological and evolutionary literature? In S. M. Reader, & K. N. Laland (Eds.), *Animal innovation* (pp. 63–82). Oxford: Oxford University Press.

Sol, D. (2009a). Revisiting the cognitive buffer hypothesis for the evolution of large brains. *Biology Letters*, *5*(1), 130–133. Available from: http://dx.doi.org/10.1098/rsbl.2008.0621.

Sol, D. (2009b). The cognitive-buffer hypothesis for the evolution of large brains. In R. Dukas, & R. M. Ratcliffe (Eds.), *Cognitive ecology* (pp. 111–134). Chicago, IL: Chicago University Press.

Sol, D., & Price, T. D. (2008). Brain size and body size diversification in birds. *American Naturalist*, *172*, 170–177. Available from: http://dx.doi.org/10.1086/589461.

Sol, D., Duncan, R. P., Blackburn, T. M., Cassey, P., & Lefebvre, L. (2005). Big brains, enhanced cognition, and response of birds to novel environments. *Proceedings of the National Academy of Sciences of the United States of America*, *102*, 5460–5465. Available from: http://dx.doi.org/10.1073/pnas.0501695102.

Sol, D., Griffin, A. S., & Bartomeus, I. (2012). Consumer and motor innovation in the common myna: The role of motivation and emotional responses. *Animal Behaviour*, *83*, 179–188. Available from: http://dx.doi.org/10.1016/j.anbehav.2011.10.024.

Sol, D., Griffin, A. S., Bartomeus, I., & Boyce, H. (2011). Exploring or avoiding novel food resources? The novelty conflict in an invasive bird. *PloS One*, *6*, e19535. Available from: http://dx.doi.org/10.1371/journal.pone.0019535.

Sol, D., Lefebvre, L., & Rodríguez-Teijeiro, J. D. (2005). Brain size, innovative propensity and migratory behaviour in temperate Palaearctic birds. *Proceedings of the Royal Society of London, Series B*, *272*, 1433–1441. Available from: http://dx.doi.org/10.1098/rspb.2005.3099.

Sol, D., Lapiedra, O., & Gonzalez-Lagos, C. (2013). Behavioural adjustments for a life in the city. *Animal Behaviour*, *85*, 1101–1112. Available from: http://dx.doi.org/10.1016/j.anbehav.2013.01.023.

Teschke, I., Cartmill, E. A., Stankewitz, S., & Tebbich, S. (2011). Sometimes tool use is not the key: No evidence for cognitive adaptive specializations in tool-using woodpecker finches. *Animal Behaviour*, *82*, 945–956. Available from: http://dx.doi.org/10.1016/j.anbehav.2011.07.032.

Teschke, I., Wascher, C. A. F., Scriba, M. F., Bayern, A. M. P., von, Huml, V., Siemers, B., et al. (2013). Did tool-use evolve with enhanced physical cognitive abilities? *Philosophical Transactions of the Royal Society of London. Series B, Biological Sciences*, *19*, 20120418. Available from: http://dx.doi.org/10.1098/rstb.2012.0418.

Troscianko, J., von Bayern, A. M. P., Chappell, J., Rutz, C., & Martin, G. R. (2012). Extreme binocular vision and a straight bill facilitate tool use in New Caledonian crows. *Nature Communications*, *3*, 1110. Available from: http://dx.doi.org/10.1038/ncomms2111.

Waddington, C. H. (1961). Genetic assimilation. *Advances in Genetics*, *10*, 257–293.

Webster, S. J., & Lefebvre, L. (2001). Problem solving and neophobia in a columbiform–passeriform assemblage in Barbados. *Animal Behaviour*, *62*, 23–32. Available from: http://dx.doi.org/10.1006/anbe.2000.1725.

Wyles, J. S., Kunkel, J. G., & Wilson, A. C. (1983). Birds, behavior and anatomical evolution. *Proceedings of the National Academy of Sciences of the United States of America*, *80*, 4394–4397. Available from: http://dx.doi.org/10.1073/pnas.80.14.4394.

Commentary on Chapter 6: Can Sol's Explanation for the Evolution of Animal Innovation Account for Human Innovation?

The aim of Sol's provocative chapter is to develop a comprehensive framework for the evolution of animal innovation. The framework can be summarized as follows:

- Innovation propensity is a by-product of a combination of traits including motivation, emotional responses, cognitive abilities, and morphological constraints.
- These traits initially evolved for other functions and were co-opted for innovative problem solving through exaptation.
- Through genetic assimilation, learned traits—such as those that underlie innovation propensity—may eventually become innate.

In this commentary we discuss this framework and (as instructed by the editors) take the ball and run with it.

INNOVATIVE CAPACITY AS EXAPTATION

The rationale for Sol's argument (that innovation propensity is not a specialized adaptation resulting from targeted selection but an instance of exaptation, that is, a by-product of selection for other traits) is that selection cannot act on situations that are only encountered once. (You only have to innovate once because if you encounter that situation again you can simply *remember* what you did the first time.) He proposes that all natural selection can do to prepare you to seize the moment and act on affordances for innovation when they present themselves is provide you with the following general characteristics: (i) *motivation*, for example, hunger may increase persistence in finding ways to obtain food, (ii) *emotional responses*, for example, neophilia versus neophobia, (iii) *cognitive abilities* such as attention, discrimination, and the capacity for episodic learning, (iv) *morphological and physiological constraints* on the type and diversity of motor patterns, including the ability to adjust actions in response to context, and (v) *time*, for example, a longer life span provides more time for problem solving. It is these traits, Sol argues, with their interacting constraints and trade-offs (which are in some cases functionally linked due to pleiotropic genetic effects and common physiological pathways) that have been the target of selection over the life history of an organism, not innovativeness itself.

We agree that these traits contribute to the capacity to innovate. We also believe they contribute to the capacity to find food, find a mate, care for offspring, and so forth, yet no one would argue that mate-finding or offspring care are exaptations. In exaptation (sometimes called pre-adaptation), a trait that originally evolved to solve one problem is co-opted to solve a new problem; the trait or traits in question must be necessary *and sufficient* to solve the new problem. Are persistence, neophilia, and so forth sufficient for innovation? Sol argues that this is so for innovation in animals, which he claims is generally a matter of trial and error. As researchers who focus on human innovation, we point out that while persistence, neophilia, and so forth, are necessary they are not sufficient for the strategic, intuitive, insightful, and even therapeutic processes involved in considering and reconsidering a complex idea from different real and imagined perspectives until all the bits and pieces fall into place.

Let us reconsider the starting point for Sol's argument: that selection cannot act on situations that are only encountered once. Viewed at a sufficiently fine level of granularity *all* situations are new situations that have never been encountered before (as Heraclitus said, you never step into the same river twice). Conversely, viewed at a sufficiently coarse level of granularity, all situations have been encountered previously. The issue then is: do novelty-affording situations collectively have enough in common at some intermediate level of granularity for selection to act upon? For example, is there a trait (or traits) that evolved, not as a by-product of some other function, but expressly for the purpose of coming up with innovative, adaptive responses to environmental variability itself?

We believe the answer is yes. This position is supported by experiments carried out using a computational model of cultural evolution that showed that the mean fitness of ideas across a society of artificial agents increases with the introduction of two innovation enhancing abilities: (i) *chaining*, the ability to combine simple ideas into complex ones, and (ii) *contextual focus*, the ability to shift from a convergent to a divergent processing mode when the fitness of one's current actions is low (Gabora & DiPaola, 2012; Gabora & Saberi, 2011). Moreover, both factors—chaining and contextual focus—proved most useful in times of environmental fluctuation (Gabora, Chia, & Firouzi, 2013). Of course, care must be taken in extrapolating from a simple computational model to the real world. However, the computer experiments are not the only source of support; Chrusch and Gabora (2014) synthesized these computational modeling results with findings from behavioral genetics, psychology, and anthropology to produce an integrated multilevel account of how chaining, contextual focus, and thereby human creative abilities could have evolved.

In short, these additions enable Sol's basic argument to be extended from trial-and-error innovative problem solving in animals to the more complex innovative abilities exhibited by humans.

THE ROLE OF CONTEXT

Sol stresses that the extent to which an animal expresses its capacity to innovate may depend on the relative costs and benefits of different actions *in its particular ecological context*; for example, populations of a given species living in more variable environments tend to be more innovative than those in predictable environments. Thus, he says, innovativeness hinges on not just the environment *per se*, but on the animal's *interaction* with its environment. We agree, and suggest that it has in this sense that innovation is exaptation. Gabora, Scott, and Kauffman (2013) developed a mathematical framework for exaptation with examples from both biological evolution and the evolution of cultural novelty through innovation. It is actually a quantum model, not in the sense of Penrose, but in the sense that it uses a generalization of the quantum formalism that was developed to model situations involving extreme contextuality in the macroworld. The state of a trait (or the starting point for an idea) is written as a linear superposition of a set of basis states, or possible forms the trait (or idea) could evolve into, in a complex Hilbert space. (For example, the basis states might represent possible ways of using a tire.) These basis states are represented by mutually orthogonal unit vectors, each weighted by an amplitude term. The choice of possible forms (basis states) depends on the context-specific goal or adaptive function of interest, which plays the role of an observable. (For example, in the context of wanting to create a playground someone turned a useless tire into a tire swing.) Observables are represented by self-adjoint operators on the Hilbert space. The possible forms (basis states) corresponding to this adaptive function (observable) are called eigenstates. In this model, innovative capacity did not evolve as an exaptation from some other, selected-for adaptive trait. Rather, innovation itself—or at least the retooling of an object or idea by considering it from a new point of view—is modeled as exaptation.

THE ROLE OF GENETIC ASSIMILATION

As Sol points out, when innovations are essential for survival, the nexus of traits underlying innovative capacity become canalized. A phenotypic response to an environmental condition, such as a learned innovative behavior, can over time be *genetically assimilated*, and thus innate. Some limitations of innate behavior are (i) it is rather unflexible and (ii) it operates over the course of biological generations. Thus while some kinds of innovation may be genetically assimilated it is unlikely that the innovations that fuel human cultural evolution are, given that they can unfold spontaneously over time frames of hours or minutes (e.g., humorous internet banter).

INNOVATION AS VIEWED BY THE ANIMAL BEHAVIOR LITERATURE VERSUS THE PSYCHOLOGICAL LITERATURE

We found it interesting to compare and contrast how innovation is viewed from the animal behavior literature versus the psychological literature. Sol's four stages in the innovation process—sampling, exploring, problem solving, and learning (by which mean means incorporating the solution into a behavioral repertoire)—bear some resemblance to Wallas' (1926) four stages of the creative process: preparation, incubation, illumination, and verification. The notion of "sampling" appears to be related to the notion of "problem finding" (Getzels & Csikszentmihalyi, 1976; Mumford, Reiter-Palmon, & Redmond, 1994; Runco & Chand, 1994), and the first two stages map onto the "generate" and "explore" stages of the creative cognition approach (Ward, 1995). What Sol refers to as *neophilia* seems comparable to the human personality trait of "openness to experience." Sol notes that innovativeness may be related to risk taking, and there is indeed evidence that highly creative individuals tend to take more risks (Merrifield, Guilford, Christensen, & Frick, 1961).

We note, however, that there are also differences in how innovativeness is viewed by these two fields. While Sol's focus is squarely on innovative problem solving (e.g., opening a lid to find hidden food), psychologists who seek a general scientific framework for creativity often unite innovative problem solving under the same broad umbrella as abilities such as art-making and scientific theorizing. Through Sol's chapter we came to better appreciate how by comparing and contrasting simple versus complex forms of innovation we sharpen our understanding of how new objects and forms of behavior come to be.

Acknowledgments

This research was supported in part by a grant from the Natural Sciences and Engineering Research Council of Canada.

References

Gabora, L., Chia, W.W., & Firouzi, H. (2013). A computational model of two cognitive transitions underlying cultural evolution. In *Proceedings of the 35th annual meeting of the cognitive science society* (pp. 2344–2349). Houston, TX: Cognitive Science Society.

Gabora, L., & DiPaola, S. (2012). How did humans become so creative? In *Proceedings of the international conference on computational creativity* (pp. 203–210). Dublin, Ireland.

Gabora, L. & Saberi, M. (2011). How did human creativity arise? An agent-based model of the origin of cumulative open-ended cultural evolution. In *Proceedings of the ACM conference on cognition & creativity* (pp. 299–306). Atlanta, GA.

Gabora, L., Scott, E., & Kauffman, S. (2013). A quantum model of exaptation: Incorporating potentiality into biological theory. *Progress in Biophysics & Molecular Biology, 113*(1), 108–116.

Getzels, J. W., & Csikszentmihalyi, M. (1976). *The creative vision: A longitudinal study of problem finding in art*. New York, NY: Wiley.

Merrifield, P. R., Guilford, J. P., Christensen, P. R., & Frick, J. W. (1961). Interrelationships between certain abilities and certain traits of motivation and temperament. *Journal of General Psychology, 55*, 57–74.

Mumford, M. D., Reiter-Palmon, R., & Redmond, M. R. (1994). Problem construction and cognition: Applying problem representation in ill-defined domains. In M. A. Runco (Ed.), *Problem finding, problem solving, and creativity*. Norwood, NJ: Ablex Publishing Corporation.

Runco, M., & Chand, I. (1994). Problem finding, evaluative thinking, and creativity. In M. Runco (Ed.), *Problem finding, problem solving, and creativity* (pp. 40–76). Norwood, NJ: Hampton.

Wallas, G. (1926). *The art of thought*. London: Cape.

CHAPTER

7

The Creative Cerebellum: Insight from Animal and Human Studies

Laura Petrosini[1,2], Debora Cutuli[1,2], Paola De Bartolo[1,3] and Daniela Laricchiuta[1,2]

[1]Laboratory of Experimental and Behavioral Neurophysiology, IRCCS Fondazione Santa Lucia, Rome, Italy [2]Department of Psychology, Medicine and Psychology Faculty, University Sapienza of Rome, Rome, Italy [3]Department of Sociological and Psychopedagogical Studies, Faculty of Formation Sciencies, University "Guglielmo Marconi", Rome, Italy

Commentary on Chapter 7: (How) Does the Cerebellum Contribute to Creativity?

Mathias Benedek

Department of Psychology, University of Graz, Graz, Austria

INTRODUCTION

Creativity is a process representing the development of innovative thoughts or actions, involving the generation of novelty and transformation of the existent, and ending with the creation of original product or the production of unusual response to the environment (Kaufman & Beghetto, 2009). The innovative pattern may be repeated and probably intra- and trans-generationally transmitted, when useful.

Understanding how the complex creative phenomenon occurs and what neuronal processing is involved is a challenge addressable by means of studies on neurophysiological and neurobiological aspects of creativity.

It has been demonstrated that highly creative individuals tend to have a high basal level of arousal, to be physiologically overreactive to stimulation, and to perform relatively poorly on biofeedback tasks (Martindale, 1977). During creative tasks, highly creative individuals exhibit regional cerebral blood flow (rCBF) in prefrontal cortex (PFC) which is higher than poorly creative individuals (Carlsson, Wendt, & Risberg, 2000). Positive correlations have been also found between creativity levels and rCBF in cerebellum, frontal and parietal cortices, and parahippocampal gyrus (Chávez-Eakle, Graff-Guerrero, García-Reyna, Vaugier, & Cruz-Fuentes, 2007). Furthermore, by using functional magnetic resonance imaging the activation in dorso-lateral PFC (DLPFC) and cerebellum has been demonstrated to covary with the number of solutions generated from the creative process (Goel & Vartanian, 2005), providing evidence of a distributed neural network related to creative process.

The three-level model of creativity advances specific neuronal sites related to each step of a creative process and provides a link between human creativity and animal innovation (Kaufman, Butt, Kaufman, & Colbert-White, 2011). In this model, the first level is composed by the ability to *recognize novelty*, a process mainly attributed to hippocampal function (Kumaran & Maguire, 2009), and to *search for novelty*, a process mainly attributed to dopaminergic systems (Marusich, Darna, Charnigo, Dwoskin, & Bardo, 2011; Van Gestel et al., 2002). The second level is *observational learning*, the ability to acquire a new competence by observing other's action. Observational learning, ranging from imitation to the transmission of the creative behavior, may critically depend on cerebellum, in addition to cortical regions (Graziano et al., 2002; Leggio et al., 2000; Petrosini, 2007; Petrosini et al., 2003; Torriero, Oliveri, Koch, Caltagirone, & Petrosini, 2007; Torriero et al., 2011). The third level is *innovative behavior*, including the establishment of new behaviors or the creation of new tools. Innovative behavior seems mainly to depend on PFC and balance between left and right hemisphere functions (Aupperle & Paulus, 2010).

In many years of research on animals and humans, we analyzed behaviors of novelty recognition, novelty seeking, observational learning, imitation, and cognitive flexibility by means of variegated methodological approaches based on virtual lesions or neuroimaging, surgical lesions or analysis of spontaneous individual differences. In all these behaviors we evidenced the cerebellar involvement that may represent the *fil rouge* linking multiple cognitive and emotional functions. Here, we are re-interpreting our behavioral data and their cerebellar substrate in the framework of the theory of creativity. It is intriguing to give our findings a "creative" twist.

After being connected to motor and cognitive functions, cerebellar structures have been completely reconsidered as playing new roles on a variety of domains. Because of the large number of anatomo-functional connections with PFC and basal ganglia, the cerebellum provides a fast computational system for timing, sequencing, and modeling aimed at the rapid and flexible manipulation of motor, cognitive and emotional processes (Bostan & Strick, 2010; Koziol, Budding, & Chidekel, 2010). It may render such processes increasingly more efficient and adaptive, mainly in facing functions linked to exploration and novelty. In accordance, Vandervert, Schimpf, and Liu (2007) proposed a cognitive theory of creativity which combines the working memory model, related to PFC functioning, with dynamic models of the cerebellum.

In the present chapter, the cerebellar involvement will be analyzed in the specific facets of the creative process by taking into account both animal and human studies. The integration of behavioral neuroscience into creativity theory will allow us to insert the interesting topic of animal innovation into the wider framework of neurobiological and behavioral animal studies.

CEREBELLAR INVOLVEMENT IN NOVELTY RECOGNITION AND NOVELTY SEEKING

A crucial behavior in novelty management is the explorativity; that is the proclivity to search for unfamiliar situations, making the unknown known. Cerebellar areas are highly involved in explorative functions that—by requiring close integration between environmental (sensory) information and searching (motor) acts—mimic the sensori-motor role classically attributed to cerebellar networks. In experimental studies, explorative components include behaviors such as suppressing the discomfort caused by unfamiliar spaces, leaving the known starting areas, building efficient foraging strategies and acquiring and using snapshots of the target view and of the representation-forming procedures (Foti et al., 2010; Mandolesi, Leggio, Spirito, & Petrosini, 2003). The explorative drive then represents a prerequisite of the recognition of and seeking for novelty, crucial components of creativity both in humans and animals. In rodents, a method for studying novelty recognition is to infer it from the behavior put into action by the animal in the open field (OF) task. The OF apparatus consists of an arena delimited by a wall. During the first session of the task, each animal is allowed to move freely in the empty arena and its baseline activity level is measured. During the subsequent habituation phase, some objects are placed in a specific arrangement in the arena and the animal is allowed to contact and explore them. During the spatial change phase, the object spatial configuration is changed by moving some objects so that the initial configuration is modified into another one. The animal typically contacts the displaced object more than the not-displaced ones. During the novelty phase, the configuration is altered by substituting one object with a new one. Again the animal typically contacts the new object more than the familiar ones. Thus, any response to change (renewal of exploration) indicates that the animal is reacting to the mismatch between the current new situation and the mental representation of the initial one. Notoriously, novelty recognition is linked to knowledge acquisition and declarative memory, functions depending on hippocampal, parahippocampal and cortical regions (Kumaran & Maguire, 2009). Considering the wide reciprocal projections between cortex and cerebellum (Middleton & Strick, 2001; Schmahmann, 1996), it has been advanced that cerebellar circuits also play a key role in processing information linked to exploring new environments and managing the novelty. In fact, the performances of animals submitted to a hemicerebellectomy (HCb) resulted impaired in the OF task (Mandolesi et al., 2003). The lesioned animals succeeded in moving in the arena, in contacting the objects, and in habituating to the new environment. However, they did not react to environmental changes, in particular when their impaired explorative pattern (featured by a compulsive exploration of the peripheral sectors of the arena) did not fit with task requirement. Animals submitted to HCb were not able to represent a new environment because they were not able to explore it appropriately, suggesting that no declarative spatial learning is possible without appropriate procedural spatial learning. Mice with mutations causing cerebellar vermian hypoplasia exhibit reduced tendency to explore novel objects or environments and increased tendency to compulsively move about (Fransen et al., 1998). Also rats treated with a single injection of lipopolysaccharide (LPS) to induce neuroinflammation showed specific and persistent impairment in reacting to the environment modifications in the OF task (Bossù et al., 2012). In fact, during the spatial change or novelty phase LPS-treated rats failed to recognize the new spatial arrangement and novel object. Interestingly, their impaired novelty recognition was accompanied by a marked increase of inflammatory cytokines in cerebellum, hippocampus and frontal cortex (Bossù et al., 2012).

Also neophilia- and neophobia-related behaviors may be indices of novelty recognition in animals. Greenberg (2003) defines neophilia or neophobia as behaviors rendering the animal attracted or repulsed by novel foods, objects or places, respectively. Although neophilia and neophobia might be at opposite ends of a behavioral continuum, at times they can coexist and the ability to balance the two drives may lead to a higher chance for survival in conflicting situations. An experimental conflict task singling out neophilic and neophobic behaviors is the approach/avoidance Y-maze. This test requires the subject to choose between two conflicting drives, reaching a new attractive reward (highly palatable food) placed in an aversive (white and lit) environment or reaching a familiar food (standard pellets) placed in a reassuring (black and opaque) environment (Laricchiuta, Rojo et al., 2012; Laricchiuta, Rossi et al., 2012). By using this task, we succeeded in detecting three phenotypes of mice that spontaneously responded with avoiding (neophobic), balancing or approaching (neophilic) behaviors to the reward. Interestingly, together with a different control of endocannabinoid system on striatal GABAergic neurotransmission, the approaching animals were more attracted by the new reward in the Y-maze and more active when facing a new context or an unknown object in the OF task (Laricchiuta, Rossi et al., 2012) (Figure 7.1A). Their neophilic behavior was associated with increased brain-derived neurotrophic factor (BDNF) levels just in the cerebellum. Notably, the protein BDNF, which belongs to the neurotrophin class, serves as survival factor for selected neuronal populations and plays a key role in synaptic plasticity as well as in learning, memory, and higher thinking (Greenberg, Xu, Lu, & Hempstead, 2009). Thus, the enhanced cerebellar BDNF levels could support the enhanced novelty responses displayed by approaching mice.

Increased exploration frequently results in coming across new events, and likewise the discovery of a new stimulus reinforces further exploration and detection of novelty. Thus, explorativity, recognition of, and seeking for novelty appear closely related behaviors. The connection between novelty seeking and creativity can be detected when analyzing the creative personality. Studying the human personality, Cloninger (1987) identified primary-basic personality temperament and character traits. Namely, Novelty Seeking, Harm Avoidance, Reward Dependence, and Persistence are the four temperamental traits described in his Temperament and Character Inventory. Individuals with high Novelty Seeking scores are exploratory, impulsive, fickle, excitable, quick tempered, and extravagant, whereas those with high Harm Avoidance scores are worried in anticipation of future problems, passive, fearful of uncertainty, shy, and easily fatigued. Thus, Novelty Seeking on one hand and Harm Avoidance on the other hand provide mechanisms to expand the range of stimuli and possibilities or to protect from potentially aversive contexts, respectively, supplying thus the appropriate feedback for sculpting the brain and developing interest in specific domains (Koziol et al., 2010). In a large sample of healthy subjects of both sexes of a range of ages, in recent neuroimaging studies (Laricchiuta et al., 2014; Picerni et al., 2013) we described the powerful positive association between cerebellar volumes and Novelty Seeking scores as well as the powerful

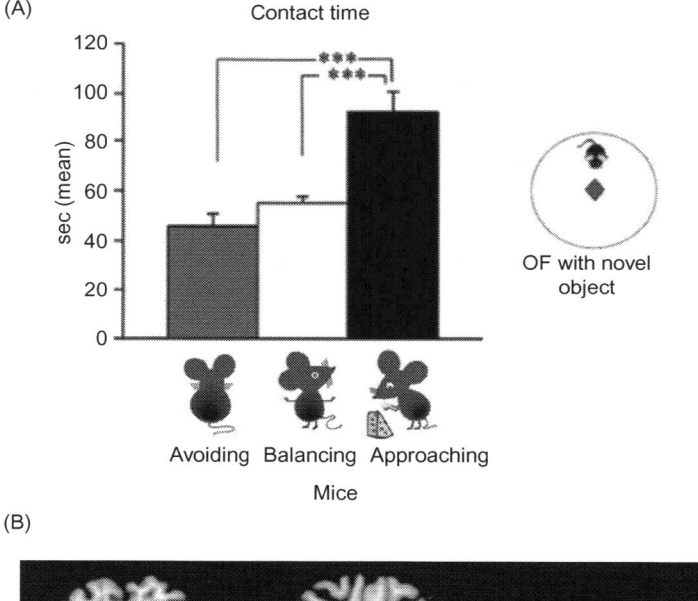

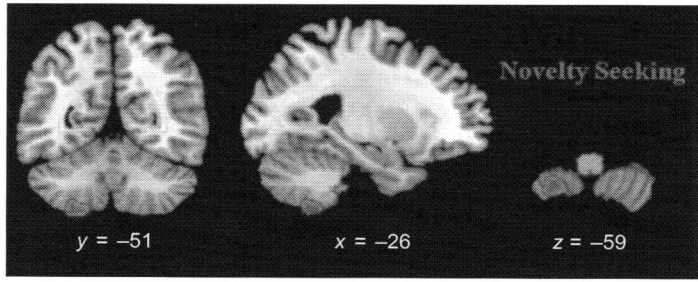

FIGURE 7.1 (A) By using the approach/avoidance Y-maze, we detected three phenotypes of mice that spontaneously responded to the reward with avoiding, balancing or
approaching behaviors. Notably, the approaching animals spent more time contacting the
new object in the OF task. (B) Representative slices (Montreal Neurological Institute template, MNI) showing that the cerebellar volumes were associated with Novelty Seeking
scores. The right side of the brain is shown on the right side. Coordinates are in MNI space.

negative association between cerebellar volumes and Harm Avoidance
scores. Namely, significant relationships between the macro- or micro-
structure of the cerebellar vermian lobules and Novelty Seeking scores
were found, emphasizing the role of posterior cerebellar vermis in the
neuroanatomical geography of novelty seeking trait (Figure 7.1B).

Furthermore, the temperamental traits are associated with emotional
processing, attentional focus, and inhibitory control, functions once
more governed by circuits involving cortex, basal ganglia and cerebellum. It appears then reasonable that even the volumes of basal ganglia,

in particular bilateral caudate and pallidum, correlated positively with Novelty Seeking scores (Laricchiuta et al., 2013).

In searching for the neuronal structure heavily involved in explorativity, recognition of and seeking for novelty, the cerebellum appears an ideal candidate, since it signals when sensory input differs from memory-driven expectations, guides exploratory drive in novel environments, allows flexible switching among multiple tasks or alternatives (Koziol et al., 2010). It was advanced that by detecting "discordances between the input from the deviant event and the sensory memory representation of the regular aspects of the preceding stimulation" (Naatanen & Michie, 1979), the cerebellum is the site where constant and deviant stimuli are compared, supporting the hypothesis that cerebellar processing is required for detecting the novelty of incoming somato-sensory stimuli (Restuccia, Della Marca, Valeriani, Leggio, & Molinari, 2007). In accordance with the cerebellar error/novelty-detection function, Ito (2008) proposed that internal models (either forward or inverse) are formed in the cerebellum to adapt motor and cognitive activities to contextual information. The internal model hypothesis for the control of novelty-related mental activities considers that the mismatch between the mentally generated solution and the new incoming information would activate the brain novelty system. Such a system consists of the hippocampal CA1 area and midbrain dopaminergic neurons in the ventral tegmental area (VTA) (Lisman & Grace, 2005). The activation of this circuit begins when the CA1 area detects new information. The resulting novelty signal is sent via subcortical pathways passing through the ventral striatum to the VTA neurons which fire in response to novelty (Bunzeck & Düzel, 2006). The novelty system in turn would activate the attentional system during the explicit phase of novelty-related thought. In triggering the new mental activity, the cerebellum could alarm the PFC about the absence of internal models matching the novel information, maintain the newly generated internal models, and incorporate them into routine schemes of thought. Thus, to successfully manage novelty, the co-activation of cerebellum and neocortical/sub-cortical areas appears needed (D'Angelo & Casali, 2012). Although this framework is largely speculative, it may be valid for supporting the cerebellar involvement in novelty recognition and novelty seeking, both in humans and animals.

CEREBELLAR INVOLVEMENT IN OBSERVATIONAL LEARNING

Humans and animals may learn new competencies through active experience or observation of others' experiences (Bandura, 1977; Meltzoff,

Kuhl, Movellan, & Sejnowski, 2009; Petrosini, 2007). In fact, although animals or humans completely isolated from others could yet reach innovative acquisitions through strategies of learning by trial and error, the observation of an experienced individual performing a complex action accelerates the observer's acquisition of the same action and limits the time-consuming process of learning by trial and error (Meltzoff et al., 2009). Observational learning not only involves copying an action, but also requires that the observer transforms the observation into an action as similar as possible to the model in terms of the goal to be reached and the motor strategies to be applied (Meltzoff & Decety, 2003). It requires the coordination of complex cognitive functions, such as action representation, attention and motivation, and at same time it requires understanding others' gestures, and making inferences about their behaviors, and thus represents a powerful social learning mechanism (Bandura, 1977; Meltzoff et al., 2009).

Since observational learning allows adding novel behaviors to one's own repertoire through the observation of a model, it is crucially involved in the creative processes leading an individual to generate original, groundbreaking ideas and behaviors (Kaufman & Kaufman, 2004; Kowatari et al., 2009). Human observational learning studies demonstrated the important effect of modeling on creativity (Bandura, 1986). For instance, a greater probability of creative behaviors has been reported following observation of creative role models (Anderson & Yates, 1999; Belcher, 1975; Groenendijk, Janssen, Rijlaarsdam, & van den Bergh, 2013; Hooker, Nakamura, & Csikszentmihalhyi, 2003; Mueller, 1978; Shalley & Perry-Smith, 2001; Simonton, 1984). Namely, research has identified "creative climates" characterized by significant factors as freedom, challenge, risk taking, safety that increase the chance for creative ideas and innovative outcomes (Jaussi & Dionne, 2003; Tierney & Farmer, 2004).

In animals, as in humans, creative behavior is promoted by the observation of innovative acts by conspecifics. Even animals exhibit creative capacities: from tactical deception to tool-use in primates, from musical aptitudes to foraging innovation in birds, from social learning in felids and cetaceans to the creative capacities of elephants (Bates & Byrne, 2007; Zuberbühler, Gygax, Harley, & Kummer, 1996).

Observational learning may occur in different cognitive domains, as simple associative learning, complex spatial learning and tool use (Guzmán et al., 2009; Mitchell, Thompson, & Miles, 1997; Ottoni, Resende, & Izar, 2005; Petrosini et al., 2003; Price, Lambeth, Schapiro, & Whiten, 2009). It is not limited to young subjects, but continues into adulthood (Russon, 2006). Furthermore, observational learning allows the transmission of "culture" within the animals' group (Whiten & Van Schaik, 2007), as reported for trans-generationally transmitted foraging behaviors in dolphins (Sargeant & Mann, 2009) or social learning in chimpanzees (Lycett, 2010).

Observing others' actions involves more than just observing a visual pattern. It implies the generation of an image of oneself performing the same action. Thus, imitating another's actions, as well as anticipating the effects of an action or representing (and feeling) the self in action, is based on the formation of motor images. Not by chance, the actual execution of a motor performance, the motor imagery of that performance, and the observation and imitation of that performance carried out by others share a number of common neural areas (Jeannerod, 1994). Thus, the mechanisms involved in learning by seeing are very similar to those involved in learning by doing. Both of them require the acquisition of varied competencies linked to the content of the action, single elements to be acquired, exact sequences of elements to be reproduced, and so on (Petrosini, 2007).

Within the neuronal structures controlling motor complex abilities, the cerebellum is also critically involved in observational learning, because of its crucial role in the internal representation of action (Petrosini et al., 2003). Through an original paradigm developed to study the observational learning in rodents (Figure 7.2), we demonstrated the cerebellar role in this facet of the "motor thought" (Leggio et al., 2000). In fact, cerebellar integrity is necessary not only to acquire spatial procedural abilities during the actual execution of a spatial task (Mandolesi et al., 2001; Petrosini, Molinari, & Dell'Anna, 1996; Petrosini et al., 2003), but also for their acquisition by observation. In our observational learning paradigm, normal rats observed another conspecific exploring a Morris water maze (MWM), a challenging task that employs a variety of navigational (procedural strategies) and mnesic (localizatory competencies) processes to search and remember the localization of a hidden platform in a water pool. To successfully navigate and locate the platform it is necessary to acquire a sequence of progressively tuned strategies (from peripherally circling to searching around to directly finding the platform) and to build a cognitive spatial map (encompassing intra- and extra-maze cues). After the observational training, the observer rat underwent a HCb and was then actually tested in the MWM. In spite of the cerebellar lesion, the observationally trained rat did not show any mnesic or explorative deficit (such as compulsive circling behavior that characterizes a hemicerebellectomized rat). The observer rat benefited from the prelesion observation, acquiring explorative competencies as efficient as those acquired by actually performing the MWM task. When the cerebellar lesion preceded the observation training, the observer rat showed a complete lack of spatial observational learning. Thus, the cerebellar circuitries provide a common neural basis for the observation of action to be reproduced and for the actual production of the same action.

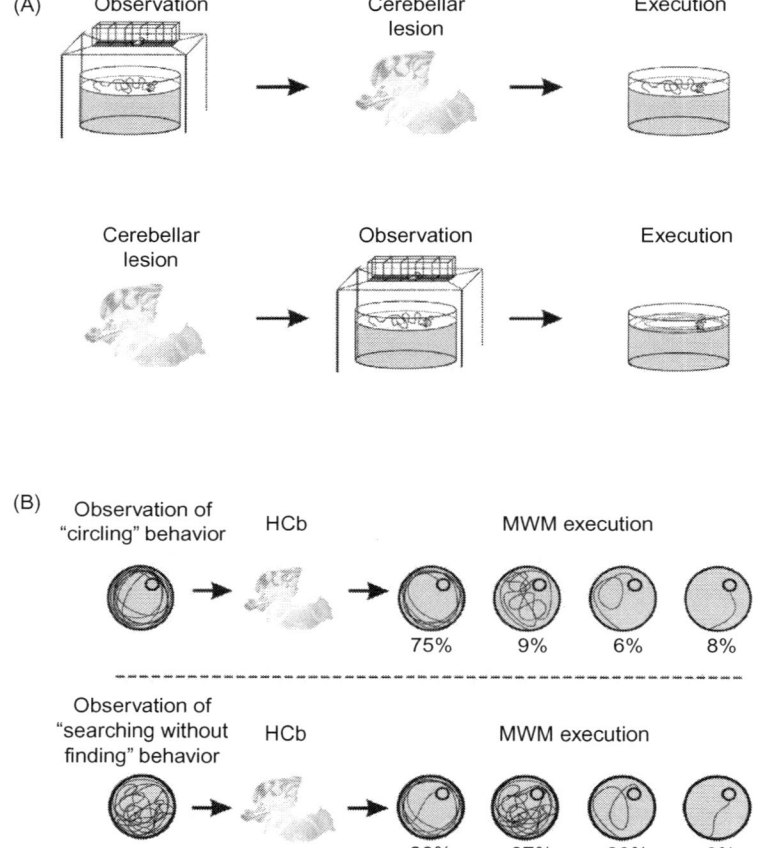

(A) Observation → Cerebellar lesion → Execution

Cerebellar lesion → Observation → Execution

(B)

Observation of "circling" behavior — HCb — MWM execution

75% 9% 6% 8%

Observation of "searching without finding" behavior — HCb — MWM execution

26% 37% 28% 8%

Observation of "finding without searching" behavior — HCb — MWM execution

61% 11% 12% 15%

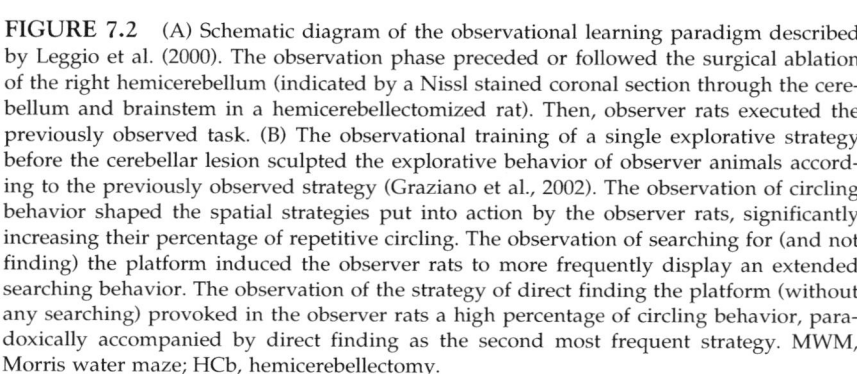

FIGURE 7.2 (A) Schematic diagram of the observational learning paradigm described by Leggio et al. (2000). The observation phase preceded or followed the surgical ablation of the right hemicerebellum (indicated by a Nissl stained coronal section through the cerebellum and brainstem in a hemicerebellectomized rat). Then, observer rats executed the previously observed task. (B) The observational training of a single explorative strategy before the cerebellar lesion sculpted the explorative behavior of observer animals according to the previously observed strategy (Graziano et al., 2002). The observation of circling behavior shaped the spatial strategies put into action by the observer rats, significantly increasing their percentage of repetitive circling. The observation of searching for (and not finding) the platform induced the observer rats to more frequently display an extended searching behavior. The observation of the strategy of direct finding the platform (without any searching) provoked in the observer rats a high percentage of circling behavior, paradoxically accompanied by direct finding as the second most frequent strategy. MWM, Morris water maze; HCb, hemicerebellectomy.

An interesting issue regarding observational learning is whether it is possible to acquire authentic competence of the meaning of the observed action or whether the behavior put into action after observation is just a copy of the observed performance. Actually, using the described observational paradigm we demonstrated that the observer did not exactly reproduce the observed behavior (Leggio et al., 2000). Indeed, during observational training, the observer viewed the entire range of explorative strategies put into action by the actor (circling, extended or restricted searching, direct finding). Thus, the observer could not acquire a fixed sequence of actions because it never observed a fixed sequence of actions. When actually undergoing the task the observer did not merely copy the previously observed behavior, but rather it exhibited highly structured spatial strategies aimed at attaining the goal.

It has been proposed that the acquisition of skills has an organizational structure that can be dissected into its simpler units of behavior (Byrne & Russon, 1998; Petrosini et al., 2003). The possibility of dissecting a complex performance allows separately verifying different hypotheses about its acquisition. On one hand, to acquire full competence of a complex behavior, it may be enough to learn its single steps. These steps thus assume the value of autonomous behavioral units that can be assembled or not as toy building blocks. On the other hand, to acquire the complex behavior, it may be necessary to learn simultaneously the entire sequence of units that are acquired as a whole and remain joined together in a unique package. By combining the aforementioned observational paradigm with the cerebellar lesion that blocks the acquisition of any new strategy, it has been possible to dissect a complex behavior into its fundamental units and to study which relationships among units have to be maintained so that the entire behavior might be acquired (Graziano et al., 2002). Normal rats were first allowed to observe actor rats performing single explorative behaviors (circling, extended searching or direct finding), then they were hemicerebellectomized and, finally, tested in the MWM. In spite of the cerebellar lesion, the observer rat displayed exploration abilities that closely matched the previously observed behaviors, indicating that the single facets that form the strategy repertoire can be independently acquired. Thus, observing the behavior of others represents a real learning process that biases the observer's behavior. To learn the entire performance, is it more favorable to observe a single strategy or the whole repertoire of strategies? In the latter case, the observer should learn "together with" the actor, taking advantage of its progressive learning of the various steps. Or, conversely, being exposed to a variety of strategies during the observational training could confound the observer, impairing its learning. Interestingly, the observer reached the highest performance level when the observational training involved the whole repertoire of strategies (Graziano et al., 2002). In fact, when actually doing the task, the observer exhibited the correct sequencing of strategies, demonstrating the learning not only of the single strategies but also of their appropriate use and sequencing. It appears that by observation it is possible to learn how to solve the task and understand the relations between specific actions and outcomes. This effect of the model observation as aid to skill acquisition may be explained by considering that learning by observation is not limited to the formation of a perceptual template, but implicates processes of "internal event generation," thus functioning as an externally guided form of mental imagery (Jeannerod, 1994; Petrosini et al., 2003).

Studies in monkeys and in humans show the existence of a widespread "mirror system" in frontal, parietal, and temporal regions (Ferrari, Rozzi, & Fogassi, 2005; Frey & Gerry, 2006; Oztop, Kawato, & Arbib, 2013; Rizzolatti, Fogassi, & Gallese, 2001). These studies have demonstrated that the wide neural network involved in motor representation, including cerebellum, is already tuned to imitating and learning actions by observation. The hypothesis that the same neural structures subserve the execution, imagination, and observation of actions is supported by studies demonstrating that learning by seeing, as learning by doing, depends on the activity of a neural network that involves the connectivity of lateral cerebellum with primary motor cortex and DLPFC (Torriero et al., 2007, 2010). Interestingly, the cerebellar involvement in the automatization of new visuo-spatial sequences acquired by observation has been recently demonstrated even in some developmental disorders (Foti et al., 2013, 2014).

Furthermore, the cerebellum is involved in learning to use tools so that this ability varies along with the cerebellar morphology across species. There is a strong correlation between tool use and the overall length of cerebellar foliation or cerebellar cortex folding in birds and mammals (Iwaniuk, Lefebvre, & Wylie, 2009). In fact, by comparing cerebellar foliation degree in a large number of birds, it has been found that birds well-known for their intelligence and tool-using tendency, such as parrots, corvids, and gulls show the greatest degree of foliation. A significant increase in cerebellar activity during tool use in monkeys (Obayski et al., 2001) and humans (Imamizu & Kawato, 2012; Imamizu, Kuroda, Miyauchi, Yoshioka, & Kawato, 2003) has also been demonstrated.

Functional neuroimaging studies in monkeys demonstrated that during a novel tool use task the rCBF increased in cerebellum, basal ganglia, pre-supplementary motor area, pre-motor cortex, and intra-parietal lobe (Obayski et al., 2001). Fronto-cerebellar and fronto-parietal circuits participate in maintaining and updating the body image for an accurate tool use. Also in humans, the dense anatomo-functional connections between the fronto-parietal and cerebellar regions indicate the crucial cerebellar contribution in tool-using acquisition (Imamizu & Kawato, 2012). Cerebellum contributes to the dexterous use of tools by providing information on the prediction of the sensory consequences of tool manipulation to the parietal regions, as well as on the motor control of tool-using to the pre-motor regions. Furthermore, by representing the kinematic or dynamic properties of tools, the cerebellum enables us to flexibly face environmental changes (Imamizu et al., 2003).

CEREBELLAR INVOLVEMENT IN COGNITIVE FLEXIBILITY

According to Kaufman's model (Kaufman et al., 2011), the third level of the process of creativity is innovation. The term "innovation" denotes mainly the product or outcome, as opposed to the term "creativity", which is more oriented to the processes engaged in. Describing innovation in animals may be difficult and subject to debate. Captive animals can be taught to be innovative by being reinforced for "doing something new" (Kaufman et al., 2011; Pryor, Haag, & O'Reilly, 1969). Within this framework, executive control (EC) may be inserted. EC is an umbrella term for the group of closely linked, high-level cognitive skills enabling animals and humans to manage different cognitive tasks to obtain a certain goal. Within the functions covered by EC, cognitive flexibility appears to be intriguingly linked to creative thinking (Pennington & Ozonoff, 1996). Cognitive flexibility (CF) allows animals and humans to adapt their behavior in response to changing environmental demands, that is, to learn how to link a changing context to a novel behavior. Thus, the ability to adjust thinking or attention in response to changing goals and/or environmental stimuli appears to be a vital component of adaptive behavior. The capacity to adjust one's thinking from old situations to new situations, as well as the ability to overcome habitual responses or thinking and shift or switch between different tasks or operations in response to a change in rules or demands may be interpreted as a part of creative thought. As such, if a subject in a new situation is able to overcome previously held beliefs or habits, then it may be considered cognitively flexible. In humans, the ability to simultaneously consider two aspects of an idea or situation and having the understanding and awareness of the possible options and alternatives simultaneously within a given situation is due to CF. All these considerations indicate how much the CF properties contribute to creativity.

To succeed in flexibly linking the context with the most appropriate response, properties of detection of novelty, use of working memory, performance monitoring, response inhibition, and selection or decision making are required (Dalley, Cardinal, & Robbins, 2004; Wolpaw & Carp, 2006). There is increasing evidence that in its different forms the CF is mediated by the medial PFC (Birrel & Brown, 2000; Miller, 2000) and the orbito-frontal cortex (Bougoulouris, Dalley, & Robbins, 2007). While medial PFC is involved in switching general rules, strategies or attentional sets (Birrel & Brown, 2000; Brown & Bowman, 2002; Ragozzino, Detrick, & Kesner, 1999), the orbito-frontal cortex has a role in stimulus-reinforcement associations.

Neuroanatomical studies have shown the pronounced fronto-cerebellar-frontal connectivity, consisting of a loop with projections between the DLPFC and cerebellum via the pontine nuclei, and projections between the cerebellum and PFC via the cerebellar dentate nucleus and thalamus (Baillieux, De Smet, Paquier, De Deyn, & Marïen, 2008; Heyder, Suchan, & Baum, 2004; Schmahmann, 1997). The reciprocal connections between the cerebellum and the PFC account for their interaction in planning new outcomes, the former by permitting acquisition of new efficient competencies, and the latter by providing flexible shifting among already acquired and stored solutions (Bellebaum & Daum, 2007; Koziol, 2013; Pochon et al., 2001). Interestingly, the phylogenetic enlargement of the cerebellar dentate parallels the enlargement of PFC, suggesting the participation of the dentate in the various facets of CF (Matano, 2001; Schmahmann & Sherman, 1998; Strick, Dum, & Fiez, 2009).

In humans, the cerebellar involvement in CF has been demonstrated by means of Wisconsin Card Sorting Task, a test assessing the ability of set-shifting (Lie, Specht, & Marshall, 2006) and the Stroop Color Word Test, a test assessing frontal attention and inhibition processes. In particular, in the Stroop Test the cerebellum was found to be activated during the incongruent part of the task, when the subjects had to name the colors of mismatching words—for example, the word red printed in green (Ravnkilde, Videbech, Rosenberg, Gjedde, & Gade, 2002). Interestingly, cerebellar involvement in CF functions has been recently demonstrated even in the presence of neurological disorders involving cerebellar circuitries (Brunamonti et al., 2014; Lalonde & Botez-Marquard, 2000).

Cerebellar involvement in CF has also been demonstrated in animals. Hemicerebellectomized and intact rats were tested daily by using a four-choice serial task in which the sequence of correct choices changed every day (De Bartolo et al., 2008, 2009). The apparatus consisted of straight alley with five compartments separated by four panels with two unidirectional swinging doors of which only one could be opened. Each animal was given a sequence of four open doors to reach the reward placed in the final compartment. The sequence of open doors remained stable for all trails of a daily sessions and was changed every day. As the correct sequence was unpredictable and different every day, the task required a continuous and flexible change in response. In the presence of a cerebellar lesion it was very difficult to forget the previous correct sequence and acquire a new one. In fact, while in the first daily sessions the lesioned animals exhibited similar performances to those of intact animals, as the session went by, they did not improve their performance. Thus, their responses were particularly defective in the final phases of the task that required "learning to learn" ability, a capacity exhibited markedly by the intact animals. This finding demonstrates the negative influence of cerebellar lesions on the acquisition, not the execution, of new responses (De Bartolo et al., 2009). This lack of flexibility appears to be due to impairment in planning intentional strategies, that is in the ability to access and use different strategies effectively to change behavior in accordance to a changing context.

The cerebellar involvement in CF was further demonstrated in a mouse model of developmental cerebellar Purkinje cell loss tested on an operant conditional visual discrimination task (Dickson et al., 2010). This test assessed the subject's ability to adapt its behavior to reversals of stimulus-response or stimulus-reward contingencies. Once more, cerebellar deficits impaired the final phases of the task when flexible responses were required. In fact, developmental cerebellar abnormalities resulted in consistent perseverative responses attributable to a failure to inhibit a strong learned response, evidencing a lack of flexibility.

CONCLUSIONS

In accordance with the wise argument of Koziol (2013), the role of the cerebellum in the various levels of the creative process could be to assist with automaticity by adapting behavior and ideas to a new context. The cerebellum works in developing both forward and inverse models (Ito, 2011). In the forward model, the cerebellum is informed by PFC about information load, plans, and intentions about what behavior to perform, as well as about the characteristics of the environment in which the behavior is to be carried out, developing progressive and anticipatory models. As the behavior and cognition are repeated and the anticipatory predicted feedback is received, the cerebellum becomes increasingly accurate in its predictive capacities. This allows behavioral execution to become more precise, fast, and independent of cortical control. With successful repetitions, behavior governed consciously by cerebellar forward model becomes more and more automated, developing the cerebellar "inverse" model that permits rapid and skilled behavior to occur at an unconscious level. The cerebellum is constantly constructing multi-pairs of models that constitute a complex modular architecture for adaptively regulating motor, cognitive and emotional material, so that the absence of these models limits creativity. Furthermore, the cerebellum may teach the PFC to predict or anticipate planning and organizational functions evidently inherent to creativity. The cerebellum may teach the PFC how to "think ahead."

In conclusion, creativity may be a fundamental characteristic of the mammalian nervous system and the development of novel and flexible responses is a luxury that only some individuals can afford (the "lure of the unknown"; Knutson & Cooper, 2006). Perceiving sameness and emitting fixed responses enhances efficiency and reliability, but at the same time promotes premature closure, perseveration and response rigidity. As Mesulam (1998) stated, "CNS has compensated for these limitations by developing specialized neural circuits for the rapid detection of unfamiliar events." In this sense, by loosening the rigid stimulus-response linkages, creative behaviors represent an antitheses to the pursuit of sameness and endow the organism with rapid change and adaptation of responses.

References

Anderson, A., & Yates, G. C. R. (1999). Clay modeling and social modeling: Effects of interactive teaching on young children's creative artmaking. *Educational Psychology, 19,* 463–470.

Aupperle, R. L., & Paulus, M. P. (2010). Neural systems underlying approach and avoidance in anxiety disorders. *Dialogues in Clinical Neuroscience, 12,* 517–531.

Baillieux, H., De Smet, H. J., Paquier, F., De Deyn, P. P., & Mariën, P. (2008). Cerebellar neurocognition: Insights into the bottom of the brain. *Clinical Neurology and Neurosurgery, 110,* 763–773.

Bandura, A. (1977). *Social learning theory.* Englewood Cliffs, NJ: Prentice Hall.

Bandura, A. (1986). *Social foundations of thought and action: A social cognitive theory.* Englewood Cliffs, NJ: Prentice Hall.

Bates, L. A., & Byrne, R. W. (2007). Creative or created: Using anecdotes to investigate animal cognition. *Methods, 42,* 12–21.

Belcher, T. L. (1975). Modeling original divergent responses: An initial investigation. *Journal of Educational Psychology, 67,* 351–358.

Bellebaum, C., & Daum, I. (2007). Cerebellar involvement in executive control. *Cerebellum, 6,* 184–192.

Birrel, J. M., & Brown, V. J. (2000). Medial frontal cortex mediates perceptual attentional set shifting in the rat. *Journal of Neuroscience, 20,* 4320–4324.

Bossù, P., Cutuli, D., Palladino, I., Caporali, P., Angelucci, F., Laricchiuta, D., et al. (2012). A single intraperitoneal injection of endotoxin in rats induces long-lasting modifications in behavior and brain protein levels of TNF-α and IL-18. *Journal of Neuroinflammation, 9,* 101.

Bostan, A. C., & Strick, P. L. (2010). The cerebellum and basal ganglia are interconnected. *Neuropsychology Review, 20,* 261–270.

Bougoulouris, V., Dalley, J. V., & Robbins, T. W. (2007). Effects of orbitofrontal, infralimbic and prelimbic cortical lesions on serial spatial reversal learning in the rat. *Behavioral Brain Research, 179*, 219–228.

Brown, V. J., & Bowman, E. M. (2002). Rodent model of prefrontal cortical function. *Trends in Neuroscience, 25*, 340–343.

Brunamonti, E., Chiricozzi, R. C., Clausi, S., Olivito, G., Giusti, M. A., Molinari, M., et al. (2014). Cerebellar damage impairs executive control and monitoring of movement generation. *PLoS One, 9*, e85997.

Bunzeck, N., & Düzel, E. (2006). Absolute coding of stimulus novelty in the human substantia nigra/VTA. *Neuron, 51*, 369–379.

Byrne, R. W., & Russon, A. E. (1998). Learning by imitation: a hierarchical approach. *Behavioral and Brain Sciences, 21*, 667–721.

Carlsson, I., Wendt, P. E., & Risberg, J. (2000). On the neurobiology of creativity. Differences in frontal activity between high and low creative subjects. *Neuropsychologia, 38*, 873–885.

Chávez-Eakle, R. A., Graff-Guerrero, A., García-Reyna, J. C., Vaugier, V., & Cruz-Fuentes, C. (2007). Cerebral blood flow associated with creative performance: A comparative study. *Neuroimage, 38*, 519–528.

Cloninger, C. R. (1987). A systematic method for clinical description and classification of personality variants. *Archives of General Psychiatry, 44*, 573–588.

Dalley, J. W., Cardinal, R. N., & Robbins, T. W. (2004). Prefrontal executive and cognitive functions in rodents: Neural and neurochemical substrates. *Neuroscience and Biobehavioral Reviews, 28*, 771–784.

D'Angelo, E., & Casali, S. (2012). Seeking a unified framework for cerebellar function and dysfunction: From circuit operations to cognition. *Frontiers in Neural Circuits, 6*, 116.

De Bartolo, P., Leggio, M. G., Mandolesi, L., Foti, F., Gelfo, F., Ferlazzo, F., et al. (2008). Environmental enrichment mitigates the effects of basal forebrain lesion on cognitive flexibility. *Neuroscience, 154*, 444–453.

De Bartolo, P., Mandolesi, L., Federico, F., Foti, F., Cutuli, D., Gelfo, F., et al. (2009). Cerebellar involvement in cognitive flexibility. *Neurobiology of Learning and Memory, 92*, 310–317.

Dickson, P. E., Rogers, T. D., Del Mar, N., Martin, L. A., Heck, D., Blaha, C. D., et al. (2010). Behavioral flexibility in a mouse model of developmental cerebellar Purkinje cell loss. *Neurobiology of Learning and Memory, 94*, 220–228.

Ferrari, P. F., Rozzi, S., & Fogassi, L. (2005). Mirror neurons responding to observation of actions made with tools in monkey ventral premotor cortex. *Journal of Cognitive Neuroscience, 17*, 212–226.

Foti, F., Mandolesi, L., Cutuli, D., Laricchiuta, D., De Bartolo, P., Gelfo, F., et al. (2010). Cerebellar damage loosens the strategic use of the spatial structure of the search space. *Cerebellum, 9*, 29–41.

Foti, F., Mazzone, L., Menghini, D., De Peppo, L., Federico, F., Postorino, V., et al. (2014). Learning by observation in children with autism spectrum disorder. *Psychological Medicine, 17*, 1–11.

Foti, F., Menghini, D., Mandolesi, L., Federico, F., Vicari, S., & Petrosini, L. (2013). Learning by observation: Insights from Williams syndrome. *PLoS One, 8*, e53782. Available from: http://dx.doi.org/10.1371/journal.pone.0053782.

Fransen, E., D'Hooge, R., Van Camp, G., Verhoye, M., Sijbers, J., Reyniers, E., et al. (1998). L1 knockout mice show dilated ventricles, vermis hypoplasia and impaired exploration patterns. *Human Molecular Genetics, 7*, 999–1009.

Frey, S. H., & Gerry, V. E. (2006). Modulation of neural activity during observational learning of actions and their sequential orders. *Journal of Neuroscience, 26*, 13194–13201.

Goel, V., & Vartanian, O. (2005). Dissociating the roles of right ventral lateral and dorsal lateral prefrontal cortex in generation and maintenance of hypothesis in set-shift problems. *Cerebral Cortex, 15*, 1170−1177.

Graziano, A., Leggio, M. G., Mandolesi, L., Neri, P., Molinari, M., & Petrosini, L. (2002). Learning power of single behavioral units in acquisition of a complex spatial behavior: An observational learning study in cerebellar-lesioned rats. *Behavioral Neuroscience, 116*, 116−125.

Greenberg, M. E., Xu, B., Lu, B., & Hempstead, B. L. (2009). New insights in the biology of BDNF synthesis and release: Implications in CNS function. *Journal of Neuroscience, 29*, 12764−12767. Available from: http://dx.doi.org/10.1523/JNEUROSCI.3566-09.2009.

Greenberg, R. (2003). The role of neophobia and neophilia in the development of innovative behaviour of birds. In S. M. Reader, & K. N. Laland (Eds.), *Animal innovation* (pp. 175−196). Oxford, England: Oxford University Press.

Groenendijk, T., Janssen, T., Rijlaarsdam, G., & van den Bergh, H. (2013). The effect of observational learning on students' performance, processes, and motivation in two creative domains. *British Journal of Educational Psychology, 83*, 3−28. Available from: http://dx.doi.org/10.1111/j.2044-8279.2011.02052.x.

Guzmán, Y. F., Tronson, N. C., Guedea, A., Huh, K. H., Gao, C., & Radulovic, J. (2009). Social modeling of conditioned fear in mice by non-fearful conspecifics. *Behavioural Brain Research, 201*, 173−178. Available from: http://dx.doi.org/10.1016/j.bbr.2009.02.024.

Heyder, K., Suchan, B., & Baum, I. (2004). Cortico-subcortical contribution to executive control. *Acta Psychologica (Amst), 115*, 271−289.

Hooker, C., Nakamura, J., & Csikszentmihalhyi, M. (2003). The group as mentor: Social capital and the systems model of creativity. In P. B. Paulus, & B. A. Nijstad (Eds.), *Group creativity*. Oxford, England: Oxford University Press.

Imamizu, H., & Kawato, M. (2012). Cerebellar internal models: Implications for the dexterous use of tools. *Cerebellum, 11*, 325−335. Available from: http://dx.doi.org/10.1007/s12311-010-0241-2.

Imamizu, H., Kuroda, T., Miyauchi, S., Yoshioka, T., & Kawato, M. (2003). Modular organization of internal models of tools in the human cerebellum. *Proceedings of the National Academy of Sciences, USA, 100*, 5461−5466.

Ito, M. (2008). Control of mental activities by internal models in the cerebellum. *Nature Reviews Neuroscience, 9*, 304−313.

Ito, M. (2011). *The cerebellum: Brain for an implicit self*. Upper Saddle River, NJ: FT Press.

Iwaniuk, A. N., Lefebvre, L., & Wylie, D. R. (2009). The comparative approach and brain-behaviour relationships: A tool for understanding tool use. *Canadian Journal of Experimental Psychology, 63*, 150−159. Available from: http://dx.doi.org/10.1037/a0015678.

Jaussi, K. S., & Dionne, S. D. (2003). Leading for creativity: The role of unconventional leader behavior. *The Leadership Quarterly, 14*, 475−498.

Jeannerod, M. (1994). The representing brain: Neural correlates of motor intention and imagery. *Behavioral and Brain Sciences, 17*, 187−245.

Kaufman, A. B., Butt, A. E., Kaufman, J. C., & Colbert-White, E. N. (2011). Towards a neurobiology of creativity in nonhuman animals. *Journal of Comparative Psychology, 125*, 255−272.

Kaufman, J. C., & Beghetto, R. A. (2009). Beyond big and little: The four C model of creativity. *Review of General Psychology, 13*, 1−12.

Kaufman, J. C., & Kaufman, A. B. (2004). Applying a creativity framework to animal cognition. *New Ideas in Psychology, 22*, 143−155.

Knutson, B., & Cooper, J. C. (2006). The lure of the unknown. *Neuron, 51*, 280−282.

Kowatari, Y., Lee, S. H., Yamamura, H., Nagamori, Y., Levy, P., Yamane, S., et al. (2009). Neural networks involved in artistic creativity. *Human Brain Mapping*, *30*, 1678—1690. Available from: http://dx.doi.org/10.1002/hbm.20633.

Koziol, L. F. (2013). From movement to thought: The development of executive function. *Applied Neuropsychology: Child*, *2*, 104—115.

Koziol, L. F., Budding, D. E., & Chidekel, D. (2010). Adaptation, expertise, and giftedness: Towards an understanding of cortical, subcortical, and cerebellar network contributions. *Cerebellum*, *9*, 499—529. Available from: http://dx.doi.org/10.1007/s12311-010-0192-7.

Kumaran, D., & Maguire, E. A. (2009). Novelty signals: A window into hippocampal information processing. *Trends in Cognitive Sciences*, *13*, 47—54. Available from: http://dx.doi.org/10.1016/j.tics.2008.11.004.

Lalonde, R., & Botez-Marquard, T. (2000). Neuropsychological deficits in patients with chronic or acute cerebellar lesions. *Journal of Neurolinguistics*, *13*, 117—128.

Laricchiuta, D., Petrosini, L., Piras, F., Cutuli, D., Macci, E., Picerni, E., et al. (2014). Linking novelty seeking and harm avoidance personality traits to basal ganglia: Volumetry and mean diffusivity. *Brain Structure and Function*, *219*, 793—803. Available from: http://dx.doi.org/10.1007/s00429-013-0535-5.

Laricchiuta, D., Petrosini, L., Piras, F., Macci, E., Cutuli, D., Chiapponi, C., et al. (2014). Linking novelty seeking and harm avoidance personality traits to cerebellar volumes. *Human Brain Mapping*, *35*, 285—296. Available from: http://dx.doi.org/10.1002/hbm.22174.

Laricchiuta, D., Rojo, M. L., Rodriguez-Gaztelumendi, A., Ferlazzo, F., Petrosini, L., & Fowler, C. J. (2012). CB1 receptor autoradiographic characterization of the individual differences in approach and avoidance motivation. *PLoS One*, *7*, e42111.

Laricchiuta, D., Rossi, S., Musella, A., De Chiara, V., Cutuli, D., Centonze, D., et al. (2012). Differences in spontaneously avoiding or approaching mice reflect differences in CB1-mediated signaling of dorsal striatal transmission. *PLoS One*, *7*, e33260.

Leggio, M. G., Molinari, M., Neri, P., Graziano, A., Mandolesi, L., & Petrosini, L. (2000). Representation of actions in rats: The role of cerebellum in learning spatial performances by observation. *Proceedings of the National Academy of Sciences, USA*, *97*, 2320—2325.

Lie, C. H., Specht, K., & Marshall, J. C. (2006). Using fMRI to decompose the neural processs underlying the Wisconsin Card Sorting Test. *Neuroimage*, *30*, 1038—1049.

Lisman, J. E., & Grace, A. A. (2005). The hippocampal-VTA loop: Controlling the entry of information into long-term memory. *Neuron*, *46*, 703—713.

Lycett, S. J. (2010). The importance of history in definitions of culture: Implications from phylogenetic approaches to the study of social learning in chimpanzees. *Learning & Behavior*, *38*, 252—264. Available from: http://dx.doi.org/10.3758/LB.38.3.252.

Mandolesi, L., Leggio. M. G., Graziano, A., Neri, P., & Petrosini, L. (2001). Cerebellar contribution to spatial event processing: Involvement in procedural and working memory components. *European Journal of Neuroscience*, *14*, 2011—2022.

Mandolesi, L., Leggio, M. G., Spirito, F., & Petrosini, L. (2003). Cerebellar contribution to spatial event processing: Do spatial procedures contribute to formation of spatial declarative knowledge? *European Journal of Neuroscience*, *18*, 2618—2626.

Martindale, C. (1977). Creativity, consciousness, and cortical arousal. *Journal of Altered States of Consciousness*, *3*, 69—87.

Marusich, J. A., Darna, M., Charnigo, R. J., Dwoskin, L. P., & Bardo, M. T. (2011). A multivariate assessment of individual differences in sensation seeking and impulsivity as predictors of amphetamine self-administration and prefrontal dopamine function in rats. *Experimental and Clinical Psychopharmacology*, *19*, 275—284. Available from: http://dx.doi.org/10.1037/a0023897.

Matano, S. (2001). Brief communication: Proportion of the ventral half of the cerebellar dentate nucleus in humans and great apes. *American Journal of Physical Anthropology*, *114*, 163−165.

Meltzoff, A. N., & Decety, J. (2003). What imitation tells us about social cognition: A rapprochement between developmental psychology and cognitive neuroscience. *Philosophical Transactions of the Royal Society of London. Series B, Biological Sciences*, *358*, 491−500.

Meltzoff, A. N., Kuhl, P. K., Movellan, J., & Sejnowski, T. J. (2009). Foundations for a new science of learning. *Science*, *325*, 284−288. Available from: http://dx.doi.org/10.1126/science.1175626.

Mesulam, M. M. (1998). From sensation to cognition. *Brain*, *121*, 1013−1052.

Middleton, F. A., & Strick, P. L. (2001). Cerebellar projections to the prefrontal cortex of the primate. *Journal of Neuroscience*, *21*, 700−712.

Miller, E. K. (2000). The prefrontal cortex and cognitive control. *Nature Reviews Neuroscience*, *1*, 59−65.

Mitchell, R. W., Thompson, N. S., & Miles, H. L. (1997). Taking anthropomorphism and anecdotes seriously. In R. W. Mitchell, N. S. Thompson, & H. L. Miles (Eds.), *Anthropomorphism, anecdotes, and animals*. Albany, NY: SUNY Press.

Mueller, L. K. (1978). Beneficial and detrimental modeling effects on creative response production. *Journal of Psychology*, *98*, 253−260.

Naatanen, R., & Michie, P. T. (1979). Early selective-attention effects on the evoked potential: A critical review and reinterpretation. *Biological Psychology*, *8*, 81−136.

Obayski, S., Suhara, T., Kawabe, K., Okauchi, T., Maeda, J., Akine, Y., et al. (2001). Functional brain mapping of monkey tool use. *Neuroimage*, *14*, 853−861.

Ottoni, E. B., Resende, B. D., & Izar, P. (2005). Watching the best nutcrackers: What capuchin monkeys (*Cebus apella*) know about others' tool-using skills. *Animal Cognition*, *24*, 215−219.

Oztop, E., Kawato, M., & Arbib, M. A. (2013). Mirror neurons: Functions, mechanisms and models. *Neuroscience Letters*, *540*, 43−55. Available from: http://dx.doi.org/10.1016/j.neulet.2012.10.005.

Pennington, B. F., & Ozonoff, S. (1996). Executive functions and developmental psychopathology. *Journal of Child Psychology and Psychiatry, and Allied Disciplines*, *37*, 51−87.

Petrosini, L. (2007). "Do what I do" and "do how I do": Different components of imitative learning are mediated by different neural structures. *Neuroscientist*, *13*, 335−348.

Petrosini, L., Graziano, A., Mandolesi, L., Neri, P., Molinari, M., & Leggio, M. G. (2003). Watch how to do it! New advances in learning by observation. *Brain Research Reviews*, *42*, 252−264.

Petrosini, L., Molinari, M., & Dell'Anna, M. E. (1996). Cerebellar contribution to spatial event processing: Morris water maze and T-maze. *European Journal of Neuroscience*, *8*, 1882−1896.

Picerni, E., Petrosini, L., Piras, F., Laricchiuta, D., Cutuli, D., Chiapponi, C., et al. (2013). New evidence for the cerebellar involvement in personality traits. *Frontiers in Behavioral Neuroscience*, *7*, 133. Available from: http://dx.doi.org/10.3389/fnbeh.2013.00133.

Pochon, J. B., Levy, R., Poline, J. B., Crozier, S., Lehéricy, S., Pillon, B., et al. (2001). The role of dorsolateral prefrontal cortex in the preparation of forthcoming actions: An fMRI study. *Cerebral Cortex*, *11*, 260−266.

Price, E. E., Lambeth, S. P., Schapiro, S. J., & Whiten, A. (2009). A potent effect of observational learning on chimpanzee tool construction. *Proceedings in Biological Science*, *276*, 3377−3383. Available from: http://dx.doi.org/10.1098/rspb.2009.0640.

Pryor, K. W., Haag, R., & O'reilly, J. (1969). The creative porpoise: Training for novel behavior. *Journal of Experimental Analysis of Behavior*, *12*, 653−661.

Ragozzino, M. E., Detrick, S., & Kesner, R. P. (1999). Involvement of the prelibic-infralimbic areas of the rodent prefrontal cortex in behavioral flexibility for place and response learning. *Journal of Neuroscience, 19*, 4585–4594.

Ravnkilde, B., Videbech, P., Gjedde, A., & Gade, A. (2002). Putative tests of frontal lobe function: A PET-study of brain activation during Stroop's test and verbal fluency. *Journal of Clinical and Experimental Neuropsychology, 24*, 534–547.

Restuccia, D., Della Marca, G., Valeriani, M., Leggio, M. G., & Molinari, M. (2007). Cerebellar damage impairs detection of somatosensory input changes. A somatosensory mismatch-negativity study. *Brain, 130*, 276–287.

Rizzolatti, G., Fogassi, L., & Gallese, V. (2001). Neurophysiological mechanisms underlying the understanding and imitation of action. *Nature Neuroscience Reviews, 2*, 661–670.

Russon, A. E. (2006). Acquisition of complex foraging skills in juvenile and adolescent orangutans *(Pongo pygmaeus)*: Developmental influences. *Aquatic Mammals, 32*, 500–510.

Sargeant, B. L., & Mann, J. (2009). Developmental evidence for foraging traditions in wild bottlenose dolphins. *Animal Behaviour, 78*, 715–721. Available from: http://dx.doi.org/10.1016/j.anbehav.2009.05.037.

Schmahmann, J. D. (1996). From movement to thought: Anatomic substrates of the cerebellar contribution to cognitive processing. *Human Brain Mapping, 4*, 174–198.

Schmahmann, J. D. (1997). *The cerebellum and cognition.* San Diego, CA: Academic Press.

Schmahmann, J. D., & Sherman, J. C. (1998). The cerebellar cognitive affective syndrome. *Brain, 121*, 161–179.

Shalley, C. E., & Perry-Smith, J. E. (2001). Effects of social-psychological factors on creative performance: The role of informational and controlling expected evaluation and modeling experience. *Organizational Behavior and Human Decision Processes, 84*, 1–22.

Simonton, D. K. (1984). Artistic creativity and interpersonal relationships across and within generations *Journal of Personality and Social Psychology, 46*, 1273–1286.

Strick, P. L., Dum, R. P., & Fiez, J. A. (2009). Cerebellum and nonmotor function. *Annual Review of Neuroscience, 32*, 413–434.

Tierney, P., & Farmer, S. M. (2004). The Pygmalion process and employee creativity. *Journal of Management, 30*, 413–432.

Torriero, S., Oliveri, M., Koch, G., Caltagirone, C., & Petrosini, L. (2007). The what and how of observational learning. *Journal of Cognitive Neuroscience, 19*, 1656–1663. Available from: http://dx.doi.org/10.1162/jocn.2007.19.10.1656.

Torriero, S., Oliveri, M., Koch, G., Lo Gerfo, E., Salerno, S., Ferlazzo, F., et al. (2011). Changes in cerebello-motor connectivity during procedural learning by actual execution and observation. *Journal of Cognitive Neuroscience, 23*, 338–348. Available from: http://dx.doi.org/10.1162/jocn.2010.21471.

Vandervert, L. R., Schimpf, P. H., & Liu, H. (2007). How working memory and the cerebellum collaborate to produce creativity and innovation. *Creativity Research Journal, 19*, 1–18.

Van Gestel, S., Forsgren, T., Claes, S., Del-Favero, J., Van Duijn, C. M., Sluijs, S., et al. (2002). Epistatic effect of genes from the dopamine and serotonin systems on the temperament traits of novelty seeking and harm avoidance. *Molecular Psychiatry, 7*, 448–450.

Whiten, A., & Van Schaik, C. (2007). The evolution of animal 'cultures' and social intelligence. *Philosophical Transactions of the Royal Society B: Biological Sciences, 362*, 603–620.

Wolpaw, J. R., & Carp, J. S. (2006). Plasticity from muscle to brain. *Progress in Neurobiology, 78*, 233–263.

Zuberbühler, K., Gygax, L., Harley, N., & Kummer, H. (1996). Stimulus enhancement and spread of spontaneous tool use in a colony of longtailed macaques. *Primates, 37*, 1–12.

Commentary on Chapter 7: (How) Does the Cerebellum Contribute to Creativity?

I am probably one of those who involuntarily raise an eyebrow when reading about the significance of the cerebellum for higher cognitive function such as creativity. Isn't the cerebellum responsible for coordination and automation of motor activity? Shouldn't creativity then be exclusively related to cerebral activity? This spontaneous response is not driven by a deep understanding about the function of the cerebellar system, but rather by the prevailing focus on the cerebrum in cognitive neuroscience research. However, it is not that one does not encounter the cerebellum in neuroscientific research on creativity—quite the opposite is true. FMRI studies from our and other labs have repeatedly revealed cerebellar brain activation during creative idea generation (e.g., Abraham et al., 2012; Aziz-Zadeh, Liew, & Dandekar, 2012; Benedek, Beaty et al., 2014; Benedek, Jauk et al., 2014), which is also evidenced in pertinent reviews (e.g., Gonen-Yaacovi et al., 2013). What these works have additionally in common is that they do not spend much or any time on discussing the cerebellar findings, but focus on the cerebral findings instead. This is certainly not due to contemptuous ignorance, but rather due to the lack of available cognitive models that would allow a clear-cut interpretation of the role of the cerebellum in creative thought. Cerebellar findings hence tend to be attributed tacitly to basic non-selective processes that are not specific for the targeted creative process (e.g., speech preparation, inner speech, or embodied cognition; Strick, Dum, & Fiez, 2009).

So far, there are only a few theoretical accounts relating the cerebellum to creativity (e.g., Vandervert, Schimpf, & Liu, 2007). I hence was very curious what Petrosini et al. (this book) have found out about the "creative cerebellum." They report sets of fascinating experimental studies demonstrating that cerebellar lesions in mice impedes observation learning of novel strategies but not the execution of those strategies. The authors conclude that the cerebellum plays a key role in the acquisition of a wide repertoire of efficient responses. Critically, it is assumed that the cerebellum not only forms models for motor actions, but also more generally for mental actions (cf. Ito, 2008), thus facilitating adaptive and flexible behavior. Further studies link the cerebellum to novelty-seeking and increased exploratory behavior. In humans, novelty-seeking is known to be positively correlated with extraversion, openness, and negatively with conscientiousness (De Fruyt, Van Wiele, &

Heeringen, 2000)—traits that are consistently associated with creativity (Feist, 1998).

It may be useful to specify what facet of the creativity construct— creative ability, or creative personality—is touched by the evidence brought together by Petrosini et al. I assume that there is still little evidence to conclude that the cerebellum is crucial for creative ability. While mental models may underlie imagination processes, the main cerebellar function of acquiring and refining response models towards efficient but highly predictive behavior does not seem to include a generative aspect (Abraham, 2007). Still, this system is thought to be involved in informing prefrontal structures when no relevant internal models are available and help to quickly incorporate new models to the repertoire (Ito, 2007). As such, and when additionally considering the evidence on novelty-seeking, the findings make a good case for a possible role of the cerebellum for creative personality.

Finally, it should be noted that the reported research approaches are very inspiring. While some experimental procedures seem out of scope for research on human creativity (e.g., hemicerebellectomy), or would need to be replaced by other approaches (e.g., inhibitory brain stimulation), the impracticability of having mice complete self-report questionnaires and tests has stimulated creative techniques for obtaining indicators of real-life creativity-related behavior. Adopting such techniques to the study of human creativity hence might open up fruitful new pathways for substantially increasing external validity in pertinent research.

References

Abraham, A. (2007). Can a neural system geared to bring about rapid, predictive, and efficient function explain creativity? *Creativity Research Journal, 19*, 19−24.

Abraham, A., Pieritz, K., Thybush, K., Rutter, B., Kröger, S., Schweckendiek, J., et al. (2012). Creativity and the brain: Uncovering the neural signature of conceptual expansion. *Neuropsychologia, 50*, 1906−1917.

Aziz-Zadeh, L., Liew, S.-L., & Dandekar, F. (2012). Exploring the neural correlates of visual creativity. *Social Cognitive Affective Neuroscience, 8*, 475−480.

Benedek, M., Beaty, R., Jauk, E., Fink, A., Silvia, P. J., Dunst, B., et al. (2014). Creating metaphors: The neural basis of figurative language production. *Neuroimage, 90*, 99−106.

Benedek, M., Jauk, E., Fink, A., Koschutnig, K., Reishofer, G., Ebner, F., et al. (2014). To create or to recall? Neural mechanisms underlying the generation of creative new ideas. *Neuroimage, 88*, 125−133.

De Fruyt, F., Van Wiele, L., & Heeringen, C. (2000). Cloninger's psychobiological model of temperament and the five-factor model of personality. *Personality and Individual Differences, 29*, 441−452.

Feist, G. J. (1998). A meta-analysis of personality in scientific and artistic creativity. *Personality and Social Psychology Review, 2*, 290−309.

Gonen-Yaacovi, G., de Souza, L. C., Levym, R., Urbanski, M., Josse, G., & Volle, E. (2013). Rostral and caudal prefrontal contribution to creativity: A meta-analysis of functional imaging data. *Frontiers in Human Neuroscience, 7*, 1–22.

Ito, M. (2007). On "How working memory and the cerebellum collaborate to produce creativity and innovation" by L.R. Vandervert, P.H. Schimpf, and H. Liu. *Creativity Research Journal, 19*, 35–38.

Ito, M. (2008). Control of mental activities by internal models in the cerebellum. *Nature Reviews Neuroscience, 9*, 304–313.

Strick, P. L., Dum, R. P., & Fiez, J. A. (2009). Cerebellum and nonmotor function. *Annual Reviews of Neuroscience, 32*, 413–434.

Vandervert, L. R., Schimpf, P. H., & Liu, H. (2007). How working memory and the cerebellum collaborate to produce creativity and innovation. *Creativity Research Journal, 19*, 1–18.

Animal Creativity: Cross-Species Studies of Cognition

*Kendra S. Knudsen[1], David S. Kaufman[2],
Stephanie A. White[3,4], Alcino J. Silva[5,6,7],
David J. Jentsch[1,3] and Robert M. Bilder[1,5,8]*

[1]Semel Institute for Neuroscience and Human Behavior, University of
California, Los Angeles, CA, USA [2]Department of Psychology, Saint Louis
University, Saint Louis, MO, USA [3]Molecular, Cellular and Integrative
Physiology Interdepartmental Program, University of California, Los
Angeles, CA, USA [4]Department of Integrative Biology and Physiology,
University of California, Los Angeles, CA, USA [5]Department of Psychiatry
and Biobehavioral Sciences, David Geffen School of Medicine at UCLA,
Los Angeles, CA, USA [6]Department of Neurobiology, David Geffen School
of Medicine at UCLA, Los Angeles, CA, USA [7]Department of Integrative
Center for Learning and Memory, David Geffen School of Medicine at
UCLA, Los Angeles, CA, USA [8]Department of Psychology, David Geffen
School of Medicine at UCLA, Los Angeles, CA, USA

Commentary on Chapter 8: Cross-Species Studies of Cognition

Oshin Vartanian

University of Toronto Scarborough, Toronto, ON, Canada

BACKGROUND

This volume epitomizes a transformation in thinking about creativity,
which has grown from a rarefied consideration of human artistic

endeavors, to a burgeoning scientific area of inquiry within which we are beginning to understand the basic biological purposes and substrates of the creative process.

Through the philanthropic vision of Michael E. Tennenbaum and his family, we at the UCLA Semel Institute have had a unique opportunity to pursue studies on the biological basis of creativity. Given the *Tennenbaum Family Center of the Biology of Creativity* (TFCBC) mandate to examine basic neuroscience bases of creativity, we needed to identify cognitive processes that could be examined across species (thereby enabling us to leverage the unique advantages of investigations that interrogate basic biological processes at the molecular and cellular levels).

We have identified three core cognitive themes that have high relevance for the study of creativity:

1. *novelty generation*—the ability to flexibly and adaptively generate products that are unique;
2. *working memory and declarative memory*—the ability to maintain, and then use relevant information to guide goal-directed performance, along with the capacity to store and retrieve this information; and
3. *response inhibition*—the ability to suppress habitual plans and substitute alternate actions in line with changing problem-solving demands.

THREE THEMES OF CREATIVE COGNITION

Novelty Generation

The ability to generate novel responses is a universal requirement for any current definition of creative thinking, namely producing new (and useful) thoughts. Novelty generation, or the ability to rapidly generate many, unique responses to a problem, is classically tied to creative cognition. The association between novelty generation and creativity originates from 1950, when J. P. Guilford made his famous APA Presidential Address that mobilized and expanded the scientific study of creativity (Guilford, 1950). In addition to flexibility, J. P. Guilford identified originality and fluency as the key components of creative thinking (Simonton, 2000). Many researchers following him have substantiated this view. This includes, notably, Simonton's (1997) empirically supported model showing scientists who produce the most tend to be the most creative. This relationship fits well with Ericsson's, Krampe, and Tesch-Römer's (1993) "10-year rule" (in which creative experts require at least 10 years of practice in order to achieve world-class creative success in any domain) and with prior work in creativity research indicating that, following an inverted-U distribution (Simonton, 1983, 1984), extensive and deliberate practice plays a strong role in creative achievement across a variety of domains—from musical performance to military leadership (e.g., see Ericsson et al., 1993; Simonton, 1980a, 1980b).

There are multiple possible reasons why increased production may result in increased creative production:

a. A trivial explanation is that it is a "base-rate" phenomenon. Those who produce more works may produce more creative works simply by chance, with the odds of any single product being considered "creative" being similar between groups. This is difficult to examine systematically in human studies because we cannot randomize individuals to be more or less productive. A less trivial interpretation is that by creating a diversity of products, the creator may be able to select the most valuable works, so the products that are presented to the world may already be extensively curated, thus reflecting "selective retention" from a broader range of "blind variations" as suggested in the be the BVSR theory (Campbell, 1960; Simonton, 2011a, 2011b).

b. The "10-year" or "10,000 h" rule is sometimes deemed important because it corresponds to extensive automatization of skills, such that even complex component processes can be executed readily as "habits" without using higher cognitive resources. In cognitive neuroscience terms this is often believed to correspond to the distinction between learning via the basal ganglia "habit" system versus "declarative" learning via the mesiotemporal lobe (hippocampal) and neocortical systems, respectively. An important but so far understudied implication, if this is true, is that more creative individuals smoothly execute or experience more complex combinations of elementary actions or ideas; this seems obvious from observation of skilled musicians or rappers, but is less clear in the creative works of scientists.

c. An unexplored (to our knowledge) hypothesis is that the creative individual may benefit from completing multiple works because it externalizes the work product, which in turn may have several consequences, including:

   i. by externalizing the work it is possible for the creator to examine it more "objectively" and determine in what ways it might be improved, or recombined with other works of the creator; and

   ii. by externalizing the work it frees the creator from the Zeigarnik Effect (Zeigarnik, 1935), insofar as incomplete actions tend to occupy mental resources.

So far the evidence about the association between overall production and creative production in humans remains largely correlational. As we will see below, it may be possible to disambiguate some of these findings in animal models.

Working Memory/Declarative Memory

Part of the rationale for targeting working and declarative memory is the consensus that to generate creative ideas, one must successfully encode, retrieve and then maintain multiple, *disparate* ideas in mind long enough to manipulate them and forge new, potentially *valuable* connections.

Many researchers believe that the more remote an idea is from its initial conception, the more likely it will be original and potentially creative (e.g., Baughman & Mumfors, 1995; Ward & Kolomyts, 2010). For novel ideas to be linked together from previously distinct ideational threads, they must be co-activated long enough to permit their association. Mednick (1962) describes this in terms of a flat association of hierarchies, which allows an individual to consider multiple items at once or a single item in great detail at once, and integrate associations for the simultaneous activation of far-flung ideas. Gabora's (2010) theory of "neurds" includes a comprehensive description of the possible relationship between creativity and the structure and activity of memory storage and retrieval in the brain.

Beyond theory to practice, the indirect association of working memory and creative cognition is supported. Strong evidence reveals working memory and fluid intelligence as very closely related constructs (e.g., Ackerman, Beier, & Boyle, 2005; Engle, Kane, & Tuholski, 1999) and research suggests a small, yet significant positive relationship of intelligence (and presumably, working memory) with cognitive tests and EEG studies of divergent thinking (Colom et al., 2007; Kim, 2008; Silvia & Beaty, 2012). The common belief that IQ influences creativity up to the "120 IQ threshold" (Barron, 1961; Guilford, 1967; MacKinnon, 1962; Simonton, 1994) may partly explain how intelligence may be necessary, but not sufficient, for creative cognition.

Psychometric data on direct correlations of cognitive constructs and working memory supports a close association between working memory capacity and the ability to perform a diversity of creative tasks. Working memory predicts performance on almost 40 different cognitive tests, including symbolic reasoning and verbal analogy (Kane et al., 2004); it benefits creative insight, musical improvisation, and novelty generation in semiprofessional musicians (De Dreu et al., 2012); and influences metaphor processing in healthy participants (Pierce et al., 2010).

Response Inhibition

The world at large and, markedly, many creativity researchers commonly view creative achievers as being uninhibited (e.g., Eysenck, 1995; Martindale, 1999, p. 143). Carson, Peterson, and Higgins (2003) reinforce this notion with their finding that decreased latent inhibition (or the decreased failure to screen out previously irrelevant stimuli) was associated with higher lifetime creative achievement.

Other work in cognitive neuroscience supports an opposite view—that response inhibition is a key component of cognitive control and the ability to *inhibit* habits and *overcome* prepotent responses may be as important as or possibly even more important than cognitive disinhibition. In a study of cognitive inhibition, high school students' performance in the color-word Stroop task (which requires participants to inhibit incongruous semantic meanings of words while naming the word's font color) was positively associated with performance on divergent thinking tasks and teacher ratings of creativity (Golden, 1975). Similarly Groborz and Nęcka (2003) found that creativity, as measured by a divergent figural production task, was linked with cognitive inhibition, as measured by the Stroop and the Navon task (which requires participants to inhibit incongruent features on local or global features of a stimulus).

More recently, Benedek, Franz, Heene, & Neubauer (2012) found a positive relationship between response inhibition (as measured by random motor generation) and tasks of insight and both figural and verbal divergent thinking; follow up latent variable analyses additionally revealed that response inhibition may predominantly enhance novelty generation (ideational fluency). This further supports our view that in order to generate new ideas, individuals must inhibit old ideas—perhaps most notably, they might need to inhibit the very first (mundane, old) idea that comes to mind.

Creative experimentation requires inhibition of habits and old ideas to access unrelated concepts and reach for more distant, unique ideas—that is, to pluck "higher hanging fruit" on the "cognitive tree." Further substantiating this view is the constructive forgetting of interfering information during incubation (Smith, 1995) and the ability to inhibit prepotent response to easily switch between tasks in the Stroop task (Zabelina & Robinson, 2010).

NOVELTY GENERATION IN SONG BIRDS

Formal assessments of creativity in humans often include measures of verbal fluency, which rely on proficient output from complex language systems in the brain. Neural networks that mediate language have been extensively studied in biomedical research, and recent developments have found important clues about the genetic basis of vocal learning and expression in non-human species.

These clues are based in part on findings about a gene known as *FOXP2*, which has been a target of intense investigation since the discovery that a mutation in this gene produces a severe human speech and language disorder in affected individuals (Lai, Fisher, Hurst, Vargha-Khadem, & Monaco, 2001). *FOXP2* is very unique in its direct link to language. As a transcription factor, FOXP2 nMRNA is not effective by itself, and can only exert its functions indirectly, through regulating target genes. FOXP2 is a master control molecule, meaning that it governs the expression of many other molecules—like a conductor coordinating the diverse musicians of an orchestra. When FOXP2 changes, so do all the other levels of gene expression in the molecular networks it affects. The mutation in *FOXP2* produced a change in protein structure, thereby affecting its function, which in turn affected the way the brain developed, resulting in altered brain morphology, and ultimately causing a direct and profound alteration of speech.

Spoken language is usually considered to be uniquely human (we lack sufficient space here to consider the contrarian views about the language capacities of some other species). This makes it very difficult to conduct basic science investigations of the brain mechanisms giving rise to novel, verbal production. Fortunately, other species engage in analogous if not homologous vocal behaviors—including vocal learning, which is the ability to modify the sounds made like when learning to speak or sing. Among primates, humans are the only species capable of greatly modifying the sounds they make (Fitch, 2000; Knornschild, Nagy, Metz, Mayer, & von Helversen, 2010; Stoeger et al., 2012). However, research has shown that dolphins, elephants, parrots, songbirds, hummingbirds, seals, and bats have this ability as well (Boughman, 1998; Gahr, 2000; Nottebohm, 1972; Pepperberg, 1994; Poole, Tyack, Stoeger-Horwath, & Watwood, 2005; Reiss & McCowan, 1993; Sanvito, Galimberti, & Miller, 2007). Of all these animals, the vocal learning of songbirds is the best characterized, and exhibits the most parallels to human speech (Jarvis, 2004).

Although distinct from humans in many ways, songbirds undergo similar developmental phases when learning vocalizations. Like humans, songbirds learn vocalizations best early in development and they first depend on listening to the sounds and productions of those around them. Later on, songbirds start making their own new sounds (which is called babbling in humans).

Zebra finch songbirds, *Taeniopygia guttata*, are an experimentally tractable species, for they are easily bred in the lab where they can be reared under controlled experimental conditions (Scharff & White, 2004). Moreover, the brain pathways involved with song learning and production can be easily distinguished, particularly because males are the only ones who learn to sing (i.e., for courting females), and their underlying neuroanatomical pathways reflect this difference (Nottebohm & Arnold, 1976) and are well-characterized—which is less the case with humans (Jarvis, 2006).

There is a strong molecular basis for genetic comparisons between zebra finches and humans. Songbird learning appears to be mediated by cortico-striatal circuitry that is similar to that of humans. Research has also linked *FoxP2* genes to zebra finch songbirds; the FoxP2 protein in zebra finches differs from that in humans at only eight amino acid positions, yielding proteins that are more than 98% identical (Hesler et al., 2004). Overall, the study of the *FoxP2* gene in novel birdsong learning may provide a neuromolecular model for procedurally-learned motor bases of novelty generation in creativity.

Zebra finch vocal learning is sexually dimorphic (Nottebohm & Arnold, 1976); males acquire their song during critical developmental phases by imitating adult males. When the male zebra finch reaches puberty, he can produce incredibly diverse, novel songs. During sexual maturity, a male zebra finch maintains the quality of his vocalization by engaging in cycles of two distinct types of singing: undirected singing (practice) and directed singing (performance) (Zann, 1996). Undirected singing is believed to consist of continuous action-based learning (i.e., practice) that contributes to song maintenance, while performance singing tends to remain consistent across different performances (Jarvis, Scharff, Grossman, Ramos, & Nottebohm, 1998; Nelson & Marler, 1994). These two distinct forms of vocal behaviors in zebra finches allow the opportunity to study and measure *FoxP2* gene expression in relation to different behavioral contexts.

Creative Expression in Songbirds and FoxP2 Genetics

Not only is *FoxP2* involved in the *development* of brain structures dedicated for learning vocal production; the *FoxP2* gene also undergoes differential expression *over the course of development*. The levels of gene expression change during vocalization and vary across different social contexts.

Importantly, while the levels of FoxP2 are similar when the male finch is not singing and when he is performing, the levels of FoxP2 drop dramatically in precisely the part of the brain responsible for these vocalizations. Expression occurs in a striatal region known as area X, which is found only in birds that exhibit vocal learning and is dedicated to song development and maintenance (Miller et al., 2008; Scharff & White, 2004). When male finches practice, FoxP2 mRNA is down-regulated within area X; however, this effect is not seen when males perform directly to females (Teramitsu & White, 2006).

The White lab discovered this by recording and analyzing the sonogram of zebra finch birdsong under experimentally altered social contexts of practice versus performance. They had two conditions, (i) practice before performance and (ii) no practice before performance. Both birds received the two conditions, but counterbalanced and in a different order, to balance out potential order effects.

In the first condition, the White lab allowed male birds to practice (which is known to attenuate area X FoxP2 levels) for 2 h before performing directed to females the next day. In the second condition, the same male birds were *not* allowed to practice (which is known to keep area X FoxP2 levels high) before giving a 2-h long serenade to females the following day. The White lab then compared bird songs' phonology and sequence under the different conditions and found that after song practice, during a time when FoxP2 levels are low, vocal song variability was high. They found that when the same birds do not practice, coincidentally when their FoxP2 levels are high, their songs were more stable.

This discovery provides evidence that FoxP2 functions as a "plasticity gate"; it behaviorally-manipulates downregulation of *FoxP2* expression during song learning and is associated with greater levels of novelty generation in bird song. Furthermore, over 2000 other genes have been identified that are co-expressed with *FoxP2* in cortico-basal ganglia pathways, suggesting that learned vocal communication is mediated by gene ensembles that are "turned on" to regulate vocal behavior (Hilliard, Miller, Horvath, & White, 2012).

FoxP2 and its genetic mechanisms in the brains of songbirds provide a neuromolecular model for procedurally-learned motor exploration, which may help to better understand the genetic basis of verbal novelty generation in humans.

Particularly exciting is the White lab's additional finding, which suggests that hearing links FoxP2 levels to the amount of vocal practice (Teramitsu, Poopatanapong, Torrisi, & White, 2010). As juvenile birds spend more time practicing than adults, their FoxP2 levels are likely to be low more often. Behaviorally-driven reductions in the mRNA encoding this transcription factor could ultimately affect downstream molecules that function in vocal exploration, especially during sensorimotor learning. The finding that behavior may actively be affecting levels of gene expression fits well with prior work in creativity research, suggesting that extensive practice (i.e., Ericsson's "10-year rule") may not only interact with genetic background, but may also be leveraged in *overcoming* certain genetic predispositions through individual converted effort. Perhaps most important is that these experiments are revealing not only that this *can* happen, but are also pointing to the specific brain regions and gene networks in which these effects are taking place.

Neural Circuitry and Patterns of Gene Expression

Further important White lab discoveries suggest the specific neural circuitry that is involved in mediating the effects of novelty generation in song birds, and interestingly, the findings point to the functioning of specific cells (medium spiny neurons) and brain regions (in the basal ganglia) (e.g., Hilliard et al., 2012) that they have linked in separate mouse studies to modulation of inhibitory control (Grant, Richter, Basken, Miller, & White, 2014; Teramitsu & White, 2008; White, Fisher, Geschwind, Scharff, & Holy, 2006). This may help us identify links between two themes of the TFCBC—namely the generation of novelty and the modulation of inhibitory control.

Following up on the finding that the transcription factor FoxP is required for normal language function in humans and for song learning in birds, the White laboratory noted an interesting convergence, namely that signals activating the expression of *FoxP* (e.g., BDNF) are also known to activate Ras/MAPK signaling. This suggests that Ras activation may lead to an increase in *FoxP*-dependent transcription (Hilliard et al., 2012).

Resultantly, the White lab tested the involvement of this gene in song learning and whether its expression is associated with cortical learning and memory, and the signaling mechanisms and timing of activation of FoxP (Teramitsu, Kudo, London, Geschwind, & White, 2004).

WORKING MEMORY IN "SMART" MICE

Neuroanatomy of Working/Declarative Memory

Despite regional specialization of working and declarative memory, the neural systems for long-term memory and working memory are not functionally isolated from each other and are believed to share both a common evolutionary cytoarchitectonic history (Bilder, 2012; Bilder & Knudsen, 2014) and multiple anatomic substrates (Gazzaley, Rissman, & Desposito, 2004).

Extensive investigations over the last four decades have provided insights into the neuroanatomical substrates of working memory. For example, electrophysiological studies of delayed response tasks in monkeys have shown that prefrontal neurons have memory fields that coordinate activity of specific neurons with consistent target locations in the visual field (Funahashi, Bruce, & Goldman-Rakic, 1989).

Accordingly, prefrontal cortex (PFC) regions are activated during encoding and retrieval in long-term memory (Henson, Rugg, Shallice, Josephs, & Dolan, 1999; Tomita et al., 1999). Additionally, manipulation in working memory is associated with an activation state of PFC neurons that appears to be neurophysiologically distinct from that which is associated with stable working memory maintenance (e.g., Cohen et al., 1997) (Bilder, 2012; Bilder & Knudsen, 2014). More specifically, tonic dopamine release and D1 transmission have been associated with stable neural network activation states (giving rise to more robust working memory maintenance), while phasic dopamine release and D2 transmission have been linked with networks exhibiting more flexibility (underlying working memory manipulation and updating) (Bilder, Volavka, Lachman, & Grace, 2004). Genetic and pharmacological investigation of catechol-O-methyltransferase, an enzyme responsible for catecholamine catabolism, has provided an avenue for exploring the influence of tonic and phasic dopamine release on higher cognitive functioning (Bilder, 2012; Bilder et al., 2002; Gogos et al., 1998), which may provide a molecular basis on which to continue investigating the complex relationship between working memory and creativity (Bilder & Knudsen, 2014). Few studies have examined contributions of long-term memory to creative thinking; however, recent work has clarified the cellular and molecular mechanisms of enhanced memory functioning in animal models.

Molecular Mechanisms Underlying Enhancements in Learning and Memory

Of the known mutations that target signaling pathways related to enhanced memory performance, most appear to boost long-term potentiation (LTP), a form of synaptic plasticity. In general, these changes occur as a result of genetic manipulation of N-methyl-D-aspartate receptors (NMDARs) signaling pathways (e.g., Kiyama et al., 1998; Sakimura et al., 1995), although other cellular mechanisms have also been implicated, including the activation of transcription factors that activate genes to produce proteins that strengthen synaptic connections. Transcription factors of interest include cyclic-AMP response-element-binding protein (Suzuki et al., 2004; Yin & Tully, 1996) and CCAAT/enhancer-binding protein (C/EBP) (Sterneck & Johnson, 1998).

One of the most widely publicized accounts of enhanced memory function was the case of the "Doogie" mice, which overexpressed the NR2B subunit of NMDAR in the adult forebrain (Tang et al., 1999). This increase in NMDAR function led to higher levels of LTP in the hippocampal CA1 region, which is known to support long-term memory formation. *Doogie* mice showed enhanced performance on several different memory tasks, including novel-object recognition

(Tang et al., 1999), spatial memory (Morris, Garrud, Rawlins, & O'Keefe, 1982), and fear conditioning (Walker, Ressler, Lu, & Davis, 2002). Remarkably, these mice maintained these memory enhancements over time and continued to outperform their age-matched counterparts in old age (Cao et al., 2007).

As past studies using mouse models have illustrated, considerable enhancement in memory can result from genetic alterations in NMDAR signaling. Manipulations of other neurotransmitter systems also result in changes in synaptic plasticity and memory. Several pharmacological agents currently target memory functioning, including donepezil and modafinil, which are believed to affect catecholamines, serotonin, glutamate, GABA, and histamine systems (Minzenberg, Watrous, Yoon, Ursu, & Carter, 2008).

Working Memory and "Smart Mice"

In prior research on smart mice, the Silva lab has been drilling into the fundamental mechanisms of learning, memory, and plasticity in the brain. This lab has developed new lines of genetically engineered "smart" mice, with working memory capacities significantly exceeding those of the wild type animals (Silva, Zhou, Rogerson, Shobe, & Balaji, 2009; Won & Silva, 2008). They have accomplished this through transgenic and knockout approaches that have enabled them to manipulate the mechanistic controls over how cells encode environmental stimuli; their related actions; and how these cells modify their own internal machinery to enable the storage, retention, and reactivation of memories (Huynh, Maalouf, Silva, Schweizer, & Pulst, 2009; Josselyn, Kida, & Silva, 2004; Matynia et al., 2008; Silva, Paylor, Wehner, & Tonegawa, 1992; Silva, Stevens, Tonegawa, & Wang, 1992). Their findings in well-characterized behavioral tasks (such as the Morris water maze and fear conditioning) implicate key synaptic and nuclear signaling events that can be manipulated to facilitate the induction or enhancement of stability of LTP, and as a result, promote the acquisition or retention of information.

More recently, Silva and colleagues have built upon this research on "smart" mice and have begun to reveal the specific changes in gene expression that are caused by memory-enhancing genetic mutations— most notably, the Silva lab has found that a genetic mutation which increases Ras/MAPK signaling leads to dramatic enhancements in hippocampal learning and memory measured across a number of tasks. The Silva lab published an important review describing the potential of this area (Lee & Silva, 2009).

II. REQUIREMENTS FOR CREATIVITY

RESPONSE INHIBITION IN RODENTS

Numerous lines of experimentation have investigated abilities of response inhibition in animal and human models, and some of the most promising investigations have focused on reversal learning tasks. These paradigms involve learning an initial novel discrimination (e.g., an association between cue A and a reward, while cues B and C are neutral), which is then followed by a reversal in contingencies (e.g., cue B or C is now associated with a reward, not A). Proficiency in this task requires the ability to notice the shift in reinforcement, inhibit a learned response to the previously rewarded cue, overcome learned irrelevance of the cues that were previously not rewarded, and form a new association that is consistent with the new response (Lee, Groman, London, & Jentsch, 2007). The strength of reversal learning tasks as measures of response inhibition is that they provide a method for examining the ability to overcome a prepotent response in the context of an individual's overall learning ability (i.e., performance in the novel discrimination condition).

Noradrenergic Modulation of Cognitive Control and its Neuroanatomy

Neuroanatomical and pharmacological approaches have revealed significant insights into the neural substrates of reversal learning. Studies in non-human primates have identified a network of brain structures that appear to regulate reversal learning, including the orbitofrontal cortex and ventral and medial striatum (Butter, McDonald, & Snyder, 1969; Dias, Robbins, & Roberts, 1996; Iverson & Mishkin, 1970; Izquierdo, Suda, & Murray, 2004; Rolls, Everitt, & Roberts, 1996). In additional studies, serotonin depletion has been shown to impair reversal learning (Clarke, Cools, & Robbins, 2004; Clarke, Walker, Dalley, Robbins, & Roberts, 2007). Blockade of dopaminergic activity through D2/D3 receptors negatively impacts reversal learning in monkeys, while manipulation of D1/D5 receptors has no effect (Lee et al., 2007).

The Jentsch lab has headed our examination of the molecular basis of response inhibition in research involving the study of rats and monkeys. Their experiments have now shown that specific drugs are capable of enhancing behavioral flexibility (Jentsch, Aarde, & Seu, 2009; Seu, Lang, Rivera, & Jentsch, 2009). The common action of these drugs is their inhibition of the norepinephrine (NE) transporter (i.e., molecular agents in cell membranes that carry this neurotransmitter from the synapse back into the cell). The drug affecting the NE transporter resulted in a significant *decrease* in perseverative responding (and thus an increase in flexibility) selectively following Atomoxetine treatment. Taken together, these results suggest a unique role for monoamine transmitter activity (dopamine, NE, and serotonin) within an orbitfrontal-striatal network that regulates response inhibition.

Patterns of Gene Expression and Cognitive Control

Jentsch and colleagues continued our examination of cognitive control in research involving the study of recombinant inbred mouse strains. They found that ability to *reverse* a previously acquired rule (i.e., to show the flexibility to shift to a new rule) was associated with dopamine D2 receptor levels in the ventral midbrain (Laughlin, Grant, Williams, & Jentsch, 2011), and was consistent with previous and ongoing observations in humans (Ghahremani, Monterosso, Jentsch, Bilder, & Poldrack, 2010).

Further investigation of reversal learning in mice has provided a novel genetic discovery that may have important implications for our understanding of response inhibition (Laughlin et al., 2011). After acquiring a learned response set (selecting from an array of five nose-poke apertures), BXD recombinant mice had to reverse their response tendencies to maintain reinforcement. The number of trials needed for reversal was correlated with DRD2 expression in the PFC and ventral midbrain, and significantly linked to a genetic locus on mouse chromosome 10, including the region for *Syn3*, *Nt5dc3*, and *Hcfc2* genes. *Syn3* expression in the nucleus accumbens was correlated with reversal learning performance, which is thought to regulate dopamine release. Interestingly, human *Syn3* is located on chromosome 22, near a genomic locus that has been implicated in psychosis. *Nt5dc3* maps onto human chromosome 12, and has been implicated in ADHD in previous genome-wide association studies. These genetic targets not only have strong implications for clinical syndromes characteristic of poor response inhibition, but may lead to a fuller understanding of the genetic and neural mechanisms of disinhibition in creativity. This work offers exciting leads that have been followed up with parallel studies of reversal learning in humans (Nestor, Ghahremani, Monterosso, & London, 2011).

CONCLUSION

Implications Now

We have identified and discussed three cardinal themes that integrate human and basic cross-species research on creative cognition: novelty generation, working/declarative memory, and response inhibition. We have hypothesized that the combination of all three processes is important in contributing to creative cognition. Novelty generation is the active driving force for deliberate and extensive production of ideas. Working memory maintains these multiple ideas in mind long enough to manipulate them in novel ways. Response inhibition actively filters and selects from all the possible products generated to enable the pursuit of truly valuable ideas. In studies of birds, mice, and nonhuman primates, the TFCBC has uncovered core dimensions of creative cognition including the brain circuits that are engaged, the gene networks that are activated, and drug treatments that influence these processes.

Research from the White lab on *novelty generation* in zebra finches that produce songs highlights the interplay of genetic mechanisms that underlie critical periods for song development with the actual behavioral practice effects. This sheds light on the specific genes that are switched on, on the specific brain circuits in which these genes operate, maps the patterns of gene expression in the brain during development, and shows that there appears to be a synergy between these influences that promotes variability in bird song to enable more new songs to be created, modified and "tuned" to attract mates.

White has further developed from her work on the genetics of bird song an entirely new line of research focused a specific gene (CNTNAP2) that plays a role in autism and probably other neurodevelopmental disorders (see Condro & White, 2014a,b; Panaitof et al., 2010; Scott-Van Zeeland, McNealy, et al., 2010; Scott-Van Zeeland, Abrahams, et al., 2010).

Silva's research on *memory* has opened up entire new areas of research on the fundamental signaling processes involved in memory formation, with implications for normal development of memory and how enhancing memory may give rise to optimal cognitive performance and creativity. Continued examination of the genetic contributions to memory enhancement provides a powerful approach to better characterizing the cellular and molecular mechanisms of memory functioning in the brain. With increased awareness of the neurophysiological processes underlying these memory systems, it will be possible to better understand the nature of enhanced cognitive states and their impact on creativity, and it will also be possible to address and provide treatments for a wide range of learning and memory disorders. Silva's research team has already brought some of these discoveries to the level of clinical trials in syndromes like neurofibromatosis (Krab, de Goede-Bolder, & Aarsen, 2008), which takes the knowledge of how these cellular operations and molecular operations work and manipulates them to enhance the plasticity and stabilization of circuits in the brain.

Jentsch and colleagues' studies on *response inhibition* have shown that the NE transporter is associated with reversal learning in rats and monkeys (Seu et al., 2009). More important, however, is their finding that the reversal behavior was specifically linked to a locus on mouse chromosome 10 (i.e., a specific genetic region, similar to the human chromosome 22 where some variants affecting human behavior are already suspected). They followed up these experiments to show that molecular expression in specific brain regions (neocortex, hippocampus, and striatum) was also correlated with the reversal learning phenotype, implicating several specific genes *Syn3*, *Nt5dc3*, and *Hcfc2*.

Building upon these findings, through an association with D1 and D2, we found all three themes of cognitive processes tie in with the theory of "on the edge of chaos" creative cognition and systems biology (Bilder & Knudsen, 2014)—which describes creative cognition as being "on the edge" of complementary cognitive systems and neural network activation states that vary from regimes that are ordered (predictable) to chaotic (unpredictable) through archicortical and paleocortical trends and actions of tonic and phasic D1- and D2-dopamine transmission, respectively (Bilder, 2012; Christensen & Bilder, 2000). The White lab research on novelty generation in zebra finches reveals *FoxP2* and its other components interacting with molecular switching via dopamine on its spiny neurons. We see a possible connection in the degree cascade of neurons to changes associated with the onset of song diversity shift in phasic D2 transmission. More generally, we also see a possible connection with the male zebra finch song bird cycles of practices and performances (which may promote vocal learning motor skills by changing expression levels of molecules that balance between enhancing plasticity and stabilization). Furthermore, this theory is connected with dopamine's crucial role in working memory and LTP (which is ultimately produced through the stability of memory traces to be reused in the future) and with the Jentsch lab's findings of the association between reversal learning and dopamine D2 receptor levels in the ventral midbrain.

The Future of Cross-Species Study of Animal Cognition

An exciting development from this work will be seen in future efforts that may link the gene networks found in birds and mice to gene networks that may underlie the same unique abilities in humans.

Given that the best characterized function of Cntnap2 is to cluster voltage-gated potassium channels to the juxtaparanodes of nerves (Poliak & Peles, 2003), and evidence suggesting its possible influence on synaptic connectivity (Konishi & Akutagawa, 1985; Mooney & Rao, 1994),

Cntnap2 might have additional, far reaching effects. Cntnap2 may be important for microcircuit connectivity in nuclei across species, by establishing and maintaining local connections within each nucleus through increasing dendritic arborization and the number of active postsynaptic connections. Further interrogation of Cntnap2 and the role it plays in vocal learning in songbirds will certainly benefit our understanding of novelty generation, and possibly the neurobiology of creativity more broadly.

There is also a connection between White and Silva's work and the molecular expression studies being executed in the Geschwind lab, and studies that aim to carry out a microarray analysis of genes whose expression is modified by enhancing Ras signaling in the brain.

Future translational work has the potential to clarify our notion of creative abilities and align them with a wider range of scientific discoveries from animal models to determine their patterns of association, and relate them to their genetic bases and specific brain mechanisms. While creativity is often considered to be a uniquely human ability, there is much that can be learned by studying animals who display core foundations of creative cognition. Animal models of cognition often confer benefits by stripping out some of the complexity that humans bring to the laboratory. Advances in genomics and neuroscience have shown that humans can learn much about their own biology from animal research. Studying how systems can be altered across species to enhance creativity can give us a unique window into how we can better design good neuropsychological treatment options for individuals affected by mental health issues. In addition to treating people who have the most severe mental health issues, by focusing on the entire dimensions of creative abilities across species, we may advance our understanding of the brain to more directly target prevention-focused and positive psychology interventions to help everyone in society live better, more fruitful, lives in the future.

Conflict of Interest Statement

The authors declare that the research was conducted in the absence of any commercial or financial relationships that could be construed as a potential conflict of interest.

Acknowledgments

This work was supported by the Michael E. Tennenbaum Family Center for the Biology of Creativity, a grant from the John Templeton Foundation, and a grant from the National Institute of Mental Health (R01MH101478).

References

Ackerman, P. L., Beier, M. E., & Boyle, M. O. (2005). Working memory and intelligence: The same or different constructs? *Psychological Bulletin, 131*(1), 30.

Barron, F. (1961). Creative vision and expression in writing and painting. In D. W. MacKinnon (Ed.), *The creative person* (pp. 237—251). Berkeley, CA: Institute of Personality Assessment Research, University of California.

Baughman, W. A., & Mumfors, M. D. (1995). Process analytic models of creative capacities: Operations involved in the combination and reorganization process. *Creativity Research Journal, 8*, 32—67.

Benedek, M., Franz, F., Heene, M., & Neubauer, A. C. (2012). Differential effects of cognitive inhibition and intelligence on creativity. *Personality and Individual Differences, 53*(4), 480—485.

Bilder, R. M. (2012). Executive control: Balancing stability and flexibility via the duality of evolutionary neurcanatomical trends. *Dialogues in Clinical Neuroscience, 14*(1), 39—47.

Bilder, R. M., & Knudsen, K. S. (2014). Creative cognition and systems biology on the edge of chaos. *Psychopathology, 5*, 1104.

Bilder, R. M., Volavka, J., Czobor, P. Á., Malhotra, A. K., Kennedy, J. L., Ni, X., et al. (2002). Neurocognitive correlates of the COMT Val[158] Met polymorphism in chronic schizophrenia. *Biological Psychiatry, 52*(7), 701—707.

Bilder, R. M., Volavka, J., Lachman, H. M., & Grace, A. A. (2004). The catechol-O-methyltransferase polymorphism: Relations to the tonic-phasic dopamine hypothesis and neuropsychiatric phenotypes. *Neuropsychopharmacology, 29*(11), 1943—1961. Available from: http://dx.doi.org/10.1038/sj.npp.1300542.

Boughman, J. W. (1998). Vocal learning by greater spear—nosed bats. *Proceedings of the Royal Society of London. Series B: Biological Sciences, 265*(1392), 227—233.

Butter, C. M., McDonald, J. A., & Snyder, D. R. (1969). Orality, preference behavior, and reinforcement value of nonfood object in monkeys with orbital frontal lesions. *Science, 164*, 1306—1307.

Campbell, D. T. (1960). Blind variation and selective retention in creative thought as in other knowledge processes. *Psychological Review, 67*, 380—400.

Cao, X., Cui, Z., Feng, R., Tang, Y. P., Qin, Z., Mei, B., et al. (2007). Maintenance of superior learning and memory function in NR2B transgenic mice during ageing. *European Journal of Neuroscience, 25*(6), 1815—1822. Available from: http://dx.doi.org/10.1111/j.1460-9568.2007.05431.x.

Carson, S. H., Peterson, J. B., & Higgins, D. M. (2003). Decreased latent inhibition is associated with increased creative achievement in high-functioning individuals. *Journal of Personality and Social Psychology, 85*(3), 499—506. Available from: http://dx.doi.org/10.1037/00223514.85.3.499.

Christensen, B. K., & Bilder, R. M. (2000). Dual cytoarchitectonic trends: An evolutionary model of frontal lobe functioning and its application to psychopathology. *The Canadian Journal of Psychiatry/La Revue Canadienne de Psychiatrie, 45*(3), 247—256.

Clark, L., Cools, R., & Robbins, T. W. (2004). The neuropsychology of ventral prefrontal cortex: decision-making and reversal learning. *Brain and Cognition, 55*(1), 41—53.

Clarke, H. F., Walker, S. C., Dalley, J. W., Robbins, T. W., & Roberts, A. C. (2007). Cognitive inflexibility after prefrontal serotonin depletion is behaviorally and neurochemically specific. *Cerebral Cortex, 17*(1), 18—27.

Cohen, J. D., Perlstein, W. M., Braver, T. S., Nystrom, L. E., Noll, D. C., Jonides, J., et al. (1997). Temporal dynamics of brain activation during a working memory task. *Nature, 386*(6625), 604—608. Available from: http://dx.doi.org/10.1038/386604a0.

Colom, R., Jung, R. E., Haier, R. J, et al. (2007). General intelligence and memory span: Evidence for a common neuroanatomic framework. *Cognitive Neuropsychology, 24*(8), 867−878.

Condro, M. C., & White, S. A. (2014a). Recent advances in the genetics of vocal learning. *Comparative Cognition Behavior Reviews, 9*, 1−24. Available from: http://dx.doi.org/10.3819/ccbr.2014.90003 *In press.*

Condro, M. C., & White, S. A. (2014b). Distribution of language-related Cntnap2 protein in neural circuitry dedicated to vocal learning. *The Journal of Comparative Neurology, 522* (1), 169−185. Available from: http://dx.doi.org/10.1002/cne.23394.

De Dreu, C. K., et al. (2012). Working memory benefits creative insight, musical improvisation, and original ideation through maintained task-focused attention. *Personality and Social Psychology Bulletin, 38*(5), 656−669.

Dias, R., Robbins, T. W., & Roberts, A. C. (1996). Dissociation in prefrontal cortex of affective and attentional shifts. *Nature, 380*(1996), 69−72.

Engle, R. W., Kane, M. J., & Tuholski, S. W. (1999). Individual differences in working memory capacity and what they tell us about controlled attention, general fluid intelligence, and functions of the prefrontal cortex. In A. Miyake, & P. Shah (Eds.), *Models of Working Memory: Mechanisms of Active Maintenance and Executive Control* (1st ed., pp. 102−134). Cambridge, England: Cambridge University Press.

Ericsson, K. A., Krampe, R. T., & Tesch-Römer, C. (1993). The role of deliberate practice in the acquisition of expert performance. *Psychological Review, 100*, 363−406.

Eysenck, H. J. (1995). *Genius. The natural history of creativity.* Cambridge, UK: Cambridge University Press.

Fitch, WT (2000). The evolution of speech: A comparative review. *Trends in Cognitive Science, 4*, 258−267. Available from: http://dx.doi.org/10.1016/s1364-6613(00)01494-7.

Funahashi, S., Bruce, C. J., & Goldman-Rakic, P. S. (1989). Mnemonic coding of visual space in the monkey's dorsolateral prefrontal cortex. *Journal of Neurophysiology, 61* (1989), 331−349.

Gabora, L. (2010). Revenge of the "Neurds": Characterizing creative thought in terms of the structure and dynamics of memory. *Creativity Research Journal, 22*(1), 1−13.

Gahr, M. (2000). Neural song control system of hummingbirds: Comparison to swifts, vocal learning (songbirds) and nonlearning (suboscines) passerines, and vocal learning (budgerigars) and nonlearning (dove, owl, gull, quail, chicken) nonpasserines. *Journal of Comparative Neurology, 426*(2), 182−196.

Gazzaley, A., Rissman, J., & D'esposito, M. (2004). Functional connectivity during working memory maintenance. *Cognitive, Affective & Behavioral Neuroscience, 4*(4), 580−599.

Ghahremani, D. G., Monterosso, J., Jentsch, J. D., Bilder, R. M., & Poldrack, R. A. (2010). Neural components underlying behavioral flexibility in human reversal learning. *Cerebral Cortex, 20*(8), 1843−1852.

Gogos, J. A., Morgan, M., Luine, V., Santha, M., Ogawa, S., Pfaff, D., et al. (1998). Catechol-O-methyltransferase-deficient mice exhibit sexually dimorphic changes in catecholamine levels and behavior. *Proceedings of the National Academy of Sciences, 95* (17), 9991−9996.

Golden, C. J. (1975). The measurement of creativity by the Stroop color and word test. *Journal of Personality Assessment, 39*, 502−506.

Grant, L. M., Richter, F., Basken, J. N., Miller, J. E., White, S. A., Fox, C. M., et al. (2014). Early vocalization deficits in a transgenic mouse model of Parkinson's Disease. *Behavioral Neuroscience, 128*(2), 110−121.

Groborz, M., & Nęcka, E. (2003). Creativity and cognitive control: Explorations of generation and evaluation skills. *Creativity Research Journal, 15*, 183−197.

Guilford, J. P. (1950). Creativity. *American Psychologist, 5*, 444−454.

Guilford, J. P. (1967). *The nature of human intelligence.* New York: McGraw-Hill.

Haesler, S., Wada, K., Nshdejan, A., Morrisey, E. E., Lints, T., Jarvis, E. D., et al. (2004). FoxP2 expression in avian vocal learners and non-learners. *The Journal of Neuroscience*, 24(13), 3164–3175.

Henson, R. N., Rugg, M. D., Shallice, T., Josephs, O., & Dolan, R. J. (1999). Recollection and familiarity in recognition memory: An event-related functional magnetic resonance imaging study. *The Journal of Neuroscience*, 19(10), 3962–3972.

Hilliard, A. T., Miller, J. E., Horvath, S., & White, S. A. (2012). Distinct neurogenomic states in basal ganglia subregions relate differently to singing behavior in songbirds. *PLoS Comparative Biology*.

Huynh, D. P., Maalouf, M., Silva, A. J., Schweizer, F. E., & Pulst, S. M. (2009). Dissociated fear and spatial learning in mice with deficiency of ataxin-2. *PLoS One*, 4 (7), e6235.

Iverson, S. D., & Mishkin, M. (1970). Perseverative interference in monkeys following selective lesions of the inferior prefrontal convexity. *Experimental Brain Research*, 11 (1970), 376–386.

Izquierdo, A., Suda, R. K., & Murray, E. A. (2004). Bilateral orbital prefrontal cortex lesions in rhesus monkeys disrupt choices guided by both reward value and reward contingency. *The Journal of Neuroscience*, 24(34), 7540–7548.

Jarvis, E. D. (2004). Learned birdsong and the neurobiology of human language. *Annals of the New York Academy of Sciences*, 1016(1), 749–777.

Jarvis, E. D. (2006). Selection for and against vocal learning in birds and mammals. *Ornithological Science*, 5(1), 5–14.

Jarvis, E. D., Scharff, C., Grossman, M. R., Ramos, J. A., & Nottebohm, F. (1998). For whom the bird sings: Context-dependent gene expression. *Neuron*, 21(4), 775–788.

Jentsch, J. D., Aarde, S. M., & Seu, E. (2009). Effects of atomoxetine and methylphenidate on performance of a lateralized reaction time task in rats. *Psychopharmacology (Berl)*, 202(1–3), 497–504.

Josselyn, S. A., Kida, S., & Silva, A. J. (2004). Inducible repression of CREB function disrupts amygdala-dependent memory. *Neurobiology of Learning and Memory*, 82(2), 159–163.

Kane, M. J., Hambrick, D. Z., Tuholski, S. W., Wilhelm, O., Payne, T. W., & Engle, R. W. (2004). The generality of working memory capacity: A latent-variable approach to verbal and visuospatial memory span and reasoning. *Journal of Experimental Psychology: General*, 133(2), 189.

Kim, K. H. (2008). Meta-analyses of the relationship of creative achievement to both IQ and divergent thinking test scores. *The Journal of Creative Behavior*, 42(2), 106–130.

Kiyama, Y., Manabe, T., Sakimura, K., Kawakami, F., Mori, H., & Mishina, M. (1998). Increased thresholds for long-term potentiation and contextual learning in mice lacking the NMDA-type glutamate receptor ε1 subunit. *The Journal of Neuroscience*, 18(17), 6704–6712.

Knörnschild, M., Nagy, M., Metz, M., Mayer, F., & von Helversen, O. (2010). Complex vocal imitation during ontogeny in a bat. *Biology Letters*, 6(2), 156–159.

Konishi, M., & Akutagawa, E. (1985). Neuronal growth, atrophy and death in a sexually dimorphic song nucleus in the zebra finch brain. *Nature*, 315, 145–147.

Krab, L. C., de Goede-Bolder, A., Aarsen, F. K., et al. (2008). Effect of simvastatin on cognitive functioning in children with neurofibromatosis type 1: A randomized controlled trial. *JAMA*, 300(3), 287–294 May 2009;33(5):690–698.

Lai, C. S., Fisher, S. E., Hurst, J. A., Vargha-Khadem, F., & Monaco, A. P. (2001). A forkhead-domain gene is mutated in a severe speech and language disorder. *Nature*, 413(6855), 519–523.

Laughlin, R. E., Grant, T. L., Williams, R. W., & Jentsch, J. D. (2011). Genetic dissection of behavioral flexibility: reversal learning in mice. *Biological Psychiatry*, *69*(11) 1109−1116.

Lee, B., Groman, S., London, E. D., & Jentsch, J. D. (2007). Dopamine D(2)/D(3) receptors play a specific role in the reversal of a learned visual discrimination in monkeys. *Neuropsychopharmacology*.

Lee, Y.-S., & Silva, A. J. (2009). Molecular and cellular mechanisms of memory allocation in neuronetworks. *Nature Reviews Neuroscience*, *10*(2), 126−140.

MacKinnon, D. W. (1962). Creativity in architects. In D. W. MacKinnon (Ed.), *The creative person* (pp. 291−320). Berkeley, CA: Institute of Personality Assessment Research, University of California.

Martindale, C. (1999). Biological bases of creativity. In R. J. Sternberg (Ed.), *Handbook of creativity* (pp. 137−152). Cambridge, UK: Cambridge University Press.

Matynia, A., Anagnostaras, S. G., Wiltgen, B. J., Lacuesta, M., Fanselow, M. S., & Silva, A. J. (2008). A high through-put reverse genetic screen identifies two genes involved in remote memory in mice. *PLoS One*, *3*(5), e2121.

Mednick, S. (1962). The associative basis of the creative process. *Psychological Review*, *69*, 220−232.

Miller, J. E., Spiteri, E., Condro, M., Dosumu-Johnson, R., Geschwind, D. H., & White, S. A. (2008). Birdsong decreases protein levels of FoxP2, a molecule required for human speech. *Journal of Neurophysiology*, *100*, 2015−2025.

Minzenberg, M. J., Watrous, A. J., Yoon, J. H., Ursu, S., & Carter, C. S. (2008). Modafinil shifts human locus coeruleus to low-tonic, high-phasic activity during functional MRI. *Science*, *322*(5908), 1700−1702.

Mooney, R., & Rao, M. (1994). Waiting periods versus early innervation: The development of axonal connections in the zebra finch song system. *The Journal of Neuroscience*, *14*(11), 6532−6543.

Morris, R. G. M., Garrud, P., Rawlins, J. N. P., & O'Keefe, J. (1982). Place navigation impaired in rats with hippocampal lesions. *Nature*, *297*(5868), 681−683.

Nelson, D. A., & Marler, P. (1994). Selection-based learning in bird song development. *Proceedings of the National Academy of Sciences*, *91*(22), 10498−10501.

Nestor, L. J., Ghahremani, D. G., Monterosso, J., & London, E. D. (2011). Prefrontal hypoactivation during cognitive control in early abstinent methamphetamine-dependent subjects. *Psychiatry Research*, *194*(3), 287−295.

Nottebohm, F. (1972). The origins of vocal learning. *American Naturalist*116−140.

Nottebohm, F., & Arnold, A. P. (1976). Sexual dimorphism in vocal control areas of the songbird brain. *Science*, *194*(4261), 211−213.

Panaitof, S. C., Abrahams, B. S., Dong, H., Geschwind, D. H., & White, S. A. (2010). Language-related Cntnap2 gene is differentially expressed in sexually dimorphic song nuclei essential for vocal learning in songbirds. *Journal of Comparative Neurology*, *518* (11), 1995−2018.

Pepperberg, I. M. (1994). Vocal learning in grey parrots (*Psittacus erithacus*): Effects of social interaction, reference, and context. *The Auk*300−313.

Pierce, R. S., et al. (2010). The role of working memory in the metaphor interference effect. *Psychonomic Bulletin & Review*, *17*(3), 400−404.

Poliak, S., & Peles, E. (2003). The local differentiation of myelinated axons at nodes of Ranvier. *Nature Reviews Neuroscience*, *4*, 968−980.

Poole, J. H., Tyack, P. L., Stoeger-Horwath, A. S., & Watwood, S. (2005). Animal behaviour: Elephants are capable of vocal learning. *Nature*, *434*(7032), 455−456.

Reiss, D., & McCowan, B. (1993). Spontaneous vocal mimicry and production by bottlenose dolphins (*Tursiops truncates*): Evidence for vocal learning. *Journal of Comparative Psychology*, *107*(3), 301.

Rolls, E. T., Everitt, B. J., & Roberts, A. (1996). The orbitofrontal cortex [and discussion]. *Philosophical Transactions of the Royal Society of London B: Biological Sciences, 351*(1346), 1433–1444.

Sakimura, K., Kutsuwada, T., Itot, I., Manabel, T., Takayama, C., Suglyamat, H., et al. (1995). Reduced hippocampal LTP and spatial learning in mice lacking NMDA receptor si subunit. *Nature, 373,* 151.

Sanvito, S., Galimberti, F., & Miller, E. H. (2007). Observational evidences of vocal learning in southern elephant seals: A longitudinal study. *Ethology, 113,* 137–146.

Scharff, C., & White, S. A. (2004). Genetic components of vocal learning. In: H. P. Zeigler, & P. Marler (Eds.), *The behavioral neurobiology of birdsong. Annals of the New York Academy of Sciences.* Vol. 1016, pp 325–347 [pdf].

Scott-Van Zeeland, A. A., Abrahams, B. S., Alvarez-Retuerto, A. I., Sonnenblick, L. I., Rudie, J. D., Ghahremani, D., et al. (2010). Altered functional connectivity in frontal lobe circuits is associated with variation in the autism risk gene CNTNAP2. *Science Translational Medicine, 2.* (56)56ra80-56ra80.

Scott-Van Zeeland, A. A., McNealy, K., Wang, A. T., Sigman, M., Bookheimer, S. Y., & Dapretto, M (2010). No neural evidence of statistical learning during exposure to artificial languages in children with autism spectrum disorders. *Biological Psychiatry, 68*(4), 345–351. Available from: http://dx.doi.org/10.1016/j.biopsych.2010.01.011.

Seu, E., Lang, A., Rivera, R. J., & Jentsch, J. D. (2009). Inhibition of the norepinephrine transporter improves behavioral flexibility in rats and monkeys. *Psychopharmacology (Berl), 202*(1–3), 505–519.

Silva, A. J., Paylor, R., Wehner, J. M., & Tonegawa, S. (1992). Impaired spatial learning in alpha-calcium-calmodulin kinase II mutant mice. *Science, 257*(5067), 206–211.

Silva, A. J., Stevens, C. F., Tonegawa, S., & Wang, Y. (1992). Deficient hippocampal long-term potentiation in alpha-calcium-calmodulin kinase II mutant mice. *Science, 257* (5067), 201–206.

Silva, A. J., Zhou, Y., Rogerson, T., Shobe, J., & Balaji, J. (2009). Molecular and cellular approaches to memory allocation in neural circuits. *Science, 326*(5951), 391–395.

Silvia, P. J., & Beaty, R. E. (2012). Making creative metaphors: The importance of fluid intelligence for creative thought. *Intelligence, 40*(4), 343–351.

Simonton, D. K. (1980a). Land battles, generals, and armies: Individual and situational determinants of victory and casualties. *Journal of Personality and Social Psychology, 38*(1), 110.

Simonton, D. K. (1980b). Thematic fame, melodic originality, and musical zeitgeist: A biographical and transhistorical content analysis. *Journal of Personality and Social Psychology, 38*(6), 972.

Simonton, D. K. (1983). Formal education, eminence and dogmatism: The curvilinear relationship. *The Journal of Creative Behavior, 17*(3), 149–162.

Simonton, D. K. (1984). Creative productivity and age: A mathematical model based on a two-step cognitive process. *Developmental Review, 4*(1), 77–111.

Simonton, D. K. (1994). *Greatness: Who makes history and why.* New York, NY: Guilford.

Simonton, D. K. (1997). Creative productivity: A predictive and explanatory model of career trajectories and landmarks. *Psychological Review, 104*(1), 66.

Simonton, D. K. (2000). Creativity: Cognitive, personal, developmental, and social aspects. *American Psychologist, 55*(1), 151.

Simonton, D. K. (2011a). Creativity and discovery as blind variation and selective retention: Multiple-variant definition and blind-sighted integration. *Psychology of Aesthetics, Creativity, and the Arts, 5,* 222. Available from: http://dx.doi.org/10.1037/a0023144.

Simonton, D. K. (2011b). Creativity and discovery as blind variation: Campbell's 1960. BVSR model after the half-century mark. *Review General Psychology, 15,* 158. Available from: http://dx.doi.org/10.1037/a0022912.

Smith, S. M. (1995). Fixation, Incubation, and insight in memory and creative thinking. In S. M. Smith, T. B. Ward, & R. A. Finke (Eds.), *The creative cognition approach* (pp. 135–156). Cambridge, MA: MIT Press.

Sterneck, E., & Johnson, P. F. (1998). CCAAT/enhancer binding protein β is a neuronal transcriptional regulator activated by nerve growth factor receptor signaling. *Journal of Neurochemistry, 70*(6), 2424–2433.

Stoeger, A. S., Heilmann, G., Zeppelzauer, M., Ganswindt, A., Hensman, S., & Charlton, B. D. (2012). Visualizing sound emission of elephant vocalizations: Evidence for two rumble production types. *PloS One, 7*(11), e48907.

Suzuki, A., Josselyn, S. A., Frankland, P. W., Masushige, S., Silva, A. J., & Kida, S. (2004). Memory reconsolidation and extinction have distinct temporal and biochemical signatures. *The Journal of Neuroscience, 24*(20), 4787–4795.

Tang, Y. P., Shimizu, E., Dube, G. R., Rampon, C., Kerchner, G. A., Zhuo, M., et al. (1999). Genetic enhancement of learning and memory in mice. *Nature, 401*(6748), 63–69.

Teramitsu, I., & White, S. A. (2006). FoxP2 regulation during undirected singing in adult songbirds. *The Journal of Neuroscience, 26*(28), 7390–7394.

Teramitsu, I, & White, SA (2008). Motor learning: The FoxP2 puzzle piece. *Current Biology, 18*(6), R335–337.

Teramitsu, I., Kudo, L. C., London, S. E., Geschwind, D. H., & White, S. A. (2004). Parallel FoxP1 and FoxP2 expression in songbird and human brain predicts functional interaction. *The Journal of Neuroscience, 24*(13), 3152–3163.

Teramitsu, I., Poopatanapong, A., Torrisi, S., & White, S. A. (2010). Striatal FoxP2 is actively regulated during songbird sensorimotor learning. *PLoS One, 5*(1), e8548.

Tomita, S., Li, R. K., Weisel, R. D., Mickle, D. A., Kim, E. J., Sakai, T., et al. (1999). Autologous transplantation of bone marrow cells improves damaged heart function. *Circulation, 100*(suppl 2), II-247.

Walker, D. L., Ressler, K. J., Lu, K. T., & Davis, M. (2002). Facilitation of conditioned fear extinction by systemic administration or intra-amygdala infusions of D-cycloserine as assessed with fear-potentiated startle in rats. *The Journal of Neuroscience, 22*(6), 2343–2351.

Ward, T. B., & Kolomyts, Y. (2010). Cognition and creativity. In J. C. Kaufman, & R. J. Sternberg (Eds.), *The Cambridge handbook of creativity* (pp. 93–112). New York, NY: Cambridge University Press.

White, S. A., Fisher, S. E., Geschwind, D. H., Scharff, C., & Holy, T. E. (2006). Singing mice, songbirds, and more: Models for FOXP2 function and dysfunction in human speech and language. *Journal of Neuroscience, 26*, 10376–10379.

Won, J., & Silva, A. J. (2008). Molecular and cellular mechanisms of memory allocation in neuronetworks. *Neurobiology of Learning and Memory, 89*(3), 285–292.

Yin, J. C., & Tully, T. (1996). CREB and the formation of long-term memory. *Current Opinion in Neurobiology, 6*(2), 264–268.

Zabelina, D. L., & Robinson, M. D. (2010). Creativity as flexible cognitive control. *Psychology of Aesthetics, Creativity, and the Arts, 4*, 136–143.

Zann, R. A. (1996). *The zebra finch: A synthesis of field and laboratory studies* (Vol. 5). Oxford: Oxford University Press.

Zeigarnik, B. (1935). On finished and unfinished tasks. In K. Lewin (Ed.), *A dynamic theory of personality* (pp. 300–314). New York, NY: McGraw-Hill.

II. REQUIREMENTS FOR CREATIVITY

Commentary on Chapter 8: Cross-Species Studies of Cognition

Knudsen et al.'s fascinating chapter on cross-species studies of creativity offers many insights into the ways in which findings from animal research can further our understanding of creative processes in humans. Fundamentally, the three core cognitive themes identified by the authors as highly relevant for the study of (animal) creativity (i.e., novelty generation, working memory and declarative memory, and response inhibition) constitute major foci of current research in human creativity as well. For our purposes here I will focus on the latter two because historically—and certainly at least since Campbell (1960)—the mechanisms that underlie novelty generation (conceptualized as cognitive and/or behavioral variability) have been studied more extensively than the other two.

WORKING MEMORY

The idea that greater working memory (WM) capacity likely facilitates creativity has received substantial empirical support recently. Interestingly, however, there are two competing hypotheses as to why this might be the case. De Dreu, Nijstad, Baas, Wolsink, and Roskes (2012) instructed participants to engage in jazz improvisation sequentially involving three different themes (e.g., summer, spring). Prior to engagement in jazz improvisation, WM capacity was assessed. They found that WM capacity predicted the creativity of the third improvised piece, but not the earlier pieces. Therefore, the authors argued that the advantage a greater WM capacity affords to creativity is in the form of sustained attention or persistence (i.e., task motivation). In turn, the competing hypothesis suggests that greater WM capacity enables one to maintain and manipulate more concepts in the span of attention at any given time, thereby increasing the likelihood that the combinatorial process will be able to fuse concepts not usually associated with each other. Of course these two hypotheses are not mutually exclusive, and may in fact be causally linked. For example, it is possible that in developmentally earlier phases a genetic advantage for higher WM capacity enables greater maintenance and manipulation in the span of attention, which through repetition develops into dispositional gains in persistence and task engagement.

With respect to individual differences, of great relevance to creativity researchers would appear to be Silva lab's development of the new lines of genetically engineered "smart" mice with greater WM capacities than wild types. Knudsen et al. discuss recent work in the Silva lab showing that a genetic mutation that increases Ras/MAPK signaling leads to dramatic enhancements in hippocampal learning and memory measured across a number of tasks. The idea that human WM capacity can be enhanced has been a topic of intense study recently. Specifically, there is now good reason to believe that rather than being fixed, WM capacity is malleable and experience-dependent. In addition, appropriate cognitive training can increase its limits (Morrison & Chein, 2011) and alter related brain function (Klingberg, 2010). The key question now is determining the extent to which such experience-dependent increases in WM capacity and skills transfer to other untrained (novel) tasks that draw on WM.

COGNITIVE CONTROL

Knudsen et al. discuss very interesting findings relating patterns of gene expression to cognitive control. Specifically, they report that consistent with previous observations in humans, the ability to reverse a previously acquired rule—defined as the flexibility to shift to a new rule—was associated with dopamine D2 receptor levels in the ventral midbrain. Recent work in humans has explicitly defined creativity as *flexible* cognitive control (Zabelina & Robinson, 2010). In other words, in contrast to earlier research that had associated creativity with either focused or defocused attention, the current data suggest that creativity is related to *variable* attention, modulated flexibly as a function of task demands (Vartanian, 2009). Yet, the question remains as to how this variability and flexibility in cognitive control and/or attention is achieved. On the one hand, Martindale (1999) argued that this adjustment is automatic or reactive rather than involving self-control. For example, a bottom-up process may be sensitive to the granularity or ambiguity of features in the problem space, such that attention is defocused when ambiguity is high and focused when ambiguity is low. However, it is also possible that this adjustment might be driven by controlled as well as spontaneous processes in relation to WM capacity, or exclusively by top-down processes in the service of strategy change (Haider, Frensch, & Joram, 2005). Either way, Knudsen et al.'s chapter makes it clear that understanding the genetic and neural mechanisms of reversal learning in animals can make strong contributions to sharpening the lens with respect to understanding the neurological and psychological bases of cognitive control in humans. To the extent that cognitive control contributes to creativity, this should in turn give us new insights into the role that the fundamental process of attention plays in creative cognition.

References

Campbell, D. T. (1960). Blind variation and selective retention in creative thought as in other knowledge processes. *Psychological Review, 67,* 380–400.

De Dreu, C. K. W., Nijstad, B. A., Baas, M., Wolsink, I., & Roskes, M. (2012). Working memory benefits creative insight, musical improvisation, and original ideation through maintained task-focused attention. *Personality and Social Psychology Bulletin, 38,* 656–669.

Haider, H., Frensch, P. A., & Joram, D. (2005). Are strategy shifts caused by data-driven processes or by voluntary processes?. *Consciousness and Cognition, 14,* 495–519.

Klingberg, T. (2010). Training and plasticity of working memory. *Trends in Cognitive Sciences, 14,* 317–324.

Martindale, C. (1999). Biological bases of creativity. In R. J. Sternberg (Ed.), *Handbook of creativity* (pp. 137–152). New York, NY: Cambridge University Press.

Morrison, A., & Chein, J. (2011). Does working memory training work? The promise and challenges of enhancing cognition by training working memory. *Psychonomic Bulletin & Review, 18,* 46–60.

Vartanian, O. (2009). Variable attention facilitates creative problem solving. *Psychology of Aesthetics, Creativity, and the Arts, 3,* 57–59.

Zabelina, D. L., & Robinson, M. D. (2010). Creativity as flexible cognitive control. *Psychology of Aesthetics, Creativity, and the Arts, 4,* 136–143.

PART III

THE STRUGGLE FOR CREATIVITY

Brain Size and Innovation in Primates

Ana Navarrete and Kevin Laland

University of St. Andrews, Behavioural and Evolutionary Biology,
St. Andrews, Fife, Scotland, UK

Commentary on Chapter 9: Innovation and the Value of Building on What We Know

Thomas B. Ward

University of Alabama, Department of Psychology, Tuscaloosa, AL, USA

For centuries, we humans have considered ourselves extraordinary relative to other animals. We possess unusual morphological traits, such as hairless skin, bipedal locomotion, and unusually large brains. We are smart, we are very innovative. We are social to the extreme of being able to live in densities that exceed those of any other mammal of similar body size (Currie & Fritz, 1993). In particular, our ability to create new behaviors and tools, and to transmit and accumulate this knowledge, have been considered traits which make us a particularly successful species. In the last decades, though, scientific exploration of the particularities of humanity has shifted to a less anthropocentric view: humans still can consider themselves a rarity in the animal world, but we are much more aware that some animals, including some not closely related to us, share some of our cognitive and behavioral traits to some degree. Instead of focusing exclusively on how the traits that define us evolved in our lineage, researchers can explore under which circumstances key traits emerged in other animal lineages. Comparative

Animal Creativity and Innovation.
DOI: http://dx.doi.org/10.1016/B978-0-12-800648-1.00009-7 241

evolutionary methods can provide insight into how the abilities and peculiarities that characterize our species evolved.

One human characteristic that is widely regarded as a major contributor to the success of our species is our adaptability, which allows us to respond to, control, and regulate our immediate environment. Thanks to our adaptability, we humans have been able to spread all over the world and thrive. It is through the generation of new ideas, and their accumulation and propagation through social learning and culture, that we have managed our greatest achievements. Our capability to innovate might be one of our most spectacular traits, but there is evidence that this ability is not evolutionarily recent or exclusive to humans. Innovativeness is a trait that is widespread amongst animals, such as birds and mammals. Amongst primates we can find some of the most impressive performers of cognitive feats in the animal kingdom. Imo, a female Japanese macaque that attracted the attention of primatologists in the 1970s, was identified as an animal pioneer due to her invention of two new foraging behaviors that quickly spread amongst her group (Kawai, 1965): she learned to wash potatoes before eating them, avoiding the dirt attached to their skin; and she devised a method to separate grain from dirt by throwing both in water ponds. For years, Imo was paraded as "the" primate innovator (Wilson, 1975). With the increase of attention of the media to animal issues and the intensification of primate research in the wild in the last decades, Imo has been joined by other celebrated primate innovators: in 2003, for example, the wild-living gorilla Leah became famous after researchers were able to film her wading a river with the help of a stick, which was considered the first ever photographic evidence of gorilla tool use (Breuer, Ndoundou-Hockemba, & Fishlock, 2005).

PRIMATE INNOVATION

Many animals will invent new behaviors or modify existing ones in order to improve performance. Innovation is considered a key component of most definitions of intelligence and culture, and in an evolutionary context, it influences ecology, macroevolution, culture, and intelligence (Ramsey, Bastian, & van Schaik, 2007a). Because we share a common ancestor with all other primates, a comparative perspective on innovation in primates is a productive avenue for investigating the evolution of human creativity, as it is for other aspects of human and primate cognition (e.g., Tomasello & Call, 1997), even when the consanguinity of primate and human innovation is still a matter of debate (Reader & Laland, 2003).

There has been considerable debate over which behaviors can sensibly be termed "innovation" (Biro et al., 2003; Kummer & Goodall, 1985; Lefebvre, Whittle, Lascaris, & Finkelstein, 1997; Reader & Laland, 2001, 2002). In general, an innovation is either regarded as a new or modified behavior pattern or as a process that results in a new or modified behavior (Reader & Laland, 2003). To facilitate the recognition of innovation in the wild, however, innovation is identified either as a new or modified learned behavior not previously found in the population (innovation *sensu product*), or as a process that results in new or modified learned behavior and that introduces novel behavioral variants into a population's repertoire (innovation *sensu process*).

Kummer and Goodall (1985) suggest three kinds of innovation: first, some innovations could derive from the ability of the individual to profit from an accidental happening. Second, innovation could result from the ability of the higher primates to use existing behavior patterns for new purposes. And, third, some innovation would involve the performance of a completely new pattern of behavior. Additionally, innovation is often only relevant to the population or the species if it is socially transmitted. Many species could be potential innovative, but only when the environmental conditions demand behavioral flexibility. The more improbable the innovation, the more likely social learning is involved in this acquisition by maturing individuals (cultural component). The more demanding observational forms of social learning are concentrated in species with large brains. Taking a more functional approach, Lee (1991) suggested that innovation may originate as a solution to a specific ecological problem. The most classic examples of innovation are those involving tool use, that is, behavioral patterns concerned with the extraction, preparation and processing of food. Such innovations make the intake of food either enhanced or energetically more efficient than when that food is exploited using simpler techniques.

Comparative measures of primate innovation are difficult to obtain. A few studies have investigated the capability and occurrence of innovation in several primate species from a comparative perspective, using behavioral tests (Auersperg, von Bayern, Gajdon, Huber, & Kacelnik, 2011; Marin Manrique, Voelter, & Call, 2013), but most comparative research in this field is done using innovation frequency as a proxy. In the literature, reports on animal innovations are stretched over decades, and are both numerous and diverse. Innovations range from the incorporation of new items or techniques into foraging repertoires, to novel courtship displays, vocalizations, deceptive acts, and tool use (Casanova & Tillquist, 2008; Lefebvre, Reader, & Sol, 2004; Reader & Laland, 2003). These records allow the frequency or rate of innovations in a given species to be computed. The methodology for obtaining this measure was originally described to assess innovativeness in birds

III. THE STRUGGLE FOR CREATIVITY

(Lefebvre et al., 1997), but it also allows us to obtain similar, reliable measures in other groups, such as primates (Reader, Hager, & Laland, 2011; Reader & Laland, 2001, 2002; Reader & MacDonald, 2003). The measure of innovation frequency involves reviewing the literature for the species of interest, and counting the reported incidences of innovation as a representative approximation of the innovative abilities of a given species. Several questions have been raised using this technique, as has been reported by Sol (2003). Innovation frequencies might be influenced by up to thirteen confounding variables (Lefebvre, 2011; Lefebvre, Juretic, Nicolakakis, & Timmermans, 2001). For example, reports on innovation depend on the number of hours of observation of a population of a particular species, on how well-known is the behavioral repertoire of that species, on the researchers recognizing innovation as such, and on the researchers being able to publish records of the innovation (Ramsey, Bastian, & van Schaik, 2007b). In primates, one important bias that needs to be controlled for is research effort. The amount of research effort put into studying primates is extremely biased, with a small number of species being extensively studied, such as chimpanzees, rhesus macaques or capuchin monkeys, compared to very small effort into investigating, say, nocturnal prosimians. This has an impact on the quality of the data that has been collected for species that have been rarely studied. It is possible that a few poorly-studied species are capable of innovating, but registers of these innovations are non-existent because the behavior of the species is not well known, innovations are overlooked or the observers were not present under the conditions that these species innovate. There is also a phylogenetic component biasing the research efforts on primate species: great apes accumulate by far most of the literature in comparison with the rest of the taxonomic groups; Old Work monkeys (catarrhines) get more attention than New World monkeys (platyrrhines); and lorises and lemurs (strepsirrhines) are proportionately the less researched group. Luckily enough, there are methods to control for this bias, such as correcting the number of observations of innovation by the number of total publications on a given species (Lefebvre, 2013; Lefebvre et al., 2004; Sol, 2003). Confidence in this measure is lent by the observation that a species' corrected innovation rate correlates strongly with its performance in laboratory tests of learning and cognition (Reader et al., 2011), supporting the idea that innovation frequency is a reliable cognitive measure. Using records on innovation frequency, we can observe that some species are more innovative than others, and that innovativeness can differ greatly even in closely-related species.

The capability to innovate seems to be taxonomically widespread in animals, and not restricted to mammals only (Laland & Reader, 2009; Reader & Laland, 2003). Together with primates, many of the most successful innovators in the animal kingdom are birds, but innovation in other groups has also been reported (Fisher & Hinde, 1949; Hinde & Fisher, 1951). Within groups, significant inter- and intraspecific variation in innovativeness has been noticed (Laland & Reader, 2009; Reader & Laland, 2003). In primates, some species have much higher innovative rates than others. A closer look into the distribution of the innovation in this group shows that, like in birds, innovative species are found in unrelated branches of the phylogenetic tree, implying convergent evolution. The most innovative lineages in primates are the great apes (*Pan, Pongo, Gorilla*), the macaques (*Macaca*), baboons (*Papio*) and the cebids or capuchin monkeys (*Cebus, Sapajus*) (Reader et al., 2011; Reader & Laland, 2002). Approximately, 60% of the innovations recorded in the primate literature occur in one of the great ape species, the common chimpanzee (*Pan troglodytes*). When we pool the great apes together (chimpanzees, gorillas, and orangutans), these species alone account for up to 75% of our innovation records. The percentage increases to 95% if we add the records from macaques, baboons, and capuchins (Lefebvre, 2013). High innovativeness thus seems to have evolved four or five times independently in primates. A possible fifth evolutionary event could have occurred in the hominin lineage, if great ape and hominin innovation does not have had a common origin (Reader & MacDonald, 2003); however, we suspect that this is unlikely.

Even within these "innovative" lineages, innovativeness does not seem to be a trait that is evenly distributed. Differences can be so striking to the point that one species in a branch can be highly innovative, as could be the case of the Japanese macaque (*Macaca fuscata*), meanwhile its closest-related species has never been observed performing a novel behavior, as is the case with the Formosan rock macaque (*Macaca cyclopis*), the sister species of the Japanese macaque (Reader et al., 2011). The distribution of innovation in the primate tree seems to indicate that innovativeness is a trait that appears in unrelated groups, and perhaps that innovativeness has been favored by very specific selection pressures. This leads us to ask which factors might have favored innovative behavior.

A few studies have compared differences in individual innovativeness between males and females, juveniles and adults, and high-ranking and low-ranking individuals (Reader & Laland, 2001). In the wild, we can observe a trend that indicates that males innovate significantly more than females, but this result has been shown to be driven by the observations in chimpanzees, where innovations are more frequently displayed by males, and it is influenced by a female-biased sex ratio in the populations. When we do not control by the sex ratio, females and males innovate equally frequently. Across age categories, though, there is evidence showing that adults innovate more often than juveniles, and low-ranking individuals innovate more frequently than their high-ranking peers (Reader & Laland, 2001).

Some researchers argue that innovativeness might be a state-dependent variable (Kummer & Goodall, 1985; Lee, Majluf, & Gordon, 1991). In this scenario, innovativeness might not be a characteristic of individuals at all: it would result from the exposure to pertinent eco-logical stimuli, such as a sudden change in the environment. Little is known about which individuals might be capable of generating new behavioral patterns, what causes them to do so, and what ecological variables influence innovation (Kummer & Goodall, 1985; Lee et al., 1991). Innovation potentially offers long-term benefits, in terms of new resources to utilize, more efficient exploitation of the environment, or increased status or mating success. Animal innovations can facilitate survival in changed circumstances (Sol, 2003), they might also be of crit-ical importance to those endangered species forced to adjust to impo-verished environments (Greenberg & Mettke-Hofmann, 2001). Species characterized as innovative are more likely to survive and establish themselves when introduced to new locations (Sol, 2003; Sol, Duncan, Blackburn, Cassey, & Lefebvre, 2005; Sol & Lefebvre, 2000; Wright, Eberhard, Hobson, Avery, & Russello, 2010). Evidence is mounting that innovation plays an important role in ecology (e.g., range expansion), in species and subspecies diversification (Tebbich, Sterelny, & Teschke, 2010), and in cultural diversification (Laland & Reader, 2009). However, if it is so advantageous to be innovative, it is striking, and significant, that innovation is not observed more often or in more species.

Innovation, together with social learning, could also be a mechanism of local adaptation (van Schaik, 2013). Innovation should typically be ini-tiated by a single individual and spread through an animal population when individuals learn from each other. Kummer and Goodall (1985) noted that the majority of innovations in primate populations do not appear to spread, and controlled studies may be the only route to resolv-ing these confounds (Lefebvre & Giraldeau, 1996). Once we calculate the frequencies of social transmission (as an approximation of social learn-ing) using the same methodology as in innovation frequency, we find a significant positive correlation between innovation and social learning (Lefebvre et al., 2004; Reader & Laland, 2002; Figure 9.1). Frequencies of reports of social learning and innovation gathered from published litera-ture seem likely to be both ecologically relevant and measurable indices of learning propensities (Lefebvre et al., 1997). Innovation and social learning rates both also show a strong correlation with tool use fre-quency (Lefebvre et al., 2004; Reader & Laland, 2002). Both tool use and social transmission have skewed distributions which overlap that of innovation, meaning that those primate species that are the most innova-tive are also the ones that show more tool use and have more cases of

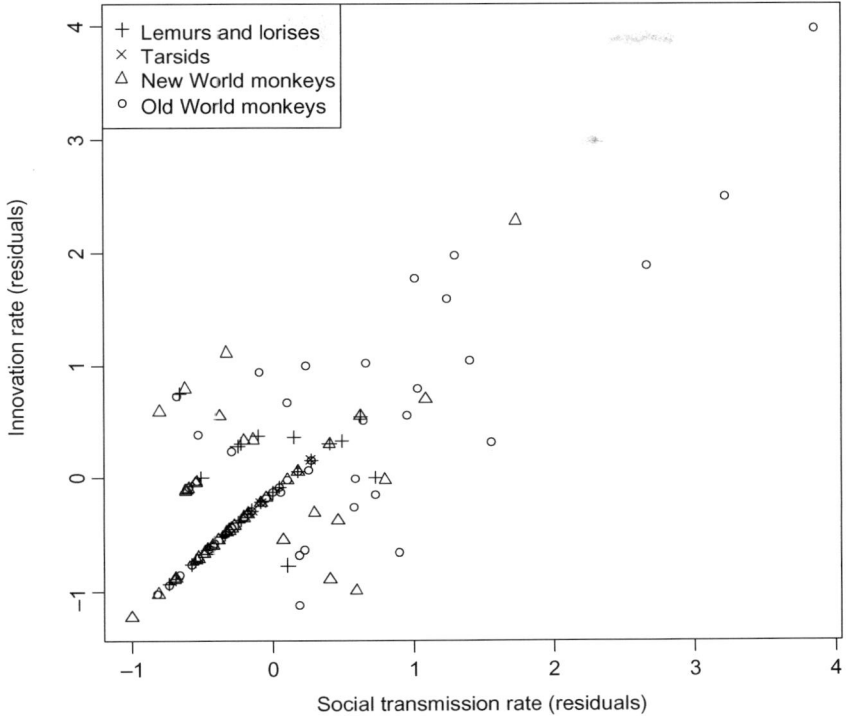

FIGURE 9.1 Correlation between social transmission and innovation rate in primates, after correcting for research effort ($N = 167$ species, lambda $= 0$, $r^2 = 0.56$, $p < 0.00001$). Behavioral data was compiled by Reader et al. (2011).

social transmission. Innovation rate also shows a strong correlation with reversal learning (Lefebvre et al., 2004; Reader & Laland, 2002) and nine types of cognitive tasks (Deaner, van Schaik, & Johnson, 2006; Reader et al., 2011). In a principal component analysis with socio-ecological measures (such as diet breadth, percentage of frugivory, and group size) and other cognitive variables (Reader et al., 2011), innovation, tool use and social transmission cluster in a principal component with five measures of cognition, such as tactical deception (Byrne & Whiten, 1992) and extractive foraging (Gibson, 1986; Parker & Gibson, 1979). This suggests that some form of general intelligence (Colom, Jung, & Haier, 2006) might underlie the evolution of primate cognition, and that these different cognitive abilities have co-evolved together (Deaner et al., 2006; Reader et al., 2011). Some researchers have argued that the selection of modular specializations requires repeated encounters with environmental conditions that favor them (Chiappe & MacDonald, 2005). As innovation constantly deals with new problems rather than repetitions of the same one, rather than being an adaptive specialization it may be heavily reliant on general intelligence (Lefebvre, 2013).

In the case of primates, arboreality could have a negative impact on innovativeness. Changes towards more terrestrial habits would have had an influence on innovation, tool use and, ultimately, in accumulative culture. An extensive body of work suggests that terrestriality may be a driver for greater technological complexity by favoring innovation, acquisition and maintenance of technological skills in primates (McGrew, 2004; Meulman, Sanz, Visalberghi, & van Schaik, 2012; Meulman, Schuppli, & van Schaik, 2013; Meulman & van Schaik, 2013; Visalberghi, Fragaszy, Izar, & Ottoni, 2005; Westergaard, Kuhn, & Suomi, 1998; Westergaard, Wagner, & Suomi, 1999). Increased tool complexity in more terrestrial species may be explained by an increase of opportunities for social learning, because of the increased visibility and/or enhanced opportunities for hand use and hand specialization, as being on the ground liberates the hands for manipulation. As terrestriality would affect the availability of food and raw material to be used as tools, a population with an intrinsic predisposition for tool innovation and social learning, which would additionally be more terrestrial, would be more likely to discover and exhibit habitual tool use that is of higher complexity than an exclusively arboreal population of the same species (Meulman et al., 2012). Although many arboreal primates are able to use tools, this only tends to happen at the substrate level, and their tool use is typically of low complexity or of incidental occurrence (Shumaker, Walkup, & Beck, 2011).

An effect of terrestriality on innovation also explains differences in innovativeness both between species and between populations of the same species, which differ in degree of terrestriality. Populations of semi-terrestrial great apes, like chimpanzees (*Pan troglodytes*), show more impressive tool kits than arboreal great apes, like orangutans (Meulman et al., 2012), but other primate populations which display frequently tool use seem to be more terrestrial than other populations from the same species (Canale, Guidorizzi, Martins Kierulff, & Ferreira Rodrigues Gatto, 2009; Ferreira et al., 2009; Spagnoletti, Izar, & Visalberghi, 2009; van Schaik, van Noordwijk, de Boer, & Dentonkelaar, 1983; Visalberghi et al., 2005). An effect of terrestriality on innovation, tool use and social learning would also explain the striking differences in performance and tool repertoire between wild and captive primate individuals of the same species, as captive primate individuals spend more time at ground level than wild individuals (Meulman et al., 2012, 2013; Meulman & van Schaik, 2013). However, we regard terrestriality as a facilitating condition for innovation and tool use, rather than a direct cause.

COSTS OF INNOVATION

The explanation for the rarity of innovation records amongst primates might be that these benefits are probably offset by associated costs, which can be diverse and may vary with factors such as the physical condition and age of the innovator, and the foraging ecology and life history of the species (Greenberg & Mettke-Hofmann, 2001). The costs of innovation range from increases in predation risk or in the risks of consuming hazardous foods, to excessive investments of time in problem solving that the animal could be using to forage, roam, or reproduce. The latter may be a significant cost where the new behavior requires long periods of time to refine the innovation in order for it to generate a successful outcome. In some species, a balance between exploratory and non-exploratory behavior appears to be important for survival. Even when assuring success, innovative behaviors might be deterred by the presence of other individuals that might take advantage of the innovator's performance (Morand-Ferron, Lefebvre, Reader, Sol, & Elvin, 2004; Morand-Ferron, Sol, & Lefebvre, 2007). If innovation is a costly behavior, it may disproportionately be present in extreme situations, when perception of the costs associated with their performance are reduced, or less acute than the costs of not performing such behavior (Godin & Smith, 1988; Kummer & Goodall, 1985; Lee, Smith, & Grimm, 2003). Alternatively, innovation may disproportionately be produced in stable environments, where an animal discovers a new method of exploiting the environment (Reader & Laland, 2001). Those circumstances that reduce the costs of innovative behavior include, amongst others, a complex social life, greater intelligence, excess energy, and free time (Kummer & Goodall, 1985).

The costs associated with innovative behavior are often invoked to explain differences in rates of innovation between animals in captivity and in the wild. Captive specimens show innovative behaviors spontaneously and more frequently than their conspecifics in the wild, and this tendency to generate innovative behaviors is consistent in several groups (Marin Manrique et al., 2013). There has been much discussion as to whether innovations in wild and captive environments are comparable. Some animals in captivity could be perceived to be more innovative than those in the wild merely because they live in small enclosures where their behavior is easier to observe. In addition, frequently in captivity, chances for innovation are presented by the researcher, in the form of puzzles, games and food, which can bias the recorded innovation frequency (Reader & Laland, 2002; Reader & MacDonald, 2003). The comparison of wild and captive conditions, though, can help researchers to assess environmental differences that might influence the incidence of novel behavior. Some animals reared in captivity might become more habituated to novelty and less neophobic, and may be conditioned to assess the risks of predation and hazard as quite low, compared with animals in the wild. Also, in captivity conditions, animals do not need to spend long periods of time finding enough food for their daily requirements, so they have more time available to explore and generate new behavior. Food, usually in excess, is provided regularly, so the risk of starvation and food competition are minimized or non-existent. Additionally, as captive animals are normally secluded in areas that are smaller than their habitat range, exploration is limited, daily locomotive activity is minimal, and interaction with other conspecifics is more intense and frequent, which increases the chances of social learning. On the other hand, it is unlikely that differences in performance in wild and captive primates are due to long-term captivity effects, as performances in the wild and in captivity seem to be consistent within the same species (Reader et al., 2011). This would indicate that captivity effects that cause anatomical and behavioral changes in different mammalian species (O'Regan & Kitchener, 2005) do not affect innovative performance in primates.

PRIMATE BRAIN EVOLUTION

Human brain size is roughly three times that of our closest relatives, the common and pygmy chimpanzees, and of early hominins, the australopithecines. This spectacular increase, which happened in a rather short period of time, is one of the core research trends in human evolution. We attribute most of our cognitive abilities to the dimensions of our brains. However, we have to wonder under what circumstances this enhanced encephalization (i.e., selection for larger brains) occurred in our lineage. If human oversized brains are associated with such advantages, and our success as a species gives credit to it, why do not we share our status as highly encephalized champions with other species? It is clear that the degree of encephalization in humans is the result of a chain of events that has only occurred once in evolution. Unraveling our own brains' evolutionary story is one of the biggest challenges that we face in our field.

In mammals and birds, the existence of an evolutionary sign trend towards larger brains has been argued (Shultz & Dunbar, 2010). Nevertheless, a large proportion of the variation of the size of the brain is caused by allometric effects related to body size: animals with larger body masses have proportionally larger brain masses (Montgomery, Capellini, Barton, & Mundy, 2010; Smaers, Dechmann, Goswami, Soligo, & Safi, 2012). This allometric relationship is not general: each mammalian order shows a different grade shift between brain mass and body mass, which suggests that different mechanisms have driven this relationship in the different lineages (Jerison, 1955, 1961, 1973; Smaers et al., 2012). A few orders, like primates, show positive grade shifts: for mammals of their body range, primate species show relatively larger brains than predicted by the mammal allometric relationship between body and brain mass. Moreover, in primates, variation in brain-body relationships seems to have been driven primarily by variability in body mass, in contrast to other groups (e.g., birds, Sol & Price, 2008), where body mass variability is driven by brain size (Smaers et al., 2012). In general terms, while body size seems to be driven primarily by birth rate, brain size is mostly dependent on behavioral pressures (Smaers et al., 2012). Among all mammals, anthropoid primates and odontocete cetaceans have significantly greater variance in encephalization quotients, that is, differences in relative brain size, suggesting that evolutionary constraints that result in a strict correlation between brain and body mass have independently become relaxed (Boddy et al., 2012; Boddy, Sherwood, Goodman, & Wildman, 2011). Recently, evidence revealed that metabolic adaptations which would be exclusive for primates could have been behind this evolutionary relaxation (Pontzer et al., 2014), overcoming the effects of other selective pressures, such as the imposition of maintaining low- and middle-sized bodies to stay arboreal (Smaers & Soligo, 2013). On the other hand, larger brains would have been selected in the ancestors, or early members, of the primate lineage in order to facilitate more effective vision, as the "visual brain hypothesis" predicts (Barton, 1998).

It is usual that the magnitude and variance of the level of integration of brain and body mass rates differ significantly amongst subgroups within orders, and here primates are not an exception. Montgomery et al.'s (2010) reconstruction of primate brain evolution suggests that absolute and relative brain mass, but not body size, generally increased over evolutionary time. In the human lineage, both measures show episodes of greater expansion. However, in several branches (strepsirrhines, callithricids and *Cercocebus* species) a decrease in absolute and relative brain size is observed. This suggests that selection has mainly acted to increase brain size in primates, with few exceptions.

Body size effects account for most of the variation in brain size. However, other possible factors have been appointed as drivers for encephalization. Changes in relative brain size (i.e., the variation of brain size once the allometric relationship effect of body size is removed) appear to correlate with enhanced cognitive abilities and sociality (Shultz & Dunbar, 2010). The "social brain hypothesis" suggests that one of the factors reinforcing encephalization at the basis of the primate tree would have been a selection for sociality. Barton (1993) found a correlation between group size and brain measures in haplorrhines (Old and New World monkeys), but not in strepsirrhines (lemurs and lorises), the most basal group in primates. This may indicate that group living favored brain size evolution amongst haplorrhines only. Large-brained species should be able to cope more efficiently with the challenges of competition and cooperation imposed by living in social groups (Barrett & Henzi, 2005; Barton & Dunbar, 1997; Byrne & Whiten, 1988; de Waal, 2003; Dunbar, 1998, 2003; Tomasello, 2000). Relatively large brains might have been selected if this increased social tolerance for a larger number of conspecifics and allowed individuals to deal successfully with the correlated increase in interactions between individuals and more complex networks. As a consequence of more cognitively demanding social environments, more neurological connections would have been needed to cope successfully with hierarchy, social interaction and more complex mating systems. According to the "social intelligence hypothesis" (Barton & Dunbar, 1997; Whiten, Byrne, Waterman, Henzi, & McCullough, 1988), a larger cerebral cortex may have had conferred selective advantages by providing increased cognitive capacity for social dynamics. In the case of primates, this social selection pressure would have led to progressive enlargement of the cerebral cortex over time.

Encephalization might be considered a general phenomenon in primates, but no general pattern in brain size changes has yet been detected using phylogenetic analyses (Shultz & Dunbar, 2010), which would explain why highly encephalized species appear in different branches of the primate tree. Ancestral state reconstructions of absolute brain mass, body mass and encephalization quotients (i.e., relative brain size) reveals patterns of increase and decrease in encephalization quotients in haplorrhines. In primates we observe three grade shifts: strepsirrhines (lemurs and lorises), platyrrhines (New World monkeys) and catarrhines (Old World monkeys, which include humans). These grade shifts suggest that encephalization pressures caused diversification very early in the primate evolution. The two primate suborders, strepsirrhines (lemurs and lorises) and haplorrhines (monkeys and apes), show greater brain sizes than other mammalian groups, but haplorrhines are more encephalized than strepsirrhines. Some authors have suggested that constraints on encephalization relaxed at the origins of the primate order (Boddy et al., 2012, 2011). Overall, primate lineages have a tendency to develop larger brains than other mammalian orders, with humans as the most extreme example. Only a few lineages oppose this trend, such as strepsirrhines and callithricids, where brain reduction is interpreted as a side-effect of dwarfism or exposure to seasonality (Montgomery et al., 2010; van Woerden, van Schaik, & Isler, 2010).

Encephalization correlates strongly with several lifestyle variables: group living, large home range, high-quality diet, strong reliance on vision, and arboreal/forest-dwelling lifestyles (Lefebvre, 2012). Large brains also correlate with slow-paced life history traits. Selection for longer juvenile periods and lifespan may be related to the selection of cognitive abilities, such as the evolution of innovativeness or the reinforcement of social learning (Kaplan, Hill, Lancaster, & Hurtado, 2000; Kummer & Goodall, 1985; Reader & Laland, 2001; Whiten & van Schaik, 2007). For instance, longer lifespans and juvenile periods may have been favored because they provided further opportunities for social learning and innovation. Encephalization might have been facilitated by life history changes in some lineages, but other pathways based on selection favoring behavioral characteristics independent of life history may also have played a relevant role in brain size evolution.

Once we eliminate the effects of body size on brain size (i.e., we consider relative brain size), we can explore how the remaining brain size variation covaries with differences in intellectual or cognitive performance between species (Deaner, Isler, Burkart, & van Schaik, 2007; Iwaniuk, Nelson, & Pellis, 2001; Lefebvre et al., 2004; Sol, Duncan et al., 2005). Different measures of brain size (absolute and relative) co-vary with differences in intellectual or cognitive performance between species (Deaner et al., 2007; Iwaniuk et al., 2001; Lefebvre et al., 2004; Sol, Duncan et al., 2005), extractive foraging (Byrne, 1997; Parker, 1990), and the ability to navigate in the landscape to exploit the spatio-temporally varying and ephemeral food items efficiently (Milton, 1981).

Innovation rate has been shown to correlate positively with absolute and relative measures of brain size in primates (Lefebvre et al., 2004; Reader et al., 2011; Reader & Laland, 2002). The results fit with similar results linking relative brain size with deception (Byrne, 1993; Byrne & Whiten, 1992), mating competition (Sawaguchi, 1997), environmental complexity (Jolicoeur, Pirlot, Baron, & Stephan, 1984), social group size (Dunbar, 1992), and frugivory (Barton, 1999).

The correlation between brain size and innovativeness can be interpreted in two ways. First, larger brain sizes that evolved for some other reason might incidentally favor innovation and social learning, as enhanced neural volumes are required to allow the information processing required for these cognitive abilities (Jerison, 1973; Madden, 2001; Wilson, 1991). However, it seems that, although relatively large brains are necessary to display high rates of these cognitive abilities, they may not be sufficient to guarantee these behaviors, and the complex relationship between brain mass, body mass, and intelligence has thus been the subject of considerable debate (Smaers et al., 2012). The second option is that, at some point in the primate evolutionary history, innovativeness may have yielded a selective advantage, thereby driving the evolution of brain enlargement (Lefebvre et al., 2004; Reader et al., 2011; Reader & Laland, 2002). Based on this premise, Wyles and colleagues developed the "behavioral drive hypothesis" (Wilson, 1985; Wyles, Kunkel, & Wilson, 1983). The ability to invent new behaviors may have played a pivotal role in primate brain evolution. Individuals capable of inventing solutions to new challenges, or exploiting the discoveries and inventions of others, may have had a selective advantage over less able conspecifics, which generated selection for the brain regions that facilitate such behavior. In large-brained species, episodes of innovation and cultural transmission would be more frequent, and they would expose species to novel selective pressures, accelerate the rate of encephalization, and favor further social learning and innovative behavior. This would also relate to the introduction of large-brained mammals to novel environments (Sol, Bacher, Reader, & Lefebvre, 2008). Following this hypothesis, it has been argued that innovation would have been the driver of periods of accelerated rates of encephalization in more recent human evolution (Smaers & Vinicius, 2009). Combined with cultural transmission, innovation may have led humans to exploit the environment in new ways, exposing them to novel selection pressures and increasing the rate of genetic evolution. Studies of primate innovation support the key assumption that brain size, innovation rate, and social learning rate are linked (Reader et al., 2011; Reader & Laland, 2002; see also Whiten & van Schaik, 2007), as does related work in birds (Lefebvre et al., 1997). Moreover, innovation rate and brain size have been shown to correlate with avian species and subspecies richness, suggesting that evolutionary rates are accelerated in large-brained, innovative taxa (Nicolakakis, Sol, & Lefebvre, 2003; Sol, 2003; Sol, Stirling, & Lefebvre, 2005), although whether this applies to primates has yet to be tested.

However, for innovation and social learning to acquire a more influential role in the encephalization processes, a relatively large brain may still be pre-requisite. New analyses, including a large sample size of primate species, show that innovation, social transmission and other cognitive behaviors increase disproportionately after a certain threshold in brain size has been achieved (Figure 9.2).

BRAIN MEASURES

Most studies on encephalization focus on the evolution of overall brain size rather than on the evolution of specific brain regions. This is partly justified because brain composition across species of a given lineage shows obvious regularities which suggest that changes in one region must necessarily be liked to a fairly large extent to changes in others (Finlay & Darlington, 1995; Jerison, 2001). Using overall brain size instead of relative brain measures has been justified in most studies exploring both behavioral and energetic factors. This is because

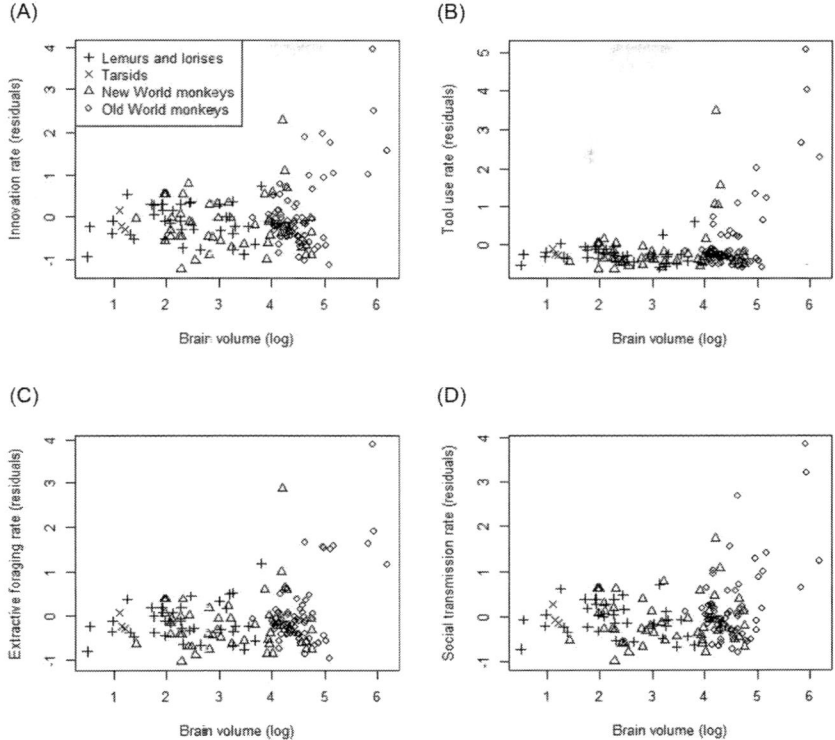

FIGURE 9.2   Correlations between brain volume (log-transformed) and behavioral variables in primates (corrected by research effort): (A) innovation rate, (B) tool use rate, (C) extractive foraging rate, (D) social transmission rate. Behavioral data was compiled by Reader et al. (2011), while brain data was obtained from the compilation of Isler et al. (2008).

cognitive performance has been linked to the size of the whole brain, not only between species (Deaner et al., 2007; Jerison, 1973), but also within species, as observed in humans (McDaniel, 2005).

However, we can also find broad and informative similarities in brain structure, internal connectivity, and anatomical composition across primates (Nieuwenhuys, Donkelaar, & Nicholson, 1998). Examining brain architecture to obtain functional principles is one great challenge to comparative neuroanatomy. The most usual approach to this problem is to make a quantitative comparison of different brain structures to describe statistical trends and to find explanations for these trends (Armstrong, 1983; Barton & Dunbar, 1997; Finlay & Darlington, 1995; Jerison, 1973; Stephan, Frahm, & Baron, 1981). However, to compare brain anatomy evolutionarily, the brain component sizes need to be normalized to whole brain volume (Burish, Kueh, & Wang, 2004). Another usual method for normalization has been to use fractions of brain size (de Winter & Oxnard, 2001; Deaner, Nunn, & van Schaik, 2000; Lefebvre et al., 1997). However, that approach differs because it tends to emphasize extreme values of volume fraction and does not yield distance metrics with clear physical interpretations. The volume fraction of cerebral cortex, for example, varies strongly in mammals, while the cerebellum volume fraction remains almost constant and the volume fractions of the remaining areas, diencephalon and brainstem, are supplanted by increases of the cerebral cortex (Clark, Mitra, & Wang, 2001). Differences in component sizes across species tend nevertheless to be correlated to each other, and are consistent with a developmental principle of concerted brain growth (Finlay & Darlington, 1995; Finlay, Darlington, & Nicastro, 2001), and growth patterns of multiple brain regions (Finlay et al., 2001). However, the results are usually dominated by the effects of absolute size differences among species. More recently, some researchers were able to control for these effects and demonstrated cerebrotypes, a composite measure based in the sizes of the different brain components (Clark et al., 2001). Cerebrotypes support the idea that primate brain architecture has been driven by directed selection pressures, and, in primates, cerebrotype shifts are associated with increases in relative neocortical size and distinct changes in the distribution of brain volume among other regions (Clark et al., 2001).

Brain anatomy in primates has been more extensively studied than in other orders, but complete volumetric measurements of specific brain regions in this group are only available for a small number of species, and few of these measurements are pooled from various individuals per species (Reader & Laland, 2003). The most complete dataset on primate brain anatomy was published by Stephan and colleagues in the 1980s (Stephan, Baron, Frahm, & Stephan, 1986; Stephan, Bauchot, & Andy, 1970; Stephan et al., 1981; Stephan, Frahm, & Baron, 1984, 1987), which includes fundamental brain components' measures obtained from serial sections of primate brains belonging to 44 non-human primate species (11% of the known primate species (Wilson & Reeder, 2005)). This dataset has been the only standard source of brain area volumes for comparative analyses in primate brain evolution. Many of these brain estimates rely only on one or two specimens, although Stephan and colleagues corrected the brain volumes to species-typical means, based on the assumption that the ratio of the different brain components is constant between individuals of the same species.

Even since the publications of Stephan's dataset, few studies have been carried out to increase the number of species with known measurements of different brain structures (Zilles, Amunts, & Smaers, 2011; see Reader & MacDonald, 2003). This lack of data is understandable when we consider that primate brain material for these studies is scarce and difficult to locate. Further publications on brain size and anatomy have mainly focused on overall brain mass (Clutton-Brock & Harvey, 1980; Harvey &

Krebs, 1990; Pagel & Harvey, 1990), endocraneal volumes as a proxy of brain size (Isler et al., 2008) and the publication of invaluable brain atlases (Black, Koller, Snyder, & Perlmutter, 2004; Ely et al., 2001; Frey et al., 2011; Hikishima et al., 2011; McLaren et al., 2009; Mikula, Graziano, Stone, & Jones, 2009; Palazzi & Bordier, 2008; Quallo et al., 2010; Sultan, Hamodeh, Murayama, Saleem, & Logothetis, 2010; Swindler & Wood, 1982; Tammer, Hofer, Merboldt, & Frahm, 2009; Woods et al., 2011). This new brain size data, combined with the publication of new primate phylogenetical trees (Arnold, Matthews, & Nunn, 2010), has allowed us to increase sample size in our analyses and test the robusticity of our results.

The lack of new, more detailed brain morphology data is a consequence of the difficulties of sampling primate brains. First, access to new primate brains is limited to a few brain collections freely available for research (Zilles et al., 2011), which are mostly provisioned with captive individuals. This is relevant because there is some controversy over matching behavioral data from wild individuals and morphological data from captive individuals, as the groups display morphological and behavioral differences. Studies comparing the morphology of wild and captive animals (O'Regan & Kitchener, 2005) have shown that rearing conditions may influence body mass, bone structure, muscle development, changes in the digestive tract, and incidence of obesity, or that different populations that have been reared under different captivity conditions differ in brain anatomy (Bogart, Bennett, Schapiro, Reamer, & Hopkins, 2014). Additionally, in some groups, but not in primates, brain size reduction occurs after a few generations in captivity. On the other hand, until recently, the only available methodology for obtaining reliable measurements required the preparation of serial sections of those brains (Stephan et al., 1981), which is extremely time-consuming and demands adequate infrastructure investment.

One alternative option for getting data is to obtain magnetic resonance images (MRIs) from primate brains. In recent years, this technology has become accessible and common in the study of brain anatomy and functionality. Moreover, because the brain-imaging technology in primates is often the same as the technology used to image human brains, direct comparisons of human and non-human neuroimaging data have become possible. Also, magnetic resonance imaging provides a better scenario for museums and brain collections to facilitate access for neurobiologists and brain evolution researchers to precious specimens without having to use very invasive techniques. High-resolution, high-quality MRIs from primate brains are, however, still difficult to obtain. Images obtained using in vivo techniques, where the animal is sedated for a short period of time while scanning their brains, might be easier to obtain, but image quality and resolution is poorer than in images obtained post-mortem. Post-mortem MRIs can have a higher resolution and are therefore more suited to extracting volumes, but there are concerns that the methods used to extract the brains from their skulls, the post-mortem delay between death and extraction and preservation, and the "age" of the brain (how long has a brain has been stored) might influence the findings. Nevertheless, we recently gained access to an extensive primate brain collection in The Netherlands, under the care of the Primate Brain Bank, and plan to publish brain volumetric measures from 65 individuals from 39 primate species, 26 of which had not previously been available.

Most efforts in the field of primate neurobiology have been focused on other aspects of the quantification of multiple brain regions, such as functionality, development and neural diseases. This research is really useful for our interests: studies on functional magnetic resonance imaging have given new insights into which brain areas might mediate responses involved with innovative, social and learning behavior in primate species, and we can use this information to infer which changes in brain anatomy may be accountable for differences in behavior. The availability and more common use of MRI technology has been of great value in the study of brain morphology in certain groups of primates (Semendeferi, 2001) and has made possible the publication of numerous new primate brain atlases (Black et al., 2004; Black, Snyder, Koller, Gado, & Perlmutter, 2001; Ely et al., 2001; Frey et al., 2011; Ghosh, Odell, Narasimhan, Fraser, & Jacobs, 1994; Hikishima et al., 2011; McLaren et al., 2009; Mikula et al., 2009; Newman et al., 2009; Quallo et al., 2010; Sultan et al., 2010; van Horn & Toga, 2009; Wisco et al., 2008; Woods et al., 2011), some of them constructed after scanning large samples of primate brains from the same species. The literature on comparative data in primate anatomy has also increased enormously in recent years (Evrard & Craig, 2012; Evrard, Logothetis, & Craig, 2014; Freeman, Cantalupo, & Hopkins, 2004; Hopkins, Lyn, & Cantalupo, 2009; Hopkins & Marino, 2000; Hopkins, Marino, Rilling, & MacGregor, 1998; Rilling & Seligman, 2000; Semendeferi, Armstrong, Schleicher, Zilles, & van Hoesen, 2001; Semendeferi & Damasio, 2000; Semendeferi, Damasio, & Frank, 1997; Sherwood et al., 2004).

The number of studies on behavioral neuroanatomy is already allowing us to have new insights into the evolution of behavioral flexibility. Using neuroimaging, we can actually associate specific brain areas to specific functions, and extrapolate which changes in brain anatomy may have accounted for the behavioral changes. For example, we now know how age affects brain size and composition (Erwin et al., 2001; Hopkins, Taglialatela, & Cantalupo, 2007; Sherwood et al., 2001, 2011). Of course, we still have to be cautious in the use of this large amount of empirical data for comparative analyses. Studies on functional neuroimaging are mostly focused on human experimental subjects and a few highly encephalized species, such as great apes or cebids. Several publications have already established that for particular behaviors (task-solving, social interaction, etc.) similar regions are activated in human and other non-primate species. For example, in macaques and humans, evidence has been collected that demonstrated that individual differences in social network size correlate with amygdala volume and areas related to this structure (Bickart, Wright, Dautoff, Dickerson, & Barrett, 2011; Kanai, Bahrami, Roylance, & Rees, 2012; Sallet et al., 2011), which could be interpreted as further evidence for the validity of the "social brain hypothesis." However, we remain very conscious of the fact that most studies of functional neuroimaging are based on species with large brains. We nonetheless hypothesize that the activation of these areas follows the same pattern in all primates.

Differences in brain anatomy across primates are noticeable at many levels. Primate species differ in degrees of gyrencephaly (i.e., brain surface folding), brain asymmetry, brain connectivity, neuron density, and distribution of grey and white matter in the cortex (Hopkins & Nir, 2010; Hopkins et al., 2008; Sherwood et al., 2006). These differences might explain better the distribution of certain cognitive abilities amongst primates than overall changes in brain size or brain composition. Primates, in comparison with other mammals, possess a larger number of neurons per unit volume in the brain (Herculano-Houzel, 2012; Herculano-Houzel & Kaas, 2011; Herculano-Houzel, Mota, Wong, & Kaas, 2010), although densities may vary depending on the brain region. In the cortex, certain regions have a higher neuron packing density than predicted by the cellular scaling model (Gabi et al., 2010). Research on cellular scaling also shows that humans have more varied neuron types and cortical areas than any other primates, suggesting that neuron specialization and gyrification, that is, the increase in brain folding, might be partially responsible for our outstanding behavioral performance (Kaas, 2012). Furthermore, it has been demonstrated that species-specific differences in neuron density in specific areas, such as the visual cortex, are not closely linked to brain or body size, as much as to the size of these brain areas (de Sousa et al., 2010). In human individuals, general intelligence is associated with increased gray matter volume throughout the brain (Colom et al., 2006).

Primate brain evolution has been characterized by an expansion of the relative size of the cerebral cortex. In humans, the cortex alone constitutes 80% of total brain size, whereas for bushbabies (*Galago*), this figure is 46% (Striedter, 2005). One of the most frequent assumptions about human brain evolution is that it involved relative enlargement of the frontal lobes of the brain, and that this enlargement is related to our outstanding cognitive abilities. Innovation and neocortical size have been found to be positively correlated in primates (Reader & Laland,

2002; Rilling & Insel, 1999b) and, within this group, great apes possess both the highest innovation rates and the largest relative brain size in associative areas (Reader & Laland, 2002). Hence, it has been argued that neocortical development might provide the basis of innovative behavior amongst primates. Barton & Venditti (2013) have shown that the size of human frontal lobes was the expected size, relative to other brain structures using five independent datasets of primate brain measures. They also found that the rate of change in the relative frontal cortex in the human branch of their evolutionary tree had not changed over time. Also, the increase of frontal lobe sizes in humans was tightly correlated with increases in other brain areas and whole brain size, and decreases in frontal neuron densities. This finding relates to the rate of increase in frontal lobe size with absolute brain size, and not to the absolute size of these structures. Importantly, this does *not* mean that human brains are just "scaled-up" primate brains, as (i) the proportional size of neocortex is greater in humans than in other primate species, and (ii) with increases in the absolute size of brains come changes in organization, connectivity and modularity (Striedter, 2005). According to (Striedter, 2005), the ratio of neocortex grey matter to medulla is 30:1 in chimps and 60:1 in humans. Nor do Barton and Venditti's (2013) findings mean that substructures of the neocortex are equivalent in size amongst primates. For instance, the lateral prefrontal cortex (a brain region associated with innovative behavior) occupies about 29% of the neocortex in humans but only 17% in chimpanzees (Striedter, 2005). Finally, this evidence does not rule out the possibility that selection on innovation and social learning has driven brain evolution, but implies that, if it has, such selection has operated on the whole brain. However, since the neocortex makes up such a large proportion of the hominin brain, selection for a larger neocortex would have a greater effect on this structure than on other brain regions. Nonetheless, the findings suggest that, in addition to focusing in cortical areas, we should focus on neural networks.

In this line of thought, interest in a cortico-cerebellar neurological system, connecting cognitive functions in the cortex and the cerebellum, has gained interest in recent years. On its own, the cerebellum makes up a significant proportion of the primate brain, from 10.9–11.5% in anthropoid primates to 13.9–14.5% in lemurs and lorises (Stephan et al., 1981; Zilles, Armstrong, Schleicher, & Kretschmann, 1988), and evidence shows that selection can act on its volume independently of selection on overall brain size or neocortex volume, which confers a larger degree of flexibility on the cerebellum (Barton & Harvey, 2000; de Winter & Oxnard, 2001; Rilling & Insel, 1999a). Also, on a cytological level, while a larger neocortex size means a larger percentage of its mass is devoted to white matter, in the cerebellum, grey and white matter increase isometrically with its size. Additionally, the concentration of neurons in the cerebellum is higher than in the cortex and increases in cerebellum size are more relevant in terms of number of neurons than in terms of volumetry (Herculano-Houzel, 2009, 2011, 2012; Herculano-Houzel & Kaas, 2011; Herculano-Houzel et al., 2010). Also, the cerebellum seems to have expanded in apes and other extractive foragers. The confluence of these comparative patterns, studies of ape foraging skills and social learning, and recent evidence on the cognitive neuroscience of the cerebellum, suggest an important role for the cerebellum in the evolution of the capacity for planning, execution, and understanding of complex behavioral sequences, including tool use and language (Barton, 2012; Day, Westcott, & Olster, 2005). As with the cerebral cortex, greater degrees of infolding of the cerebellar cortex provide more surface area and consequently greater potential for information processing.

The cerebellum is also responsible, together with the cortex, for fine sensory-motor control. In order to innovate successfully, motor coordination needs to be refined. Many areas controlling for handedness while using tools have been located in the cortex and corpus callosum (Hopkins, Dunham, Cantalupo, & Taglialatela, 2007; Hopkins & Pilcher, 2001; Hopkins, Russell, & Schaeffer, 2012; Hopkins, Taglialatela, & Russell, 2006; Hopkins, Taglialatela, Russell, Nir, & Schaeffer, 2010; Phillips & Sherwood, 2005, 2007; Phillips, Sobieski, Thompson, & Sherwood, 2008), and it is associated with gyrification of the cortex (i.e., folding of the cortical surface) and larger processing areas (Hopkins, Cantalupo, & Taglialatela, 2007) and cortical differences are detectable between tool-using and non-tool-using individuals (Phillips et al., 2008). Also, handedness might be influenced by asymmetries in cortical connectivity associated with differences in the distribution of white matter (Hopkins, Taglialatela, Dunham, & Pierre, 2007). Again, in studies using functional magnetic resonance in primates, we find evidence for a cortico-cerebellar system integrating information. In comparison with changes in the neocortex, cerebellum changes better explain other variables such as locomotion mode, group size, tool use, and extractive foraging in primates (Barton & Venditti, 2013; Cantalupo, Freeman, Rodes, & Hopkins, 2008; Hopkins, Dunham et al., 2007; Phillips & Hopkins, 2007; Rilling & Insel, 1998; Stephan & Pirlot, 1970). In great apes, the evolutionary growth rate in the cerebellum is more pronounced that in the neocortex (Barton & Venditti, 2013). Moreover, the cerebellum is also involved in internal representations of action, and correlates with the ability to understand and learn skills by observation (Petrosini et al., 2003).

On other aspects, innovation requires the inhibition of previously learned solution strategies before a novel solution can be found. Inhibitory centers have been placed in the prefrontal cortex (Miller &

Cohen, 2001). For most primates, inventing novel solutions for an old problem is considered to be particularly difficult, and would deter their innovativeness. Great apes are more eager to solve new problems and overcome obsolete techniques quickly and efficiently, indicating high degrees of behavioral flexibility and inhibitory control (Barton & Venditti, 2013; Marin Manrique et al., 2013), though not all species seem to be equally efficient in inhibiting programmed responses and producing novel, creative solutions (Amici, Aureli, & Call, 2008; Boysen & Berntson, 1995; Uher & Call, 2008; Vlamings, Hare, & Call, 2010).

Other neural structures that could have an influence on innovation would be areas that are related to spatial learning (Healy, de Kort, & Clayton, 2005), such as the hippocampus, which is related to food-storing and home-range size, and other cortical areas such as the medial temporal lobe, which participates in the formation and storage of declarative memories (Zola-Morgan & Squire, 1993).

However, we still need to be cautious about the interpretation of this large amount of empirical data. Studies on functional neuroimaging are mostly focused on human experimental subjects and a small number of highly encephalized species, such as great apes or cebids, which are kept in captivity, and there is evidence that primate populations show differences in cortical organization under different rearing conditions (Bogart et al., 2014). Although this homology has been until now only found in those few primate species that are commonly used in studies in primate cognition, we can deduce that, because of the regularities that make primate brains so characteristic, these areas also might play a role in less researched species which display the same type of behaviors.

COSTS OF ENCEPHALIZATION

The study of brain size evolution in birds and mammals focuses on explaining variation in brain size between species, but ultimately much of this research is motivated by attempts to understand the conditions under which the evolution of the human brain occurred. Many researchers attribute our cognitive abilities and our success as a species to our large brains. However, if human oversized brains are associated with such advantages, and our success as a species gives credit to it, why do not we share our status as highly encephalized champions with other species?

A major part of the answer to that question is that increases in brain size are costly. Larger brains might be associated with enhanced cognitive performances, but investing in relatively large brains can be a risky strategy. To begin with, brain tissue requires a high supply of energy

(Mink, Blumenschine, & Adams, 1981) and, in contrast to other meta-bolically demanding tissues, its needs cannot be temporarily reduced (Bauchinger, Wohlmann, & Biebach, 2005; Karasov, Pinshow, Starck, & Afik, 2004). Moreover, serious energy deficits, as happens in the case of starvation, can lead to permanent brain damage (Lukas & Campbell, 2000). Maintaining brain mass seems to be prioritized physiologically: during periods of starvation, the brain is the last organ to reduce its mass, only after adipose depots, viscera and muscle mass have been consumed (Lukas & Campbell, 2000). Brain consumption differs for each species. Chimpanzees shunt 13% of their metabolic energy to their brains in their resting state (Mink et al., 1981). Humans, as the extreme example, devote one-fifth of their daily energy budget to maintain a brain that only constitutes 2% of their total body mass (Holliday, 1986; Mink et al., 1981). This consumption is higher for immature individuals, who have to add extra energy for development to the maintenance costs of a larger brain for the body weight. Human neonates devote over 60% of their metabolism to their brains (Holliday, 1986).

Taking into consideration the energetic bargain of encephalization, we would expect that the high proportion of energy necessarily allo-cated to developing and maintaining brain size must impose serious constraints on the evolution of increasing encephalization. Moreover, any constraints on brain size should also limit the associated cognitive abilities. Taking into consideration both costs and benefits of encephali-zation would complement ideas positing benefits to larger brains and reveal the conditions under which positive selection pressures would actually be able to produce increases in brain size and the appearance of innovative behavior. In short, increases in brain size must pay for themselves by enhancing cognitive capabilities in a way that leads to reliable increases in energy intake. It is about time that the cost side of brain evolution became integrated into the mainstream theory of cogni-tive evolution.

Until recently, the costs of brain evolution had not been well inte-grated into the mainstream theory of cognitive evolution. However, a spurt in interest in the energetics of brain evolution in recent years has led to a better understanding of the conditions under which encephali-zation might occur.

We would, from the energetic perspective, only expect an increase in brain size in a context where the extra requirements of developing and maintaining enlarged brains could be compensated by the benefits of enhanced cognition (Aiello & Wheeler, 1995; Isler & van Schaik, 2006; Navarrete, van Schaik, & Isler, 2011). Hypotheses on the energetic side of encephalization in mammals go back as far as the 1980s. They have been recently summarized in the "expensive brain framework" (Isler & van Schaik, 2009b; Navarrete et al., 2011), which proposes that the costs of an increase in brain size can be paid by a combination of an increase of the total energy turnover of the species or a reduction of energy allocation to other expensive functions, such as maintenance, locomotion, or production.

Metabolism

Recently, evidence was collected to demonstrate that primates are distinct among placental mammals in having exceptionally slow rates of growth, reproduction and aging. This could reflect an energetic strategy consisting of allocating energy away from growth and reproduction and to somatic investment in the development and maintenance of large brains (Pontzer et al., 2014). After this study, primates would be only expending half of the daily energy budget expected for a placental mammal of similar body mass. As these metabolic expenses are not reflected in physical activity, they imply metabolic adaptations. As these adaptations seem to be exclusive to primates, they might have been behind the loosening of encephalization constraints at the origin of the primate lineage.

Life History

Part of these pathways to reduce constraints on encephalization might be mediated by changes in life history traits, selecting for slower life paces and extended juveniles periods (Barton & Capellini, 2011; Dean et al., 2001; Isler & van Schaik, 2009a). Notably, Hillard Kaplan and colleagues argue that increases in longevity coevolved with increases in brain size in humans, mediated by our ability to exploit difficult to access, high-nutrient foods (Kaplan et al., 2000; Kaplan & Robson, 2002). According to these authors, longer lifespans are favored, as that gives greater opportunity to benefit from difficult-to-acquire food-procurement skills learned earlier in life. Additionally a slower pace of life and a longer lifespan would enable mobile and long-lived individuals to achieve local adaptation using geographical variation in morphology or in behavior varying selection of local genotypes (van Schaik, 2013). The only strategy for a long-lived individual to survive unpredictable changes in the environmental is being behaviorally flexible. Extended longevity can have its disadvantages, however. Isler and colleagues have suggested that non-human primates with a hominoid lifestyle (high longevity, long weaning periods, belated sexual maturity and long interbirth intervals) might encounter a gray ceiling of demographic non-viability, when the reduction of costs of reproduction in encephalized species and the low number of produced offspring cannot compensate for increases in mortality (Isler, van Woerden, Navarrete, & van Schaik, 2012).

Physiological Buffers

In a seasonal habitat where the energy supply is periodically low, brain size should be constrained, even if the species is able to buffer seasonality physiologically, using strategies such as storing fat, reducing activity, or hibernating. Relative brain size and the relative size of fat deposits are negatively correlated in mammals, which suggests that accumulating adipose tissue and growing larger brains might have been alternative strategies to avoid the effects of seasonal food starvation (Navarrete et al., 2011). Although no evidence of this correlation can be found in primates, we attribute the lack of correlation to the fact that the sampled specimens had been reared in captivity and their adipose depots had to be estimated (see supplementary information in Navarrete et al., 2011).

Each species is adapted to its preferred or staple diet morphologically. If a species would be able to fully compensate (or even overcompensate) for the change in diet during lean periods, for example, by increasing foraging effort, and thus be better adapted to fall-back foods, these foods would become its staple diet during the good periods as well (as is the case in many folivorous primates). Of course, some differences in brain size may result from the primary adaptation to staple diet during the lean periods, and, therefore, it is important to control for diet type when testing the correlations between seasonality and brain size. During periods of scarcity, many animals change their diet to fall-back foods which are abundant during the season, though of lower dietary quality (Hemingway & Bynum, 2005). Diet shifts are also considered physiological buffers, although foraging on these non-seasonal items can rarely provide the same amount of net energy that is available during the periods of plenty and still exposes individuals to habitat seasonality.

Cognitive Buffers

As body and brain size increases, local adaptation can increasingly be achieved through selection of plasticity. However, animals with very large brains may be particularly vulnerable to drastic forms of environmental change if and when their brains do not confer a significantly enhanced ability to cope with these challenges (van Schaik, 2013). It is not a surprise that one of the relevant mechanisms that animals use to provide extra energy for developing and maintaining larger brains is behavior, particularly behavior that functions to increase energy intake or foraging efficiency. From the energetic perspective, we would expect an increase in brain size in a context where the extra requirements of developing and of maintaining enlarged brains could be compensated by the benefits of enhanced cognition (Aiello & Wheeler, 1995; Isler & van Schaik, 2006; Navarrete et al., 2011).

In birds, there is substantial evidence showing that there is a link between seasonality and behavioral flexibility. Bird species living in temperate habitats, subjected to seasonal conditions, have significantly larger brains that birds species living in tropical, more stable habitats (Schuck-Paim, Alonso, & Ottoni, 2008), or migrating birds (Sol, Lefebvre, & Rodriguez-Teijeiro, 2005; Winkler, Leisler, & Bernroider, 2004). This correlation between seasonality and large brains has been explained by the "cognitive buffer hypothesis" (Allman, McLaughlin, & Hakeem, 1993): seasonal habitats are likely to be more cognitively demanding than non-seasonal habitats because preferred food sources are more variable or more dispersed in space and over time. Animals that are more capable of flexible behavior might be better placed to take advantage of variable opportunities to exploit difficult to access resources and to survive periods of scarcity successfully. If innovative species are able to exploit new sources of difficult-to-access, high-quality food by means of tool use or extractive foraging, this would give them an energetic advantage, allowing the selection of larger brains. In these species, behavioral flexibility, in the means of innovation and (social) learning, could be a by-product of actual or historical seasonality in a lineage's evolutionary past. The "cognitive buffer hypothesis" opposes the central prediction of the "expensive brain framework" on seasonality: the average brain size within a population should be negatively related to the duration and/or frequency of periods of low food availability, when daily energetic requirements cannot be fully compensated for by increasing foraging effort or shifting diet (Isler & van Schaik, 2009b; Navarrete et al., 2011). Note that the above reasoning does not require seasonality. Innovation may earn its keep in energetic terms if it allows animals to exploit more resources, or higher-quality resources, even in a non-seasonal context.

Recent research has provided evidence for the "cognitive buffer hypothesis" and the "expensive brain framework" in primates: the differences between climatic seasonality and experienced seasonality indicate that cognitive buffering occurs in this group, but its effects are counterbalanced by the energetic demands of larger and more metabolically expensive brains (van Woerden et al., 2010; van Woerden, Willems, van Schaik, & Isler, 2012). In catarrhines (Old World monkeys), relative brain sizes correlate negatively with seasonal effects, but they also show a positive correlation between relative brain size and cognitive buffering, which means that the actual net energy intake of primates species is higher than expected for the harshness of the seasonal changes on resources (van Woerden et al., 2012). However, buffering effects can only compensate to an extent. In orangutans, populations in areas exposed more frequently to "El Niño"-induced droughts had relatively small brains compared to populations living in more stable environments (Taylor & van Schaik, 2007). When periods of unavoidable starvation become too unpredictable or too frequent, cognitive buffering becomes less effective and the constraints on brain size get reinforced.

The more extreme example of how seasonality can constrain brains can be found in strepsirrhines (lemurs and lorises), a group of animals whose brain metabolism is already high *per se* (Isler & van Schaik, 2009b; Mink et al., 1981). Here, experienced seasonality correlates negatively with relative brain size, after controlling for phylogenetic effects and other confounding variables as the extent of folivory. Large-brained lemurs seem to experience less variation in dietary intake than would be expected by the seasonality of their habitat (van Woerden et al., 2010). Seasonal energetic constraints would seem to be responsible for most of the physiological and life history adaptations of this group to harsh climates (Ganzhorn, Fietz, Rakotovao, Schwab, & Zinner, 1999; Wright, 1999). Strepsirrhines are the only primates that have adjusted to extreme birth seasonality (Janson & Verdolin, 2005). The basal metabolic rates of most strepsirrhines are below those of other primate groups (Genoud, 2002) and they include the only two species that show torpor or hibernation (Dausmann, Glos, Ganzhorn, & Heldmaier, 2004; Schülke & Ostner, 2007). In strepsirrhines, we do not find the usual correlates of encephalization with group size that are observed in other primate groups (MacLean, Barrickman, Johnson, & Wall, 2009; Shultz & Dunbar, 2007) or diet quality (Fish & Lockwood, 2003). The exception to this pattern in this group is the aye-aye (*Daubentonia madagascarensis*), the single extant member of a family that split off from the rest at the base of the strepsirrhine tree, almost 60 millions of years ago (Martin, 2000). The aye-aye's brain is exceptionally large, within the range of anthropoid primates, and is matched by a relatively high basal metabolic rate for a lemur (Barrickman & Lin, 2010; Isler et al., 2008). The experienced seasonality of the aye-aye seems to be low, but this might be due to its condition as an extractive forager who is able to exploit high-quality food sources throughout the year (Sterling, 1994). We might conclude that extractive foraging could be one of the best strategies to buffer seasonality. It is still not clear, though, if exposure to seasonality could, under specific circumstances, have been an effective driver of encephalization in primates; moreover, cognitive buffering may operate in non-seasonal contexts, too.

III. THE STRUGGLE FOR CREATIVITY

NEW PERSPECTIVES

Until now, research on brain and cognition evolution has focused in identifying factors that might have facilitated the emergence of large brains, rather than the evolution of innovation, social learning and accumulative culture. Now, researchers are finally in a position to correlate the evolution of brain mass and innovation to a variety of different cognitive, life history, physical, social, and ecological variables. However, as biologists, we have to remember that (unless sophisticated statistical tools are brought to bear) correlation does not typically imply causation (Shipley, 2000). Even if the newest comparative statistical methods can be deployed to identify hypotheses consistent with the data, it remains a major challenge to know whether these relationships are causal and not the result of confounds or co-dependencies. It remains difficult to determine with certainty whether innovation and brain size are causally related, and it is possible that the observed correlation only occurs because of intermediate variables connecting both factors.

Of course, the difficulties of establishing cause-effect relationships is not specific to the evolution of innovation and brain size, but is a feature of comparative statistical methods in general (Harvey & Pagel, 1991). Researchers in fields unrelated to biology have elaborated methods that might ascertain cause-effect relationships from empirical data. Although these methods were originally developed in the study of artificial intelligence (Pearl, 1988; Shipley, 2000; Spirtes, Glymour, & Scheines, 1993), there have been several attempts to apply causal statistics to biological systems. Shipley's guide to the application of causal statistics for biologists (Shipley, 2000) is the most relevant source for researchers interested in evaluating models of causality, while Kalisch, Maechler, Colombo, Maathuis, and Buehlmann (2012) recently developed an R package on causal models, *pcalg*.

This methodology allows inferring causality from good quality correlation data, using so-called causal or directed graphs (Pearl, 1988; Shipley, 2000). A causal graph is a formal means of specifying the causal relationship between variables, which can be used to generate a list of conditional and unconditional independence relations that would be found if the graph were supported by the data (Shipley, 2000). This allows researchers to test specific causal models against observational data, or to search for a set of causal graphs that might be consistent with the observational data (Spirtes, Glymour, & Scheines, 2001).

The application of this methodology might help to resolve the long-standing controversies on the most likely cause of primate brain evolution, and to integrate and/or sort between the available alternative explanations for the evolution of innovation. Moreover, causal statistics would allow us to integrate, in the same model, all possible factors that have been linked to the evolution of either encephalization or cognition. However, application of this methodology to evolutionary models is not straightforward. First, the methods must be adapted to allow for phylogenetic control. Second, causal graphs imply that, in order to obtain reliable results, all the relevant variables in the system appear in the graph, and that these variables are linked unidirectionally. Causal graphs struggle to consider loops or feed-backs between variables. Therefore, any attempt to apply causal statistics to biological systems, where we observe constant feedback over time, needs to reconcile the fact that the best model might itself be biased. This is relevant to the study of the evolution of innovation and social learning, as Wilson's "behavioral drive" hypothesis emphasizes an autocatalytic mechanism. Nevertheless, our preliminary findings in applying these new methodologies to the evolution of innovation and social learning imply that both behavioral variables are not directly related to brain size in primates, and that interactions between brain size and innovation only occur when mediated by other ecological variables, such as group size and diet breadth (unpublished). Such findings do not rule out the possibility that social learning and innovation have driven the evolution of the primate brain, but suggest that if they have they operated via intermediate variables.

SUMMARY

Primates are outstanding, relative to most other mammals, with respect to their cognitive abilities, especially their ability to show innovative behavior. Research into this group has thrived in the last decades, as how and why primate innovation occurs, and how innovation relates to brain size, can potentially provide new insights into how humans have achieved outstanding levels of innovativeness. The body of information that we have accumulated referring to the evolution of both brain size and cognitive abilities has increased rapidly. Every day, more knowledge is being added, thanks to new technologies that allow researchers to associate changes of brain size, composition, and connectivity with behavioral measures. This new information has generated large datasets on brain morphology and behavior that, together with a more reliable primate phylogeny, allow researchers to test hypotheses that could not be tested before.

Evidence for encephalization begins early in the primate lineage. Initially, primates probably experienced constraints on their body size due to their arboreality, but larger brains were selected nevertheless, facilitated by appearance of metabolic traits that lifted energetic constraints on encephalization. Whether or not larger brains, in comparison to other mammals, were selected in ancestral primates because of their associated benefits is a question that researchers may struggle to answer. However, there is now reasonable evidence that sociality may have driven encephalization further in haplorrhines, but not in strepsirrhines. This could account for the lower grade shift in the relation between body and brain mass in the latter group. Additionally, strepsirrhines most probably did not evolve larger brains and enhanced cognitive abilities due to strong energetic constraints caused by constant exposure to seasonality, which has strongly affected their biology. Although strepsirrhines' relatively large brains for a mammal apparently confers the possibility of buffering seasonal effect to a reduced degree, these benefits seemingly did not overcome the costs associated with large brains. The only exception in this group is the case of the aye-aye, a very encephalized species whose success might be related to it being an extractive forager.

Within haplorrhines, platyrrhines and catarrhines show different grade shifts in the relationship between body mass and brain mass. In catarrhines, there exists further evidence that some species are capable of partially buffering seasonal effects, and again this ability is strongly correlated with extractive foraging and access to high-quality dietary habits. However, energetic constraints on brain size can seemingly be very severe, such that, if seasonal periods are too frequent or unpredictable, cognitive buffers cannot compensate for the energetic demand, leading to selection for smaller brains. This suggests that, unlike in birds, seasonality might not be a major driver of primate innovation and encephalization.

Amongst haplorrhines, there are several lineages in unrelated branches of the primate tree that are responsible for most of our records on innovation: capuchins, baboons, macaques, and great apes. This distribution of innovative behavior indicates that innovation might be a trait that only appears under specific selection pressures, or when specific pre-conditions (such as a sufficiently large brain and group size for innovations to propagate) are met. A strong correlation between large brains and innovativeness is apparent, which, together with the assessment of the evolutionary rates, might provide evidence for the "behavioral drive hypothesis" in some primate lineages. Other factors such as terrestriality, social group size, and diet breadth might also facilitate primate innovation. Such factors need not be regarded as alternatives to behavioral drive, and may be wrapped up in the same mechanism.

There is also evidence that those conditions that do not select for larger brain size also do not favor innovation: having a small body size, being unable to achieve stable net energy intake, or unable to improve survival through cognitive performance would deter selection for behavioral flexibility and innovation. A fruitful means to proceed is by way of a cost-benefit analysis that considers how the energetic costs of a larger brain can be outweighed by the potential to increase energy gathering efficiency through enhanced cognition. When researchers consider these costs and benefits, they might be better placed to explain a larger amount of the co-variation in brain size and innovation. The evolution of early hominins may well be an example of how cognitive abilities have undermined these energetic constrains (Kaplan et al., 2000; Kaplan & Rcbson, 2002).

Currently, the question of why some species are more innovative than others of similar brain size remains a puzzle. One possibility is that certain regions, like the prefrontal cortex, may be critical to effective innovation (Striedter, 2005). A second possibility is that innovation may require extensive neural networks connecting executive and per-ceptual/motor brain regions, and only certain ecological circumstances or life histories favor the evolution of this connectivity, or confer the experience necessary for these networks to develop during ontogeny. Currently, there is insufficient data to identify these structure-functional correspondences across primates.

Acknowledgments

Research supported in part by a grant from the John Templeton Foundation to KNL.

References

Aiello, L. C., & Wheeler, P. (1995). The expensive-tissue hypothesis—the brain and the digestive-system in human and primate evolution. *Current Anthropology, 36*(2), 199–221.

Allman, J. M., McLaughlin, T., & Hakeem, A. (1993). Brain weight and life span in primate species. *Proceedings of the National Academy of Sciences of the United States of America, 90* (1), 118–122.

Amici, F., Aureli, F., & Call, J. (2008). Fission-fusion dynamics, behavioral flexibility, and inhibitory control in primates. *Current Biology, 18*(18), 1415–1419. Available from: http://dx.doi.org/10.1016/j.cub.2008.08.020.

Armstrong, E. (1983). Relative brain size and metabolism in mammals. *Science, 220*(4603), 1302–1304.

Arnold, C., Matthews, L. J., & Nunn, C. L. (2010). The 10 k trees website: A new online resource for primate phylogeny. *Evolutionary Anthropology, 19*(3), 114–118. Available from: http://dx.doi.org/10.1002/evan.20251.

Auersperg, A. M. I., von Bayern, A. M. P., Gajdon, G. K., Huber, L., & Kacelnik, A. (2011). Flexibility in problem solving and tool use of kea and new caledonian crows in a multi access box paradigm. *PLoS One, 6*(6). Available from: http://dx.doi.org/10.1371/journal.pone.0020231.

Barrett, L., & Henzi, P. (2005). The social nature of primate cognition. *Proceedings of the Royal Society of London. Series B: Biological Sciences, 272*(1575), 1865–1875.

Barrickman, N. L., & Lin, M. J. (2010). Encephalization, expensive tissues, and energetics: An examination of the relative costs of brain size in strepsirrhines. *American Journal of Physical Anthropology*, 143(4), 579–590. Available from: http://dx.doi.org/10.1002/ajpa.21354.

Barton, R. A. (1993). Independent contrasts analysis of neocortical size and socioecology in primates. *Behavioral and Brain Sciences*, 16(04), 694–695.

Barton, R. A. (1998). Visual specialization and brain evolution in primates. *Proceedings of the Royal Society of London Series B—Biological Sciences*, 265(1409), 1933–1937.

Barton, R. A. (1999). The evolutionary ecology of the primate brain. In P. C. Lee (Ed.), *Comparative primate socioecology* (pp. 167–194). Cambridge: Cambridge University Press.

Barton, R. A. (2012). Embodied cognitive evolution and the cerebellum. *Philosophical Transactions of the Royal Society B: Biological Sciences*, 367(1599), 2097–2107.

Barton, R. A., & Capellini, I. (2011). Maternal investment, life histories, and the costs of brain growth in mammals. *Proceedings of the National Academy of Sciences of the United States of America*, 108(15), 6169–6174. Available from: http://dx.doi.org/10.1073/pnas.1019140108.

Barton, R. A., & Dunbar, R. I. M. (1997). Evolution of the social brain. In A. Whiten, & R. Byrne (Eds.), *Machiavellian intelligence II: Extensions and evaluations* (pp. 240–263). Cambridge: Cambridge University Press.

Barton, R. A., & Harvey, P. H. (2000). Mosaic evolution of brain structure in mammals. *Nature*, 405(6790), 1055–1058. Available from: http://dx.doi.org/10.1038/35016580.

Barton, R. A., & Venditti, C. (2013). Human frontal lobes are not relatively large. *Proceedings of the National Academy of Sciences of the United States of America*, 110(22), 9001–9006. Available from: http://dx.doi.org/10.1073/pnas.1215723110.

Bauchinger, U., Wohlmann, A., & Biebach, H. (2005). Flexible remodeling of organ size during spring migration of the garden warbler (*Sylvia borin*). *Zoology*, 108(2), 97–106.

Bickart, K. C., Wright, C. I., Dautoff, R. J., Dickerson, B. C., & Barrett, L. F. (2011). Amygdala volume and social network size in humans. *Nature Neuroscience*, 14(2), 163–164. Available from: http://dx.doi.org/10.1038/nn.2724.

Biro, D., Inoue-Nakamura, N., Tonooka, R., Yamakoshi, G., Sousa, C., & Matsuzawa, T. (2003). Cultural innovation and transmission of tool use in wild chimpanzees: Evidence from field experiments. *Animal Cognition*, 6(4), 213–223. Available from: http://dx.doi.org/10.1007/s10071-003-0183-x.

Black, K. J., Koller, J. M., Snyder, A. Z., & Perlmutter, J. S. (2004). Atlas template images for nonhuman primate neuroimaging: Baboon and macaque. *Imaging in Biological Research, Pt A*, 385, 91–102.

Black, K. J., Snyder, A. Z., Koller, J. M., Gado, M. H., & Perlmutter, J. S. (2001). Template images for nonhuman primate neuroimaging: 1. Baboon. *Neuroimage*, 14(3), 736–743. Available from: http://dx.doi.org/10.1006/nimg.2001.0752.

Boddy, A. M., McGowen, M. R., Sherwood, C. C., Grossman, L. I., Goodman, M., & Wildman, D. E. (2012). Comparative analysis of encephalization in mammals reveals relaxed constraints on anthropoid primate and cetacean brain scaling. *Journal of Evolutionary Biology*, 25(5), 981–994. Available from: http://dx.doi.org/10.1111/j.1420-9101.2012.02491.x.

Boddy, A. M., Sherwood, C. C., Goodman, M., & Wildman, D. E. (2011). Phylogenetic analysis reveals relaxed constraints in primate encephalization during mammalian descent. *American Journal of Physical Anthropology*, 144. 93–93

Bogart, S. L., Bennett, A. J., Schapiro, S. J., Reamer, L. A., & Hopkins, W. D. (2014). Different early rearing experiences have long-term effects on cortical organization in captive chimpanzees (*Pan troglodytes*). *Developmental Science*, 17(2), 161–174. Available from: http://dx.doi.org/10.1111/desc.12106.

Boysen, S. T., & Berntson, G. G. (1995). Responses to quantity—perceptual versus cognitive mechanisms in chimpanzees (*Pan troglodytes*). *Journal of Experimental Psychology—Animal Behavior Processes*, 21(1), 82–86. Available from: http://dx.doi.org/10.1037/0097-7403.21.1.82.

Breuer, T., Ndoundou-Hockemba, M., & Fishlock, V. (2005). First observation of tool use in wild gorillas. *PLoS Biology*, 3(11), 2041–2043. Available from: http://dx.doi.org/10.1371/journal.pbio.0030380.

Burish, M. J., Kueh, H. Y., & Wang, S. S. H. (2004). Brain architecture and social complexity in modern and ancient birds. *Brain Behavior and Evolution*, 63(2), 107–124. Available from: http://dx.doi.org/10.1159/000075674.

Byrne, R. W. (1993). Do larger brains mean greater intelligence? *Behavioral and Brain Sciences*, 16(4), 696–697.

Byrne, R. W. (1997). The technical intelligence hypothesis: An additional evolutionary stimulus to intelligence? In A. Whiten, & R. W. Byrne (Eds.), *Machiavellian intelligence II: Extension and evaluations* (pp. 289–311). Cambridge: Cambridge University Press.

Byrne, R. W., & Whiten, A. (1992). Cognitive evolution in primates—evidence from tactical deception. *Man*, 27(3), 609–627. Available from: http://dx.doi.org/10.2307/2803931.

Byrne, R. W., & Whiten, A. (1988). *Machiavellian intelligence: Social expertise and the evolution of intellect in monkeys, apes, and humans*. Oxford: Clarendon Press.

Canale, G. R., Guidorizzi, C. E., Martins Kierulff, M. C., & Ferreira Rodrigues Gatto, C. A. (2009). First record of tool use by wild populations of the yellow-breasted capuchin monkey (*Cebus xanthosternos*) and new records for the bearded capuchin (*Cebus libidinosus*). *American Journal of Primatology*, 71(5), 366–372. Available from: http://dx.doi.org/10.1002/ajp.20648.

Cantalupo, C., Freemar, H. D., Rodes, W., & Hopkins, W. D. (2008). Handedness for tool use correlates with cerebellar asymmetries in chimpanzees (*Pan troglodytes*). *Behavioral Neuroscience*, 122(1), 191–198. Available from: http://dx.doi.org/10.1037/0735-7044.122.1.191.

Casanova, M. F., & Tillquist, C. R. (2008). Encephalization, emergent properties, and psychiatry: A minicolumnar perspective. *Neuroscientist*, 14, 101–118. Available from: http://dx.doi.org/10.1177/1073858407309091 | issn 1073-8584

Chiappe, D., & MacDonald, K. (2005). The evolution of domain-general mechanisms in intelligence and learning. *Journal of General Psychology*, 132(1), 5–40. Available from: http://dx.doi.org/10.3200/genp.132.1.5-40.

Clark, D. A., Mitra, P. P., & Wang, S. S. H. (2001). Scalable architecture in mammalian brains. *Nature*, 411, 189–193.

Clutton-Brock, T. H., & Harvey, P. H. (1980). Primates, brains and ecology. *Journal of Zoology*, 190(MAR), 309–323.

Colom, R., Jung, R. E., & Haier, R. J. (2006). Distributed brain sites for the g-factor of intelligence. *Neuroimage*, 31(3), 1359–1365. Available from: http://dx.doi.org/10.1016/j.neuroimage.2006.01.006.

Currie, D. J., & Fritz, J. T. (1993). Global patterns of animal abundance and species energy use. *Oikos*, 67(1), 56–68. Available from: http://dx.doi.org/10.2307/3545095.

Dausmann, K. H., Glos, J., Ganzhorn, J. U., & Heldmaier, G. (2004). Physiology: Hibernation in a tropical primate. *Nature*, 429(6994), 825–826. Available from:http://dx.org/doi:10.1038/429825a; http://www.nature.com/nature/journal/v429/n6994/suppinfo/429825a_S1.html

Day, L. B., Westcott, D. A., & Olster, D. H. (2005). Evolution of bower complexity and cerebellum size in bowerbirds. *Brain, Behavior and Evolution*, 66(1), 62–72.

de Sousa, A. A., Sherwood, C. C., Mohlberg, H., Amunts, K., Schleicher, A., MacLeod, C. E., et al. (2010) Hominoid visual brain structure volumes and the position of the lunate sulcus. *Journal of Human Evolution*, 58(4), 281–292. Available from: http://dx.doi.org/10.1016/j.jhevol.2009.11.011.

III. THE STRUGGLE FOR CREATIVITY

de Waal, F. B. M. (2003). Social syntax: The if-then structure of social problem solving. In F. B. M. de Waal, & P. L. Tyack (Eds.), *Animal social complexity: Intelligence, culture, and individualized societies* (pp. 230−248). Cambridge, MA: Harvard University Press.

de Winter, W., & Oxnard, C. E. (2001). Evolutionary radiations and convergences in the structural organization of mammalian brains. *Nature, 409*(6821), 710−714.

Dean, C., Leakey, M. G., Reid, D., Schrenk, F., Schwartz, G. T., Stringer, C., et al. (2001). Growth processes in teeth distinguish modern humans from *Homo erectus* and earlier hominins. *Nature, 414*(6864), 628−631. Available from: http://dx.doi.org/10.1038/414628a.

Deaner, R. O., Isler, K., Burkart, J. M., & van Schaik, C. P. (2007). Overall brain size, and not encephalization quotient, best predicts cognitive ability across non-human primates. *Brain Behavior and Evolution, 70*(2), 115−124. Available from: http://dx.doi.org/10.1159/000102973.

Deaner, R. O., Nunn, C. L., & van Schaik, C. P. (2000). Comparative tests of primate cognition: Different scaling methods produce different results. *Brain, Behavior and Evolution, 55*(1), 44−52.

Deaner, R. O., van Schaik, C. P., & Johnson, V. E. (2006). Do some taxa have better domain-general cognition than others? A meta-analysis of nonhuman primate studies. *Evolutionary Psychology, 4*, 149−196.

Dunbar, R. I. M. (1992). Neocortex size as a constraint on group size in primates. *Journal of Human Evolution, 22*(6), 469−493.

Dunbar, R. I. M. (1998). The social brain hypothesis. *Evolutionary Anthropology, 6*(5), 178−190. Available from: http://dx.doi.org/10.1002/(sici)1520-6505(1998)6:5 < 178:: aid-evan5 > 3.0.co;2-8

Dunbar, R. I. M. (2003). The social brain: Mind, language, and society in evolutionary perspective. *Annual Review of Anthropology, 32*, 163−181. Available from: http://dx.doi.org/10.1146/annurev.anthro.32.061002.093158.

Ely, J. J., Sherwood, C. C., Delman, B. N., Gentile, J. C., Naidich, T. P., Perl, D. P., et al. (2001). Comparative atlases of great ape brains from magnetic resonance images. *Society for Neuroscience Abstracts, 27*(2), 2270.

Erwin, J. M., Sherwood, C. C., Delman, B. N., Naidich, T. P., Gentile, J. C., Bruner, H. J., et al. (2001). The aging great ape brain: A volumetric MRI study of hippocampus and striatum. *Society for Neuroscience Abstracts, 27*(2), 2270.

Evrard, H. C., & Craig, A. D. (2012). Modular architectonic parcellation of the insula in the macaque monkey. *Society for Neuroscience Abstract Viewer and Itinerary Planner, 42*.

Evrard, H. C., Logothetis, N. K., & Craig, A. D. (2014). Modular architectonic organization of the insula in the macaque monkey. *Journal of Comparative Neurology, 522*(1), 64−97. Available from: http://dx.doi.org/10.1002/cne.23436.

Ferreira, R. G., Jerusalinsky, L., Farias Silva, T. C., Fialho, M. d. S., Roque, A. d. A., Fernandes, A., et al. (2009). On the occurrence of *Cebus flavius* (Schreber 1774) in the Caatinga, and the use of semi-arid environments by *Cebus* species in the Brazilian state of Rio Grande do Norte. *Primates, 50*(4), 357−362. Available from: http://dx.doi.org/10.1007/s10329-009-0156-z.

Finlay, B. L., & Darlington, R. B. (1995). Linked regularities in the development and evolution of mammalian brains. *Science, 268*(5217), 1578−1584.

Finlay, B. L., Darlington, R. B., & Nicastro, N. (2001). Developmental structure in brain evolution. *Behavioral and Brain Sciences, 24*(2), 298−308.

Fish, J. L., & Lockwood, C. A. (2003). Dietary constraints on encephalization in primates. *American Journal of Physical Anthropology, 120*(2), 171−181. Available from: http://dx. doi.org/10.1002/ajpa.10136.

Fisher, J., & Hinde, R. A. (1949). The opening of milk bottles by birds. *British Birds, 42*(2), 347−357.

Freeman, H. D., Canta_upo, C., & Hopkins, W. D. (2004). Asymmetries in the hippocampus and amygdala of chimpanzees (*Pan troglodytes*). *Behavioral Neuroscience*, *118*(6), 1460–1465. Available from: http://dx.doi.org/10.1037/0735-7044.118.6.1460.

Frey, S., Pandya, D. N., Chakravarty, M. M., Bailey, L., Petrides, M., & Collins, D. L. (2011). An MRI based average macaque monkey stereotaxic atlas and space (MNI monkey space). *Neuroimage*, *55*(4), 1435–1442. Available from: http://dx.doi.org/10.1016/j.neuroimage.2011.01.040.

Gabi, M., Collins, C. E., Wong, P., Torres, L. B., Kaas, J. H., & Herculano-Houzel, S. (2010). Cellular scaling rules for the brains of an extended number of primate species. *Brain Behavior and Evolution*, *76*(1), 32–44. Available from: http://dx.doi.org/10.1159/000319872.

Ganzhorn, J. U., Fietz, J., Rakotovao, E., Schwab, D., & Zinner, D. (1999). Lemurs and the regeneration of dry deciduous forest in madagascar lemures y la regeneración de bosques deciduos secos en madagascar. *Conservation Biology*, *13*(4), 794–804. Available from: http://dx.do_.org/10.1046/j.1523-1739.1999.98245.x.

Genoud, M. (2002). Comparative studies of basal rate of metabolism in primates. *Evolutionary Anthropology*, *11*, 108–111.

Ghosh, P., Odell, M., Narasimhan, P. T., Fraser, S. E., & Jacobs, R. E. (1994). Mouse lemur microscopic MRI b_ain atlas. *Neuroimage*, *1*(4), 345–349. Available from: http://dx.doi.org/10.1006/nimg.1994.1019.

Gibson, K. R. (1986). Cognition, brain size and the extraction of embedded food resources. In J. G. Else, & P. C. Lee (Eds.), *Primate ontogeny, cognition, and social behavior* (pp. 93–105). Cambridge: Cambridge University Press.

Godin, J. G. J., & Smith, S. A. (1988). A fitness cost of foraging in the guppy. *Nature*, *333* (6168), 69–71. Available from: http://dx.doi.org/10.1038/333069a0.

Greenberg, R., & Mettke-Hofmann, C. (2001). Ecological aspects of neophobia and neophilia in birds. *Current ornithology*, *16*, 119–178.

Harvey, P. H., & Krebs, J. R. (1990). Comparing brains. *Science*, *249*, 140–146.

Harvey, P. H., & Pagel, M. D. (1991). *The comparative method in evolutionary biology*. Oxford: Oxford University Press.

Healy, S. D., de Kort, S. R., & Clayton, N. S. (2005). The hippocampus, spatial memory and food hoarding: A puzzle revisited. *Trends in Ecology & Evolution*, *20*(1), 17–22.

Hemingway, C. A., & Bynum, N. (2005). The influence of seasonality on primate diet and ranging. In D. K. Erockman, & C. P. van Schaik (Eds.), *Seasonality in primates: Studies of living and extinct human and non-human primates* (pp. 57–104). Cambridge: Cambridge University Press.

Herculano-Houzel, S. (2009). The human brain in numbers: A linearly scaled-up primate brain. *Frontiers in Human Neuroscience*, *3*. Available from: http://dx.doi.org/10.3389/neuro.09.031.2009.

Herculano-Houzel, S. (2011). Not all brains are made the same: New views on brain scaling in evolution. *Brain, Behavior and Evolution*, *78*(1), 22–36.

Herculano-Houzel, S. (2012). Neuronal scaling rules for primate brains: The primate advantage. *Progress in Brain Research*, *195*, 325–340.

Herculano-Houzel, S., & Kaas, J. H. (2011). Gorilla and orangutan brains conform to the primate cellular scaling rules: Implications for human evolution. *Brain, Behavior and Evolution*, *77*(1), 33–44.

Herculano-Houzel, S., Mota, B., Wong, P., & Kaas, J. H. (2010). Connectivity-driven white matter scaling and folding in primate cerebral cortex. *Proceedings of the National Academy of Sciences of the United States of America*, *107*(44), 19008–19013. Available from: http://dx.doi.org/10.1073/pnas.1012590107.

Hikishima, K., Quallo, M. M., Komaki, Y., Yamada, M., Kawai, K., Momoshima, S., et al. (2011). Population-averaged standard template brain atlas for the common marmoset

(*Callithrix jacchus*). *Neuroimage*, *54*(4), 2741–2749. Available from: http://dx.doi.org/10.1016/j.neuroimage.2010.10.061.

Hinde, R. A., & Fisher, J. (1951). Further observations on the opening of milk bottles by birds. *British Birds*, *44*, 395–396.

Holliday, M. A. (1986). Body composition and energy needs during human growth. In (2nd ed.F. Falkner, & J. M. Tanner (Eds.), *Human growth: A comprehensive treatise* (Vol. 2New York, NY: Plenum Press.

Hopkins, W. D., Cantalupo, C., & Taglialatela, J. P. (2007). Handedness is associated with asymmetries in gyrification of the cerebral cortex of chimpanzees. *Cerebral Cortex*, *17* (8), 1750–1756. Available from: http://dx.doi.org/10.1093/cercor/bhl085.

Hopkins, W. D., Dunham, L., Cantalupo, C., & Taglialatela, J. P. (2007). The association between handedness, brain asymmetries, and corpus callosum size in chimpanzees (*Pan troglodytes*). *Cerebral Cortex*, *17*(8), 1757–1765. Available from: http://dx.doi.org/10.1093/cercor/bhl086.

Hopkins, W. D., Lyn, H., & Cantalupo, C. (2009). Volumetric and lateralized differences in selected brain regions of chimpanzees (*Pan troglodytes*) and bonobos (*Pan paniscus*). *American Journal of Primatology*, *71*(12), 988–997. Available from: http://dx.doi.org/10.1002/ajp.20741.

Hopkins, W. D., & Marino, L. (2000). Asymmetries in cerebral width in nonhuman primate brains as revealed by magnetic resonance imaging (MRI). *Neuropsychologia*, *38*(4), 493–499. Available from: http://dx.doi.org/10.1016/s0028-3932(99)00090-1

Hopkins, W. D., Marino, L., Rilling, J. K., & MacGregor, L. A. (1998). Planum temporale asymmetries in great apes as revealed by magnetic resonance imaging (MRI). *Neuroreport*, *9*(12), 2913–2918. Available from: http://dx.doi.org/10.1097/00001756-199808240-00043.

Hopkins, W. D., & Nir, T. M. (2010). Planum temporale surface area and grey matter asymmetries in chimpanzees (*Pan troglodytes*): The effect of handedness and comparison with findings in humans. *Behavioral Brain Research*, *208*(2), 436–443. Available from: http://dx.doi.org/10.1016/j.bbr.2009.12.012.

Hopkins, W. D., & Pilcher, D. L. (2001). Neuroanatomical localization of the motor hand area with magnetic resonance imaging: The left hemisphere is larger in great apes. *Behavioral Neuroscience*, *115*(5), 1159–1164. Available from: http://dx.doi.org/10.1037/0735-7044.115.5.1159.

Hopkins, W. D., Russell, J. L., & Schaeffer, J. A. (2012). The neural and cognitive correlates of aimed throwing in chimpanzees: A magnetic resonance image and behavioral study on a unique form of social tool use. *Philosophical Transactions of the Royal Society B—Biological Sciences*, *367*(1585), 37–47. Available from: http://dx.doi.org/10.1098/rstb.2011.0195.

Hopkins, W. D., Taglialatela, J. P., & Cantalupo, C. (2007). Normative data of the chimpanzee brain as revealed by magnetic resonance imaging: Age and sex effects. *American Journal of Primatology*, *69*, 46–47.

Hopkins, W. D., Taglialatela, J. P., Dunham, L., & Pierre, P. (2007). Behavioral and neuroanatomical correlates of white matter asymmetries in chimpanzees (*Pan troglodytes*). *European Journal of Neuroscience*, *25*(8), 2565–2570. Available from: http://dx.doi.org/10.1111/j.1460-9568.2007.05502.x.

Hopkins, W. D., Taglialatela, J. P., Meguerditchian, A., Nir, T. M., Schenker, N. M., & Sherwood, C. C. (2008). Gray matter asymmetries in chimpanzees as revealed by voxel-based morphometry. *Neuroimage*, *42*(2), 491–497. Available from: http://dx.doi.org/10.1016/j.neuroimage.2008.05.014.

Hopkins, W. D., Taglialatela, J. P., & Russell, J. L. (2006). Localizing handedness in the chimpanzee brain: A combined MRI and PET study. *American Journal of Primatology*, *68*, 120.

Hopkins, W. D., Taglialatela, J. P., Russell, J. L., Nir, T. M., & Schaeffer, J. A. (2010). Cortical representation of lateralized grasping in chimpanzees (*Pan troglodytes*): A combined MRI and PET study. *PLoS One, 5*(10). Available from: http://dx.doi.org/10.1371/journal.pone.0013383.

Isler, K., Kirk, C. E., Miller, J. M. A., Albrecht, G. A., Gelvin, B. R., & Martin, R. D. (2008). Endocranial volumes of primate species: Scaling analyses using a comprehensive and reliable data set. *Journal of Human Evolution, 55*(6), 967—978. Available from: http://dx.doi.org/10.1016/j.jhevol.2008.08.004.

Isler, K., & van Schaik, C. P. (2006). Costs of encephalisation: The energy trade-off hypothesis tested on birds *Journal of Human Evolution, 51*(3), 228—243.

Isler, K., & van Schaik, C. P. (2009a). Cooperative breeding and brain size in mammals and birds. *Folia Primatologica, 80*(2), 121.

Isler, K., & van Schaik, C. P. (2009b). The expensive brain: A framework for explaining evolutionary changes in brain size. *Journal of Human Evolution, 57*(4), 392—400. Available from: http://dx.doi.org/10.1016/j.jhevol.2009.04.009.

Isler, K., van Woerden, J. T., Navarrete, A. F., & van Schaik, C. P. (2012). The "gray ceiling": Why apes are not as large-brained as humans. *American Journal of Physical Anthropology, 147*, 173.

Iwaniuk, A. N., Nelson, J. E., & Pellis, S. M. (2001). Do big-brained animals play more? Comparative analyses of play and relative brain size in mammals. *Journal of Comparative Psychology, 115*(1), 29—41. Available from: http://dx.doi.org/10.1037/0735-7036.115.1.29.

Janson, C. H., & Verdolin, J. (2005). Seasonality of primate births in relation to climate. In D. A. Brockman, & C. P. van Schaik (Eds.), *Seasonality in primates* (pp. 307—350). New York, NY: Cambridge University Press.

Jerison, H. J. (1955). Brain to body ratios and the evolution of intelligence. *Science, 121* (3144), 447—449.

Jerison, H. J. (1961). Quantitative analysis of evolution of brain in mammals. *Science, 133* (345), 1012.

Jerison, H. J. (1973). *Evolution of the brain and intelligence*. New York, NY: Academic Press.

Jerison, H. J. (2001). Archaeological implications of paleoneurology. In A. Nowell (Ed.), *In the mind's eye: Multidisciplinary approaches to the evolution of human cognition* (pp. 83—96). Ann Arbor, MI: International Monographs in Prehistory.

Jolicoeur, P., Pirlot, P., Baron, G., & Stephan, H. (1984). Brain structure and correlation patterns in insectivora, chiroptera, and primates. *Systematic Zoology, 33*(1), 14—29.

Kaas, J. H. (2012). The evolution of neocortex in primates. *Evolution of the Primate Brain: From Neuron to Behavior, 195*, 91—102. Available from: http://dx.doi.org/10.1016/b978-0-444-53860-4.00005-2.

Kalisch, M., Maechler, M., Colombo, D., Maathuis, M. H., & Buehlmann, P. (2012). Causal inference using graphical models with the R package *pcalg*. *Journal of Statistical Software, 47*(11), 1—26.

Kanai, R., Bahrami, B., Roylance, R., & Rees, G. (2012). Online social network size is reflected in human brain structure. *Proceedings of the Royal Society B—Biological Sciences, 279*(1732), 1327—1334. Available from: http://dx.doi.org/10.1098/rspb.2011.1959.

Kaplan, H. S., Hill, K., Lancaster, J., & Hurtado, A. M. (2000). A theory of human life history evolution: Diet, intelligence, and longevity. *Evolutionary Anthropology, 9*(4), 156—185. Available from: http://dx.doi.org/10.1002/1520-6505(2000)9:4 < 156::aid-evan5 > 3.0.co;2-7

Kaplan, H. S., & Robson, A. J. (2002). The emergence of humans: The coevolution of intelligence and longevity with intergenerational transfers. *Proceedings of the National Academy of Sciences of the United States of America, 99*(15), 10221—10226.

Karasov, W. H., Pinshow, B., Starck, J. M., & Afik, D. (2004). Anatomical and histological changes in the alimentary tract of migrating blackcaps (*Sylvia atricapilla*): A comparison among fed, fasted, food-restricted, and refed birds. *Physiological and Biochemical Zoology*, 77(1), 149–160.

Kawai, M. (1965). Newly-acquired pre-cultural behavior of the natural troop of Japanese monkeys on Koshima islet. *Primates*, 6(1), 1–30. Available from: http://dx.doi.org/10.1007/bf01794457.

Kummer, H., & Goodall, J. (1985). Conditions of innovative behavior in primates. *Philosophical Transactions of the Royal Society of London Series B—Biological Sciences*, 308 (1135), 203–214. Available from: http://dx.doi.org/10.1098/rstb.1985.0020.

Laland, K. N., & Reader, S. M. (2009). Comparative perspectives on human innovation. In M. J. O'Brien, & S. J. Shennan (Eds.), *Innovation in cultural systems: Contributions from evolutionary anthropology* (pp. 37–51). Cambridge, MA: Massachusetts Institute of Technology(eds).

Lee, H., Smith, K. G., & Grimm, C. M. (2003). The effect of new product radicality and scope on the extent and speed of innovation diffusion. *Journal of Management*, 29(5), 753–768.

Lee, P. C (1991). Adaptations to environmental change: an evolutionary perspective. In H. O. Box (Ed.), *Primate responses to environmental change* (pp. 39–56). New York, NY: Chapman & Hall.

Lee, P. C., Majluf, P., & Gordon, I. J. (1991). Growth, weaning and maternal investment from a comparative perspective. *Journal of Zoology*, 225, 99–114.

Lefebvre, L. (2011). Taxonomic counts of cognition in the wild. *Biology Letters*, 7(4), 631–633. Available from: http://dx.doi.org/10.1098/rsbl.2010.0556.

Lefebvre, L. (2012). Primate encephalization. In M. A. Hofman & D. Falk (Eds.), *Evolution of the primate brain: From neuron to behavior* (Vol. 195, pp. 393–412).

Lefebvre, L. (2013). Brains, innovations, tools and cultural transmission in birds, non-human primates, and fossil hominins. *Frontiers in Human Neuroscience*, 7. Available from: http://dx.doi.org/10.3389/fnhum.2013.00245.

Lefebvre, L., & Giraldeau, L. A. (1996). Is social learning an adaptive specialization? In C. M. Heyes, & B. G. Galef, Jr. (Eds.), *Social learning in animals: The roots of culture* (pp. . 107–128). San Diego, CA: Academic Press, Inc..

Lefebvre, L., Juretic, N., Nicolakakis, N., & Timmermans, S. (2001). Is the link between forebrain size and feeding innovations caused by confounding variables? A study of Australian and North American birds. *Animal Cognition*, 4(2), 91–97. Available from: http://dx.doi.org/10.1007/s100710100102.

Lefebvre, L., Reader, S. M., & Sol, D. (2004). Brains, innovations and evolution in birds and primates. *Brain, Behavior and Evolution*, 63(4), 233–246.

Lefebvre, L., Whittle, P., Lascaris, E., & Finkelstein, A. (1997). Feeding innovations and forebrain size in birds. *Animal Behavior*, 53, 549–560. Available from: http://dx.doi.org/10.1006/anbe.1996.0330.

Lukas, W. D., & Campbell, B. C. (2000). Evolutionary and ecological aspects of early brain malnutrition in humans. *Human Nature—An Interdisciplinary Biosocial Perspective*, 11(1), 1–26. Available from: http://dx.doi.org/10.1007/s12110-000-1000-8.

MacLean, E. L., Barrickman, N. L., Johnson, E. M., & Wall, C. E. (2009). Sociality, ecology, and relative brain size in lemurs. *Journal of Human Evolution*, 56(5), 471–478.

Madden, J. (2001). Sex, bowers and brains. *Proceedings of the Royal Society of London Series B—Biological Sciences*, 268(1469), 833–838.

Marin Manrique, H., Voelter, C. J., & Call, J. (2013). Repeated innovation in great apes. *Animal Behavior*, 85(1), 195–202. Available from: http://dx.doi.org/10.1016/j.anbehav.2012.10.026.

Martin, R. D. (2000). Origins, diversity and relationships of lemurs. *International Journal of Primatology*, 21(6), 1021–1049. Available from: http://dx.doi.org/10.1023/A:1005563113546.

McDaniel, M. A. (2005). Big-brained people are smarter: A meta-analysis of the relationship between *in vivo* brain volume and intelligence. *Intelligence, 33*(4), 337–346. Available from: http://dx.doi.org/10.1016/j.intell.2004.11.005.

McGrew, W. C. (2004). *The cultured chimpanzee: Reflections on cultural primatology.* Cambridge, UK: Cambridge University Press.

McLaren, D. G., Kosmatka, K. J., Oakes, T. R., Kroenke, C. D., Kohama, S. G., Matochik, J. A., et al. (2009). A population-average MRI-based atlas collection of the rhesus macaque. *Neuroimage, 45*(1), 52–59. Available from: http://dx.doi.org/10.1016/j.neuroimage.2008.10.058.

Meulman, E. J. M., Sanz, C. M., Visalberghi, E., & van Schaik, C. P. (2012). The role of terrestriality in promoting primate technology. *Evolutionary Anthropology, 21*(2), 58–68. Available from: http://dx.doi.org/10.1002/evan.21304.

Meulman, E. J. M., Schuppli, C., & van Schaik, C. P. (2013). How ecology may affect orangutan innovation and culture. *Folia Primatologica, 84*(3–5), 303.

Meulman, E. J. M., & van Schaik, C. P. (2013). Orangutan tool use and the evolution of technology. In C. M. Sanz, J. Call, & C. Boesch (Eds.), *Tool Use in Animals: Cognition and Ecology* (pp. 176–202). Cambridge, UK: Cambridge University Press.

Mikula, S., Graziano, A., Stone, J. M., & Jones, E. G. (2009). A three-dimensional atlas of neural connectivity in the macaque brain. *Society for Neuroscience Abstract Viewer and Itinerary Planner, 39.*

Miller, E. K., & Cohen, J. D. (2001). An integrative theory of prefrontal cortex function. *Annual Review of Neuroscience, 24*, 167–202. Available from: http://dx.doi.org/10.1146/annurev.neuro.24.1.167.

Milton, K. (1981). Distribution patterns of tropical plant foods as an evolutionary stimulus to primate mental development. *American Anthropologist, 83*(3), 534–548.

Mink, J. W., Blumenschine, R. J., & Adams, D. B. (1981). Ratio of central nervous system to body metabolism in vertebrates—its constancy and functional basis. *American Journal of Physiology, 241*(3), R203–R212.

Montgomery, S. H., Capellini, I., Barton, R. A., & Mundy, N. I. (2010). Reconstructing the ups and downs of primate brain evolution: Implications for adaptive hypotheses and *Homo floresiensis. BMC Biology, 8.* Available from: http://dx.doi.org/10.1186/1741-7007-8-9.

Morand-Ferron, J., Lefebvre, L., Reader, S. M., Sol, D., & Elvin, S. (2004). Dunking behavior in Carib grackles. *Animal Behavior, 68*(Journal Article), 1267–1274.

Morand-Ferron, J., Sol, D., & Lefebvre, L. (2007). Food stealing in birds: Brain or brawn? *Animal Behavior, 74*, 1725–1734. Available from: http://dx.doi.org/10.1016/j.anbehav.2007.04.031.

Navarrete, A. F., van Schaik, C. P., & Isler, K. (2011). Energetics and the evolution of human brain size. *Nature, 480*(7375), 91–U252. Available from: http://dx.doi.org/10.1038/nature10629.

Newman, J. D., Kenkel, W. M., Aronoff, E. C., Bock, N. A., Zametkin, M. R., & Silva, A. C. (2009). A combined histological and MRI brain atlas of the common marmoset monkey, *Callithrix jacchus. Brain Research Reviews, 62*(1), 1–18. Available from: http://dx.doi.org/10.1016/j.brainresrev.2009.09.001.

Nicolakakis, N., Sol, D., & Lefebvre, L. (2003). Behavioral flexibility predicts species richness in birds, but not extinction risk. *Animal Behavior, 65*(Journal Article), 445–452.

Nieuwenhuys, R., Donkelaar, H. J., & Nicholson, C. (1998). *The central nervous system of vertebrates.* Berlin: Springer.

O'Regan, H. J., & Kitchener, A. C. (2005). The effects of captivity on the morphology of captive, domesticated and feral mammals. *Mammal Review, 35*(3–4), 215–230.

Pagel, M. D., & Harvey, P. H. (1990). Diversity in the brain sizes of newborn mammals. *Bioscience, 40*(2), 116–122.

Palazzi, X., & Bordier, N. (2008). *The marmoset brain in stereotaxic coordinates the marmoset brain in stereotaxic coordinates.* New York, NY: Springer, (pp. 1–59).

Parker, S. T. (1990). Why big brains are so rare: Energy costs of intelligence and brain size in anthropoid Primates. In S. T. Parker, & K. R. Gibson (Eds.), *'Language' and intelligence in monkeys and apes: Comparative developmental perspectives* (pp. 129–154). Cambridge, UK: Cambridge University Press.

Parker, S. T., & Gibson, K. R. (1979). A developmental model for the evolution of language and intelligence in early hominids. *Behavioral and Brain Sciences, 2*(3), 367–381.

Pearl, J. (1988). *Probabilistic reasoning in intelligent systems: Networks of plausible inference.* San Francisco, CA: Morgan Kaufmann Publishers Inc.

Petrosini, L., Graziano, A., Mandolesi, L., Neri, P., Molinari, M., & Leggio, M. G. (2003). Watch how to do it! New advances in learning by observation. *Brain Research Reviews, 42*(3), 252–264.

Phillips, K. A., & Hopkins, W. D. (2007). Exploring the relationship between cerebellar asymmetry and handedness in chimpanzees (*Pan troglodytes*) and capuchins (*Cebus apella*). *Neuropsychologia, 45*(10), 2333–2339. Available from: http://dx.doi.org/10.1016/j.neuropsychologia.2007.02.010.

Phillips, K. A., & Sherwood, C. C. (2005). Primary motor cortex asymmetry is correlated with handedness in Capuchin monkeys (*Cebus apella*). *Behavioral Neuroscience, 119*(6), 1701–1704. Available from: http://dx.doi.org/10.1037/0735-7044.119.6.1701.

Phillips, K. A., & Sherwood, C. C. (2007). Cerebral petalias and their relationship to handedness in capuchin monkeys (*Cebus apella*). *Neuropsychologia, 45*(10), 2398–2401. Available from: http://dx.doi.org/10.1016/j.neuropsychologia.2007.02.021.

Phillips, K. A., Sobieski, C. A., Thompson, C. R., & Sherwood, C. C. (2008). Brain structures differ among tool using and non-tool using capuchin monkeys. *Society for Neuroscience Abstract Viewer and Itinerary Planner, 38.*

Pontzer, H., Raichlen, D. A., Gordon, A. D., Schroepfer-Walker, K. K., Hare, B. A., O'Neill, M. C., et al. (2014). Primate energy expenditure and life history. *Proceedings of the National Academy of Sciences of the United States of America, 111*(4), 1433–1437. Available from: http://dx.doi.org/10.1073/pnas.1316940111.

Quallo, M. M., Price, C. J., Ueno, K., Asamizuya, T., Cheng, K., Lemon, R. N., et al. (2010). Creating a population-averaged standard brain template for Japanese macaques (*M. fuscata*). *Neuroimage, 52*(4), 1328–1333. Available from: http://dx.doi.org/10.1016/j.neuroimage.2010.05.006.

Ramsey, G., Bastian, M. L., & van Schaik, C. P. (2007a). Animal innovation defined and operationalized. *Behavioral and Brain Sciences, 30*(4), 393–407. Available from: http://dx.doi.org/10.1017/s0140525x07002373.

Ramsey, G., Bastian, M. L., & van Schaik, C. P. (2007b). On the concept of animal innovation and the challenge of studying innovation in the wild. *Behavioral and Brain Sciences, 30*(4), 425–437. Available from: http://dx.doi.org/10.1017/s0140525x07002567.

Reader, S. M., Hager, Y., & Laland, K. N. (2011). The evolution of primate general and cultural intelligence. *Philosophical Transactions of the Royal Society B: Biological Sciences, 366*(1567), 1017–1027. Available from: http://dx.doi.org/10.1098/rstb.2010.0342.

Reader, S. M., & Laland, K. N. (2001). Primate innovation: Sex, age and social rank differences. *International Journal of Primatology, 22*(5), 787–805. Available from: http://dx.doi.org/10.1023/a:1012069500899.

Reader, S. M., & Laland, K. N. (2002). Social intelligence, innovation, and enhanced brain size in primates. *Proceedings of the National Academy of Sciences of the United States of America, 99*(7), 4436–4441. Available from: http://dx.doi.org/10.1073/pnas.062041299.

Reader, S. M., & Laland, K. N. (2003). *Animal innovation.* New York, NY: Oxford University Press.

Reader, S. M., & MacDonald, K. (2003). Environmental variability and primate behavioral flexibility. In S. M. Reader, & K. N. Laland (Eds.), *Animal innovation* (pp. 83–116). New York, NY: Oxford University Press.

Rilling, J. K., & Insel, T. R. (1998). Evolution of the cerebellum in primates: Differences in relative volume among monkeys, apes and humans. *Brain, Behavior and Evolution, 52* (6), 308–314.

Rilling, J. K., & Insel, T. R. (1999a). Differential expansion of neural projection systems in primate brain evolution. *Neuroreport, 10*(7), 1453–1459. Available from: http://dx.doi. org/10.1097/00001756-199905140-00012.

Rilling, J. K., & Insel, T. R. (1999b). The primate neocortex in comparative perspective using magnetic resonance imaging. *Journal of Human Evolution, 37*(2), 191–223. Available from: http://dx.doi.org/10.1006/jhev.1999.0313.

Rilling, J. K., & Seligman, R. A. (2000). A quantitative morphometric comparative analysis of the primate temporal lobe. *Society for Neuroscience Abstracts, 26.* (1—2)Abstract No. -71.11

Sallet, J., Mars, R. B., Noonan, M. P., Andersson, J. L., O'Reilly, J. X., Jbabdi, S., et al. (2011). Social network size affects neural circuits in macaques. *Science, 334*(6056), 697–700. Available from: http://dx.doi.org/10.1126/science.1210027.

Sawaguchi, T. (1997). Possible involvement of sexual selection in neocortical evolution of monkeys and apes. *Folia Primatologica, 68*(2), 95–99. Available from: http://dx.doi. org/10.1159/000157236.

Schuck-Paim, C., Alonso, W. J., & Ottoni, E. B. (2008). Cognition in an ever-changing world: Climatic variability is associated with brain size in neotropical parrots. *Brain Behavior and Evolution, 71*(3), 200–215. Available from: http://dx.doi.org/10.1159/ 000119710.

Schülke, O., & Ostner, J. (2007). Physiological ecology of cheirogaleid primates: Variation in hibernation and torpor. *Acta Ethologica, 10*(1), 13–21. Available from: http://dx.doi. org/10.1007/s10211-006-0023-5.

Semendeferi, K. (2001). Advances in the study of hominoid brain evolution: magnetic resonance imaging (MRI) and 3-D reconstruction. In D. Falk, & K. R. Gibson (Eds.), *Evolutionary Anatomy of the Primate Cerebral Cortex* (pp. 257–289). New York, NY: Cambridge Universitary Press.

Semendeferi, K., Armstrong, E., Schleicher, A., Zilles, K., & van Hoesen, G. W. (2001). Prefrontal cortex in humans and apes: A comparative study of area 10. *American Journal of Physical Anthropology, 114*(3), 224–241. Available from: http://dx.doi.org/ 10.1002/1096-8644(200103)114:3 < 224::aid-ajpa1022 > 3.0.co;2-i

Semendeferi, K., & Damasio, H. (2000). The brain and its main anatomical subdivisions in living hominoids using magnetic resonance imaging. *Journal of Human Evolution, 38*(2), 317–332. Available from: http://dx.doi.org/10.1006/jhev.1999.0381.

Semendeferi, K., Damasio, H., & Frank, R. (1997). The evolution of the frontal lobes: A volumetric analysis based on three-dimensional reconstructions of magnetic resonance scans of human and ape brains. *Journal of Human Evolution, 32*(4), 375–388. Available from: http://dx.doi.org/10.1006/jhev.1996.0099.

Sherwood, C. C., Cranfield, M. R., Mehlman, P. T., Lilly, A. A., Garbe, J. A. L., Whittier, C. A., et al. (2004) Brain structure variation in great apes, with attention to the mountain gorilla (*Gorilla beringei beringei*). *American Journal of Primatology, 63*(3), 149–164. Available from: http://dx.doi.org/10.1002/ajp.20048.

Sherwood, C. C., Erwin, J. M., Delman, B. N., Naidich, T. P., Bruner, H. J., Braun, A. R., et al. (2001). Brain volume in aging great apes: A postmortem MRI study. *American Journal of Primatology, 54*(Suppl. 1), 45–46.

Sherwood, C. C., Gordon, A. D., Allen, J. S., Phillips, K. A., Erwin, J. M., Hof, P. R., et al. (2011). Aging of the cerebral cortex differs between humans and chimpanzees.

Proceedings of the National Academy of Sciences of the United States of America, 108(32), 13029–13034. Available from: http://dx.doi.org/10.1073/pnas.1016709108.

Sherwood, C. C., Stimpson, C. D., Raghanti, M. A., Wildmand, D. E., Uddin, M., Grossman, L. I., et al. (2006). Evolution of increased glia-neuron ratios in the human frontal cortex. *Proceedings of the National Academy of Sciences of the United States of America, 103*(37), 13606–13611.

Shipley, B. (2000). *Cause and correlation in biology: A user's guide to path analysis, structural equations and causal inference*. Cambridge, UK: Cambridge University Press.

Shultz, S., & Dunbar, R. I. M. (2007). The evolution of the social brain: Anthropoid primates contrast with other vertebrates. *Proceedings of the Royal Society B—Biological Sciences, 274*(1624), 2429–2436.

Shultz, S., & Dunbar, R. I. M. (2010). Encephalization is not a universal macroevolutionary phenomenon in mammals but is associated with sociality. *Proceedings of the National Academy of Sciences of the United States of America, 107*(50), 21582–21586. Available from: http://dx.doi.org/10.1073/pnas.1005246107.

Shumaker, R. W., Walkup, K. R., & Beck, B. B. (2011). *Animal tool behavior: The use and manufacture of tools by animals*. Baltimore, MD: Johns Hopkins University Press, Revised and updated edition.

Smaers, J. B., Dechmann, D. K. N., Goswami, A., Soligo, C., & Safi, K. (2012). Comparative analyses of evolutionary rates reveal different pathways to encephalization in bats, carnivorans, and primates. *Proceedings of the National Academy of Sciences of the United States of America, 109*(44), 18006–18011. Available from: http://dx.doi.org/10.1073/pnas.1212181109.

Smaers, J. B., & Soligo, C. (2013). Brain reorganization, not relative brain size, primarily characterizes anthropoid brain evolution. *Proceedings of the Royal Society B—Biological Sciences, 280*(1759), 20130269. Available from: http://dx.doi.org/10.1098/rspb.2013.0269.

Smaers, J. B., & Vinicius, L. (2009). Inferring macro-evolutionary patterns using an adaptive peak model of evolution. *Evolutionary Ecology Research, 11*(7), 991–1015.

Sol, D. (2003). Behavioral innovation: A neglected issue in the ecological and evolutionary literature? In S. M. Reader, & K. N. Laland (Eds.), *Animal innovation* (pp. 63–82). Oxford, NY: Oxford University Press.

Sol, D., Bacher, S., Reader, S. M., & Lefebvre, L. (2008). Brain size predicts the success of mammal species introduced into novel environments. *American Naturalist, 172*, S63–S71. Available from: http://dx.doi.org/10.1086/588304.

Sol, D., Duncan, R. P., Blackburn, T. M., Cassey, P., & Lefebvre, L. (2005). Big brains, enhanced cognition, and response of birds to novel environments. *Proceedings of the National Academy of Sciences of the United States of America, 102*(15), 5460–5465.

Sol, D., & Lefebvre, L. (2000). Behavioral flexibility predicts invasion success in birds introduced to New Zealand. *Oikos, 90*(3), 599–605. Available from: http://dx.doi.org/10.1034/j.1600-0706.2000.900317.x.

Sol, D., Lefebvre, L., & Rodriguez-Teijeiro, J. D. (2005). Brain size, innovative propensity and migratory behavior in temperate Palaearctic birds. *Proceedings of the Royal Society of London. Series B: Biological Sciences, 272*, 1433–1441.

Sol, D., & Price, T. D. (2008). Brain size and the diversification of body size in birds. *American Naturalist, 172*(2), 170–177. Available from: http://dx.doi.org/10.1086/589461.

Sol, D., Stirling, D. G., & Lefebvre, L. (2005). Behavioral drive or behavioral inhibition in evolution: Subspecific diversification in holarctic passerines. *Evolution, 59*(12), 2669–2677.

Spagnoletti, N., Izar, P., & Visalberghi, E. (2009). Tool use and terrestriality in wild bearded capuchin monkey (*Cebus libidinosus*). *Folia Primatologica, 80*(2), 142.

Spirtes, P., Glymour, C., & Scheines, R. (1993). *Causation, prediction, and search.* Cambridge, MA: Massachusetts Institute of Technology.

Spirtes, P., Glymour, C., & Scheines, R. (2001). *Causation, prediction, and search* (2nd ed.). Cambridge, MA: Massachusetts Institute of Technology.

Stephan, H., Baron, G., Frahm, H. D., & Stephan, M. (1986). Comparison of size in brains and brain structures of mammals. *Zeitschrift Fur Mikroskopisch-Anatomische Forschung, 100*(2), 189–212.

Stephan, H., Bauchot, R., & Andy, O. J. (1970). Data on size of the brain and various brain parts in insectivores and primates. In C. R. Nobak, & W. Montagna (Eds.), *The primate brain* (pp. 289–297). New York, NY: Appleton.

Stephan, H., Frahm, H. D., & Baron, G. (1981). New and revised data on volumes of brain structures in Insectivores and Primates. *Folia Primatologica, 35*(1), 1–29.

Stephan, H., Frahm, H. D., & Baron, G. (1984). Comparison of brain structure volumes in insectivora and primates.4. Non-cortical visual structures. *Journal Fur Hirnforschung, 25* (4), 385–403.

Stephan, H., Frahm, H. D., & Baron, G. (1987). Comparison of brain structure volumes in insectivora and primates.7. Amygdaloid components. *Journal Fur Hirnforschung, 28*(5), 571–584.

Stephan, H., & Pirlot, P. (1970). Volumetric comparisons of brain structures in bats 1. *Journal of Zoological Systematics and Evolutionary Research, 8*(1), 200–236.

Sterling, E. J. (1994). Taxonomy and distribution of daubentonia—a historical-perspective. *Folia Primatologica, 62*(1–3), 8–13.

Striedter, G. F. (2005). *Principles of brain evolution.* Sunderland, MA: Sinauer Associates.

Sultan, F., Hamodeh, S., Murayama, Y., Saleem, K. S., & Logothetis, N. K. (2010). Flat map areal topography in *Macaca mulatta* based on combined MRI and histology. *Magnetic Resonance Imaging, 28*(8), 1159–1164. Available from: http://dx.doi.org/10.1016/j.mri.2010.03.023.

Swindler, D. R., & Wood, C. D. (1982). *An atlas of primate gross anatomy: Baboon, chimpanzee, and man.* Malabar, FL: Robert E. Krieger.

Tammer, R., Hofer, S., Merboldt, J. D., & Frahm, J. (2009). *Magnetic resonance imaging of the rhesus monkey brain.* Göttingen, Germany: Vandenhoeck & Ruprecht.

Taylor, A. B., & van Schaik, C. P. (2007). Variation in brain size and ecology in Pongo. *Journal of Human Evolution, 52*(1), 59–71.

Tebbich, S., Sterelny, K., & Teschke, I. (2010). The tale of the finch: Adaptive radiation and behavioral flexibility. *Philosophical Transactions of the Royal Society B—Biological Sciences, 365*(1543), 1099–1109. Available from: http://dx.doi.org/10.1098/rstb.2009.0291.

Tomasello, M. (2000). Two hypotheses about primate cognition. In C. Heyes, & L. Huber (Eds.), *The evolution of cognition* (pp. 165–183). Cambridge, MA: MIT Press.

Tomasello, M., & Call, J. (1997). Primate cognition.

Uher, J., & Call, J. (2008). How the great apes (*Pan troglodytes, Pongo pygmaeus, Pan paniscus, Gorilla gorilla*) perform on the reversed reward contingency task II: Transfer to new quantities, long-term retention, and the impact of quantity ratios. *Journal of Comparative Psychology, 122*(2), 204–212. Available from: http://dx.doi.org/10.1037/0735-7036.122.2.204.

van Horn, J. D., & Toga, A. W. (2009). Brain atlases: Their development and role in functional inference. In M. Filippi (Ed.), *FMRI techniques and protocols* (Vol. 41, pp. 263–281).

van Schaik, C. P. (2013). The costs and benefits of flexibility as an expression of behavioral plasticity: A primate perspective. *Philosophical Transactions of the Royal Society B—Biological Sciences, 368*(1618). Available from: http://dx.doi.org/10.1098/rstb.2012.0339.

III. THE STRUGGLE FOR CREATIVITY

van Schaik, C. P., van Noordwijk, M. A., de Boer, R. J., & Dentonkelaar, I. (1983). The effect of group size on time budget and social-behavior in wild long-tailed macaques (*Macaca fascicularis*). *Behavioral Ecology and Sociobiology*, 13(3), 173−181.

van Woerden, J. T., van Schaik, C. P., & Isler, K. (2010). Effects of seasonality on brain size evolution: Evidence from strepsirrhine primates. *American Naturalist*, 176(6), 758−767. Available from: http://dx.doi.org/10.1086/657045.

van Woerden, J. T., Willems, E. P., van Schaik, C. P., & Isler, K. (2012). Large brains buffer energetic effects of seasonal habitats in catarrhine primates. *Evolution*, 66(1), 191−199. Available from: http://dx.doi.org/10.1111/j.1558-5646.2011.01434.x.

Visalberghi, E., Fragaszy, D. M., Izar, P., & Ottoni, E. B. (2005). Terrestriality and tool use. *Science*, 308(5724), 951.

Vlamings, P. H. J. M., Hare, B. A., & Call, J. (2010). Reaching around barriers: The performance of the great apes and 3−5-year-old children. *Animal Cognition*, 13(2), 273−285. Available from: http://dx.doi.org/10.1007/s10071-009-0265-5.

Westergaard, G. C., Kuhn, H. E., & Suomi, S. J. (1998). Effects of upright posture on hand preference for reaching vs. the use of probing tools by tufted capuchins (*Cebus apella*). *American Journal of Primatology*, 44(2), 147−153. Available from: http://dx.doi.org/10.1002/(sici)1098-2345(1998)44:2 < 147::aid-ajp5 > 3.0.co;2-w.

Westergaard, G. C., Wagner, J. L., & Suomi, S. J. (1999). Manipulative tendencies of captive *Cebus albifrons*. *International Journal of Primatology*, 20(5), 751−759. Available from: http://dx.doi.org/10.1023/a:1020756803437.

Whiten, A., Byrne, R. W., Waterman, P. G., Henzi, S. P., & McCullough, F. M. (1990). Specifying the rules underlying selective foraging in wild mountain baboons, P. ursinus. In M. Thiago de Mello, A. Whiten, & R. W. Byrne (Eds.), *Baboons. Behaviour and ecology, use and care. Selected proceedings of the 12th Congress of the International Primatological Society* (pp. 5−22). Brazil: International Primatological Society.

Whiten, A., & van Schaik, C. P. (2007). The evolution of animal 'cultures' and social intelligence. *Philosophical Transactions of the Royal Society B—Biological Sciences*, 362(1480), 603−620. Available from: http://dx.doi.org/10.1098/rstb.2006.1998.

Wilson, A. C. (1985). The molecular basis of evolution. *Scientific American*, 253(4), 148−157.

Wilson, A. C. (1991). From molecular evolution to body and brain evolution. In J. Campisi, D. Cunningham, M. Inouye, & M. Riley (Eds.), *Perspectives on cellular regulation: From bacteria to cancer* (pp. 331−340). New York, NY: Wiley-Liss.

Wilson, D. E., & Reeder, D. M. (2005). *Mammal species of the world. A taxonomic and geographic reference* (3rd ed.). Baltimore, MD: Johns Hopkins University Press.

Wilson, E. O. (1975). *Sociobiology*. Cambridge, MA: Harvard University Press.

Winkler, H., Leisler, B., & Bernroider, G. (2004). Ecological constraints on the evolution of avian brains. *Journal of Ornithology*, 145(3), 238−244.

Wisco, J. J., Rosene, D. L., Killiany, R. J., Moss, M. B., Warfield, S. K., Egorova, S., et al. (2008). A rhesus monkey reference label atlas for template driven segmentation. *Journal of Medical Primatology*, 37(5), 250−260. Available from: http://dx.doi.org/10.1111/j.1600-0684.2008.00288.x.

Woods, R. P., Fears, S. C., Jorgensen, M. J., Fairbanks, L. A., Toga, A. W., & Freimer, N. B. (2011). A web-based brain atlas of the vervet monkey, *Chlorocebus aethiops*. *Neuroimage*, 54(3), 1872−1880. Available from: http://dx.doi.org/10.1016/j.neuroimage.2010.09.070.

Wright, P. C. (1999). Lemur traits and Madagascar ecology: Coping with an island environment. *American Journal of Physical Anthropology*, 110(S29), 31−72.

Wright, T. F., Eberhard, J. R., Hobson, E. A., Avery, M. L., & Russello, M. A. (2010). Behavioral flexibility and species invasions: The adaptive flexibility hypothesis. *Ethology Ecology & Evolution*, 22(4), 393−404. Available from: http://dx.doi.org/10.1080/03949370.2010.505580.

Wyles, J. S., Kunkel, J. G., & Wilson, A. C. (1983). Birds, behavior, and anatomical evolution. *Proceedings of the National Academy of Sciences of the United States of America—
Biological Sciences*, *80*(14), 4394–4397. Available from: http://dx.doi.org/10.1073/
pnas.80.14.4394.

Zilles, K., Amunts, K., & Smaers, J. B. (2011). Three brain collections for comparative neuroanatomy and neuroimaging. *Annals of the New York Academy of Sciences*, *1225*(S1),
E94–E104. Available from: http://dx.doi.org/10.1111/j.1749-6632.2011.05978.x.

Zilles, K., Armstrong, E., Schleicher, A., & Kretschmann, H. J. (1988). The human pattern
of gyrification in the cerebral cortex. *Anatomy and Embryology*, *179*(2), 173–179.
Available from: http://dx.doi.org/10.1007/bf00304699.

Zola-Morgan, S., & Squire, L. R. (1993). Neuroanatomy of memory. *Annual Review of
Neuroscience*, *16*(1), 547–563.

Commentary on Chapter 9: Innovation and the Value of Building on What We Know

Reflecting on the ways in which the mental capacities of humans resemble and differ from those of other species is useful in at least two respects. Perhaps most importantly, it helps us to understand ourselves; where do we fit within the broader context of the world around us? Second, as a side benefit, it forces us to be specific about what we mean by particular types of mental operations, and hence, clarify how best to conceptualize them. I want to consider these points in reverse order and then, hopefully, tie them together.

A hallmark of human cognition our creativity, that is, our capacity to generate novel and useful ideas that we bring to fruition in the form of behaviors or tangible products. Key ingredients in the recipe for creativity are the production of novel behavioral variations ("innovation" in the sense used by Navarrete, this volume), and their social transmission. Although humans are often thought of as being especially innovative, there is little doubt that members of other species are capable of being innovative. This includes novel behaviors, such as new foraging techniques, as well as tools, found objects that are modified in some way to increase their utility for a particular purpose. It is also clear from the Navarrete review that innovation in this sense is more frequent in primates, linked to brain size, especially in cortical regions, more common in conjunction with social transmission and other indicators of general intellectual function, and perhaps associated with the ability to inhibit old responses, thus allowing the production of new ones.

However, this careful characterization of innovative behavior and its occurrence in a wide range of species also raises the question of what is different about the ways in which humans innovate. We have only been observing and tallying innovative behavior in other species for the tiniest fraction of time in the geologic sense. Thus we have no way of knowing if those behaviors are newly emerging in those species or have been present in the long-term sense, much as human innovative behavior has been present for ages. For argument sake I will propose that innovative tendencies in other species are similarly long-standing rather than recent.

The assumption that innovation has been going on in other species for as long as it has been in humans is important because it brings into perspective that human innovation is a cumulative process in a way that is true for no other species. Put differently, when an individual human introduces a new idea, and it gets transmitted to others, we use it as a starting point for continued innovation, not an end point; we continue to build on it. For example, rather than sticking with the innovation of chipping away at a rock to form a simple scraping tool, we continued to refine that type of tool, with our capacity for cumulative innovation ultimately culminating in "tools" that took us to the Moon and Mars to observe the rocks there, and most recently to land on a comet. No other species, no matter what their brain size, does this. I say this not to wave the banner of humankind or to demean the capacities of other species, but to highlight something I see as vital to our sense of where humans fit in the bigger picture and how best to think about creativity.

Our cumulative innovation is of course linked to the fact that we develop products that continue to exist and that we have developed various means of preserving a record of them in the form of painting, carving, oral communication, writing, and so on. The physical presence of objects or their descriptions certainly provides an advantage, but it is also our mental capacity to hold things in mind that is also vital. It may seem ironic, but it is being able to mentally hold onto the past that allows us to innovate in ways that continue to build. Much of the focus on creativity is on the side of processes that introduce the novel variations, but in my view it our capacity to physically and mentally hold onto the old that allows us to innovate in the extraordinary way that we do. That is, creativity and the human capacity for it are as much about what is old about new ideas as what is new about them.

10

Minding the Gap: A Comparative Approach to Studying the Development of Innovation

Jackie Chappell[1], Nicola Cutting[2], Emma C. Tecwyn[1,3], Ian A. Apperly[2], Sarah R. Beck[2] and Susannah K.S. Thorpe[1]

[1]School of Biosciences, University of Birmingham, Birmingham, UK
[2]School of Psychology, University of Birmingham, UK [3]Department of Psychology, University of Toronto, CA

Commentary on Chapter 10: Minding the Gap: Problem Construction and Ill-Defined Problems

Roni Reiter-Palmon

Department of Psychology, University of Nebraska Omaha, NE, USA

INTRODUCTION

What is Innovation?

Imagine that you are performing some maintenance on the engine of your car. You remove a small screw from a component, but it falls down into the depths of the engine, well beyond the reach of your fingers. It is

287

a particular size and shape of screw, and you have no spares, so you must retrieve it. You might start by trying to extend your reach with tools such as pliers, but the gap into which the screw has fallen is too narrow and deep to permit this. You know what materials you have on hand, and you know what you want to achieve (retrieving the screw from the gap), but what can you do to achieve that goal? What you need is something long and thin to "fish" for the lost screw. But how could you get the screw to stick to the end of this long, thin object? You have a roll of adhesive tape nearby, and it occurs to you that you could perhaps unroll a length of tape, attach a small weight near the end and dangle the length of tape into the gap, hoping that the free end of the tape will stick to the screw and allow you to lift it out. You put this plan into action, and after a great deal of frustration and careful adjustment of your technique, you retrieve the lost screw. This vignette captures the kind of everyday, *ad hoc* innovative solution that we have all produced at one time or another. We are faced with a frustrating problem, for which our usual solutions do not suffice. We understand the problem with which we are faced, and know what we want to achieve, but there is a gap between the two that we must bridge by generating appropriate actions or transforming materials. Understanding this process in both humans and nonhuman animals is the main focus of this chapter.

Innovation is easy to identify when you encounter it, but difficult to define precisely. A loose definition that captures the everyday meaning of "innovation" might be "something new and different" (O'Brien & Shennan, 2010a, p. 3). However, this is too broad and difficult to operationalize. A more specific definition that has gained popularity in the literature is "a solution to a novel problem or a novel solution to an old problem" (Benson-Amram & Holekamp, 2012; Manrique & Call, 2011; Manrique, Völter, & Call, 2013). One important distinction to make is whether innovation is considered the end *product*, for example a new behavior, idea or artifact (see Kummer & Goodall, 1985), or the *process* that results in a novel behavior (see discussion in Osvath & Karvonen, 2012; Reader & Laland, 2003). Reader and Laland (2003) proposed making this distinction explicit, with "innovation *sensu* product" referring to "a new or modified learned behavior not previously found in the population", and "innovation *sensu* process" referring to "a process that results in new or modified learned behavior and that introduces novel behavioral variants into a population's repertoire" (Reader & Laland, 2003, p. 14).

Different forms of innovation may involve different processes and problems. Solving a novel problem may involve applying previously learned and generated actions or transformations to a novel problem, through processes of generalization, analogy and so on. In contrast, producing a novel solution to an old problem involves inhibiting previously generated actions that may no longer be appropriate. Manrique and colleagues (2013) argue that this process of inhibiting inappropriate actions may be more difficult (at least for nonhuman animals) than applying pre-existing actions to a novel problem. In either case, individuals must select and organize new and pre-existing behaviors appropriately. Epstein (1999) coined the term "Generativity Theory" to describe the process of innovation in which pre-existing behaviors compete together in a dynamic way, the outcome of which is a new set of behaviors formed by resequencing or "blending" the pre-existing behaviors. He proposed that this process may be initiated by multiple previous behavioral repertoires being elicited at once, often because the usual outcome is thwarted, as in the vignette at the start of this chapter. To summarize then, the process of innovation may vary, but it is a generative process, and usually involves selecting between alternative courses of action on the basis of their fit to the context presented. The fact that we subjectively experience innovation as "difficult" probably arises from the fact that it involves generating something new.

Role of Novelty

The concept of "novelty" is an important one in the definitions of innovation (and indeed creativity). Thus, it is important to establish what is meant by "new." There are two main dimensions along which novelty can vary. In the first, the novelty of the components of the solution might vary, from an entirely new solution to one in which one or two preexisting components have been modified. In the second dimension, the degree of novelty may vary from a solution new to the individual to one novel for the population. At one extreme, it could be a solution that has *never* been produced before, by individuals of any species. At the other extreme, it might be an innovation that is novel for the particular individual involved, but one that has been produced by many others in the population or species before (see Boden, 1996; Vaesen, 2012 for extensive discussion). Both dimensions of novelty are important, because they determine what information is available to the individual, and can act as components or building blocks of the new behavior, artifact or idea. Innovation rarely comes from nowhere: even the most startling new idea emerges out of analogies made with other problems, or insights about the structure of a problem, that provides a new approach.

It is difficult—even when studying humans—to determine what is genuinely novel at a population level, as complex patent laws, designed to legislate on the novelty of ideas, imply. Moreover, it is impossible to establish population-level novelty definitively for nonhuman animals, where observation and documentation of behaviors is likely to be incomplete. Given this difficulty, and since our research focuses on understanding the processes involved within individuals, in this chapter we will focus on instances of innovation where the behavior is novel to the individual concerned.

Creativity and Innovation

In everyday language, "creativity" and "innovation" are used to mean broadly similar things, although in English there is a tendency to use "creativity" when referring to artistic products and processes, and "innovation" when referring to technological or scientific products and processes. However, some researchers maintain a subtle but important distinction between the two. Both may result in new behaviors, artifacts or ideas, but innovation requires there to be a particular goal in mind (e.g., a chimpanzee concealing rocks for the first time to throw at visitors later, Osvath & Karvonen, 2012), while creativity involves generating novel actions with no specific end goal. However, both categories of activity may result in trying out similar actions and transformations.

This presents a problem. If the outcome of both processes is similar, how can we distinguish between them? In the case of experiments designed to pose a problem with a defined goal, it may be clearer, but—particularly with nonhuman subjects—we do not know for certain whether the *subject's* goal is the one we have designed. Interacting with a problem in a playful, exploratory way may be inherently rewarding (Ellis & Siegler, 1997; Miyata, Gajdon, Huber, & Fujita, 2011), and may have in important role during ontogeny in building capacity for innovation in adulthood. Similarly, we do not know the extent to which different kinds of abilities or cognitive processes are involved in creativity and innovation. Indeed, exploration of objects in the environment may be a very important component of innovation (Chappell, Cutting, Apperly, & Beck, 2013; Chappell, Demery, Arriola-Rios, & Sloman, 2012), thus creativity and innovation are linked. Experiments on a diverse range of taxa (Benson-Amram & Holekamp, 2012; Caruso, 1993; Parker, 1974; Sol & Lefebvre, 2000) point to the fact that exploratory tendencies are important predictors of innovativeness at both the species and individual level.

In this chapter, we will restrict our discussion to situations in which a specific problem is posed to the subject, and the actions and transformations are assumed to be generated with a particular end goal in mind, because these are the most amenable to analysis. However it should be remembered that even in these situations, subjects might have a different goal, or additional motivations for generating the actions and transformations, than those intended by the experimenter. Human adults may be able to communicate their goal, but young children and nonhuman animals cannot. We will discuss why innovation is important, and review the ways in which innovation has been tested experimentally across a variety of species. We will then discuss the cognitive abilities and environmental conditions which may promote and support innovation, before discussing whether considering such problems as "ill-structured" problems (see section "Ill-structured problems") may help us to tease apart what makes innovation so difficult.

WHY IS INNOVATION IMPORTANT?

Innovation (*sensu* product) has an important role in the way that individuals adapt to variability in their environment and also allows species to invade and adapt to novel environments. This can avoid the need for biological evolution of morphological traits to solve key habitat problems, and therefore allows animals to adapt quickly to more extreme environments than would otherwise be possible. Furthermore, innovation (*sensu* process) provides the raw material for new cultural variants, differentiating one population from another through their material culture and behavioral repertoire.

Adaptation to Novel Environments

One of the key determinants of success in solving novel problems (such as those encountered in novel environments) is simply trying a lot of different behaviors (Greenberg & Mettke-Hofmann, 2001). Observations of the diversity and complexity of exploratory manipulatory behaviors in 11−12 month old human infants found correlations within individuals between these measures, and that diversity of exploration predicted success on problem-solving tasks (Caruso, 1993). A similar, earlier study on a range of nonhuman primate taxa showed that great apes perform a significantly greater diversity and complexity of exploratory manipulations than the monkey and prosimian species tested (Parker, 1974). These tantalizing links between behavioral diversity or flexibility, problem-solving ability and ability to innovate have motivated many studies that have attempted to show correlations between brain size and other "cognitive" attributes, by examining reports in scientific journals of novel behavior in the chosen taxa (Sol, Duncan, Blackburn, Cassey, & Lefebvre, 2005; Sol & Lefebvre, 2000; Sol, Timmermans, & Lefebvre, 2002). Indeed, it has been suggested that the need to produce novel behavioral patterns to overcome ecological challenges (the "cognitive buffer hypothesis") has been an important driver of encephalization in birds and mammals (Sol, 2009).

These studies have shown significant correlations between various measures of brain size and innovation in birds (Lefebvre, Reader, & Sol, 2004; Lefebvre, Whittle, Lascaris, & Finkelstein, 1997; Overington, Morand-Ferron, Boogert, & Lefebvre, 2009; Reader & Laland, 2002) and primates (Lefebvre et al., 2004). Furthermore, studies examining the outcome of human introductions of bird species to nonnative locations showed a relationship between innovation (or behavioral flexibility: the number of innovations per taxon) and invasion success in the new location (Sol & Lefebvre, 2000; Sol et al., 2002, 2005). However, while intriguing, these comparative studies have been criticized because of a lack of clarity and unstated assumptions about the cognitive attributes being measured, and because of inconsistencies in the way that brain size is measured (Healy & Rowe, 2007). Nevertheless, the weight of evidence suggests there is some kind of relationship between innovative ability and success in new environments, but the tight relationship between multiple attributes and the impossibility of determining the direction of causality in correlational studies makes it difficult to be more precise.

Importance for Culture

Innovation is obviously an important source of novel behaviors or artifacts that could spread throughout a population and form the basis of a culture. Many different factors (such as the nature of the innovation itself, fidelity of transmission and population density) determine whether or not a particular innovation will be adopted by the whole group and persist over time (see review in Laland & Rendell, 2013). Furthermore, there appear to be important differences in the processes by which humans and nonhuman animals develop traditions or cultures.

Whiten et al. (1999, 2001) surveyed a variety of population-specific behaviors in chimpanzees (*Pan troglodytes*), and concluded that there was evidence that many of the behaviors were culturally transmitted, and not due to differences in the ecological circumstances of the populations (which might, for example, determine the availability of suitable tools for nut cracking). However, other authors (Tennie, Call, & Tomasello, 2009; Tomasello, Kruger, & Ratner, 1993) have suggested that, while there are many similarities in the mechanisms of human and chimpanzee cultural transmission, there are also important differences. Experiments have shown that when chimpanzees observe another individual performing a new behavior, they tend to reproduce the effect of the behavior on the environment, rather than copying the behavior or process itself as humans do, particularly when causal information about the relevance of the action is available (Horner & Whiten, 2005; however, see Whiten, Horner, & de Waal, 2005; Whiten, McGuigan,

Marshall-Pescini, & Hopper, 2009 for an alternative view). Thus Tennie and colleagues (2009) argue that cumulative culture is not tenable in chimpanzees, because effectively each chimpanzee learns the behavior anew, through a combination of stimulus enhancement and emulative mechanisms. In contrast, humans have a strong tendency to imitate faithfully and conform to the observed behavior, even to the point of reproducing causally irrelevant actions (e.g., McGuigan, Whiten, Flynn, & Horner, 2007; Nielsen, 2013).

This highlights the conflict between individual innovation and the development of cumulative culture: innovators are needed to "seed" new variants of a behavior, but in order to retain those novel variants in the population and build upon them, precisely the opposite behavioral tendency is required, with individual variation being suppressed in favor of faithful reproduction of the behavior (Brosnan & Hopper, 2014; O'Brien & Shennan, 2010b). Cumulative culture is itself a very important process. Once it is possible, it generates novel artifacts, ideas, and behaviors, which in turn alter the environment and introduce novel selection pressures, driving further innovation. This is a form of niche construction, which has been argued as a key factor in the evolution of human intelligence (Sterelny, 2007).

APPROACHES TO STUDYING INNOVATION

As mentioned in the Introduction, there are several difficulties with studying innovation, both in humans and in nonhuman animals. First, if we are interested in innovation as distinct from creativity, we need to be reasonably confident that the subjects' actions are goal directed and intentional. Second, since novel behaviors are fundamental to innovation, we need to determine whether the behavior (or the solution generated for the problem) is novel to the individual involved, when their history may be unknown.

Both problems are often addressed experimentally in captive nonhuman animals (and "captive" humans taking part in experiments) by presenting the subject with a novel problem to solve (which may require the subject to employ previously learned behaviors in a novel context). Successful solution of the problem results in a reward of some kind, increasing confidence that subjects are behaving in a goal-directed, intentional way when they interact with the problem, and not exploring the apparatus in a nondirected manner. While it is impossible to be absolutely certain that the task is novel for the subject, such experimental paradigms often use unusual or man-made materials in a configuration that the subject is highly unlikely to have encountered before.

Occasionally, field researchers observe examples of novel behaviors in wild nonhuman animals, and can observe the context in which that behavior is shown to determine the function and therefore infer whether it is goal directed. Furthermore, observations of the whole group can determine whether (and how) the behavior spreads to other individuals in the population. However, given the constraints of field research, it is rare that researchers observe the very first time that a behavior is exhibited by an individual, or that they know the entire behavioral history of that individual.

A final approach blends the field and laboratory approach by presenting novel problems (often remotely operated and monitored) to wild individuals in the field (e.g., Morand-Ferron, Cole, Rawles, & Quinn, 2011; Morand-Ferron & Quinn, 2011). This allows investigation of both individual innovation and the spread of innovative behavior in a more ecologically valid context, where one can be reasonably confident that the problem is novel. However, since participation by subjects in such experiments is voluntary, with the schedule decided by the individual itself, there is less control over experimental variables such as trial length, order, and frequency.

Evidence from Nonhuman Animals

Famous examples of animal innovation in the wild include blue tits and great tits piercing the foil lids of milk bottles to drink the cream (Fisher & Hinde, 1949) and Japanese macaques washing sand and mud off sweet potatoes (Kawai, 1965). Both of these novel behaviors, first performed by a single innovative individual, subsequently spread through the population. In more recent examples, a single rhesus macaque has been observed using a novel throwing technique to open coconuts (Comins, Russ, Humbert, & Hauser, 2011) and a single Japanese macaque has been seen performing idiosyncratic dental flossing using hair (Leca, Gunst, & Huffman, 2010). Unlike the two aforementioned examples, other group members have not adopted these techniques, which presents the interesting question of what factors might constrain the spread of innovations within a population.

Although these are seemingly compelling examples of animal innovation, the history of wild subjects is nearly always unknown and so it is unclear what previous experience may have been necessary for an individual to "discover" a novel behavior. Therefore, to better understand innovation there is a need to present subjects with "novel" problems under controlled conditions. This makes it easier to address important questions such as: which individuals innovate? What are the characteristics of these individuals? What are the cognitive processes underlying innovative behavior? What previous experience is required for innovation of a particular solution? Which noncognitive processes influence innovation? Several studies have attempted to address some of these questions in a variety of taxa. As mentioned previously, this discussion will be restricted to studies where a specific problem is posed to a subject. The most popular approach to studying animal innovation has been to present individuals with extractive foraging tasks, or "puzzle boxes" that conceal a desirable food reward. For example, it might be necessary to open a door to gain access to the food, or the food might be accessible but out of reach, necessitating the use of a tool to obtain it (like the screw lost in the car engine at the start of this chapter).

Auersperg, von Bayern, Gajdon, Huber, and Kacelnik (2011) presented captive kea (*Nestor notabilis*) and New Caledonian crows (*Corvus moneduloides*) with a reward in the center of a transparent box (the "Multi-Access-Box" paradigm). There were four possible ways of accessing the reward (one via each wall of the box): opening a window and reaching into the box; pulling a string attached to the reward; using a stick tool; and using a ball tool. Once a technique was mastered by an individual it was blocked, so that the reward could no longer be obtained using this method. This enabled investigation of the behavioral flexibility of different individuals, in terms of their ability to abandon a previously successful strategy and explore alternative options, thus discovering additional solutions. One out of six kea and one out of five crows discovered all four solutions to the task. Kea explored the apparatus more and switched between solutions more readily, and they therefore discovered multiple solutions in fewer sessions than the crows. However, they were less efficient at the solution requiring the use of a stick tool, which is unsurprising given that the crows are natural stick-tool users in the wild. This study provides a clear demonstration of how multiple factors (e.g., exploratory behavior, natural behavioral propensities, neophobia, and flexibility) interact to affect a subject's ability to innovate (Auersperg et al., 2011).

Further evidence for the important role of exploratory behavior in innovating a solution to a novel problem comes from a study of wild hyenas (Benson-Amram & Holekamp, 2012). Individuals were presented with a puzzle box containing meat, which could be accessed by sliding a bolt across and swinging the door open. Nine out of 62 individuals that interacted with the apparatus succeeded in retrieving the meat at least once. Notably it was the less neophobic individuals who exhibited a greater diversity of exploratory behaviors and were therefore more likely to succeed (Benson-Amram & Holekamp, 2012). This study is interesting from a comparative perspective given that the vast majority of studies addressing innovation have tested primates (mainly apes) or birds (mainly corvids); indeed this is an issue in the field of animal cognition as a whole. This study also demonstrates that it is possible (albeit challenging) to run innovation experiments in the wild, and raises the question of the extent to which findings from studies with captive animals are ecologically valid. To address this, Benson-Amram, Weldele, and Holekamp (2013) compared the performance of wild and captive hyenas presented with the same extractive foraging task. Captive hyenas were more successful at retrieving the food than their wild counterparts, less neophobic, and more persistent in their first trial. This suggests that for hyenas at least, findings with captive individuals cannot be extended to their wild counterparts (Benson-Amram et al., 2013).

Manrique and colleagues (2013) adopted a similar approach to Auersperg et al. (2011) to investigate the ability of nonhuman great apes to produce a novel response to retrieve a food reward and subsequently abandon it in favor of another new response, when the previous method was blocked. Three different sets of apparatus were presented, starting with the easiest first. They were all transparent rectangular boxes with an open top and a handle at the bottom that could be used to push up the base with the reward on. The first apparatus could be solved using any of three techniques (fingering; lifting; shooting); the reward in the second apparatus could only be retrieved via two of the three methods (lifting; shooting); and the final apparatus could only be solved by using a single technique (shooting). Chimpanzees, bonobos, and gorillas efficiently switched from acquired techniques once they became obsolete. On the other hand only one out of seven orangutans discovered the shooting technique. However, the authors acknowledge that this may have been related to African apes' greater propensity for banging/pounding actions involved in the shooting technique, rather than a difference in cognitive ability between species (Manrique et al., 2013). Further evidence for the natural behavioral propensities of a species influencing their innovative ability comes from a study by Manrique and Call (2011), which found that orangutans were more likely than chimpanzees or bonobos to use a piece of cable as a straw to suck up juice, instead of using the established (but less efficient) technique of dipping and licking. The authors suggest that the observed species differences might be due to orangutans' greater reliance on oral exploration and object manipulation (Manrique & Call, 2011).

Other studies have combined serendipitous observations of tool innovation with follow-up studies presenting the task in which the individuals innovated in a more controlled manner. In a famous example, Betty, a New Caledonian crow, spontaneously bent a straight piece of wire into a hook shape to retrieve a small bucket containing food from a vertical tube (Weir, Chappell, & Kacelnik, 2002). She subsequently went on to replicate this behavior in 9/10 trials. Interestingly, this innovative hook-making behavior has since been replicated in nontool-using rooks (Bird & Emery, 2009). Similarly, a Goffin's cockatoo (not a tool-user in the wild) was observed using a piece of bamboo to attempt to retrieve a stone that he had dropped out of reach. In ten subsequent trials in which the stone was replaced with a cashew nut, the cockatoo manufactured stick tools and successfully used them to rake in the reward (Auersperg, Szabo, von Bayern & Kacelnik, 2012).

Evidence from Humans

Much of the research on human problem solving focuses on how we learn to solve problems by observing others (innovation *sensu* process: e.g., Horner & Whiten, 2005; Want & Harris, 2002; Williamson, Jaswal, & Meltzoff, 2010). In contrast there is very little research investigating how we solve problems for ourselves, that is, how we innovate new solutions (innovation *sensu* product). It has been well documented that a propensity for faithful imitation and learning from others has been the driving force behind human cultural evolution (Boyd & Richerson, 1996; Tomasello et al., 1993); however, as discussed in "Importance for culture," cumulative culture also requires new innovations to "seed" the process and spread novel behaviors throughout the population. Recently, there has been a shift towards trying to understand the role of innovation.

Some studies have directly tested children's ability to innovate. Beck, Apperly, Chappell, Guthrie, and Cutting (2011) tested children's ability to innovate a novel tool needed to solve a task. In this task, children were presented with a tall transparent tube that had a bucket containing a sticker in place in the bottom (a child-friendly version of the task that was presented to Betty the crow, see section "Evidence from nonhuman animals"). Children were instructed to retrieve the bucket from the tube in order to win the sticker. They were given materials such as pipecleaners, string and short sticks with which to solve the task. Children found it extremely difficult to innovate the solution of bending the pipecleaner into the hooked tool that was required to solve the task. Under the age of 5 children rarely innovated a pipecleaner hook, with success only reaching around 50% by the age of 8. Children's difficulty innovating novel tools has been shown to extend to other tools (Cutting, Apperly, & Beck, 2011), methods of making tools and materials (Cutting, 2013). In contrast to children's difficulty in innovating tools for themselves these studies demonstrated children's impressive aptitude for learning from others. If children failed to innovate a hook tool they were shown a demonstration of how to bend the pipecleaner into a hook. Following this demonstration children's success rates were near ceiling. It appears that children are well adapted to learning new skills from others, but have great difficulty in innovating new solutions for themselves.

Children's difficulty with innovating novel tools has been demonstrated in other tasks. Tennie et al. (2009) tested children's (and chimpanzees') ability to make a loop from wooden wool needed to loop around a screw protruding from the top of a platform. By looping the screw children could pull the platform towards them and win the reward. None of the 24 4-year-old children succeeded in innovating the loop and succeeding in the task. This provides further evidence that innovating novel tools is a difficult task for young children, even when the tools they are required to make appear relatively simple.

Other studies have investigated children's innovative use of tools and objects to solve problems. For example children have been tested on the floating peanut task (Hanus, Mendes, Tennie, & Call, 2011; Nielsen, 2013). In this task children were required to pour water into a tube in order to raise the water level so they could retrieve a reward. Children found the innovative solution of pouring water into the tube extremely difficult. Hanus and colleagues tested 4-, 6-, and 8-year-olds on this task. Success rates were low; only 8% of 4-year-olds succeeded in innovating the correct solution, rising to only 58% in 8-year-olds. Nielsen (2013) tested 4-year-old children on a similar version of the task and he too found that children had great difficulty innovating the solution. Out of the 36 children who took part only five (14%) spontaneously poured water into the tube to retrieve the reward. These studies required children to innovate a new process to retrieve their reward rather than a new product as in the tool-innovation tasks above. However, taken together this evidence suggests that innovation is generally difficult for young children, irrespective of the task presented.

Whilst innovation is not always the main focus of studies, often innovations occur that were not predicted by the authors. This has been the case in studies of social learning where children are required to solve problems after observing a model demonstrate the solution. Despite children being faithful imitators as evidenced by a large body of work in this field (e.g., McGuigan et al., 2007; Nielsen & Blank, 2011; Simpson & Riggs, 2011), occasionally children innovate new solutions that are not expected. Whiten and Flynn (2010) were investigating the social transmission of seeded techniques in an open diffusion experiment when they observed innovations. Groups of preschoolers were seeded with one of two tool-using methods (lift or poke) to retrieve a reward from an apparatus. Although the majority of children remained faithful to the method they had observed, the authors witnessed three types of innovation during their experiment. Some children innovated their own solution of lift or poke without having seen a demonstration of any kind (termed innovation-blind). Other children used the technique that was not seeded in their group (innovation-major), for example they had seen a demonstration of poke but used the lift technique. Children also innovated completely new techniques, which had not been anticipated by the authors. For example, children used elements of the poke technique they had seen to create a new technique termed T-bar by the authors (innovation-minor). Although rare, these instances of innovation have the potential to tell us a lot about the nature of innovation. In a world where learning from others and imitating behavior are the norm it is helpful for us to understand what is different about people who innovate. It is of course important for us to learn from others and successful transmission of knowledge relies on faithful imitation, but for culture to evolve it also needs innovators. Therefore these instances where children act in new and innovative ways are important to help us discover the mechanisms and motivations behind innovation.

To summarize, to date there is limited research on innovation in humans. The little research there is has shown that innovation is hard and occurs infrequently. Research has documented this phenomenon but has done little to help us understand what exactly it is about innovation that makes it so difficult. In the next section, we explore potential components of innovation in an attempt to understand why it is so difficult.

COMPONENTS OF INNOVATION

What kinds of abilities does innovation require? What conditions and attributes do successful innovators need? Several predisposing conditions for innovation have already been discussed in the preceding sections, for example, a tendency towards a lack of neophobia and exploration of objects in the environment are important predisposing factors (Chappell et al., 2012; Greenberg & Mettke-Hofmann, 2001). Furthermore, in order to spread throughout a population, innovation *sensu* process requires an ability to learn socially. However, there may be other cognitive abilities that play an important role in innovation.

Physical Cognition and Causal Reasoning

As with tool use, an understanding of the physical structure, materials, properties or causal structure of objects in the environment is not *necessary* for innovation. Just as animals can use tools competently and effectively without understanding their function, one can imagine that individuals might discover by chance that a particular novel action has a favorable outcome and repeat it: after all, this is the basis of associative learning. However, if this was all that was required for innovation to occur, one would expect it to be much more common, since so many species are capable of associative learning. In humans, at least, an understanding of the properties of objects and the causal structure of a particular situation can allow us to work backwards from the requirements of the situation, and therefore search for a novel object or action that would fit the requirements. In the example at the start of this chapter, knowledge about what kind of object was required (something long and thin that could stick to the screw) drove a local search for an item with those properties. This means that the search is far more tightly directed than it would be if an individual was simply trying out random actions or trying to use random objects. It is reasonable to suppose that nonhuman animals with such capabilities (e.g., Chappell & Kacelnik, 2002, 2004; Huber & Gajdon, 2006; Jelbert, Taylor, Cheke, Clayton, & Gray, 2014; Schuck-Paim, Borsari, & Ottoni, 2009; Seed, Tebbich, Emery, & Clayton, 2006; Weir & Kacelnik, 2006) might also gain an advantage in innovation.

One aspect of this process is the ability to abstract a particular function of an object or action away from its current function, and to recognize analogies between superficially dissimilar problems. For example the conventional function of a cup is to act as a container for liquids. However, the containment function can be abstracted to "containment" more generally. Thus a cup could also function as a container for pens, or as a trap for a spider if inverted. Intriguingly, in humans, this ability can be constrained by a strong tendency to be biased towards the designed or conventional function of an object—a process called "functional fixedness" (Defeyter & German, 2003; German & Defeyter, 2000). This tendency seems to develop in human children sometime between the ages of 5 and 6 years old (German & Defeyter, 2000). There have been suggestions recently that similar processes may even occur in some nonhuman apes (Hanus et al., 2011). Thus, there is a tension between the ability to behave flexibly with respect to the functions of objects on the one hand, and respecting the primacy of the "designed" or "intended" function of an object on the other. The former tendency supports innovation *sensu* product, while the latter supports innovation *sensu* process.

Planning and Sequencing Actions Appropriately

Innovative behavior often involves linking together behaviors (that may have been previously learned) into a novel sequence, or in a novel context (Taylor, Elliffe, Hunt, & Gray, 2010). Thus the ability to sequence or plan a series of behaviors appropriately may be important in innovation. The majority of paradigms investigating innovation of this nature have consisted of sequential tool-use tasks. In the simplest version, one tool must be used to access a second tool, which can subsequently be used to access a food reward. The number and nature of the intermediate steps can be manipulated to alter the complexity of the task.

Japanese macaques (*Macaca fuscata*; Hihara, Obayashi, Tanaka, & Iriki, 2003), gorillas (*Gorilla gorilla*) and orangutans (*Pongo pygmaeus*; Mulcahy, Call, & Dunbar, 2005) have succeeded in sequential stick-tool-use tasks with a single intermediate step, though all individuals had received pretraining of the task components. Naturally tool-using New Caledonian crows (*C. moneduloides*) solved a similar task spontaneously (Taylor, Hunt, Holzhaider, & Gray, 2007), and have since demonstrated the ability to use three stick tools sequentially (Wimpenny, Weir, Clayton, & Kacelnik, 2009). Martin-Ordas, Schumacher, and Call (2012) replicated and extended Wimpenny et al.'s (2009) study and found that chimpanzees, bonobos (*Pan paniscus*) and orangutans were able to use up to five tools in sequence to retrieve a reward.

Bird and Emery (2009) presented rooks (*Corvus frugilegus*) with three tubes with collapsible bases and a large stone. One of the tubes contained a reward, the second a large stone and the third a small stone. However, the tube containing the reward was too narrow for the large stone to be dropped in to release the reward. Therefore, to access the reward the large stone had to be dropped into the tube containing the small stone, which could subsequently be dropped into the narrow tube containing the reward. All four rooks solved this task from their initial trial (Bird & Emery, 2009). Although this is seemingly impressive, all individuals had previous experience of both operating the collapsible platforms and having to select a small stone to retrieve a reward from a narrow tube (other experiments in Bird & Emery, 2009); therefore just how novel and innovative this solution was is debatable. von Bayern, Heathcote, Rutz, and Kacelnik (2009) demonstrated that for New Caledonian crows, experience of using stones to collapse platforms was not required for success in this type of task. Two out of four crows that lacked experience of dropping stones into tubes but had experience of pushing the platform down directly with their bill went on to innovate using a stone as a tool to retrieve an item from a tube (von Bayern et al., 2009). This suggests that collapsing the platform directly provided sufficient information regarding the affordances of the apparatus for at least some individuals to succeed at the task, though the insertion of a feather into the tube by one bird is certainly indicative of incomplete causal understanding.

A comparable experiment was conducted with two bottlenose dolphins (*Tursiops truncatus*) by Kuczaj, Gory, and Xitco (2009). Individuals were presented with three tubes, each of which required a weight to be dropped into it to release a reward. Two of the tubes had open bases so that the weight fell out of the bottom and could be reused, whereas one tube (which was clearly marked) had a closed base and hence trapped the weight. Providing dolphins were able to view all of the tubes once they had picked up the weight, both subjects maximized the number of rewards obtained by going to the weight-retaining tube last in the majority of trials (Kuczaj et al., 2009).

The experiments described so far in this section have required the same behavior to be performed multiple times: namely using a stick to retrieve an out-of-reach item, or dropping a stone or weight into a tube. Taylor et al. (2007) presented New Caledonian crows with a three-stage problem that involved combining different previously established behaviors in a novel context. Solving the task required individuals to pull up a string to reach a short stick tool; use the short stick tool to access an out of reach long stick tool; and finally use the long stick tool to access an out of reach food reward. Out of four individuals who had no previous experience of using tools to access other tools (only to access food directly), two solved this task in their first trial, demonstrating the ability to use previously learned behaviors in a novel context, and to perform them in an appropriate sequence (Taylor et al., 2007).

The ability of animals to produce a novel sequence of behaviors in a nontool-using context has rarely been investigated. The advantage of presenting paradigms that do not require subjects to use tools is that it provides greater scope for comparative work, specifically by increasing the validity of comparisons with nontool-users, such as many monkey species. Tecwyn, Thorpe, and Chappell (2013) presented orangutans and bonobos with a paradigm (the "paddle-box") that required them to perform novel sequences of up to three actions to retrieve a reward. The paddle-box consisted of a transparent Perspex box containing eight rotatable seesaw-like paddles on three levels. At the bottom of the box were four possible goal locations, each of which could either be open (allowing the reward to be accessed) or blocked (resulting in the reward becoming trapped). The starting position of the reward could be manipulated, as could the position of the open goal, making it possible to present trial-unique configurations of the task. Subjects received minimal training with the apparatus; it simply ensured that they were physically able to turn a paddle and retrieve a reward from an open goal. In the "advance planning" task, all paddles apart from the one that the reward started on were set up diagonally, so subjects had to move one or two paddles into an appropriate position before turning the paddle with the reward on, otherwise the reward would become trapped. All subjects of both species performed poorly in this task, because they nearly always turned the paddle with the reward on immediately. This failure may have been related to the subjects' inability to inhibit the prepotent response to act on the reward straight away due to its salience, therefore causing them to perseverate on an incorrect behavioral response (Tecwyn et al., 2013). This leads on to another important component of innovation: inhibitory control.

Inhibitory Control

If the ability to sequence and plan actions is so important for innovation, inhibitory control over that process is vital to ensure that inappropriate or unsuccessful strategies are abandoned, and new strategies attempted. Inhibitory control has multiple features, but two kinds of inhibitory control failures are particularly relevant to innovation. First, perseveration is the inability to switch between strategies if the current strategy is not working (Chappell et al., 2013; Diamond, 2006; Frye, Zelazo, & Palfai, 1995). This is very important because—as indicated in the example given at the start of this chapter—innovative behavior often involves multiple steps or difficult behaviors that require persistence in order to succeed. However, it is important that the individual can recognize when the current actions are not succeeding, and move on to trying something else. Second, impulsivity is the inability to wait before performing an action, failure to take into account the consequences of the current action, or failure to inhibit behaviors that are inappropriate in the current context (Reynolds, Ortengren, Richards, & Wit, 2006; Schachar & Logan, 1990). If individuals cannot inhibit inappropriate actions or take the consequences of their actions into account, they may fail immediately. This is particularly relevant to innovation because certain actions may have resulted in success in a previous behavioral context, but lead to failure in the novel context.

In humans, inhibitory control seems to develop between the ages of 3 and 5 years (Diamond, 2006). For example, Vlamings, Hare, and Call (2010) tested 3- to 5-year-old children, chimpanzees, bonobos, orangutans, and gorillas on a detour reaching task, in which participants had to inhibit their prepotent response of reaching for the reward directly, but access it through an indirect path. Older children (4- and 5-year-olds) significantly outperformed younger children, and orangutans outperformed the other nonhuman great apes. However, inhibitory control problems may not be the key performance-limiting factor in all such paradigms. For example, our study employing the advance planning task of the "paddle-box" paradigm mentioned in the previous section with 4- to 10-year-old children suggested that in humans at least, reducing the inhibitory demands of the task did not improve performance (Tecwyn, Thorpe, & Chappell, 2014). It therefore seems that what makes the advance planning task so difficult may be the requirement to "think outside the box" of the most obvious option, or to innovate turning other paddles before turning the one with the reward on. Indeed, orangutans, bonobos, and children were able to perform appropriate sequences of three paddle rotations when all the paddles were flat and therefore turning the paddle with the reward on immediately was an appropriate strategy (Tecwyn et al., 2013, 2014). Supporting evidence for this distinction is provided by a reverse contingency task (not involving tool use), in which children had to point to the empty box to win the box containing the prize (Apperly & Carroll, 2009). Apperly and Carroll (2009) found evidence that children's difficulties with the task concerned "thinking outside the box" to come up with the correct response strategy, rather than with inhibition *per se*.

Taking all these elements together, it is clear that one of the key factors making innovative behavior difficult is the fact that individuals have to compose a sequence of actions, while lacking information about what those steps should be, or what order they should be in. In other words, the initial conditions are known, as is the desired goal, but the sequence of actions to get from one to the other is unclear. In the next section, we will discuss this class of problems, known as "ill-structured problems."

Ill-STRUCTURED PROBLEMS

What Are They?

Ill-structured problems are problems that are missing information required to solve them. This makes the problem particularly difficult for the solver because they must generate this information for themselves. Ill-structured problems can be missing information about the start state, the goal state or the transformation required to go between the two. This is in contrast to well-structured problems, which have defined start and goal states, and defined possible transformations to achieve the goal. For example, a well-structured version of the task described in the vignette at the start of this chapter might involve the kind of special-purpose telescopic rod with a magnet at its tip, sold specifically to help retrieve small metallic items in awkward spaces. The task would then have a clearly defined structure: the starting position and desired goal are known as before, but now a suitable tool is already available, and the required transformation is the *only* sensible one that can be performed with the rod, namely to extend it and direct it towards the lost screw.

There are two strands of research that investigate ill-structured problems. One strand suggests that ill-structured problems are difficult because domain knowledge alone is not sufficient to solve them (Chen & Bradshaw, 2007). The knowledge we possess must be well-integrated into what is termed "structural knowledge." It is only when we have structural knowledge that we can flexibly access this knowledge and manipulate it in such a way to come up with successful problem solutions (Jonassen, Beissner, & Yacci, 1993). Novices may possess the individual pieces of knowledge required to solve a problem, but only experts possess structural knowledge which can be flexibly considered and coordinated into a solution (Voss, Blais, Means, Greene, & Ahwesh, 1986; Wineburg, 1998).

Ill-structured problem solving has also been used to investigate executive deficits observed in clinical human populations. It has been observed that some patients with brain damage have difficulty carrying out simple everyday tasks, yet, when tested in the laboratory, these patients perform at normal levels on traditional tests of executive function, that is, working memory, inhibition, task switching. Based on this finding Shallice and Burgess (1991) devised new ill-structured executive tasks that more closely resemble scenarios from everyday life, and require multi-tasking and prospective memory. The difficulty of these new ill-structured tasks was due to the need for participants to use a number of executive functions in conjunction with each other, without simply reducing down to being a basic working memory or inhibition task. For example, in the Multiple Errands task, participants were taken to a shopping center and were required to retrieve items and information from a number of shops whilst following simple rules such as only entering each shop once and only entering a shop to buy something. Shallice and Burgess found that the clinical patients performed worse on these new ill-structured tasks than age and IQ-matched controls, despite performing at similar levels on traditional well-structured executive tasks. Recent brain-imaging studies have shown that patients who show this divergence in abilities have damage to an area of the brain termed Brodmann Area 10 (Dumontheil, Burgess, & Blakemore, 2008). This area of the brain has protracted maturation through childhood and adolescence, and develops rapidly between 5 and 11 years of age (Burgess, Dumontheil, & Gilbert, 2007).

Why Does This Fit the Pattern of Innovation?

Ill-structured problems provide a good framework within which to think about innovation. The very nature of innovation means that information is missing which the innovator must come up with. The innovator may have a goal in mind but needs to generate the transformations required to get there. This may involve a predefined start state where certain materials are available, or there may be no start state and the innovator must generate the boundaries of how to achieve the goal for himself.

As discussed above, innovation has been shown to be rare, which is in line with the notion that domain knowledge alone is not sufficient for innovations to occur. Many people have knowledge of certain domains, for example mechanics have a good understanding of car engines. However, not all mechanics have flexible structural knowledge of engines, which gives them the capacity to be innovators and design new components that could be used in the latest Formula One cars for instance.

The executive account of ill-structuredness also fits the pattern of innovation. Innovation requires people to multi-task, it requires the ability to hold in mind relevant ideas, switch between different strategies, and inhibit things that do not work. It seems likely that innovation would fit more closely with Shallice and Burgess' (1991) ill-structured executive tasks rather than simply being a working memory or inhibition task.

Using This Framework to Test Innovation in Children

We have used an ill-structured framework to investigate innovation in children (see Chappell et al., 2013). In our tool-innovation tasks, children are given a start state (the apparatus and materials), and a goal (retrieve a reward from the apparatus). What makes these tasks ill-structured is that children must work out how to get from the start state to the goal state for themselves. That is, they must innovate a tool that will enable them to achieve their goal.

As shown above, children find ill-structured tool-innovation tasks extremely difficult. We can compare children's performance on this task with performance on a well-structured version of the same task. In Beck et al. (2011), children were given the same start state as in the innovation task, the apparatus and materials, and the same goal, retrieve the bucket to win the sticker, but this time they were given the choice of two possible transformations. They could choose from a straight pipecleaner or a hooked pipecleaner. In this well-structured version of the task where the possible transformations are clearly defined, children's performance is extremely good. From the age of 4 children select the correct hooked pipecleaner first significantly more than chance and easily solve the problem.

The first strand of research into ill-structured problems suggests that domain knowledge alone is not sufficient to be able to solve these types of problem. Ill-structured problems require the solver to bring to mind knowledge that is relevant to the task and then coordinate these pieces of knowledge together into a useful solution. Children have been shown to have great difficulty with both of these components (Cutting, Apperly, Chappell, & Beck, 2014).

To test children's abilities to bring to mind relevant task knowledge and coordinate information we highlighted various task components to children to investigate how they helped them to solve our tool-innovation problem (Cutting et al., 2014). Half of the children in this experiment were given bending practice where they manipulated a pipecleaner before they encountered the innovation task. As in previous studies we found that highlighting information about pipecleaner properties did not help children to bring to mind the solution to the task. If children failed the innovation task we then showed them a readymade pipecleaner hook (but not how to use it). Children aged over 5 were good at innovating the solution to the task only if they had information about properties and the need for a hook highlighted for them. On their own neither piece of information was enough to induce innovation. This shows that although children had difficulty in bringing to mind the different elements of knowledge for themselves they were good at coordinating information into a solution if it was highlighted. In contrast, younger children were still poor at this task even when both pieces of information were highlighted. This demonstrates that young children not only have difficulty with bringing to mind information but also with coordinating it.

This study supports the idea that even when we have domain knowledge we are not always able to use it to come up with new solutions. This demonstrates just how difficult innovation can be. In our task children had information about the pliability of pipecleaners and were shown the hook tool they needed to make, yet they were still unable to come up with the solution of bending their own straight pipecleaner into a hook.

CONCLUSION

Innovation is clearly a complex, multi-faceted problem, and this review highlights a number of abilities that will be necessary in an innovator: physical cognition and causal reasoning; planning and sequencing; inhibitory control; and the ability to impose structure on situations with inherently many degrees of freedom. One important lesson is that while each of these components may be necessary for successful innovation, they are unlikely to all be limiting factors at the same time. An important job for researchers investigating a given innovation problem, species or age group is to understand *which* of these factors might be the critical constraint on success.

Of these components, the need to handle the ill-structured nature of innovation problems has received least attention in the literature to date. We believe that this makes an important addition to our understanding, in several ways. Firstly, as we have illustrated with our own studies, it helps us understand limitations on the abilities of human children, and leads to novel hypotheses for investigating these limitations. Secondly, it provides a concrete basis for understanding how innovation requires "creativity" and "novelty." Thirdly, since ill-structured problem solving in humans is strongly associated with prefrontal brain regions—in particular BA10—and since these structures are among those that have undergone the largest and most recent expansion in the development of modern humans (Falk, 2012; Semendeferi, Armstrong, Schleicher, Zilles, & Van Hoesen, 2001), it is a plausible hypothesis that a greater capacity for bringing structure to inherently ill-structured problems is a distinctive feature of human innovation.

While innovation in nonhuman animals clearly lacks the sophistication of that shown by humans, this review has shown that they are also capable of creativity and innovative behavior. Might the ill-structured problem framework be a similarly promising framework with which to investigate innovation and problem solving in nonhuman animals? The framework integrates many of the cognitive capacities we know to be important in innovation, but by focusing on the gap in knowledge the subject faces between the starting conditions and the goal, it allows us to experimentally provide or withhold information incrementally (to "scaffold" the subject across the gap) and to observe the effect on performance. This may help us to pin down the ways in which nonhuman animal species solve problems. For example, do they require concrete information about what object to use or how to transform it, or would more abstract, analogical information suffice? Does providing the next step from the initial condition help more than providing the penultimate step from the solution, suggesting that they work "forwards" from the current situation, rather than attempting to reverse engineer the solution from the goal? Note that in these kinds of experiments, it is important to record and analyze the errors made, as well as recording successful performance (Chappell & Hawes, 2012; Thornton & Lukas, 2012).

In summary, characterizing tasks involving innovation and creativity as ill-structured problems provides much more than a convenient term for a multi-faceted phenomenon: it provides a solid foundation for generating and testing hypotheses in order to understand the process of innovation in both humans and nonhuman animals.

References

Apperly, I. A., & Carroll, D. J. (2009). How do symbols affect 3- to 4-year-olds' executive function? Evidence from a reverse-contingency task. *Developmental Science*, *12*(6), 1070–1082. Available from: http://dx.doi.org/10.1111/j.1467-7687.2009.00856.x.

Auersperg, A. M. I., Szabo, B., von Bayern, A. M. P., & Kacelnik, A. (2012). Spontaneous innovation in tool manufacture and use in a Goffin's cockatoo. *Current Biology*, *22*(21), R903–R904. Available from: http://dx.doi.org/10.1016/j.cub.2012.09.002.

Auersperg, A. M. I., von Bayern, A. M. P., Gajdon, G. K., Huber, L., & Kacelnik, A. (2011). Flexibility in problem solving and tool use of kea and New Caledonian crows in a multi access box paradigm. *PLoS One*, *6*(6), e20231-EP. Available from: http://dx.doi.org/10.1371/journal.pone.0020231.

Beck, S. R., Apperly, I. A., Chappell, J., Guthrie, C., & Cutting, N. (2011). Making tools isn't child's play. *Cognition*, *119*(2), 301–306. Available from: http://dx.doi.org/10.1016/j.cognition.2011.01.003.

Benson-Amram, S., & Holekamp, K. E. (2012). Innovative problem solving by wild spotted hyenas. *Proceedings. Biological Sciences/The Royal Society*, *279*(1744), 4087–4095. Available from: http://dx.doi.org/10.1098/rspb.2012.1450.

Benson-Amram, S., Weldele, M. L., & Holekamp, K. E. (2013). A comparison of innovative problem-solving abilities between wild and captive spotted hyaenas, *Crocuta crocuta*. *Animal Behaviour*, *85*(2), 349−356. Available from: http://dx.doi.org/10.1016/j.anbehav.2012.11.003.

Bird, C. D., & Emery, N. J. (2009). Insightful problem solving and creative tool modification by captive nontool-using rooks. *Proceedings of the National Academy of Sciences of the United States of America*, *106*(25), 10370−10375. Available from: http://dx.doi.org/10.1073/pnas.0901008106.

Boden, M. A. (1996). What is creativity? In M. A. Boden (Ed.), *Dimensions of creativity* (pp. 75−117). Cambridge, MA: The MIT Press.

Boyd, R., & Richerson, P. J. (1996). Why culture is common but cultural evolution is rare. *Proceedings of the British Academy, 88*, 73−93.

Brosnan, S. F., & Hopper, L. M. (2014). Psychological limits on animal innovation. *Animal Behaviour*. Available from: http://dx.doi.org/10.1016/j.anbehav.2014.02.026.

Burgess, P. W., Dumontheil, I., & Gilbert, S. J. (2007). The gateway hypothesis of rostral prefrontal cortex (area 10) function. *Trends in Cognitive Sciences, 11*(7), 290−298. Available from: http://dx.doi.org/10.1016/j.tics.2007.05.004.

Caruso, D. A. (1993). Dimensions of quality in infants' exploratory behavior: relationships to problem-solving ability. *Infant Behavior and Development, 16*(4), 441−454. Available from: http://dx.doi.org/10.1016/0163-6383(93)80003-Q.

Chappell, J., Cutting, N., Apperly, I. A., & Beck, S. R. (2013). The development of tool manufacture in humans: what helps young children make innovative tools? *Philosophical Transactions of the Royal Society of London Series B, Biological Sciences, 368* (1630), 20120409. Available from: http://dx.doi.org/10.1098/rstb.2012.0409.

Chappell, J., Demery, Z. P., Arriola-Rios, V., & Sloman, A. (2012). How to build an information gathering and processing system: lessons from naturally and artificially intelligent systems. *Behavioural Processes, 89*(2), 179−186. Available from: http://dx.doi.org/10.1016/j.beproc.2011.10.001.

Chappell, J., & Hawes, N. (2012). Biological and artificial cognition: what can we learn about mechanisms by modelling physical cognition problems using artificial intelligence planning techniques? *Philosophical Transactions of the Royal Society of London Series B, Biological Sciences, 367*(1603), 2723−2732. Available from: http://dx.doi.org/10.1098/rstb.2012.0221.

Chappell, J., & Kacelnik, A. (2002). Tool selectivity in a non-primate, the New Caledonian crow (*Corvus moneduloides*). *Animal Cognition, 5*(2), 71−78. Available from: http://dx.doi.org/10.1007/s10071-002-0130-2.

Chappell, J., & Kacelnik, A. (2004). Selection of tool diameter by New Caledonian crows *Corvus moneduloides*. *Animal Cognition, 7*(2), 121−127. Available from: http://dx.doi.org/10.1007/s10071-003-0202-y.

Chen, C.-H., & Bradshaw, A. C. (2007). The effect of web-based question prompts on scaffolding knowledge integration and ill-structured problem solving. *Journal of Research on Technology in Education, 39*(4), 359−375. Available from: http://dx.doi.org/10.1080/15391523.2007.10782487.

Comins, J. A., Russ, B. E., Humbert, K. A., & Hauser, M. D. (2011). Innovative coconut-opening in a semi free-ranging rhesus monkey (*Macaca mulatta*): a case report on behavioral propensities. *Journal of Ethology, 29*(1), 187−189. Available from: http://dx.doi.org/10.1007/s10164-010-0234-0.

Cutting, N., Apperly, I. A., & Beck, S. R. (2011). Why do children lack the flexibility to innovate tools? *Journal of Experimental Child Psychology, 109*(4), 497−511. Available from: http://dx.doi.org/10.1016/j.jecp.2011.02.012.

Cutting, N., Apperly, I. A., Chappell, J., & Beck, S. R. (2014). The puzzling difficulty of tool innovation: why can't children piece their knowledge together? *Journal of*

Experimental Child Psychology, *125*, 110–117. Available from: http://dx.doi.org/10.1016/j.jecp.2013.11.010.

Cutting, N., (2013) Children's tool making: from innovation to manufacture. Ph.D. thesis, University of Birmingham. <http://etheses.bham.ac.uk/3969/>.

Defeyter, M. A., & German, T. P. (2003). Acquiring an understanding of design: evidence from childrens insight problem solving. *Cognition*, *89*, 133–155. Available from: http://dx.doi.org/10.1016/S0010-0277(03)00098-2.

Diamond, A. (2006). The early development of executive functions. In A. Diamond, E. Bialystok, & F. Craik (Eds.), *Lifespan cognition: Mechanisms of change* (pp. 70–95). New York, NY: Oxford University Press. Available from: http://dx.doi.org/10.1093/acprof:oso/9780195169539.003.0006.

Dumontheil, I., Burgess, P. W., & Blakemore, S.-J. (2008). Development of rostral prefrontal cortex and cognitive and behavioural disorders. *Developmental Medicine & Child Neurology*, *50*(3), 168–181. Available from: http://dx.doi.org/10.1111/j.1469-8749.2008.02026.x.

Ellis, S., & Siegler, R. S. (1997). In S. Friedman, & E. Scholnick (Eds.), *The developmental psychology of planning: Why, how, and when do we plan?* (pp. 183–208). Hillsdale, NJ: Lawrence Erlbaum Associates.

Epstein, R. (1999). Generativity theory. In M. A. Runco, & S. R. Pritzker (Eds.), *Encyclopedia of creativity* (Vol. 1, pp. 759–766). San Diego, CA: Academic Press.

Falk, D. (2012). Hominin paleoneurology: where are we now? In M. A. Hofman, & D. Falk (Eds.), *Progress in brain research* (Vol. 195, pp. 255–272). Netherlands: Elsevier. Available from: http://dx.doi.org/10.1016/B978-0-444-53860-4.00012-X.

Fisher, J. B., & Hinde, R. A. (1949). The opening of milk bottles by birds. *British Birds*, *42*, 347–357.

Frye, D., Zelazo, P. D., & Palfai, T. (1995). Theory of mind and rule-based reasoning. *Cognitive Development*, *10*(4), 483–527. Available from: http://dx.doi.org/10.1016/0885-2014(95)90024-1.

German, T. P., & Defeyter, M. A. (2000). Immunity to functional fixedness in young children. *Psychonomic Bulletin & Review*, *7*(4), 707–712. Available from: http://dx.doi.org/10.3758/BF03213010.

Greenberg, R., & Mettke-Hofmann, C. (2001). Ecological aspects of neophobia and neophilia in birds. *Current Ornithology*, *16*, 119–178.

Hanus, D., Mendes, N., Tennie, C., & Call, J. (2011). Comparing the performances of apes (*Gorilla gorilla, Pan troglodytes, Pongo pygmaeus*) and human children (*Homo sapiens*) in the floating peanut task. *PLoS One*, *6*(6), e19555. Available from: http://dx.doi.org/10.1371/journal.pone.0019555.

Healy, S. D., & Rowe, C. (2007). A critique of comparative studies of brain size. *Proceedings of the Royal Society B: Biological Sciences*, *274*(1609), 453–464. Available from: http://dx.doi.org/10.1098/rspb.2006.3748.

Hihara, S., Obayashi, S., Tanaka, M., & Iriki, A. (2003). Rapid learning of sequential tool use by macaque monkeys. *Physiology and Behavior*, *78*, 427–434. Available from: http://dx.doi.org/10.1016/S0031-9384(02)01006-5.

Horner, V., & Whiten, A. (2005). Causal knowledge and imitation/emulation switching in chimpanzees (*Pan troglodytes*) and children (*Homo sapiens*). *Animal Cognition*, *8*(3), 164–181. Available from: http://dx.doi.org/10.1007/s10071-004-0239-6.

Huber, L., & Gajdon, G. K. (2006). Technical intelligence in animals: the kea model. *Animal Cognition*, *9*(4), 295–305. Available from: http://dx.doi.org/10.1007/s10071-006-0033-8.

Jelbert, S. A., Taylor, A. H., Cheke, L. G., Clayton, N. S., & Gray, R. D. (2014). Using the Aesop's fable paradigm to investigate causal understanding of water displacement by New Caledonian crows. *PLoS One*, *9*(3), e92895. Available from: http://dx.doi.org/10.1371/journal.pone.0092895.

Jonassen, D. H., Beissner, K., & Yacci, M. (1993). *Structural knowledge: Techniques for representing, conveying, and acquiring structural knowledge*. Hillsdale, NJ: Lawrence Erlbaum Associates.

Kawai, M. (1965). Newly-acquired pre-cultural behaviour of the natural troop of Japanese monkeys on Koshima Islet. *Primates, 6,* 1–30. Available from: http://dx.doi.org/10.1007/BF01794457.

Kuczaj, S., Gory, J., & Xitco, M. (2009). How intelligent are dolphins? A partial answer based on their ability to plan their behavior when confronted with novel problems. *Japanese Journal of Animal Psychology, 59,* 99–115. Available from: http://dx.doi.org/10.2502/janip.59.1.9.

Kummer, H., & Goodall, J. (1985). Conditions of innovative behaviour in primates. *Philosophical Transactions of the Royal Society of London Series B, Biological Sciences, 308* (1135), 203–214. Available from: http://dx.doi.org/10.1098/rstb.1985.0020.

Laland, K. N., & Rendell, L. (2013). Cultural memory. *Current Biology, 23*(17), R736–R740. Available from: http://dx.doi.org/10.1016/j.cub.2013.07.071.

Leca, J.-B., Gunst, N., & Huffman, M. A. (2010). The first case of dental flossing by a Japanese macaque (*Macaca fuscata*): implications for the determinants of behavioral innovation and the constraints on social transmission. *Primates, 51*(1), 13–22. Available from: http://dx.doi.org/10.1007/s10329-009-0159-9.

Lefebvre, L., Reader, S. M., & Sol, D. (2004). Brains, innovations and evolution in birds and primates. *Brain, Behavior and Evolution, 63,* 223–246. Available from: http://dx.doi.org/10.1159/000076784.

Lefebvre, L., Whittle, P., Lascaris, E., & Finkelstein, A. (1997). Feeding innovations and forebrain size in birds. *Animal Behaviour, 53,* 549–560. Available from: http://dx.doi.org/10.1006/anbe.1996.0330.

Manrique, H. M., & Call, J. (2011). Spontaneous use of tools as straws in great apes. *Animal Cognition, 14*(2), 213–226. Available from: http://dx.doi.org/10.1007/s10071-010-0355-4.

Manrique, H. M., Völter, C. J., & Call, J. (2013). Repeated innovation in great apes. *Animal Behaviour, 85*(1), 195–202. Available from: http://dx.doi.org/10.1016/j.anbehav.2012.10.026.

Martin-Ordas, G., Schumacher, L., & Call, J. (2012). Sequential tool use in great apes. *PLoS One, 7*(12), e52074. Available from: http://dx.doi.org/10.1371/journal.pone.0052074.

McGuigan, N., Whiten, A., Flynn, E., & Horner, V. (2007). Imitation of causally opaque versus causally transparent tool use by 3-and 5-year-old children. *Cognitive Development, 22* (3), 353–364. Available from: http://dx.doi.org/10.1016/j.cogdev.2007.01.001.

Miyata, H., Gajdon, G. K., Huber, L., & Fujita, K. (2011). How do keas (*Nestor notabilis*) solve artificial-fruit problems with multiple locks? *Animal Cognition, 14,* 45–58. Available from: http://dx.doi.org/10.1007/s10071-010-0342-9.

Morand-Ferron, J., Cole, E. F., Rawles, J. E. C., & Quinn, J. L. (2011). Who are the innovators? A field experiment with 2 passerine species. *Behavioral Ecology, 22*(6), 1241–1248. Available from: http://dx.doi.org/10.1093/beheco/arr120.

Morand-Ferron, J., & Quinn, J. L. (2011). Larger groups of passerines are more efficient problem solvers in the wild. *Proceedings of the National Academy of Sciences of the United States of America, 108*(38), 15898–15903. Available from: http://dx.doi.org/10.1073/pnas.1111560108.

Mulcahy, N. J., Call, J., & Dunbar, R. I. M. (2005). Gorillas (*Gorilla gorilla*) and orangutans (*Pongo pygmaeus*) encode relevant problem features in a tool-using task. *Journal of Comparative Psychology, 119*(1), 23–32. Available from: http://dx.doi.org/10.1037/0735-7036.119.1.23.

Nielsen, M. (2013). Young children's imitative and innovative behaviour on the floating object task. *Infant and Child Development, 22*(1), 44–52. Available from: http://dx.doi.org/10.1002/icd.1765.

Nielsen, M., & Blank, C. (2011). Imitation in young children: when who gets copied is more important than what gets copied. *Developmental Psychology, 47*(4), 1050–1053. Available from: http://dx.doi.org/10.1037/a0023866.

III. THE STRUGGLE FOR CREATIVITY

O'Brien, M. J., & Shennan, S. (Eds.), (2010a). *Innovation in cultural systems: Contributions from evolutionary anthropology* Cambridge, MA: MIT Press.

O'Brien, M. J., & Shennan, S. (2010b). Issues in anthropological studies of innovation. In M. J. O'Brien, & S. Shennan (Eds.), *Innovation in cultural systems: Contributions from evolutionary anthropology* (pp. 3–17). Cambridge, MA: MIT Press.

Osvath, M., & Karvonen, E. (2012). Spontaneous innovation for future deception in a male chimpanzee. *PLoS One, 7*(5), e36782. Available from: http://dx.doi.org/10.1371/journal.pone.0036782.

Overington, S. E., Morand-Ferron, J., Boogert, N. J., & Lefebvre, L. (2009). Technical innovations drive the relationship between innovativeness and residual brain size in birds. *Animal Behaviour, 78,* 1001–1010. Available from: http://dx.doi.org/10.1016/j.anbehav.2009.06.033.

Parker, C. E. (1974). Behavioral diversity in ten species of nonhuman primates. *Journal of Comparative and Physiological Psychology, 87*(5), 930. Available from: http://dx.doi.org/10.1037/h0037228.

Reader, S. M., & Laland, K. N. (2002). Social intelligence, innovation, and enhanced brain size in primates. *Proceedings of the National Academy of Sciences of the United States of America, 99*(7), 4436–4441. Available from: http://dx.doi.org/10.1073/pnas.062041299.

Reader, S. M., & Laland, K. N. (2003). Animal innovation: an introduction. In S. M. Reader, & K. N. Laland (Eds.), *Animal innovation* (pp. 3–35). Oxford: Oxford University Press. Available from: http://dx.doi.org/10.1093/acprof:oso/9780198526223.003.0001.

Reynolds, B., Ortengren, A., Richards, J. B., & Wit, H. de. (2006). Dimensions of impulsive behavior: personality and behavioral measures. *Personality and Individual Differences, 40* (2), 305–315. Available from: http://dx.doi.org/10.1016/j.paid.2005.03.024.

Schachar, R., & Logan, G. D. (1990). Impulsivity and inhibitory control in normal development and childhood psychopathology. *Developmental Psychology, 26*(5), 710–720. Available from: http://dx.doi.org/10.1037/0012-1649.26.5.710.

Schuck-Paim, C., Borsari, A., & Ottoni, E. B. (2009). Means to an end: neotropical parrots manage to pull strings to meet their goals. *Animal Cognition, 12*(2), 287–301. Available from: http://dx.doi.org/10.1007/s10071-008-0190-z.

Seed, A. M., Tebbich, S., Emery, N. J., & Clayton, N. S. (2006). Investigating physical cognition in rooks, *Corvus frugilegus. Current Biology, 16,* 697–701. Available from: http://dx.doi.org/10.1016/j.cub.2006.02.066.

Semendeferi, K., Armstrong, E., Schleicher, A., Zilles, K., & Van Hoesen, G. W. (2001). Prefrontal cortex in humans and apes: a comparative study of area 10. *American Journal of Physical Anthropology, 114*(3), 224–241.

Shallice, T., & Burgess, P. (1991). Higher-order cognitive impairments and frontal lobe lesions in man. In H. S. Levin, H. M. Eisenberg, & A. L. Benton (Eds.), *Frontal lobe function and dysfunction* (pp. 125–138). New York, NY: Oxford University Press.

Simpson, A., & Riggs, K. J. (2011). Under what conditions do children have difficulty in inhibiting imitation? Evidence for the importance of planning specific responses. *Journal of Experimental Child Psychology, 109*(4), 512–524. Available from: http://dx.doi.org/10.1016/j.jecp.2011.02.015.

Sol, D. (2009). Revisiting the cognitive buffer hypothesis for the evolution of large brains. *Biology Letters, 5*(1), 130–133. Available from: http://dx.doi.org/10.1098/rsbl.2008.0621.

Sol, D., Duncan, R. P., Blackburn, T. M., Cassey, P., & Lefebvre, L. (2005). Big brains, enhanced cognition, and response of birds to novel environments. *Proceedings of the National Academy of Sciences of the United States of America, 102*(15), 5460–5465. Available from: http://dx.doi.org/10.1073/pnas.0408145102.

Sol, D., & Lefebvre, L. (2000). Behavioural flexibility predicts invasion success in birds introduced to New Zealand. *Oikos, 90*(3), 599–605. Available from: http://dx.doi.org/10.1034/j.1600-0706.2000.900317.x.

Sol, D., Timmermans, S., & Lefebvre, L. (2002). Behavioural flexibility and invasion success in birds. *Animal Behaviour*, *63*, 495–502. Available from: http://dx.doi.org/10.1006/anbe.2001.1953.

Sterelny, K. (2007). Social intelligence, human intelligence and niche construction. *Philosophical Transactions of the Royal Society of London Series B, Biological Sciences*, *362* (1480), 719–730. Available from: http://dx.doi.org/10.1098/rstb.2006.2006.

Taylor, A. H., Elliffe, D., Hunt, G. R., & Gray, R. D. (2010). Complex cognition and behavioural innovation in New Caledonian crows. *Proceedings of the Royal Society B: Biological Sciences*, *277*(1694), 2637–2643. Available from: http://dx.doi.org/10.1098/rspb.2010.0285.

Taylor, A. H., Hunt, G. R., Holzhaider, J. C., & Gray, R. D. (2007). Spontaneous metatool use by New Caledonian crows. *Current Biology*, *17*(17), 1504–1507. Available from: http://dx.doi.org/10.1016/j.cub.2007.07.057.

Tecwyn, E. C., Thorpe, S. K., & Chappell, J. (2013). A novel test of planning ability: great apes can plan step-by-step but not in advance of action. *Behavioural Processes*, *100*, 174–184. Available from: http://dx.doi.org/10.1016/j.beproc.2013.09.016.

Tecwyn, E. C., Thorpe, S. K. S., & Chappell, J. (2014). Development of planning in 4- to 10-year-old children: reducing inhibitory demands does not improve performance. *Journal of Experimental Child Psychology*, *125*, 85–101.

Tennie, C., Call, J., & Tomasello, M. (2009). Ratcheting up the ratchet: on the evolution of cumulative culture *Philosophical Transactions of the Royal Society of London Series B, Biological Sciences*, *364*(1528), 2405–2415. Available from: http://dx.doi.org/10.1098/rstb.2009.0052.

Thornton, A., & Lukas, D. (2012). Individual variation in cognitive performance: developmental and evolutionary perspectives. *Philosophical Transactions of the Royal Society of London Series B, Biological Sciences*, *367*(1603), 2773–2783. Available from: http://dx.doi.org/10.1098/rstb.2012.0214.

Tomasello, M., Kruger, A. C., & Ratner, H. H. (1993). Cultural learning. *Behavioral and Brain Sciences*, *16*(03), 495–511. Available from: http://dx.doi.org/10.1017/S0140525X0003123X.

Vaesen, K. (2012). The cognitive bases of human tool use. *Behavioral and Brain Sciences*, 1–16. Available from: http://dx.doi.org/10.1017/S0140525X11001452.

Vlamings, P. H. J. M., Hare, B., & Call, J. (2010). Reaching around barriers: the performance of the great apes and 3-5-year-old children. *Animal Cognition*, *13*(2), 273–285. Available from: http://dx.doi.org/10.1007/s10071-009-0265-5.

von Bayern, A. M. P., Heathcote, R. J. P., Rutz, C., & Kacelnik, A. (2009). The role of experience in problem solving and innovative tool use in crows. *Current Biology*, *19*(22), 1965–1968. Available from: http://dx.doi.org/10.1016/j.cub.2009.10.037.

Voss, J. F., Blais, J., Means, M. L., Greene, T. R., & Ahwesh, E. (1986). Informal reasoning and subject matter knowledge in the solving of economics problems by naive and novice individuals. *Cognition and Instruction*, *3*(3), 269–302. Available from: http://dx.doi.org/10.1207/s1532690xci0303_7.

Want, S. C., & Harris, P. L. (2002). How do children ape? Applying concepts from the study of non-human primates to the developmental study of "imitation" in children. *Developmental Science*, *5*(1), 1–14. Available from: http://dx.doi.org/10.1111/1467-7687.00194.

Weir, A. A., Chappell, J., & Kacelnik, A. (2002). Shaping of hooks in New Caledonian crows. *Science*, *297*(5583), 981. Available from: http://dx.doi.org/10.1126/science.1073433.

Weir, A. A. S., & Kacelnik, A. (2006). A New Caledonian crow (*Corvus moneduloides*) creatively re-designs tools by bending or unbending aluminium strips. *Animal Cognition*, *9* (4), 317–334. Available from: http://dx.doi.org/10.1007/s10071-006-0052-5.

Whiten, A., & Flynn, E. (2010). The transmission and evolution of experimental microcultures in groups of young children. *Developmental Psychology, 46*(6), 1694–1709. Available from: http://dx.doi.org/10.1037/a0020786.

Whiten, A., Goodall, J., McGrew, W. C., Nishida, T., Reynolds, V., Sugiyama, Y., et al. (1999). Cultures in chimpanzees. *Nature, 399*(6737), 682–685. Available from: http://dx.doi.org/10.1038/21415.

Whiten, A., Goodall, J., McGrew, W. C., Nishida, T., Reynolds, V., Sugiyama, Y., et al. (2001). Charting cultural variation in chimpanzees. *Behaviour, 138*(11/12), 1481–1516. Available from: http://dx.doi.org/10.1163/156853901317367717.

Whiten, A., Horner, V., & de Waal, F. B. M. (2005). Conformity to cultural norms of tool use in chimpanzees. *Nature, 437*, 737–740. Available from: http://dx.doi.org/10.1038/nature04047.

Whiten, A., McGuigan, N., Marshall-Pescini, S., & Hopper, L. M. (2009). Emulation, imitation, over-imitation and the scope of culture for child and chimpanzee. *Philosophical Transactions of the Royal Society of London Series B, Biological Sciences, 364*(1528), 2417–2428. Available from: http://dx.doi.org/10.1098/rstb.2009.0069.

Williamson, R. A., Jaswal, V. K., & Meltzoff, A. N. (2010). Learning the rules: observation and imitation of a sorting strategy by 36-month-old children. *Developmental Psychology, 46*(1), 57. Available from: http://dx.doi.org/10.1037/a0017473.

Wimpenny, J. H., Weir, A. A. S., Clayton, L., & Kacelnik, A. (2009). Cognitive processes associated with sequential tool use in New Caledonian crows. *PLoS One, 4*(8), e6471. Available from: http://dx.doi.org/10.1371/journal.pone.0006471.

Wineburg, S. (1998). Reading Abraham Lincoln: an expert/expert study in the interpretation of historical texts. *Cognitive Science, 22*(3), 319–346. Available from: http://dx.doi.org/10.1207/s15516709cog2203_3.

Commentary on Chapter 10: Minding the Gap: Problem Construction and Ill-Defined Problems

In this chapter, Chappell et al. focus on the gap in knowledge that requires innovation, and the ill-defined nature of the problem that results in innovation. While ill-defined problems have not been studied extensively in cognition, they have been studied by creativity researchers. It has been suggested that ill-defined problems and well-defined problems are placed on a continuum, where problems can be more or less ill-defined, and vary on a number of dimensions (Dillon, 1982). Problems may be ill-defined because they may have multiple goals, which may not always be compatible. Problems may be ill-defined because there is no one acceptable or appropriate way to reach a solution, instead, a solution may be derived from multiple processes and procedures. Problems may be ill-defined because there is no one acceptable solution, rather multiple solutions exist that may satisfy one or more of the goals of the problem in acceptable manner. Finally, ill-defined problems may be so ambiguous that not everyone will even realize that there is a problem to be solved. It is this ambiguity of the ill-defined problem that allows for creative solutions to arise (Mumford, Mobley, Uhlman, Reiter-Palmon, & Doares, 1991; Schraw, Dunkle, & Bendixen, 1995).

In the studies described in the chapter, the problems were ill-defined because there were multiple ways of successfully solving the problem, and participants may not have all the knowledge they needed to solve the problem, but in most cases the goal of the problem, retrieving a reward, was known. However, many problems we face, such as those related to public policy, relationships, and problems that face business and industry, do not have a clear goal. The goal may be more ambiguous or multiple goals may exist.

As a result, the creative problem solver must first construct the problem to be solved. Many process models of creative problem solving suggest that problem identification and construction is the first step in the creative process (e.g., Finke, Ward, & Smith, 1992; Mumford et al., 1991). The problem construction includes addressing three issues or questions: recognizing and identifying the problem to be solved (is there indeed a problem?), defining the problem (what is the nature of the problem?), and constructing the problem (what are the parameters of the problem to guide possible solutions?).

Past research suggests that creative individuals engage in problem identification and construction more than less creative individuals (Getzels & Csikszentmihalyi, 1976), and that problem identification and construction ability is linked to creative performance above and beyond intelligence or divergent thinking (e.g., Reiter-Palmon, Mumford, O'Connor Boes, & Runco, 1997). Additionally, past research suggests that active engagement in problem construction, typically induced through instruction or training, improves creative performance (Mumford, Reiter-Palmon, & Redmond, 1994; Reiter-Palmon et al., 1997). Problem construction and identification has been found to be critical for creative problem solving in a variety of settings and for a variety of occupations, such as artists, scientists, and managers.

This research indicates that problem identification and construction is a critical cognitive process for creative problem solving. This process provides the problem solver with context for the application of other processes in the creative problem-solving effort, it has been suggested that the way the problem is constructed will have a marked impact on creative production and solution generation.

References

Dillon, J. T. (1982). Problem finding and solving. *Journal of Creative Behavior, 16,* 97–111.
Finke, R. A., Ward, T. B., & Smith, S. M. (1992). *Creative cognition: Theory, research, and applications.* Cambridge, MA: MIT.

Getzels, J. W., & Csikszentmihalyi, M. (1976). *The creative vision: A longitudinal study of problem finding in art.* New York, NY: Wiley.

Mumford, M. D., Mobley, M. I., Uhlman, C. E., Reiter-Palmon, R., & Doares, L. M. (1991). Process analytic models of creative capacities. *Creativity Research Journal, 4,* 91–122.

Mumford, M. D., Reiter-Palmon, R., & Redmond, M. R. (1994). Problem construction and cognition: applying problem representations in Ill-defined domains. In M. A. Runco (Ed.), *Problem finding, problem solving, and creativity* (pp. 3–39). Norwood, NJ: Ablex.

Reiter-Palmon, R., Mumford, M. D., O'Connor Boes, J., & Runco, M. A. (1997). Problem construction and creativity: the role of ability, cue consistency, and active processing. *Creativity Research Journal, 10,* 9–23.

Schraw, G., Dunkle, M. E., & Bendixen, L. D. (1995). Cognitive processes in well-defined and ill-defined problem solving. *Applied Cognitive Psychology, 9,* 523–538.

III. THE STRUGGLE FOR CREATIVITY

Necessity, Unpredictability and Opportunity: An Exploration of Ecological and Social Drivers of Behavioral Innovation

Phyllis C. Lee[1] and Antonio C. de A. Moura[2]

[1]Behaviour and Evolution Research Group, Psychology, School of Natural Sciences, University of Stirling, Stirling, UK [2]Department of Engineering and Environment, Federal University of Paraiba, Rio Tinto, PB, Brazil

Commentary on Chapter 11: Necessity, Unpredictability, Opportunity, and Creativity

Marie J.C. Forgeard[1,2] and Eranda Jayawickreme[3]

[1]Department of Psychology, University of Pennsylvania, Philadelphia, PA, USA [2]McLean Hospital, Harvard Medical School, Boston, MA, USA [3]Department of Psychology, Wake Forest University, Winston-Salem, NC, USA

INTRODUCTION

The nature and diversity of animal innovation continues to be a topic of considerable interest to evolutionary ecologists, since it reflects the cognitive capacities of animals and relates to behavioral diversity among species. Innovation, as distinct from learning or mere exploration (e.g., Reader & Laland, 2003), is a process that results in novel behaviors being used to solve existing problems (what Reader & Laland, 2003, call "innovation as a product") or existing behaviors which are used to produce an entirely new response to problems and in effect opening up a new niche to the innovator. Does innovation as a product—for example, a newly developed specific tool type—differ from the process of using a tool in the first instance, which then enables new foods to be exploited? Both elements form an essential part of an innovation that spreads to become established in a population. In neither case do we have a clear insight into the causal variables which have enabled the initial innovation or its spread. We aim to explore these questions here using examples of foraging innovations, including tool use in nonhuman primates, alongside a perspective on the opportunities for innovation due to social dynamics.

The key traits associated with innovation have been discussed by many authors since Kummer and Goodall (1985) addressed the issue. Attraction to novelty—or curiosity—is of importance for predicting innovation in birds (Lefebvre, 2000), while boldness or other personality traits are other predictors at the individual level (Sol, 2003). Cognitive capacity and learning abilities, both for technical and social skills, are associated with propensities to innovate (Byrne, 2003; van Schaik & Pradhan, 2003), and elements of cognitive control or a capacity for deferred gratification (Addessi, Paglieri, & Focaroli, 2011) may also play a role. Life history traits and group demography—the availability of individuals across a range of ages and sex—are further influences on the capacity for plasticity in behavior (Lee, 1991, 2003). Finally, the size, structure and complexity of social networks create opportunities for learning and dissemination of innovations (Aplin, Sheldon, & Morand-Ferron, 2013; Franz & Nunn, 2009; Hoppitt, Boogert, & Laland, 2010; van Schaik, 2003; van Schaik, Deaner, & Merrill, 1999).

Our approach here is that of attempting to assess the costs and benefits of innovations, using ecological currencies of energy acquisition, time costs and mortality risks. These are not quantitative, as most of these costs have yet to be determined; they are however a useful heuristic device for establishing possibly interesting or important domains for measuring costs and benefits.

UNPREDICTABILITY

There are two key features of interest in understanding the drivers of ecological or foraging innovations such as tool use. The first of these is *unpredictability*. When returns, for example from foraging, are unpredictable, then any enhancement of returns is potentially beneficial.

Experimentation or the process of innovating will have few costs when resources are abundant but will have significantly enhanced potential benefits when resources are scarce. If these resource conditions and their variation cannot be predicted, then innovation may produce benefits during shortages that far outweigh the average costs of experimentation, and animals living in marginal habitats (challenging and costly, or with an abundance of concealed food sources; e.g., Rutz & St Clair, 2012) could benefit most from innovation (Figure 11.1). The capacity to develop resilience to unpredictability via a long developmental period and high rates of play (e.g., Joffe, 1997; Spinka, Newberry, & Bekoff, 2001; Wood-Gush & Vestergaard, 1991) as seen in many nonhuman primates could give them this Protean (in the sense of Driver & Humphries, 1988) edge in the capacity to innovate. Adaptive uncertainty produces opportunities which are associated with innovation in the larger brained species such as dolphins (Kuczaj, Gory, & Xitco, 2009), elephants (Chevalier-Skolnikoff & Liska, 1993; Foerder, Galloway, Barthel, Moore, & Reiss, 2011), corvids (Rutz & St Clair, 2012), and of course, primates.

The second key element of interest is *necessity*. Energetically challenging habitats may enhance a propensity for innovation when novel foraging strategies improve capacities to cope with periods of food shortage. Thus, food scarcity—whether due to concealed or embedded foods or unpredictable variability in supply—could induce novel innovations in the form of tool use to enable access to energetically-dense or micronutrient rich foods and thus to improve the profitability of foraging or dietary breadth (Emedio & Ferreira, 2012; Moura & Lee, 2004; Tebbich,

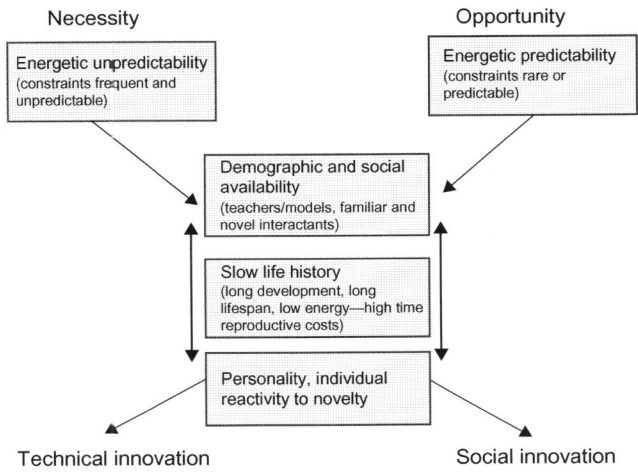

FIGURE 11.1 Potential paths for influences on the evolution of technical and social innovation in relation to ecological unpredictability, demography and life history.

Taborsky, Fessl, & Dvorak, 2002; Yamakoshi, 1998). We note that necessity in the context of coping with social constraints (e.g., limited access to sexual partners, Reader & Laland, 2001) can also be a major driver of innovation (see below). We contrast necessity, here proposed as a driver of innovation, with recent suggestions that *opportunity* alone is sufficient for the emergence and spread of tool innovations, where the availability of resources and substrates might underlie innovative tool use simply by the repeated exposure to appropriate circumstances, for example, the juxtaposition of hammer stones and nuts as seen in wild capuchins (Fragazsy et al., 2013; Spagnolleti et al., 2012; Visalberghi et al., 2007).

INNOVATION IN CAPUCHIN TOOL USE

That unpredictability, necessity and opportunity no doubt co-vary, if not co-exist, in a population makes the separation of cause and consequence difficult with respect to the origins of primate tool use. Capuchins of the genus *Sapajus* provide an appropriate model for investigating innovative behavior as they have only relatively recently been documented as habitual tool users (Fragaszy, Izar, Visalberghi, Ottoni, & Gomes de Oliveira, 2004; Moura & Lee, 2004), although capuchin capacity for foraging innovations is legendary (Fragaszy, Visalberghi, & Fedigan, 2004; Perry et al., 2003). Is tool use such as that seen in capuchins locally enabled by cognitive goal-directed activities, manual dexterity, and a varied diet (Souto et al., 2011), what Ramsey, Bastian, and van Schaik (2007) call an adaptive improvisation, or is it the result of an evolved capacity for tool use which we consider to be an adaptation that emerged in only a few species or populations to solve problems of inadequate energy balance (Moura, 2004)? Capacities for goal-directed behavior in a technological context among tufted capuchins need to be combined with ecological contexts where the benefits outweigh the costs of a behavior that has potentially low rewards for a high investment of time and energy (Emidio & Ferreira, 2012; Moura & Lee, 2010; Spagnolleti et al., 2012). Therefore, we ask if ecological factors underlie selection for these cognitive propensities, or does local ecology merely create conditions for local and immediate expression of an innovative improvisation (e.g., Fox, van Schaik, Sitompul, & Wright, 2004)? We suggest that energy and/or micro-nutrient benefits must be substantial for individuals to engage in a potentially high cost behavior (e.g., Liu et al., 2009) among tool-using capuchin populations. These behaviors are also noisy, conspicuous, time consuming, and often occur in areas exposed to predators. It is thus likely that energy constraints provided a selective bottleneck, enhancing those novelties in the form of behavioral and technological skills which then cope with energy and nutrient constraints. As expected, chimpanzee communities with a high diversity and availability of foods are less likely to develop complex extractive tool use than are those living in harsher environments (Gruber et al., 2012). The importance of the social or developmental context in enabling diverse tool use in varied contexts remains to be considered in detail for capuchins (but see Perry, 2011).

In setting up this argument, we explored the ecological context for tool use in a population of blond capuchin monkeys (*Sapajus flavius*) using tools in the Atlantic forest. A highly varied tool kit has already been reported for several populations of *Sapajus* species (Ottoni & Izar, 2008). Souto et al. (2011) have described a fascinating new feeding tool for these capuchins from Atlantic forest. The monkeys have developed sticks that are used as "screwdrivers" to capture termites and which, when combined with tapping an arboreal nest, increase capture efficiency. Only males use these termite tools, providing further evidence for sex-biased tool use in some capuchins (Moura & Lee, 2010).

A catholic diet and dexterity were proposed as the main factors explaining this form of tool use, disregarding food scarcity as a selective driver (Souto et al., 2011). Semi-free ranging capuchin groups also use tools despite extensive provisioning (i.e., high energy balance) (Ottoni & Izar, 2008). Thus positive reinforcement has been suggested as the most parsimonious explanation for tool use, since food scarcity should predict greater tool use than has yet been observed (Souto et al., 2011).

However, we lack meaningful information on the abundance or availability of resources that could potentially engender blond capuchin tool use. In most explorations of the *necessity* versus *opportunity* hypotheses, comparisons are between populations that vary in their extent of tool use or in the diversity of tool types (e.g., Emidio & Ferreira, 2012; Fox et al., 2004; Gruber et al., 2012), rather than between tool-using and non-tool-using populations of the same species in different habitats. Such comparisons urgently remain to be done. Here, we evaluated the availability of fruit resources and arboreal termites nests in the area where capuchins used sticks to "fish" for termites (*Nasutitemes*) and compared these with resource availability data from a larger and better preserved fragment of Atlantic forest considered to be representative of the norm for tree and insect species diversity and density. Although there are forests where blond capuchins have yet to be seen using tools (e.g., Montenegro, 2011), no sampling of resource availability has yet been carried out. Resources in the two forests were assessed by determining the diversity and abundance of trees from belt transects based on standardized vegetation sampling protocol (Gentry, 1982). We used a slightly modified version of this protocol (Moura, 2004) with 25 rectangular quadrats measuring 50×4 m (totaling 0.5 ha) located randomly

TABLE 11.1 Comparison of Key Fruit Tree Densities for Two Surveyed Atlantic Forests: One with Capuchin Tool Use (ASPLAN) and a Second Well Preserved, Representative Atlantic Forest (PACATUBA)

	ASPLAN (100 ha)	PACATUBA (265 ha)
Sample size	78	114
Individual trees	1336	955
Density (Fisher α)	18.67	33.75
Diversity (Shannon H)	3.071	3.757
N species producing fruits/seed known to be consumed by monkeys	31	47
Mean tree height (min−max)	9.9 (2−30 m)	10.5 (2−50 m)
Mean DBH (min−max)	7.9 (3.8−42.2 cm)	10.53 (3.8−197.8 cm)

in the sampled area. All trees with a diameter at breast height (DBH) greater than or equal to 3.8 cm were measured and their height estimated. Most trees were identified in the field by AC de AM, and specimens were collected and deposited in the Herbarium Lauro Pires Xavier (JPB) from Universidade Federal da Paraiba. Termite nest density was estimated by counting all active arboreal nests within the established transects as arboreal *Nasutitermes* can be easily identified through the external nest appearance (Thorne, 1980). We counted nests only in in the ASPLAN forest as comparative data for other Atlantic forest fragments already exist (Vasconcellos, 2010).

In the tool-using area (ASPLAN), a low availability of fruit resources and key tree foods (Table 11.1) for capuchin monkeys was combined with a significantly higher density of termite nests ($x^2 = 20.9$; df = 2; $p < 0.01$; Figure 11.2). The larger forested area of Pacatuba had more key fruit resources available during seasonal or unpredictable periods of fruit scarcity: *Cariniana legalis*, *Hymenaea* sp. and fig trees (*Ficus gomellera* and *Ficus* sp.), which were absent at ASPLAN. These trees species also occur in the Gargau forest (Moura, pers. obs.) where blond capuchins have been intensively studied but no tool use observed (Montenegro, 2011). Food scarcity could promote the innovative use of tools to capture termites (*necessity hypothesis*), while the relative abundance of termite nests could also enable the use of tools (*opportunity hypothesis*). *Nasutitemes* soldiers, however, have strong chemical protection that can repel even specialist feeders (Lubin & Montgomery, 1981). Thus the specialized use of tools to feed on termites in their nests suggests a strong need to obtain energy or nutrients, supporting the

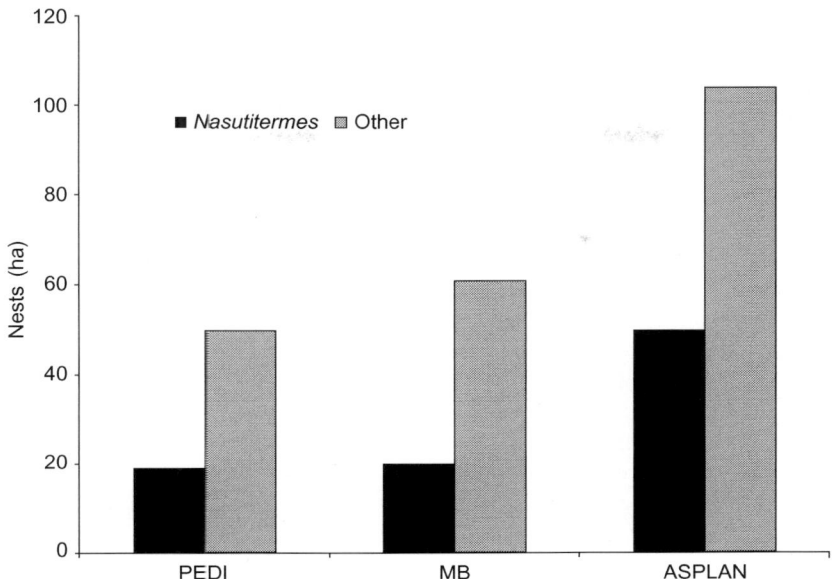

FIGURE 11.2 The abundance of arboreal termite nests in the ASPLAN forest with capuchin tool use and in two similar fragments of Atlantic forest for comparison (PEDI and MB). *Data from Vasconcellos (2010).*

necessity hypothesis. In this example, fruit scarcity is suggested to be a local trigger for innovation in the resident capuchins, resulting in the habitual use of diverse novel feeding tools while the availability of prey species dictates the design of the tools.

SOCIAL SKILL INNOVATION

While Byrne (2003) argues that tactical deception represents innovative social maneuvering, the existence of behaviors which promote conflict resolution (Aureli & de Waal, 2000) and the importance of valuable relationships (Cords & Aureli, 2000) in sustaining cohesive social groups have long suggested that affiliative social innovations are equally important drivers of behavioral diversity (e.g., Kummer & Goodall, 1985). Innovative behavior under stress has been associated with personality traits of boldness in puppies (Riemer, Müller, Virányi, Huber, & Range, 2013), highlighting the individual nature of flexible and novel behavior. New displays, or new contexts for a modified display such as seen in chimpanzees (Goodall, 1986; Nakamura, 2002), are constant reminders of the potential to translate a behavior into novel contexts with survival or reproductive consequences (see also Reader & Laland, 2001). Inter-sexual signaling often appears to incorporate high levels of innovation, possibly due to the high benefits of small shifts in behavior. Bower birds use novel (anthropogenic) objects to increase the amplitude of their signals (Frith & Frith, 2004); using a "super stimulus" may not reflect novelty over simple stimulus-response, but the intentional incorporation of these new elements into the most visible locations suggests innovation in choice both for object novelty and diversity. Innovation in bird song to enhance sexual signaling is well established (Slater, 1986), while innovative communication in cetaceans appears to be both sexual and social—producing dialects and genetic diversity (Whitehead, 1998). Signaling of emotions in self and others, in the form of consolation or empathy (Bates et al., 2008; Plotnik & de Waal, 2014), produces innovative communication and contacts between elephant friends. Innovations in social skill acquisition in the form of teaching may represent another modality for the generation and transmission of novelty (Boesch, 1991). For example, experienced elephant mothers demonstrate modes of appropriate interaction with large sexually active males to their inexperienced daughters (Bates et al., 2010). The innovation here is in the mother's capacity to time her displays of estrous behavior to the onset of estrus in her daughter, resulting in the novel linking of temporal cause and effect with a behavioral display. Such an association of cause and effect which then leads to a behavioral innovation is exceedingly rare among the other social skills that might also be classed as innovations.

As suggested elsewhere (Lee, 2003), behavioral innovators change the dynamics of interactions; they alter the outcomes of an interaction and thus may lessen its predictability. Behavioral innovators might also change the value of an interaction to either or both participants, and to bystanders or eavesdroppers (Valone, 2007). Whole new interactive, strategic and plastic arenas open up through such innovations, with implications for understanding the ways in which social innovation constructs new niches for behavior (e.g., Flack, Girvan, de Waal, & Krakauer, 2006).

DISCUSSION

Many examples of innovative behavior tend to come from the technical intelligence domains, and almost all the experimental work on transmission and acquisition of novelty has been in the context of food rewards (e.g., puzzle boxes and demonstrators, Marshall-Pescini & Whiten, 2008; van de Waal, Borgeaud, & Whiten, 2013), which represents learning modalities but not the generation of the novelty in the first place. Dissemination of foraging innovations can be shown to be a function of the social centrality of the innovator (Cambefort, 1981), as well as age, sex, and social status (Moura, 2004; Reader & Laland, 2001). Matching between the social dimensions of individuals and diffusion of technical innovations nevertheless provides important clues about the importance of the social context in the spread of innovations, as elegantly demonstrated by Aplin et al. (2013) in her experiments on the spread of foraging innovations in great tits. New analysis techniques using social networks have provided us with insights into several gaps in our understanding of the spread and persistence of innovations (Franz & Nunn, 2009; Hoppitt et al., 2010; Hoppitt & Laland, 2011).

However, understanding the ecological and social contexts of the generation of innovations needs to be contrasted with studies of dissemination. Therefore, why innovations *arise* is a specific question, which we suggest is most often related to energetic or other direct survival benefits. This may be especially the case of technical innovation, as we illustrated with the capuchin example above. Innovation as insightful problem-solving can exist in the absence of tool use (e.g., Bird & Emory, 2009) and the conditions for the expression of innovation may often differ from those for tool use.

Who is likely to innovate is a question we have not addressed in detail here. In some cases, the conditions for innovation may be simply that of an individual propensity for problem-solving (Benson-Amram & Holekamp, 2012). Reader and Laland (2001) found that male primates were the most frequent innovators (both technical and social), while the beneficiaries of innovation may be those who have the greatest gain in terms of energy or access to resources and thus are the fastest adopters of innovations (e.g., female chimpanzees using tools, McGrew, 1992; or female orangutans feeding on honey, Fox et al., 2004). Innovators may be younger individuals who have yet to have fixed behavioral expectations and actions (Lee, 2003; Santos, Hauser, & Spelke, 2001), and thus have more opportunities for experimentation at low risk. In that case, opportunity may dictate *who* innovates rather than *what* the innovation consists of (e.g., a response to necessity). Innovators could be dispersers or those who leave social groups and therefore individuals who are more likely to encounter new environments and new opportunities (Thornton & Samson, 2012). As mentioned above, individuals with flexible personality traits, those of openness or boldness, may be more prone to risk-taking and thus experimentation regardless of the costs or consequences. A schematic of the dialectic of cognitive, social, and physical conditions under which innovation might arise is illustrated in Figure 11.3. Note that the question of "when" is not addressed by the overlapping circles of underlying conditions; "when" rather than "how" or "whom" remains a function of how dynamic ecological

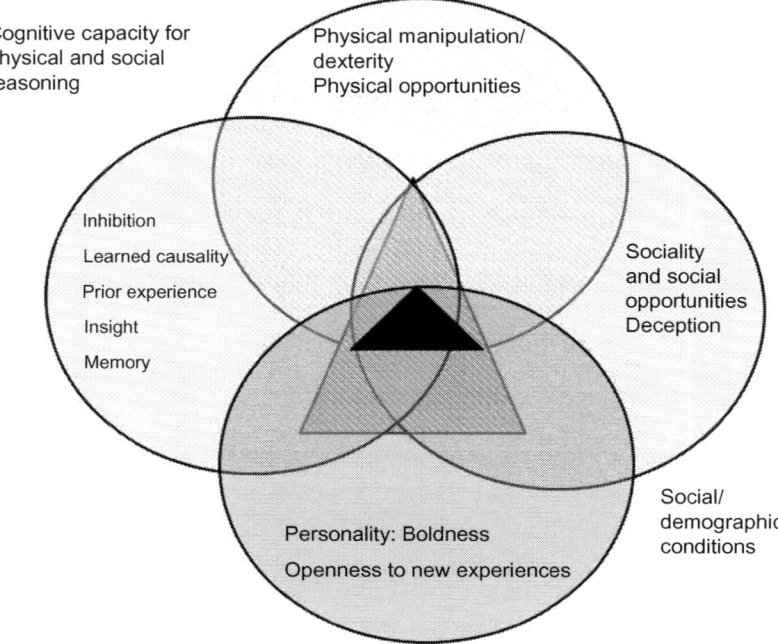

Cognitive capacity for physical and social reasoning

Physical manipulation/ dexterity
Physical opportunities

Inhibition

Learned causality

Prior experience

Insight

Memory

Sociality and social opportunities
Deception

Personality: Boldness

Openness to new experiences

Social/ demographic conditions

FIGURE 11.3 Schematic of how cognitive, social, individual, and physical traits overlap and interact in determining contexts for the generation of innovation. The central triangles illustrate areas where innovation is likely (hatched) and innovation is necessary (solid) when ecologically unpredictable contexts underlie survival and reproductive success.

opportunities (what we call *necessity*) enable creative solutions to problems.

Finally, once an innovation is present, will it persist and be replicated across individuals and generations? Again, the functionality of the innovation plays a role in its future perseverance, but persistence must also depend on opportunity, ability and motivation (e.g., Hoppitt, Samson, Laland, & Thornton, 2012). It may survive through dissemination, diffusion or imitation, or alternatively it may spread through drift, and change in non-functional ways over time across a population (producing cultural variants; e.g., Rutz & St Clair, 2012). It can be suggested that most innovations are unique events that will simply become extinct leaving no technological or behavioral trace. Thus innovation, or what Ramsey et al. (2007) called an improvisation, may actually be far more common than we think, but with a legacy in the form of a distributed and learned behavior that is only rarely spread and established in populations as a function of local social or ecological conditions (Reader

& Laland, 2003). We suggest that the twofold conditions of importance for understanding innovations remain individual propensities for experimentation and the dynamic and unpredictable social, demographic or ecological contexts within which these occur.

Acknowledgments

We would like to thank Kevin Laland for comments, critique, and very useful feedback.

References

Addessi, E., Paglieri, F , & Focaroli, V. (2011). The ecological rationality of delay tolerance: insights from capuchin monkeys. *Cognition, 119,* 142–147.

Aplin, L. M., Sheldon, B. C., & Morand-Ferron, J. (2013). Milk bottles revisited: social learning and individual variation in the blue tit, *Cyanistes caeruleus. Animal Behaviour, 85,* 1225–1232.

Aureli, F., & de Waal, F. B. (Eds.), (2000). *Natural conflict resolution* Berkeley, CA: University of California Press.

Bates, L. A., Handford, R., Lee, P. C., Njiraini, N., Poole, J. H., Sayialel, K., et al. (2010). Why do African elephants (*Loxodonta africana*) simulate oestrus? An analysis of longitudinal data. *PLoS One, 5,* e10052.

Bates, L. A., Lee, P. C., Njiraini, N., Poole, J. H., Sayialel, K., Sayialel, S., et al. (2008). Do elephants show empathy? *Journal of Consciousness Studies, 15,* 204–225.

Benson-Amram, S., & Holekamp, K. E. (2012). Innovative problem solving by wild spotted hyenas. *Proceedings of the Royal Society, Series B, 279,* 4087–4095.

Bird, C. D., & Emery, N. J. (2009). Insightful problem solving and creative tool modification by captive nontool-using rooks. *Proceedings of the National Academy of Sciences, 106,* 10370–10375.

Boesch, C. (1991). Teaching among wild chimpanzees. *Animal Behaviour, 41,* 530–532.

Byrne, R. W. (2003). Novelty in deceit. In S. M. Reader, & K. N. Laland (Eds.), *Animal innovation.* Oxford: Oxford University Press.

Cambefort, J. P. (1981). A comparative study of culturally transmitted patterns of feeding habits in the chacma baboon *Papio ursinus* and the vervet monkey *Cercopithecus aethiops. Folia Primatologica, 36,* 243–263.

Chevalier-Skolnikoff, S., & Liska, J. (1993). Tool use by wild and captive elephants. *Animal Behaviour, 46,* 209–219.

Cords, M., & Aureli, F. (2000). Reconciliation and relationship qualities. In F. Aureli, & F. B. M. Waal (Eds.), *Natural conflict resolution* (pp. 177–198). Berkeley, CA: University of California Press.

Driver, P. M., & Humphries, D. A. (1988). *Protean behaviour.* Oxford: Oxford University Press.

Emidio, R. A., & Ferreira, R. G. (2012). Energetic payoff of tool use for capuchin monkeys in the Caatinga: variation by season and habitat type. *American Journal of Primatology, 74,* 332–343.

Flack, J. C., Girvan, M., de Waal, F. B., & Krakauer, D. C. (2006). Policing stabilizes construction of social niches in primates. *Nature, 439,* 426–429.

Foerder, P., Galloway, M., Barthel, T., Moore, D. E., III, & Reiss, D. (2011). Insightful problem solving in an Asian elephant. *PLoS One, 6,* e23251.

Fox, E. A., van Schaik, C. P., Sitompul, A., & Wright, D. N. (2004). Intra- and interpopulational differences in orangutan (*Pongo pygmaeus*) activity and diet: Implications for the invention of tool use. *American Journal of Physical Anthropology, 125*, 162–174.

Fragaszy, D., Izar, P., Visalberghi, E., Ottoni, E. B., & Gomes de Oliveira, M. (2004). Wild capuchin monkeys (*Cebus libidinosus*) use anvils and stone pounding tools. *American Journal of Primatology, 64*, 359–366.

Fragaszy, D. M., Biro, D., Eshchar, Y., Humle, T., Izar, P., Resende, B., et al. (2013). The fourth dimension of tool use: Temporally enduring artefacts aid primates learning to use tools. *Philosophical Transactions of the Royal Society, Series B, 368*, 20120410.

Fragaszy, D. M., Visalberghi, E., & Fedigan, L. M. (2004). *The complete capuchin: The biology of the genus Cebus*. Cambridge: Cambridge University Press.

Franz, M., & Nunn, C. L. (2009). Network-based diffusion analysis: a new method for detecting social learning. *Proceedings of the Royal Society Series B, 276*, 1829–1836.

Frith, C. B., & Frith, D. W. (2004). *The bowerbirds*. Oxford: Oxford University Press.

Gentry, A. H. (1982). Patterns of neotropical plant species diversity. *Evolutionary Biology, 6*, 1–84.

Goodall, J. (1986). *The chimpanzees of Gombe: Patterns of behavior*. Harvard, MA: Harvard University Press.

Gruber, T., Potts, K. B., Krupenye, C., Byrne, M. R., Mackworth-Young, C., McGrew, W. C., et al. (2012). The influence of ecology on chimpanzee (*Pan troglodytes*) cultural behavior: A case study of five Ugandan chimpanzee communities. *Journal of Comparative Psychology, 126*, 446–457.

Hoppitt, W., Boogert, N. J., & Laland, K. N. (2010). Detecting social transmission in networks. *Journal of Theoretical Biology, 263*, 544–555.

Hoppitt, W., & Laland, K. N. (2011). Detecting social learning using networks: A user's guide. *American Journal of Primatology, 73*, 834–844.

Hoppitt, W., Samson, J., Laland, K. N., & Thornton, A. (2012). Identification of learning mechanisms in a wild meerkat population. *PLoS One, 7*, e42044.

Joffe, T. H. (1997). Social pressures have selected for an extended juvenile period in primates. *Journal of Human Evolution, 32*, 593–605.

Kuczaj, S. A., Gory, J. D., & Xitco, M. J. (2009). How intelligent are dolphins? A partial answer based on their ability to plan their behavior when confronted with novel problems. *Japanese Journal of Animal Psychology, 59*, 99–115.

Kummer, H., & Goodall, J. (1985). Conditions of innovative behaviour in primates. *Philosophical Transactions of the Royal Society, Series B, 308*, 203–214.

Lee, P. C. (1991). Adaptations to environmental change by primates: An evolutionary perspective. In H. O. Box (Ed.), *Primate responses to environmental change* (pp. 39–56). London: Chapman & Hall.

Lee, P. C. (2003). Innovation as a behavioural response to environmental challenges. In S. Reader, & K. Laland (Eds.), *Animal innovation* (pp. 261–278). Oxford: Oxford University Press.

Lefebvre, L. (2000). Feeding innovations and their cultural transmission in bird populations. In C. M. Heyes, & L. Huber (Eds.), *The evolution of cognition* (pp. 211–328). New York: MIT Press.

Liu, Q., Simpson, K., Izar, P., Ottoni, E., Visalberghi, E., & Fragaszy, D. (2009). Kinematics and energetics of nut-cracking in wild capuchin monkeys (*Cebus libidinosus*) in Piauí, Brazil. *American Journal of Physical Anthropology, 138*, 210–220.

Lubin, Y. D., & Montgomery, G. G. (1981). Defences of *Nasutitermes termites* (Isoptera, Termitidae) against *Tamandua anteaters* (Edentata, Myrmecophagidae). *Biotropica, 13*, 66–76.

Marshall-Pescini, S., & Whiten, A. (2008). Chimpanzees (*Pan troglodytes*) and the question of cumulative culture: An experimental approach. *Animal Cognition, 11*, 449–456.

McGrew, W. C. (1992). *Chimpanzee material culture: Implications for human evolution.* Cambridge: Cambridge University Press.

Montenegro, M.V. (2011). *Ecologia de* Cebus flavius *(Schreber, 1774) em Remanescentes de Mata Atlântica no Estado da Paraíba* (PhD Thesis). Universidade de São Paulo.

Moura, A.C. de A. 2004. *The capuchin monkey and the Caatinga dry forest: A hard life in a harsh habitat* (PhD Thesis). University of Cambridge.

Moura, A. C. de A., & Lee, P. C. (2004). Capuchin stone tool use in Caatinga dry forest. *Science, 306*, 1909.

Moura, A. C. de A., & Lee, P. C. (2010). Wild capuchins show male-biased feeding tool use. *International Journal of Primatology, 31*, 457–470.

Nakamura, M. (2002). Grooming-hand-clasp in Mahale M group chimpanzees: Implications for culture in social behaviours. In C. Boesch, G. Hohmann, & L. Marchant (Eds.), *Behavioural diversity in chimpanzees and bonobos* (pp. 71–83). Cambridge: Cambridge University Press.

Ottoni, E. B., & Izar, P. (2008). Capuchin monkey tool use: Overview and implications. *Evolutionary Anthropology, 17*, 171–178.

Perry, S. (2011). Social traditions and social learning in capuchin monkeys (*Cebus*). *Philosophical Transactions of the Royal Society, Series B, 366*, 988–996.

Perry, S., Panger, M., Rose, L. M., Baker, M., Gros-Louis, J., Jack, K., et al. (2003). Traditions in wild white-faced capuchin monkeys. In D. Fragaszy, & S. Perry (Eds.), *The biology of traditions: Models and evidence* (pp. 391–425). Cambridge: Cambridge University Press.

Plotnik, J. M., & de Waal, F. B. (2014). Asian elephants (*Elephas maximus*) reassure others in distress. *PeerJ, 2*, e278.

Ramsey, G., Bastian, M. L., & van Schaik, C. P. (2007). Animal innovation defined and operationalized. *Behavioral and Brain Sciences, 30*, 393–407.

Reader, S. M., & Laland, K. N. (2001). Primate innovation: Sex, age and social rank differences. *International Journal of Primatology, 22*, 787–805.

Reader, S. M., & Laland, K. N. (Eds.), (2003). *Animal innovation* Oxford: Oxford University Press.

Riemer, S., Müller, C., Virányi, Z., Huber, L., & Range, F. (2013). Choice of conflict resolution strategy is linked to sociability in dog puppies. *Applied Animal Behaviour Science, 149*, 36–44.

Rutz, C., & St Clair, J. J. H. (2012). The evolutionary origins and ecological context of tool use in New Caledonian crows. *Behavioural Processes, 89*, 153–165.

Santos, L. R., Hauser, M. D., & Spelke, E. S. (2001). Recognition and categorization of biologically significant objects by rhesus monkeys (*Macaca mulatta*): The domain of food. *Cognition, 82*, 127–155.

Slater, P. J. B. (1986). The cultural transmission of bird song. *Trends in Ecology & Evolution, 1*, 94–97.

Sol, D. (2003). Behavioural flexibility: A neglected issue in the ecological and evolutionary literature. In S. M. Reader, & K. N. Laland (Eds.), *Animal innovation* (pp. 63–82). Oxford: Oxford University Press.

Souto, A., Bione, C. B. C., Bastos, M., Bezerra, B. M., Fragaszy, D., & Schiel, N. (2011). Critically endangered blonde capuchins fish for termites and use new techniques to accomplish the task. *Biology Letters, 7*, 532–535.

Spagnolleti, N., Visalbergh, E., Verderane, M. P., Ottoni, E., Izar, P., & Fragaszy, D. (2012). Stone tool use in wild bearded capuchin monkeys, *Cebus libidinosus*. Is it a strategy to overcome food scarcity?. *Animal Behaviour, 83*, 1285–1294.

Spinka, M., Newberry, R. C., & Bekoff, M. (2001). Mammalian play: Training for the unexpected. *Quarterly Review of Biology, 81*, 141–168.

Tebbich, S., Taborsky, M., Fessl, B., & Dvorak, M. (2002). The ecology of tool-use in the woodpecker finch (*Cactospiza pallida*). *Ecology Letters, 5*, 656–664.

Thorne, B. (1980). Differences in nest architecture between the arboreal termites *Nasutitermes corniger* and *Nasutitermes ephratae* (Isoptera: Termitidae). *Psyche, 87,* 235–243.

Thornton, A., & Samson, J. (2012). Innovative problem solving in meerkats. *Animal Behaviour, 83,* 1459–1468.

Valone, T. J. (2007). From eavesdropping on performance to copying the behavior of others: A review of public information use. *Behavioral Ecology and Sociobiology, 62,* 1–14.

van de Waal, E., Borgeaud, C., & Whiten, A. (2013). Potent social learning and conformity shape a wild primate's foraging decisions. *Science, 340,* 483–485.

van Schaik, C. P. (2003). Local traditions in orangutans and chimpanzees: Social learning and social tolerance. In D. Fragaszy, & S. Perry (Eds.), *The biology of traditions: Models and evidence* (pp. 297–328). Cambridge: Cambridge University Press.

van Schaik, C. P., Deaner, R. O., & Merrill, M. Y. (1999). The conditions for tool use in primates: implications for the evolution of material culture. *Journal of Human Evolution, 36,* 719–741.

van Schaik, C. P., & Pradham, G. R. (2003). A model for tool-use traditions in primates: Implications for the coevolution of culture and cognition. *Journal of Human Evolution, 44,* 645–664.

Vasconcellos, A. (2010). Biomass and abundance of termites in three remnant areas of Atlantic Forest in northeastern Brazil. *Revista Brasileira Entomologia, 54,* 455–461.

Visalberghi, E., Fragaszy, D., Ottoni, E., Izar, P., de Oliveira, M. G., & Andrade, F. R. D. (2007). Characteristics of hammer stones and anvils used by wild bearded capuchin monkeys (*Cebus libidinosus*) to crack open palm nuts. *American Journal of Physical Anthropology, 132,* 426–444.

Whitehead, H. (1998). Cultural selection and genetic diversity in matrilineal whales. *Science, 282,* 1708–1711.

Wood-Gush, D. G. M., & Vestergaard, K. (1991). The seeking of novelty and its relation to play. *Animal Behaviour, 42,* 599–606.

Yamakoshi, G. (1998). Dietary response to fruit scarcity of wild chimpanzees at Bossou, Guinea: Possible implications for ecological importance of tool use. *American Journal of Physical Anthropology, 106,* 283–295.

Commentary on Chapter 11: Necessity, Unpredictability, Opportunity, and Creativity

In their thought-provoking chapter, Lee and Moura (2014) emphasize the important role that innovation plays in helping animals solve existing ecological problems. In particular, the authors discuss the influence of necessity, unpredictability, and opportunity in fostering creative problem-solving among nonhuman species. Past research examining similar questions in humans suggests that necessity can indeed be the mother of invention, as immediate problems and needs provide important occasions and motivational impetus for creative thinking (e.g., Grant & Berry, 2011; Roskes, De Dreu, & Nijstad, 2012).

Taking the case of necessity to its extreme, we have been interested in examining the degree to which adverse, highly stressful, and challenging life events are associated with adaptive or positive changes (including changes in creativity), a phenomenon referred to as *growth through adversity* or *post-traumatic growth* in the psychological literature (Jayawickreme & Blackie, 2014; Tedeschi & Calhoun, 1995). Preliminary research showed that among a sample of English-speaking adults, feelings of distress associated with experiences of adversity predicted perceptions of increased creativity; these perceptions were also associated with a more general tendency to notice new possibilities for one's life, as well as both positive and negative changes in interpersonal relationships (Forgeard, 2013a). Thus, individuals in this sample commonly discerned a link between their suffering and their ability to generate novel and useful ideas in various areas of life (e.g., at work, interpersonally, during artistic activities). In addition, individuals high in the personality trait of openness to experience who experienced high levels of distress were more likely than others to report perceptions of increased creativity (Forgeard, 2013b), echoing Lee and Moura's proposition that "individuals with flexible personality traits, those of openness or boldness, may be more prone to risk-taking and thus experimentation regardless of the costs or consequences."

One of the many interesting strengths of the research described by Lee and Moura is the necessary reliance on behavioral observation methods to detect creative problem-solving in nonhuman species. In humans, research dedicated to examining growth through adversity has for the most part used self-report methods. While understanding how humans subjectively construe changes in their lives is important, recent scholarship suggests that investigating whether and how perceptions of change translate into behavioral manifestations may provide a more nuanced and comprehensive understanding of the effect of adversity on human functioning (Jayawickreme & Blackie, 2014). Such research can help scientists tease out the nature and function of perceptions of increased creativity following adversity. These perceptions may reflect changes in behaviors (including the ability to come up with creative cognitions and products), and beliefs used to cope with the situation, which may or may not correspond to objective changes in creative cognition (Forgeard, Mecklenburg, Lacasse, & Jayawickreme, 2014; Taylor, 1983).

Research investigating these questions can also help shed light on the universality of this phenomenon, and more generally, the circumstances under which necessity may stimulate human creativity. We briefly discuss here one context in human history (among many others) which presents challenges paralleling those discussed by Lee and Moura (2014)—the experience of refugee populations. The refugee

III. THE STRUGGLE FOR CREATIVITY

experience is part of the wider history of human migration, which has stressed the important and positive role of this process in the growth and evolution of human civilization, while at the same time acknowledging the tremendous challenges, suffering, and losses associated with this experience. As Jablensky et al. (1994, p. 327) explained:

> Throughout human history, individual and group migration has played an important role in the social evaluation of human society by contributing to cultural and biological diversity. This diversity has permitted human beings to thrive in the face of adversity by providing an ever-changing set of perspectives and solutions to pressing environmental demands.

Forced population movements have been a commonplace practice since the beginning of human history, as seen in the expulsion of Moses and his followers from Egypt and the "social death" of banishment in classical Greece. Interesting, the phenomenon of being "cast-out" of one's community is even seen in monkeys such as the rhesus macaque (Maestripieri, 2007). The concept of the "refugee" as a social category and global legal problem was however only formed in the wake of World War II, when more than seven million Europeans were unable to return to their homelands (Malkki, 1995).

Given the challenges faced by refugees, scientists can examine predictors and manifestations of innovation and creativity in these populations. Preliminary historical evaluations suggest a wide range of outcomes reflecting different ways to react to stressful and ever-changing political circumstances including returning to the home country, completely assimilating in a new country, continuing to live in refugee settlements, adopting a nomadic existence, or founding a new country (Jayawickreme, 2010). Further historical and psychological scholarship can help determine the degree to which individuals' ability to flexibly adapt is influenced by environmental circumstances, pre-existing individual dispositions, and how these interact to shape creative behavior and cognition.

Thus, the work described by Lee and Moura has important parallels and implications for the study of creativity and innovation in humans. It is important to note here that such investigations do not seek to show that adversity or necessity is "good," and all efforts to eradicate the circumstances producing human suffering are of course of paramount importance. Given this caveat, the perspectives and methods used by researchers interested in the creativity of nonhuman species may provide important inspiration and information for psychological scientists dedicated to understanding how human suffering, if inevitable, can be harnessed towards adaptive change.

References

Forgeard, M. J. C. (2013a). Perceiving benefits after adversity: The relationship between self-reported post-traumatic growth and creativity. *Psychology of Aesthetics, Creativity, and the Arts, 7*, 245–264.

Forgeard, M. J. C. (2013b). The role of openness to experience in growth through adversity. In E. Jayawickreme (Chair), *The most consequential trait? New directions in openness to experience research. Symposium conducted at the biennial conference of the Association for Research in Personality*. Charlotte, CT.

Forgeard, M. J. C., Mecklenburg, A. C., Lacasse, J. J., & Jayawickreme, E. (2014). "Bringing the whole universe to order:" Creativity, healing, and post-traumatic growth. In J. C. Kaufman (Ed.), *New ideas about an old topic: Creativity and mental illness* (pp. 321–342). New York, NY: Cambridge University Press.

Grant, A. M., & Berry, J. W. (2011). The necessity of others is the mother of invention: Intrinsic and prosocial motivations, perspective taking, and creativity. *Academy of Management Journal, 54*, 73–96.

Jablensky, A., Marsella, A. J., Ekbald, S., Jansson, B., Levi, L., & Bornemann, T. (1994). Refugee mental health and well-being: Conclusions and recommendations. In A. J. Marsella, et al. (Eds.), *Amidst peril and pain: The mental health and well-being of the world's refugees* (pp. 327–340). Washington, DC: American Psychological Association.

Jayawickreme, E. (2010). *Well-being, growth and war: Mental health in war-affected regions of Sri Lanka* (Doctoral dissertation). Retrieved from ProQuest Dissertations and Theses (Accession Order No. AAI3447162).

Jayawickreme, E., & Blackie, L. E. R. (2014). Post-traumatic growth as positive personality change: Evidence, controversies and future directions. *European Journal of Personality, 28*, 312–331.

Lee, P. C., & Moura, A. C. de A. (2014). Necessity, unpredictability and opportunity: An exploration of ecological and social drivers of behavioral innovation. In A. B. Kaufman, & J. C. Kaufman (Eds.), *Animal creativity and innovation*. Waltham, MA: Academic Press.

Maestripieri, D. (2007). *Machiavellian intelligence: How rhesus macaques and humans Have conquered the world*. Chicago: University of Chicago Press.

Malkki, L. H. (1995). *Purity and exile: Violence, memory, and national cosmology among Hutu refugees in Tanzania*. Chicago: University of Chicago Press.

Roskes, M., De Dreu, C. K., & Nijstad, B. A. (2012). Necessity is the mother of invention: Avoidance motivation stimulates creativity through cognitive effort. *Journal of Personality and Social Psychology, 103*, 242–256.

Taylor, S. E. (1983). Adjustment to threatening events: A theory of cognitive adaptation. *American Psychologist, 38*, 1161–1173.

Tedeschi, R. G., & Calhoun, L. G. (1995). *Trauma and transformation: Growing in the aftermath of suffering*. Thousand Oaks, CA: Sage.

Cognitive and Noncognitive Aspects of Social Learning

Thomas R. Zentall

Department of Psychology, University of Kentucky, Lexington, KY, USA

Commentary on Chapter 12: Imitation and Creativity

John Baer

Rider University, Education Department, Lawrence Township, NJ, USA

It might seem paradoxical to propose that social learning is at the heart of innovation, because learning from others implies the copying of existing behavior. But in general, humans can be creative only in domains that they know quite well (Gardner, 1993). To be creative, one must have a well-developed base of knowledge (Sternberg & Lubart, 1991). That is, one must first become an expert in a field before one can make a creative contribution. As Louis Pasteur observed, "chance favors only the prepared mind" (Vallery-Radot, 1916). But how is that expertise acquired? Historically it has been through education and apprenticeship; i.e., social learning. Thus, the ability to learn from the behavior of others, social learning, is a necessary prerequisite of innovation. As I will make clear, social learning itself consists of several different psychological mechanisms, the most complex of which is true imitation or what Piaget (1962) interpreted as perspective taking, the ability to imagine oneself as others would see one.

335

Social learning can be defined as the degree to which an organism is more likely to acquire a response from observation of the performance of that response by an organism of the same species (a conspecific) than on its own. This simple definition belies the complexity involved in the psychological analysis of the phenomenon. Depending on the goal of the researcher social learning has been of interest for two quite different reasons.

Behavioral ecologists have reasoned that social learning occupies an important position between genetically predisposed behavior and trial and error leaning (Boyd & Richerson, 1988). Genetically predisposed behavior has the advantage of being produced reliably when appropriate, based on the demands of the environment (e.g., nest building by birds) but it suffers from being relatively inflexible when environments change in unpredictable ways. On the other hand, trial and error learning can allow an animal to adapt to a rapidly changing environment. It has the advantage of being quite flexible (e.g., learning by rats of the consequences of eating new foods) but it suffers from the potentially lethal consequences of making errors (e.g., the possibility of consuming poison). Social learning may occupy an important middle ground. It can have much of the flexibility of trial and error learning by benefitting from the trial and error learning of others but without the potential negative consequences of the social learner making errors. Thus, behavioral ecologists recognize the adaptive value of social learning and are interested in which species can acquire behavior by social means. For these researchers, the means by which the information is transmitted is not as important as the degree to which the information is useful (for survival or reproductive success). For example, if a fish finds itself on one side of a transparent partition with food on the other side, following another fish to the food side through a small hole in the partition is sufficient to demonstrate the value to the observing fish of the social context, if later the original observing fish will swim through the hole on its own (Laland & Williams, 1997).

Interestingly, an approach that focuses on adaptive value would predict that there are conditions under which social learning would have less value than individual (trial and error) learning (Laland, 2004). For example, social learning might be less valuable when individual learning is not costly (errors do not have serious negative consequences), when the current behavior of the observer is productive, and when outcomes are certain. Furthermore, this approach predicts that the identity of the model is likely to be important. Thus, social learning is more likely to occur when the model is related to the observer, when the model is successful, or when the model is older (see Rendell et al., 2011, for additional conditions under which social learning would be advantageous over individual learning).

Psychologists, on the other hand, have been more interested in the mechanisms by which social learning occurs. When the acquisition of a response is facilitated by observation of that response by another animal, several possible mechanisms may be involved. Given the above scenario of following by fish, a psychologist would typically be interested in *how* the observer fish learned to get to the food. Did the observer have a natural tendency to affiliate with others and thus, follow others through the opening in the partition? If it did, it may have learned incidentally where the hole was and that the other side was safe. Alternatively, the psychologist might ask whether learning would occur if the observer was not permitted to follow immediately. Would it still learn about the opening by attending to the other fish as they passed through the opening such that it could find the opening at a later time? This would suggest more than a following response accompanied by incidental learning. But even if such learning could be demonstrated, the psychologist would want to distinguish between learning facilitated by local enhancement (attention to a location) or stimulus enhancement (attention to an object, the hole) and learning to copy the behavior of demonstrated (imitation).

But why does a psychologist care about the nature of the process by which the observer learns? To answer this question one must see how imitation has been viewed in the human psychological literature. For example, imitation by young children has been viewed by some as an instrumental conditioning process in which behavior that happens to match the behavior of a model is selectively reinforced (e.g., Horne & Erjavec, 2007; Zukow-Goldring & Arbib, 2007), and by others as a cognitive process reflecting an understanding or assimilation of the relation between one's own body parts and those of others (e.g., Guillaume, 1926/1971; Mitchell, 1987). For example, imagine an adult human walking with his hand clasped behind his back and being followed by a young child with his hands also clasped behind his back. According to this view, imitation can be an intentional, conscious process involving the ability to take the perspective of another (Whiten, 2000). Kurdek and Rodgon (1975) call this kind of process *perceptual perspective taking*, especially when the observer cannot easily see itself performing the behavior. When imitation is this of kind, it is has been referred to as imitation of invisible actions (Piaget, 1962) or as opaque imitation (Heyes, 2002).

If it can be shown that animals are capable of such imitation, it might be argued that those animals have a relatively advanced representational system. However, it should be noted that a simpler learning model, that will be discussed later, has been proposed to account for opaque imitation (Heyes & Ray, 2000; Ray & Heyes, 2011).

Another view of the cognitive implications of imitation has been proposed by Bandura (1969), who noted a difference between simple imitation, the copying of behavior occurring at about the same time as it is observed, and deferred imitation or observational learning, in which performance of the observed behavior occurs at a later time. Bandura's distinction is based on the premise that for one to be able to defer imitation requires one to have a representation of the observed behavior that can be retrieved and performed at a later time when the model is no longer present. The potential value of deferred imitation is that it can function as a tool that can be used flexibly at an appropriate time after observation of the behavior. Of course, in a sense, all imitation can be thought of as deferred because it always occurs after observation, but a useful if not precise distinction can be made between a delay that is no longer than working memory (seconds) and one that is minutes long or longer (Cowan, Wood, Nugent, & Treisman, 1997).

Bandura's distinction between deferred imitation and imitation that occurs at or about the time of observation, what Byrne and Russon (1998) have referred to as response facilitation, rests on the assumption that immediate imitation involves a simpler process. However, referring to imitation as response facilitation implies that the ability to copy the behavior of another is automatic and reflexive. But describing the immediate copying of behavior as reflexive fails to explain how the observation of the model's behavior translates into the performed behavior by the observer, especially if the observer's own behavior is opaque. If one is interested in how organisms understand and reproduce observed behavior, distinguishing among the various contributions to leaning from others is a meaningful endeavor.

Several reviews of observational learning have appeared in the past 20 years including those by Galef (1988b), Whiten and Ham (1992), Zentall (1996), Whiten, Horner, Litchfield, and Marshall-Pescini (2004), Huber et al. (2009), and Hoppitt and Laland (2008). In the present review, I will attempt not only to distinguish among the various kinds of social influence, social learning, and learning by nonsocial means (e.g., emulation), but I will also to examine several variables that appear to be important in determining whether a particular behavior will be copied. In addition, I will attempt to identify several kinds of complex social learning that may be unique to humans.

In examining the literature on social learning, I will start with several examples of social influence in which behavior is influenced by the presence of others but for which either learning is not involved or learning is a side effect of the observed behavior. I will then consider several examples of simple social learning in which the other animal plays an important role in facilitating the same behavior in the observing animal but associative learning processes by the observer are sufficient to account for the learning. Finally, I will address the possibility that under certain conditions animals may perform an observed behavior that requires a more cognitive explanation. Such a process, by which the behavior of others is translated into one's own behavior, may not involve what Piaget (1962) referred to as perspective taking nor what Bandura (1969) viewed as a case of mental representation. Yet it would be difficult to account for such behavior as predisposed, induced by increased motivation, enhanced by attention, or readily subsumed under the rubric of trial and error learning involving a social stimulus. Although the processes involved in these more complex examples of imitation may be difficult to disentangle from simpler motivational, attentional, and trial and error processes because in nature they do not typically occur in isolation, they do have theoretical implications for the conceptual capacities of animals.

SOCIAL INFLUENCE

Social influence occurs when behavior is influenced by the presence of other members of the same species that is part of the animals species typical behavior (contagion), that affects the motivation of the observing animal (mere presence), or that causes the observing animal to direct its attention to a place or object (local or stimulus enhancement).

Contagion

Predisposed tendencies to match specific behaviors of a conspecific are often referred to as contagious behaviors, mimesis, response facilitation, or response priming. Contagion can be used to describe certain elicited courtship displays, antipredator behavior (such as mobbing), and social eating. These behaviors are often reflexive, species-typical responses to the behavior of another animal. For example, a chicken that is provided with food until it has eaten its fill and has stopped eating will often begin eating again if one introduces a hungry chicken that begins eating (Tolman, 1964). In this case, the behavior of one animal appears to serve as a releaser for the natural unlearned behavior of others (Thorpe, 1963) and is unrelated to the copying of more instrumental behavior.

Motivational Influences

Social factors that affect the general arousal or motivation of an animal may affect its general activity, which may in turn affect the probability that it will make a matching response.

Social Facilitation/Social Enhancement

The mere presence of another animal, irrespective of that animal's behavior, may increase (or decrease) arousal, a phenomenon called social facilitation or social enhancement. Increased arousal can lead to increased activity, leading to increased contact with environmental contingencies. For example, if the presence of another animal increases a rat's general exploratory activity, that rat may discover (on its own) a lever that when pressed leads to reinforcement (Zajonc, 1965). Alternatively, in a novel environment, the presence of a conspecific may lead to a decrease in fear which may lead to a decrease in arousal and increased general exploratory activity (Moore, Byers, & Baron, 1981) or possibly following behavior.

Incentive Motivation

If the observer's behavior is reinforced during the demonstration of the response, observation of the reinforcer may play a role in the rate at which the response is acquired by the observer by way of incentive motivation (see, Caldwell & Whiten, 2003). That is, if an animal observes inaccessible reinforcement or sees another animal eating, it may increase incentive motivation that may lead to an increase in activity sufficient to explain facilitated learning.

Observation of Aversive Conditioning

The copying of a response being acquired or being performed by a demonstrator that is motivated by the avoidance of painful stimulation (e.g., electric shock) may result from the induction of motivation in the observer. Emotional cues of pain or fear provided by a conspecific, either escaping from or avoiding shock, may instill fear in the observer. For example, John, Chesler, Bartlett, and Victor (1968) found that cats that had observed a demonstrator being trained to jump over a hurdle to avoid foot shock learned the hurdle-jumping response faster than controls that did not observe the demonstrators. It may be, however, that being in the presence of a cat being shocked was sufficient to increase the observers' fear (motivation) associated with the conditioning context. Under such conditions, the increase in motivation may account for the facilitated acquisition.

Perceptual Factors

When the observation of a demonstrator draws attention to a response (e.g., a lever press), it may alter the salience of the lever (stimulus enhancement) or the place where the lever is located (local enhancement).

Local Enhancement

Local enhancement refers to the facilitation of learning that results from drawing attention to a locale or place associated with reinforcement (Roberts, 1941). For example, Lorenz (1935) found that ducks enclosed in a pen did not react to a hole in the pen that was large enough for them to escape, unless they happen to be near another duck as it was escaping. The sight of a duck passing through the hole in the pen may merely draw attention to the hole. Thus, one could ask, for example, if observing a ball roll through the hole would produce a similar effect. If so, it would not need to be interpreted as social learning.

Local enhancement may also be involved in John's et al. (1968, Exp. 1) finding of facilitated acquisition of an aversively motivated hurdle jump response. Attributing the matching behavior to local enhancement in this case may not be obvious, but observation of the demonstrator jumping over the hurdle may draw the observer's attention to the top of the hurdle. In other words, it might be sufficient to see a ball bounce over the hurdle to find facilitation of the hurdle jumping response. Similarly, an attentional mechanism may be responsible for the rapid mastery of a V-shaped-fence-detour problem when demonstrated by a human (Pongrácz et al., 2001). Seeing the demonstrator pass around the end of the fence may draw the observer's attention to the place where it can gain access to the goal on the other side. In general, whenever the performance observed involves an object (e.g., a manipulandum, a hurdle, a barrier) to which the observer must later respond, local enhancement may play a role and appropriate control conditions should be included.

Although performance by the demonstrator may draw the attention of the observer to a location, the outcome of the demonstrator's behavior may also play a role in the observer's tendency to copy. For example, Lefebvre and Palameta (1988) found that pigeons that observed a model pierce the paper cover on a food well to obtain hidden grain, later acquired that response on their own, whereas those that observed that same response but with no grain in the well (the model performed in extinction), failed to acquire the response. In this case the observed behaviors by the two groups were quite similar but the consequences of the observed behavior were quite different (see also Akins & Zentall, 1998). Although one might be inclined to interpret this result cognitively, as the observer's understanding of the consequences of the demonstrator's behavior, a Pavlovian conditioning account in terms of the pairing of attention to a location and the appearance of a reinforcer may be sufficient to explain the more rapid acquisition of the target behavior by the observer when the demonstrator's behavior is reinforced. This point will be addressed again shortly.

Stimulus Enhancement

In the case of local enhancement, the attention of an observer is drawn to a particular *place* by the activity of the demonstrator. The term stimulus enhancement is used when the activity of the demonstrator draws the attention of the observer to a particular *object* (e.g., a manipulandum). Quite often in the study of imitative learning, the object in question is at a fixed location so local enhancement and stimulus enhancement are indistinguishable. In the duplicate-chamber procedure (see Gardner & Engel, 1971; Warden & Jackson, 1935), however, a manipulandum (e.g., a lever) is present in both the demonstration chamber and in the observation chamber. Under these conditions, drawing attention to the demonstrator's lever might not be expected to enhance the observer's lever-pressing behavior. In fact, one could argue that it should retard acquisition of lever pressing by the observer because it should draw the observer's attention away from its own lever. However, the similarity between the demonstrator's lever and that of the observer may make it more likely that the observer would notice its own lever after having its attention drawn to the demonstrator's lever. Thus, stimulus enhancement can refer to the combination of a perceptual, attention-getting process resulting from the activity of the demonstrator in the presence of the lever and stimulus generalization between the demonstrator's and observer's levers. Because it subsumes the effects of local enhancement, the term stimulus enhancement may be more inclusive and thus, is often preferred (Galef, 1988b).

Stimulus enhancement may also play a role in mate-choice copying by animals (Dugatkin, 1996; Galef, Lim, & Gilbert, 2008). For example, female guppies that see a demonstrator or model female in the presence of a courting male will prefer that male over an alternative male (Dugatkin, 1992; Dugatkin & Godin, 1992). In this case, it may not be possible to separate the observer's attention to the model female, and therefore to the male nearby, from the putative attraction of the model to the adjacent male.

The facilitation of learning through perceptual factors presents a difficult problem for the study of imitation in animals. If the similarity between the demonstrator's location or the demonstrator's manipulandum and that of the observer presents an interpretational problem because of perceptual factors, making the location or the nature of the manipulandum for the observer different from that of the demonstrator is likely to interfere with the observer's potential interpretation of the relation between the two tasks. This problem, which will be addressed later, requires a new approach to defining adequate control procedures.

SIMPLE LEARNING IN THE CONTEXT OF SOCIAL CUES

A number of cases of learning in a social context may be mediated by simple nonsocial learning mechanisms. Although social stimuli are present and those social stimuli may play a role in facilitating acquisition of the target behavior (perhaps because often social stimuli are more salient than nonsocial alternatives), the processes by which the observer acquires the behavior may be more parsimoniously explained in terms of simpler individual (trial and error) learning processes.

Discriminated Following (or Matched Dependent) Behavior

Perhaps the clearest case of learning in a social context for which the behavior is likely to involve simple associations is when the observer is reinforced for following the model. For example, rats can learn to follow a trained conspecific to food in a T maze in the absence of any other discriminative stimulus (Haruki & Tsuzuki, 1967). Although the leader rat in these experiments is clearly a social stimulus, the data are more parsimoniously interpreted in terms of simple discriminative learning. If, for example, the demonstrator were replaced with a block of wood pulled along by a string, or even an arrow at the choice point, directing the rat to turn left or right, it is clear that one would identify the cue (i.e., the demonstrator, the block of wood, or the arrow) as a simple discriminative stimulus. Even if following a demonstrator led to faster learning than following a passive signal, it might indicate merely that the social cue was more salient than either a static or a moving nonliving cue.

Observational Conditioning

As noted in the section on local enhancement, the observation of a performing demonstrator may not merely draw attention to the object being manipulated (e.g., the lever), but because the observer's interaction with the object is often followed immediately by presentation of food to the demonstrator, a Pavlovian association may be established. This form of conditioning has been called observational conditioning (Whiten & Ham, 1992) or valence transformation (Hogan, 1988) and it occurs when the observer learns the relation between some event in the environment and the reinforcer (e.g., a rat approaches a lever that has appeared shortly before the demonstrator—that has pressed the lever—has been fed). Although such conditioning would have to take the form of higher-order conditioning (because the observer would not actually experience the unconditional stimulus), there is evidence that such higher-order conditioning can occur even in the absence of a demonstrator. For example, pigeons that were presented with a localized light, followed shortly by the presentation of inaccessible grain, spontaneously initiated pecking to the light (Zentall & Hogan, 1975). The presence of a demonstrator drawing additional attention to the light (by pecking at it) and to the reinforcer (by eating) may further enhance associative processes in the absence of imitative learning.

With regard to the nature of the conditioning process, it is of interest that when reinforcement of the demonstrator's response cannot be observed (or the response-reinforcer association is difficult to make) acquisition may be impaired (Akins & Zentall, 1998; Heyes, Jaldow, & Dawson, 1994). Furthermore, rats appear to acquire a lever-pressing response faster following observation of a lever-pressing demonstrator if they are fed at the same time as the performing demonstrator (Del Russo, 1971). Although that result was mentioned earlier in the context of increased motivation on the part of the observer, it is also possible that feeding the observer following the demonstrator's response may result in simple Pavlovian conditioning (i.e., the pairing of movement of the lever with food).

Socially-transmitted food preferences (e.g., Galef, 1988a; Strupp & Levitsky, 1984) represent a special case of observational conditioning. Although food preference may appear to fall into the category of unlearned behavior, subject to elicitation through contagion, consuming food with a *novel* taste should be thought of as an acquired behavior. The mechanisms responsible for socially-acquired food preferences appear to have strong simple associative learning components (e.g., learned safety or the habituation of neophobia to the novel taste), for which the presence of a conspecific may serve as a catalyst. Furthermore, these specialized mechanisms may be unique to foraging and feeding systems.

One of the best examples of observational conditioning is in the acquisition of fear of snakes by laboratory-reared monkeys exposed to a wild-born conspecific in the presence of a snake (Mineka & Cook, 1988). Presumably, the fearful conspecific serves as the unconditioned stimulus and the snake serves as the conditioned stimulus. This interpretation is supported by the finding that exposure to a fearful conspecific or to a snake alone is insufficient to produce fear of snakes in the observer. Interestingly, not all stimuli are as easily associated with a fear response. For example, a fearful wild-born conspecific in the presence of a flower is not sufficient for observational conditioning. For an excellent discussion of the various forms of observational conditioning, see Heyes (1994).

Goal-Emulation, Object-Movement-Reenactment, and Emulation via Affordance Learning

When subjects learn about aspects of their environment and use this information to achieve their own goals, it may not involve demonstration by another organism. Instead, the learning can be defined in terms of the occurrence of events in the environment that typically lead to reinforcement or the products of the behavior of a demonstrator. Although the terms emulation, end-state emulation, goal-emulation, object-movement-reenactment, and emulation via affordance learning cannot always be clearly differentiated, they have been used in somewhat different contexts (Hopper, 2010; Huang, Heyes, & Charman, 2006).

End-state or *outcome emulation* is used when the presence of an outcome motivates an observer to replicate the result. An observer may see a demonstrator obtain food (by making a response) and thus, the observer may be induced to explore because it is motivated to obtain food itself. The induced exploration may be what facilitates making the target response. *Goal emulation*, or more properly *outcome emulation*, has been used to describe, for example, an observer's attention to a tool that has been used to obtain a reward, but not exactly how that tool should be used (Tomasello, 1990). *Object-movement-reenactment* refers to copying what an object does (e.g., a door moves toward the animal or away from the animal to allow it to gain access to reinforcement), without regard to the specific actions of the demonstrator to move the door (Whiten et al., 2004). *Emulation* or *emulation via affordance learning* refers to learning how the environment works (Byrne, 1998). For example, learning that a door can be opened by seeing the door knob turn to the right and then seeing the door move away from the observer.

For purposes of the present review, the term emulation will be used to indicate learning about those changes in the environment that are necessary to obtain a goal, independent of the actions of a demonstrator. When observation of a demonstrator allows an animal to learn how the environment functions, a sophisticated form of learning may be involved. For example, if a pigeon observes a screen (that is capable of moving to the left or to the right) move to the left to allow access to food, when the pigeon is given access to the screen, it is more likely to move the screen in the same direction (Campbell, Heyes, & Goldsmith, 1999; Klein & Zentall, 2003). However, because learning how the environment works may occur in the absence of the behavior of another animal, one would not want to view such learning as social learning. Emulation resembles observational conditioning in the sense that both involve the relation between environmental events and an outcome, and the two may not be easy to disentangle (Byrne, 1998). The difference is that in observational conditioning the response to be accounted for is closely related to the unconditioned response to the reinforcer; for example, a pigeon pecking a light that has been followed by reinforcement or a rat approaching a lever, the appearance of which has been closely followed by the delivery of food. Emulation, on the other hand, could involve a more arbitrary, means-end, instrumental relation, such as learning that a swinging door swings out (rather than in) to allow the animal to obtain food on the other side (Bugnyar & Huber, 1997).

Emulation may be involved in a procedure used with chimpanzees in which they learn to open a box to obtain a reward (Whiten & Custance, 1996). In that study, some demonstrator chimpanzees were trained to poke a bolt to open the box, whereas other demonstrators were trained to twist and pull the bolt to achieve the same result. Observers given access to the box tended to remove the bolt the same way that they had seen it removed. However, because the bolt moved differently in the two cases, it is possible that the observers learned how the bolt moved (by emulation) rather than learning to copy the actions of their demonstrators (by imitation).

Emulation may also have played a role in an experiment in which observation of experienced demonstrators facilitated the opening of hickory nuts by red squirrels, relative to trial-and-error learning (Weigle & Hanson, 1980). In this case, differential local enhancement can be ruled out because animals in both groups quickly approached and handled the nuts, and the observers actually handled the nuts less than control animals (perhaps because observers were more efficient at opening them). However, those animals that observed demonstrators opening nuts were able to see the open nuts (end-state emulation) and they had the opportunity to associate open nuts with eating by the demonstrator (Heyes & Ray, 2000).

Similarly, chimpanzees that observed a demonstrator spit water into a cylinder to raise a floating peanut learned to do the same (Tennie, Call, & Tomasello, 2010). But emulation may be involved here as well, because observers that watched a human experimenter pour water into the cylinder were just as likely to spit water into the cylinder as those that observed the demonstrator spit water into the cylinder. Thus, the observers in the second group learned that water inserted into the cylinder would raise the level of the peanut. That is, they learned to emulate via the affordances of the task.

Although emulation typically takes place in a social context, it may not be social learning because it does not require learning the *actions* of a demonstrator. To help make this distinction, learning involving the actions of a demonstrator is often compared with a "ghost" control which does not involve a social stimulus.

Although emulation should be distinguished from social learning or imitation, it is a phenomenon of interest in its own right. Learning about how the environment works by observation has important implications for cognitive learning. For example, there is evidence not only that chimpanzees can emulate the movement of a tool (Nagell, Olguin, & Tomasello, 1993), as well as increasing the level of water in a cylinder to gain access to a reward (Tennie et al., 2010), but also evidence that pigeons (Klein & Zentall, 2003) and dogs (Miller, Rayburn-Reeves, & Zentall, 2009) can emulate the direction of movement of a screen which permits access to food.

Bird Song

A special case of matching behavior by animals is the acquisition of bird song (Hinde, 1969; Marler, 1970; Nottebohm, 1970; Thorpe, 1961; see also vocal mimicry; e.g., Pepperberg, 1986, 2002; Thorpe, 1967). Although for a few species of songbird the development of species-typical song is regulated to a large extent by maturation and the seasonally fluctuating release of hormones, for others, social interaction plays a large role (Saar, Mitra, Derégnaucourt, & Tchernichovski, 2008; White, Gros-Louis, King, Papakhian, & West, 2007). Notably, regional variations in bird song appear to depend on the bird's early experience with conspecifics (Baptista & Petrinovitch, 1984). That is, young song birds sometimes learn their regional dialect by copying the song of more mature conspecifics.

But acquisition of bird-song dialect is a special case of social learning. First, although it is learned, bird song is a variation of a species-typical behavior and thus, it is relatively constrained, although acquisition of other species' songs is possible (see Pepperberg, 1988 for a review). Second, and most importantly, bird song takes place in the auditory modality and a characteristic of animal-produced auditory events is that the stimulus produced by the demonstrator and that produced by the "observer" can be a close match, not only to a third party (i.e., the experimenter) but also to the observer itself (Thorpe, 1961). Thus, verbal behavior, for which comparisons between one's own behavior and that of others may be relatively easy to acquire because one can hear one's own utterances with relative fidelity, may be a special "prepared" case of generalized stimulus identity learning (e.g., animals that have been trained to match shape stimuli can now use the principle of stimulus matching to match novel hue stimuli; see Zentall, Edwards, & Hogan, 1983). Byrne (2002) suggests that matching a vocalization is likely to be mediated by copying the final result and thus, should be considered an example of outcome emulation.

This analysis of the copying of verbal behavior can also be applied to certain examples of visual behavior copying. Any behavior that produces a clear change in the environment, such that *from the perspective of the observer* there is a match between the stimulus produced by the demonstrator and that produced by the observer, may be a case of stimulus matching (e.g., observing someone turning up the volume of a radio— when the knob turns to the right, the volume increases). Such cases of visual stimulus matching can be distinguished from the perhaps more abstract and interesting case in which no visual stimulus match is possible (opaque imitation, e.g., the imitation of hands clasped behind the back when observing a person who has his hands clasped behind his back) which Piaget (1962) called imitation of invisible actions.

Traditions

A tradition is a behavior that may be acquired socially but is then passed on to other individuals, presumably by social means (Laland & Galef, 2009). Traditions can be thought of as the building blocks of cumulative culture, characteristic of human populations. When traditions can be modified and improved by learning, it permits the accumulation of knowledge in a population or what has been referred to by Tomasello (1994) as a "ratchet effect" (see Tennie, Call, & Tomasello, 2009). This is where innovation plays an important role.

Supposedly, if one population of animals engages in a particular behavior that other nearby populations does not, one might conclude that a tradition was involved. The problem with this interpretation is that the evidence for traditions in natural settings requires that one distinguish the social transmission of behavior from other differences between the populations that might be responsible for the behavioral differences, such as genetic and environmental differences, and it is very difficult to isolate social transmission from nonsocial means by which behavior can spread throughout a population (Langergraber et al., 2010; van Schaik et al., 2003).

An alternative approach has been to study traditions experimentally (i.e., by introducing a new behavior and studying its spread or diffusion through a population; Tomasello et al., 1997). Although considerable laboratory research on the diffusion of behavior has been conducted with human groups (see Mesoudi & Whiten, 2008), only a few studies of this kind have been done with nonhuman animals. For example, evidence for the diffusion of either of two techniques for extracting trapped food has been demonstrated in chimpanzees (see Whiten, McGuigan, Marshall-Pescini, & Hopper, 2009), meerkats (Thornton & Malapert, 2009), and mongooses (Müller & Cant, 2010).

There is even evidence for the diffusion of an acquired taste preference in rats (Galef & Allen, 1995). But Claidière and Sperber (2010) have argued that the diffusion of behavior that has been demonstrated (e.g., Whiten et al.) is insufficiently stable to qualify as a tradition.

Although traditions may involve the kinds of complex social learning that will be discussed in the next section, I have chosen to include a brief discussion of traditions here because there are likely to be a variety of mechanisms involved in the spread of traditions including stimulus enhancement, observational conditioning, and emulation.

THE SOCIAL LEARNING OF BEHAVIOR

To researchers interested in the possibility that under certain conditions there may be cognitive processes involved in social learning, the most interesting forms of social learning are those that cannot be explained easily by any of the previously described mechanisms. Such complex social learning is generally referred to as imitation or true imitation.

Imitation

The term imitation is used to indicate behavior of an observer that matches the behavior of a demonstrator but that cannot be accounted for with any of the motivational, attentional, or simple learning processes described earlier. Under appropriate conditions, the bidirectional control and two-action procedures are accepted methods for demonstrating imitation.

One issue that often comes up when imitation is considered is whether the behavior that is observed and later performed must be novel to the observer. In principle, one would think that novelty of the behavior would be a prerequisite (see Thorpe, 1961). But in practice, novelty is a difficult requirement to assess. Birds may have a number of different behaviors available to them but which of these has never been performed before? A bird may peck at a red light that one may presume it has never seen before but it has surely pecked at objects before. Even in the case of a sequence of responses, it can be argued that each component of the sequence is likely to have been performed earlier. In fact, one could argue that it would be very difficult for an animal to perform a response sequence, if the individual components of the sequence had never been performed before. Thus, a requirement that is more tractable than novelty is that the behavior should be otherwise improbable (Thorpe, 1961). That is, if the response is unlikely to occur in the absence of its demonstration, one can consider it improbable and thus, if it should occur following conditions of observation, and it cannot be explained easily by any of the previously described mechanisms, it can be attributed to the behavior demonstrated. To be considered imitation, it is important that there be evidence of behavior transmission.

The Bidirectional Control Procedure

When an overhead pole can be pushed to the left or to the right, observer rats tend to push the pole in the same direction that they saw a demonstrator push it (Heyes & Dawson, 1990; see also Klein & Zentall, 2003, for similar results with pigeons and Miller et al., 2009, with dogs). To distinguish this learning from emulation, a control condition (sometimes referred to as a "ghost" control) is needed in which the manipulandum appears to move by itself without a demonstrator pushing it (see Hopper, 2010; Hopper, Lambeth, Schapiro, & Whiten, 2008). Although the ghost control provides a necessary comparison condition, it can be argued that it does not control for social facilitation (the mere presence of a conspecific, see Klein & Zentall, 2003) nor does it control adequately for observation of the outcome or goal (the observed "demonstrator" should be provided with reinforcement to control for the motivation induced by seeing an conspecific eating as well as potential associations between the moving manipulandum and the sight a conspecific eating—what Hopper et al. call an enhanced ghost control). But if one controls for social facilitation by including the presence of an inactive conspecific (Hopper, 2010), the conspecific may be distracting, or imitation of the conspecific could result in the absence of emulation, but for the wrong reason. Thus, if one uses the bidirectional control procedure, there may not be an ideal control procedure.

The Two-Action Procedure

If the demonstrator produces the same effect on a manipulandum in one of two different ways (e.g., by a bird stepping on or pecking at a manipulandum) it is known as the two-action procedure (Figure 12.1). It should be noted that in some cases in which the bidirectional control has been used it has been referred to as the two-action procedure (e.g., Bugnyar & Huber, 1997; Whiten, 1998). However, the two-action procedure differs from the bidirectional control in that with the two-action procedure, the movement of the manipulandum is the same with either action, whereas with the bidirectional control, the manipulandum moves differently (e.g., in the case of Heyes & Dawson, 1990, to the left or to the right).

With the two-action procedure, because the two responses have the same effect on the manipulandum, the two-action procedure controls for emulation as well as local and stimulus enhancement. It also

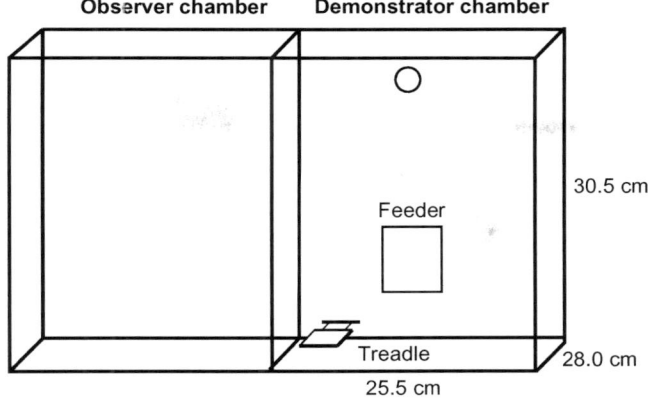

FIGURE 12.1   The two-action apparatus: Observer birds could see a demonstrator bird either stepping on the treadle or pecking at the treadle. Observer birds could then step or peck for food.

controls for social facilitation because a conspecific is present in both conditions.

Akins and Zentall (1996) trained Japanese quail to activate a treadle (a small metal plate near the floor of the chamber) for food, either by pecking at the treadle or by stepping on it. When later given access to the treadle, observers used the same part of their body (their foot or their beak) to make the response as had their respective demonstrators (see also Zentall, Sutton, & Sherburne, 1996, for similar results with pigeons). Of course, one could argue that pecking and stepping are responses that are predisposed and thus, are neither novel nor of low probability, however, selectively directing those responses to a never before seen treadle, following the observation of those specific responses made by a demonstrator, would not normally be considered already acquired responses. Furthermore, Kaiser, Zentall, and Galef (1997) have found that in the absence of a treadle-pecking or treadle-stepping demonstrator (no demonstrator or a merely-present "demonstrator") the probability of the occurrence of either response is very low.

It is important to note, first, that the environmental consequences of stepping and pecking were the same (i.e., everything was the same except the actions or response topographies of the demonstrators). Second, there was little if any similarity between the visual stimulus the observer saw during observation and the visual stimulus it saw during its own performance of either response. That is, the appearance of the demonstrator's beak on the treadle must have appeared quite different to the observer from the sight of its own beak on the treadle. Similarly, although perhaps not so obviously, when the quail stepped on the treadle (located near the corner of the chamber between the feeder and the observer) it had to pull its head back and thrust its chest forward and for this reason it could not see its foot making contact with the treadle. Once again, to the observer, the demonstrator's response to the treadle must have appeared quite different from the observer's own response to the treadle. For these reasons, in such an experiment, the imitated response can be thought of as opaque to the observer and any account of imitation based on visual-visual stimulus matching is implausible.

A similar example of imitation involving two actions was reported in marmosets by Voelkl and Huber (2000). Demonstrators opened a plastic (photographic film) canister either by using their hands or by using their teeth. All of the observers that watched the canister being opened with the teeth did the same, whereas almost half of the observers that watched the canister being opened with the hand did so as well.

The two-action procedure allows one to assess a special case of social learning called imitation of invisible actions or opaque imitation that directly controls for emulation because, from the perspective of the observer, the observer's behavior does not match that of the demonstrator. Opaque imitation is of particular interest to comparative psychologists because the mechanisms responsible for the transmission of information from the demonstrator to the observer are not well understood (see Whiten, 2005) and when it occurs in children, traditionally, it has been attributed to the ability to take the perspective of a third person (Piaget, 1962). That is, one must imagine what one looks like and what the demonstrator looks like to a third individual, and respond such that the observer would note that the two behaviors look alike. However, as perspective taking does not appear in children before the age of 3 years (Selman, 1980, says it occurs between the ages of 3 and 6), whereas other animals (e.g., Japanese quail, pigeons, and marmosets) show evidence of imitation, it seems unlikely that perspective taking is the mechanism responsible for imitation (see Ray & Heyes, 2011).

Variables that May Influence Opaque Imitation

Several variables have been found to influence whether imitation will be found or not, and those variables may be of interest not only because of their practical implications but also because they may help to identify the nature of the cognitive processes that are involved in social learning.

Demonstrator Reinforcement

A cognitive account of imitation implies that the observer understands what the demonstrator is doing and perhaps even why it is doing it. If such an interpretation is correct, whether evidence of imitation is found may depend on the consequences of the demonstrated response for the demonstrator. Alternatively, animals may have a species-typical tendency to imitate regardless of the consequence of the behavior for the demonstrator. It is also possible that the copying of a demonstrator's behavior is reinforcing in its own right. For example, children play a game called "follow the leader" which involves doing everything that the leader does and there is no apparent external reinforcement for the copying behavior. It should be clear, however, that even if the tendency to imitate is predisposed or self-reinforcing and independent of demonstrator reinforcement, it does not explain how the observer understands what it should do to replicate the behavior of the demonstrator.

Interestingly, Akins and Zentall (1998) found that quail imitated when they observed demonstrators receiving a reward after they pecked or stepped on a treadle but not unless they had observed the demonstrator's response being reinforced (see also, Palameta & Lefebvre, 1985; but see McGregor, Saggerson, Pearce, & Heyes, 2006). Although the effects of the presence versus the absence of demonstrator reinforcement suggests a cognitive account (i.e., there is no reason to imitate if reinforcement does not follow the response), it is possible to explain the effect of demonstrator reward on observer imitation by appealing to observational conditioning (the simpler form of learning described earlier). In observational conditioning, an observer's attention may be drawn to a stimulus (in this case, to the demonstrator quail depressing the treadle) because this action precedes demonstrator reinforcement (for the observer, a secondary reinforcer). Although observational conditioning might account for the effect of reinforcement on observation, observational conditioning cannot account for the *correspondence* between observer's and demonstrator's response topographies. Thus, the effect of demonstrator reinforcement may be to act as a catalyst to bring out imitative learning in an observer.

The effect of the outcome for the demonstrator may play an even more important role when responses are demonstrated that are more complex than pressing a treadle. For example, in one experiment, humans performed a task in the presence of a chimpanzee in which first one response (poking a stick in a hole in the top of a box) did not lead to obtaining food but another response (poking the stick in a hole in the side of the box) did (Horner & Whiten, 2005). Next, subjects were given access to the box and the stick. Interestingly, when the box was opaque, so that subjects could not see that the top hole did not provide access to food, the subjects often started by poking the stick into the top hole. However, when the box was transparent and subjects could see that the top hole did not provide access to food, the subjects generally avoided poking the stick in the top hole and instead poked the stick directly into the side hole that had produced the food.

Horner and Whiten (2005) proposed that when the box was transparent, subjects recognized the causal structure of the task and avoided the response that did not lead to reward, however, when the box was opaque, it was not clear that inserting the stick in the top hole was not a necessary prerequisite to inserting the stick in the side hole. Thus, the chimpanzees could acquire the entire sequence of responses through observation but they omitted part of the sequence when it was apparent that one of the demonstrated responses was not necessary to achieve the goal. It has been argued, however, that this stick-poking behavior may not actually involve imitation because the chimpanzees were adept at stick poking and only need to learn where to poke the stick (Tennie, Call, & Tomasello, 2006). Furthermore, in the case of the transparent box, it may have been that causal information (the result of the first stick-poke) overrode learning where to poke (local enhancement).

Surprisingly, although the chimpanzees omitted the initial unnecessary response when the box was transparent, children in the Horner and Whiten (2005) study did not. This result suggests that children are more prone to "blind" copying, that is copying that is intrinsically reinforcing and has no other goal, than are chimpanzees. For example, children will often imitate the posture of an adult when there is no extrinsic reinforcement, that is, even if they are not visible to the adult or to others, and as noted earlier, children appear to enjoy playing follow the leader without any obvious external reinforcement for doing so. Blind imitation, sometimes referred to as over-imitation, may be the major distinction between the copying behavior of humans (especially children) and other animals.

On the other hand, there is evidence that children do not always blindly imitate (see Gergely, Bekkering, & Király, 2002; Meltzoff, 1995; Nielsen, 2008; Uzgiris, 1981; Whiten et al., 2009). The conditions under which blind imitation and selective imitation will be found are likely to depend on contextual factors such as whether imitation is reinforcing in its own right and whether causal information is judged to be important. Thus, there is evidence that older children are more prone to over-imitate than younger children (Huang, Heyes, & Charman, 2006), perhaps because the older children see imitation as a game or an end in itself, rather than as a means to an end. With regard to the acquired tendency to imitate observed behavior by children, it would be interesting to know if this ability varies across cultures depending on the importance that the culture places on learning by imitation.

Observer Motivation

If at the time of observation, observers are not motivated to obtain the reinforcers for which the demonstrators are working, one can ask if they will acquire the response for use later when they are motivated. Such learning might be expected if animals are predisposed to learn from a demonstrator and blindly imitate. Alternatively, it could imply a higher level of cognitive functioning. For example, it could indicate the ability of the observer to retrieve an earlier representation of an observed behavior or even the ability of the observer to plan for the future (Piaget, 1936; Tulving, 2004) if the observer recognized that although the information was not useful at the time of observation, it may be useful at a later time.

The hypothesis that observer motivation may affect imitation was tested in quail by comparing imitative learning by quail that were either hungry or sated at the time of observation (Dorrance & Zentall, 2001). It was found that hungry quail matched the demonstrator's reinforced behavior, whereas sated quail, when later tested hungry, did not. Animals that are not hungry may not be motivated to learn from a conspecific or they may not attend as well to the behavior of the conspecific. Thus, differential motivation or differential attention during the time of observation could account for differential learning by the two groups. One might also propose that a representation of the observed behavior does not survive the delay between observation and performance required by the need to test the animal when sufficiently hungry. That possibility will be addressed in the next section.

Deferred Imitation

As noted earlier, Bandura (1969) proposed that there is an important cognitive difference between immediate imitation (that Bandura called imitation) and deferred imitation (that he called observational learning) in which some time passes between the time of observation and performance by the observer (see also Piaget, 1936, for an earlier version of this theory). For Bandura, immediate imitation was viewed as a reflexive response akin to contagious behavior, whereas deferred imitation indicated a more cognitive process in which an observer had to *represent* the response at the time of observation for later retrieval when performance was assessed.

To determine whether animals are capable of deferred imitation, as part of a study already cited (Dorrance & Zentall, 2001), hungry quail observed either a treadle pecking or treadle stepping demonstrator. When one group of quail was tested 30 min later, they imitated as frequently as observers tested immediately following observation. If, as Bandura proposed, deferred imitation is evidence of a more cognitive process, then quail show good evidence of the cognitive representation of the earlier observed behavior.

Enculturation

One of the variables that may play a role in imitative learning by primates appears to be the degree to which the animals have had extensive interactions with humans—what Tomasello (1990) refers to as enculturation. Enculturated chimpanzees and orangutans readily show signs of imitative learning (Russon & Gladikas, 1993, 1995; Tomasello, Gust, & Frost, 1989; Tomasello, Savage-Rumbaugh, & Kruger, 1993), whereas lab housed/reared chimpanzees often do not (Whiten & Custance, 1996; but see Tomasello & Call, 2004). Furthermore, Tomasello et al. (1993) have suggested that enculturated apes may develop an understanding of intentionality (see Searle, 1983).

Enculturation may produce its effect in a number of ways. First, it may reduce the apes' anxiety (response to novelty) during tests. Second, it could increase their attentiveness to social cues (see, e.g., Bering, 2004). Third, it could give them prior reinforced experience with imitating (i.e., a generalized imitation response). Fourth, there could be a general kind of learning-to-learn (i.e., learning that a task will be presented in which reinforcement can be obtained). Fifth, enculturation may actually improve the general cognitive ability of the animal.

Although Whiten (1993) has suggested the failure to observe imitation in non-enculturated apes may be related to the fact that human demonstrators have been used, the use of conspecifics in such experiments does not guarantee success (Tennie et al., 2006, 2010; Tomasello et al., 1997). A better understanding of the various components of enculturation might provide important insights into the mechanisms involved in imitation by apes.

Gestural Single-Response Imitation

The two-action procedure provides the best control for non-imitative learning, however, if the behavior to be demonstrated is a gesture and it is sufficiently unlikely to occur by chance because the number of possible gestures is very large, it may not be necessary to have a control group. Instead, one can use a within-subject design and observe the behavior of the observing animal for some time before it observes the gesture of the demonstrator, using that as a baseline from which to judge performance of the observed behavior (Tomasello et al., 1997).

More Complex Forms of Imitation

Program Level Imitation

Byrne (1994) has distinguished action level imitation, involving a single response, for example, pressing a lever or poking at a bolt, from program level imitation that involves learning a coordinated sequence of actions leading to reward. Byrne and Russon (1998) describe the sequence of behaviors needed by gorillas to consume leaves that have stinging nettles on one side. Although they have argued that the sequence of actions was socially acquired, the results of an experiment by Tennie, Hedwig, Call, and Tomasello (2008) suggest that gorilla nettle feeding derives mostly from genetic predispositions and individual learning of plant affordances.

Nonetheless, one can ask if it is possible for animals to acquire a sequence of actions through observation. Evidence for the imitation of a response sequence by chimpanzees has been demonstrated by Whiten and Custance (1996) using the artificial fruit task in which a box containing a treat can be opened only by performing a sequence of actions on the boxes "defenses" and observers appear to learn how to do this through observation.

There is also evidence that pigeons will imitate a sequence of two quite different response alternatives. Nguyen, Klein, and Zentall (2005) trained demonstrators to either step on a treadle or peck at the treadle to present a feeder, access to which was blocked by a screen. The demonstrator then had to push the screen in the assigned direction (some demonstrators learned to push the screen to the left others to the right). Observers could then step on or peck at the treadle and then push the screen in either direction. Results indicated that there was a significant correlation between the response sequence performed by the demonstrators and the response sequence performed by the observers. Thus, although program level imitation likely involves a greater memory load, contrary to Byrne's distinction, program level imitation does not appear to be conceptually different from imitation of a single response. Instead, the difference appears to be quantitative and likely influenced by the animal's ability to remember the sequence of events.

Generalized Gestural Imitation

A form of imitative learning that is conceptually related to the two-action procedure involves copying the gestures of a model on command (e.g., "Do this!"). Successful do-as-I-do performance has been reported in chimpanzees (Custance, Whiten, & Bard, 1995; Myowa-Yamakoshi & Matsuzawa, 1999), orangutans (Call, 2001; Miles, Mitchell, & Harper, 1996), dolphins (Herman, Matus, Herman, Ivancic, & Pack, 2001), dogs (Huber et al., 2009; Topal, Byrne, Miklosi, & Csanyi, 2006), and parrots (Pepperberg, 1988), but only to a limited extent in monkeys (Fragaszy, Deputte, Cooper, Colbert-White, & Hémery, 2011). Remarkably, because the imitated models were humans, in the case of dolphins and parrots there would be little similarity between the corresponding body parts of observer and the human demonstrators.

With regard to generalized imitation, Custance et al. (1995) found that chimpanzees learned to respond to the command "Do this!" by imitating a broad class of behaviors demonstrated by humans including touching the back of the head and other actions that could not be seen as they were performed. Thus, such imitated opaque actions cannot be explained as some form of visual stimulus matching. Furthermore, because objects are not involved in this kind of imitation, accounts based on local and stimulus enhancement are irrelevant. Finally, each imitated gesture serves as a control for other imitated gestures, and the broad range of gestures that have been imitated within a few seconds of demonstration suggests that differential motivation does not play a role. Success in such do-as-I-do experiments shows not only that (enculturated) chimpanzees can imitate, but also that they are capable of forming a generalized *concept* of imitation because they selectively imitate any of a broad class of gestures when cued to do so.

Byrne and Tanner (2006) have offered a different interpretation of the positive results of do-as-I-do studies. They propose that the behaviors imitated were not novel, that is, they were already in the animals' repertoire and the sight of the demonstration merely evoked a similar response (response enhancement). But as noted earlier, truly novel responses are very difficult to define, especially if one considers that the presumed novel behavior is likely to be similar to some past behavior by the observer. Thus, it is more reasonable to require that the response have a very low probability of occurrence in the absence of observation of its demonstration by the model and that they cannot be explained by an alternative account. Furthermore, the notion of response enhancement fails to deal with the most perplexing question of the correspondence problem, how it is that the seen response of another comes to match the felt response of the observer. In Byrne and Tanner's view, those responses are "prewired" and reflexive. The possibility that an observed response will be reproduced automatically, perhaps by means of the mirror system (Rizzolatti, Fadiga, Fogassi, & Gallese, 2002), will be addressed shortly.

Intentionality

Interest in imitation research can be traced, at least in part, to the possibility that imitation involves some degree of purposiveness or goal directedness. But when an individual imitates the behavior of demonstrator it is not clear that the goal is always to obtain the same outcome as the demonstrator. In the case of blind imitation, the goal of the observer is generally not the same as the goal of the demonstrator. However, when behavioral copying depends on the outcome of the demonstrator's behavior, it does suggest that the observer's intention is responsible (but see Akins & Zentall, 1998). Intentionality (Searle, 1983), is surely involved in many higher order forms of imitation by humans, such as the student dancer who repeats the movements of the teacher. Interestingly, when it comes to the precise movements involved in dance or in sports, we humans are not particularly good at repeating them from demonstration. Instead, it is typically necessary to practice the movements many times and to learn from their consequences. In fact, the use of mirrors by dancers to perfect their movements suggests that trial and error stimulus matching plays an important role in the learning process (a process that Galef, 2010, refers to as *performance emulation*).

Intentionality is difficult enough to study in humans because, although humans generally have language, they may not always be aware of their intentions. It is even more difficult to study in animals and evidence for intentionality appears most often in the literature in the form of anecdote rather than experiment.

Mitchell (1987), for example, provides a number of examples of imitation in animals at these higher levels that imply intentionality. For example, Mitchell discusses observations of a young female rhesus monkey who after seeing her mother carrying a sibling, walked around carrying a coconut shell at the same location on her own body. If there were some way to conduct experiments involving the manipulation of intentionality, the credibility of these anecdotes would be greatly increased.

Understanding the Intentions of Others

Evidence suggests that 14-month-old children are able to understand the intentions of another person and use this understanding to mediate their imitative behavior (Gergely, Nadasdy, Csibra, & Biro, 1995). When young children watched a demonstrator, whose hands were occupied, turn on a light by touching it with her forehead, they subsequently turned on the light more efficiently by using their hands. However, when the demonstrator's hands were not occupied, so that the observing children might assume that it was necessary to use their forehead to turn on the light, children showed a greater tendency to copy the demonstrator by using the forehead.

Curiously, as children get older, they may engage in more acts of blind imitation. Thus Huang et al. (2006) reported that when 41-month-old children were shown a failed attempt to complete a response, they were more likely to copy the failed attempt than 31-month olds. That is, in this case, younger children appeared to understand the intentions of the adult better than the older children. It may be, however, that the older children were more likely to have social copying (repeating the actions of the model, a kind of follow the leader game) as a goal rather than merely obtaining the outcome.

Results similar to those reported by Gergely et al. (1995) also have been found with enculturated chimpanzees (Buttelmann, Carpenter, Call, & Tomasello, 2007). More surprising, there is evidence that dogs may be able to make similar inferences (Range, Viranyi, & Huber, 2007). When dogs watched a dog demonstrator with a ball in its mouth pull a rod with its paw to obtain a treat, the observer dogs pulled the rod more efficiently with the mouth. However, if the demonstrator's mouth was not occupied and it pulled the rod with its paw, the observers also pulled the rod with the paw, suggesting that dogs, like human children, can not only imitate but it appears that they can also understand the intentions of the demonstrator (but see Kaminski et al., 2011, for an alternative explanation).

Symbolic Imitation

At the highest level of imitative behavior, what Mitchell (1987) refers to as fifth-level imitation, the behavior of the observer does not actually match the behavior of the demonstrator. In fact, the differences between the actions of the demonstrator and those of the observer are explicit and they are produced for the purpose of drawing attention to specific characteristics of the demonstrator. Examples of such symbolic imitation can be found in the human use of parody and caricature—exaggerating someone's limp or their facial expression. Such forms of imitation are mentioned primarily for completeness and to note the degree of subtlety that can be involved in imitation.

POSSIBLE BEHAVIORAL AND BIOLOGICAL MECHANISMS

Recently, there have been several attempts to account for imitation using simpler behavioral and biological mechanisms. These are addressed here because they purport to provide simpler accounts of imitative processes.

An Associative Learning Account of Imitation

Ray and Heyes (2011, see also Heyes & Ray, 2000) have proposed an associative learning account of imitation based on Pavlovian and instrumental conditioning processes. According to their *associative sequence learning model*, in the case of humans, imitation is learned through prior experience with (i) direct self-observation (the correspondence of one's body parts, e.g., fingers with those of others), (ii) mirror self-observation (the correspondence of felt actions with the image of those actions in a mirror), (iii) synchronous action (two individuals happen to be making the same response), (iv) acquired equivalence experience (experiencing a similar reaction when one is making a response, someone saying "you look angry" and when someone else is making the same response, someone saying "he looks angry"), and (v) being imitated (mothers imitate their babies and they reinforce imitation by their babies).

In the case of imitation by animals, for example, the two-action procedure with Japanese quail (Akins & Zentall, 1996), stepping and pecking have been reinforced in the past when they have occurred in the presence of other animals engaged in similar behavior. According to this theory, it is assumed that before participating in any experiment, observers ate at the same time as others and consequently learned to peck when others were pecking, and they fed from a similar feeder and consequently learned to step toward the feeder when others were doing so. As a result, seeing others pecking or stepping in the context of an experiment would become a discriminative stimulus for engaging in the same behavior.

However, there are several problems with this account. First, it is highly speculative and is not supported by data. For example, in our laboratory, birds are typically not fed at the same time, so pecking and stepping at the same time as other birds would not have been reinforced. Second, the theory requires that the home-cage context will generalize to the experimental context despite the fact that the two contexts are quite different. Third, in the two-action procedure, the treadle is usually not actually located in front of the feeder. In our apparatus it is located in the front corner nearest to the observer and in a direction away from the feeder (see Figure 12.1). Finally, it is not clear how associative sequence learning can account for results of bidirectional control experiments in which a screen encountered for the first time is pushed in the same direction as a demonstrator pushed it. As Whiten (2005) suggests, if the processes responsible for imitation involve basic learning processes that are present in many animal species, why is it that only humans, certain great apes, dolphins, and birds have been found to show clear evidence of imitative learning?

Kinesthetic-Visual Matching

Another solution to the correspondence problem was proposed by Guillaume (1926/1971) and expanded on by Mitchell (1997b). The idea is that in addition to visual-visual matching (the ability to have a generalized concept of sameness—the concept that two things look the same), some animals also are able to match across modalities and recognize the similarity between something they can see and something that they can feel. It is presumed that this ability is acquired through experience with objects and parts of one's body that can be both seen and felt. More specifically, given considerable experience seeing and feeling objects and parts of one's body that can be seen, one should then be able to generalize to parts of one's body that cannot be seen. To accomplish this, one would have to be able to form an image or representation of the parts of one's body that one cannot see. For older children and adult humans from cultures with mirrors, it would not seem unreasonable to learn the correspondence between how an object looks and how it feels. And there is evidence that chimpanzees and several other species that have had some exposure to a mirror may show evidence that they can recognize themselves in a mirror. However, there is little evidence (except in human children) that generalized bodily imitation and self-recognition occur in the same individual (Mitchell, 1997a).

POSSIBLE BIOLOGICAL MECHANISMS

Response Facilitation

Byrne has proposed that response facilitation can account for reports of imitation found in animals. As noted earlier, response facilitation implies that observation of a response elicits a similar response in an observer. By this account, the observed behavior is already in the repertoire of the observer and observation of it automatically primes the representation of the behavior in the brain, increasing the probability that the behavior will occur. Although this view provides a noncognitive mechanism for the kinds of response imitation most frequently studied in animals, it does not provide a particularly convincing account of imitation of behaviors that an animal encounters for the first time in an experimental setting, for example, pushing a screen to left or right to access food (when accompanied by the appropriate controls for emulation). Furthermore, it is not clear how the observed behavior became connected to the observer's motor response. Of course one could propose that it evolved as an adaptation to an unpredictable environment by means of something like the mirror system (see below). However, if this were the case, one would have to posit the evolution of predisposed connections for each behavior that has been found to be imitated; an unlikely and unparsimonious account given the range of behavior and contexts in which it has been found. In addition, as we will see, the mirror system is constrained in the kinds of imitation that it is able to explain.

The Mirror System

There is evidence that neurons found in the premotor cortex of monkeys are activated not only when the monkey picks up an object but also when it sees either a human or another monkey pick up an object (Rizzolatti et al., 2002). These so called *mirror neurons* have been proposed to be responsible for imitation, and their presence in the premotor cortex rather than the visual cortex suggests that they may have a preparatory cognitive function. However, it is important to know whether these mirror neurons belong to "prewired" neural pathways that evolved to facilitate imitation or they have to be trained to behave the way they do. If learning is required, mirror neurons may *result* from imitation, rather than be its *cause* (Csibra, 2007; Heyes, 2010). Furthermore, although mirror neurons may be involved in stimulus matching, it is not clear that they can account for perceptually opaque imitation in which there is little similarity between the visual input animals receive from watching the behavior of another and what they can see of their own behavior. But perhaps most important, although evidence for mirror neurons was originally reported in monkeys, with the exception of the study by Voelkl and Huber (2000) with marmosets, there is surprisingly little evidence that monkeys show evidence of imitation (Fragaszy & Visalberghi, 2004; Visalberghi & Fragaszy, 1990).

Ferrari, Bonini, and Fogassi (2009) have described an innovative solution to this paradox. They suggest that there are two pathways that are involved in copying the behavior of others. The first is a direct parieto-premotor pathway that exerts a direct influence on the motor output during action observation. This pathway is involved in simple motor acts such as grasping (Rizzolatti et al., 2002) and does not require learning. It is responsible for neonatal imitation such as facial imitation described by Meltzoff and Moore (1977) and Lepage and Théoret (2007) in humans and by Ferrari et al. (2006) in monkeys. However, Heyes, Bird, Johnson, and Haggard (2005) showed that non-matching, or incompatible sensorimotor training—in which the participant repeatedly performs one action while observing another—can abolish and even reverse both imitative behavior and the action matching properties of the mirror neuron system.

The second pathway is indirect, linking parietal and premotor areas with ventro-lateral prefrontal cortex (see Tanji & Hoshi, 2008). This pathway could exploit the sensory-motor representations provided by the direct pathway for more complex cognitive and behavioral functions, such as those required for delayed imitative behaviors and opaque imitation. But it is precisely those processes that presumably take place in the prefrontal context that are of greatest interest to those who are interested in the processes responsible for complex imitative processes.

Others have argued that the behavioral aspect of mirror neurons are a by-product of the simulation theory of emotion recognition (Goldman & Sripada, 2005) which allows humans to have empathy and also prepares them to act similarly to the behavior of others (Gallese, 2001). Humans then learn when to engage in similar behavior (imitate) and when to inhibit that behavior. Furthermore, they can also learn to engage in quite arbitrary behavior when presented with the behavior of another person (Catmur, Walsh, & Heyes, 2007). This simulation model can account not only for perspective taking but also for opaque imitation by humans. However, it is not clear that it can account for the opaque imitation shown by Japanese quail (Akins & Zentall, 1996) and pigeons (Zentall et al., 1996). So far, possible avian correlates of mirror neurons have been found only in the song system of vocal learning birds (Prather, Peters, Nowicki, & Mooney, 2008).

CONCLUSIONS

Procedures have recently been developed that separate imitation from other forms of social influence and social learning, and the results of initial studies indicate that species from chimpanzees to quail can imitate. Such findings should not be surprising because social learning, whether by imitation or some other process, often provides greater benefits than genetically predisposed behavior or trial-and-error learning (Boyd & Richerson, 1988). However, the processes involved that enable animals to match their behavior to that of a demonstrator are poorly understood. Imitation may involve some form of coordination of visual and tactile sensory modalities, perspective taking, or response facilitation. However, the role of such processes in opaque imitation is still unknown. A reasonable strategy to better understand the mechanisms involved in imitation would be to determine the necessary and sufficient conditions for opaque imitation to occur and to explore the range of behaviors that animals can imitate. Finally, the ability to learn from others may not only be adaptive in making more efficient the acquisition of behavior important for survival and reproduction but it may also free the animal from trial and error learning to engage in innovative variants that may be adopted by others (see, e.g., Kawai, 1965).

Acknowledgments

Preparation of this article was supported by National Institute of Mental Health Grant 63726 and by National Institute of Child Health and Development Grant 60996.

References

Akins, C. K., & Zentall, T. R. (1996). Imitative learning in male Japanese quail (*Coturnix japonica*) using the two-action method. *Journal of Comparative Psychology, 110*, 316—320.

Akins, C. K., & Zentall. T. R. (1998). Imitation in Japanese quail: The role of reinforcement of demonstrator responding. *Psychonomic Bulletin & Review, 5*, 694—697.

Bandura, A. (1969). Social learning theory of identificatory processes. In D. A. Goslin (Ed.), *Handbook of socialization theory and research* (pp. 213—262). Chicago, IL: Rand-McNally.

Baptista, L. F., & Petrinovich, L. (1984). Social interaction, sensitive phases, and the song template hypothesis in the white-crowned sparrow. *Animal Behaviour, 32*, 172—181.

Bering, J. M. (2004). A critical review of the "enculturation hypothesis": The effects of human rearing on great ape social cognition. *Animal Cognition, 7*, 201—212.

Boyd, R., & Richerson. P. J. (1988). An evolutionary model of social learning: The effect of spatial and temporal variation. In T. R. Zentall, & B. G. Galef, Jr. (Eds.), *Social learning: Psychological and biological perspectives* (pp. 29—48). Hillsdale, NJ: Erlbaum.

Bugnyar, T., & Huber L. (1997). Push or pull: An experimental study on imitation in marmosets. *Animal Behaviour, 54*, 817—831.

Buttelmann, D., Carpenter, M., Call, J., & Tomasello, M. (2007). Enculturated chimpanzees imitate rationally. *Developmental Science, 10*, F31—F38.

Byrne, R. W. (1994). The evolution of intelligence. In P. J. B. Slater, & T. R. Halliday (Eds.), *Behavior and evolution* (pp. 223—265). Cambridge: Cambridge University Press.

Byrne, R. W. (1998). Comment on chimpanzee and human cultures. *Current Anthropology, 39*, 591—614.

Byrne, R. W. (2002). Emulation in apes: Verdict 'not proven'. *Developmental Science, 5*, 20—22.

Byrne, R. W., & Russon, A. E. (1998). Learning by imitation: A hierarchical approach. *Behavioral and Brain Sciences, 21*, 667—721.

Byrne, R. W., & Tanner, J. E. (2006). Gestural imitation by a gorilla: Evidence and nature of the capacity. *International Journal of Psychology and Psychological Therapy, 6*, 215—231.

Caldwell, C., & Whiten, W. (2003). Scrounging facilitates social learning in common marmosets, *Callithrix jacchus*. *Animal Behaviour, 65*, 1085—1092.

Call, J. (2001). Body imitation in an enculturated orangutan (*Pongo pigmaeus*). *Cybernetics and Systems, 32*, 97—119.

Campbell, F. M., Heyes, C. M., & Goldsmith, A. R. (1999). Stimulus learning and response learning by observation in the European starling, in a two-object/two-action test. *Animal Behaviour. 58*, 151—158.

Catmur, C., Walsh, V., & Heyes, C. (2007). Sensorimotor learning configures the human mirror system. *Current Biology, 17*, 1527—1531.

Claidière, N., & Sperber, D. (2010). Imitation explains the propagation, not the stability of animal culture. *Proceedings of the Royal Society of London: B, 277*, 651—659.

Cowan, N., Wood, N. L., Nugent, L. D., & Treisman, M. (1997). There are two word-length effects in verbal short-term memory: Opposed effects of duration and complexity. *Psychological Science, 8*, 290—295.

Csibra, G. (2007). Action mirroring and action interpretation: An alternative account. In Sensorimotor foundations of higher cognition. In P. Haggard, Y. Rossetti, & M. Kawato (Eds.), *Sensorimotor foundations of higher cognition. Attention and performance XXII* (pp. 435—459). New York, NY: Oxford University Press.

Custance, D. M., Whiten, A., & Bard, K. A. (1995). Can young chimpanzees imitate arbitrary actions? Hayes and Hayes revisited. *Behaviour, 132*, 839—858.

Del Russo, J. E. (1971). Observational learning in hooded rats. *Psychonomic Science, 24*, 37—45.

Dorrance, B. R., & Zentall, T. R. (2001). Imitative learning in Japanese quail depends on the motivational state of the observer at the time of observation. *Journal of Comparative Psychology, 115,* 62–67.

Dugatkin, L. A. (1992). Sexual selection and imitation: Females copy the mate choice of others. *American Nature, 139,* 1384–1389.

Dugatkin, L. A. (1996). Copying and mate choice. In C. M. Heyes, & B. G. Galef (Eds.), *Social learning in animals: The roots of culture* (pp. 85–105). San Diego, CA: Academic Press.

Dugatkin, L. A., & Godin, J.-G. J. (1992). Reversal of female mate choice by copying. *Proceedings of the Royal Society of London: B, 249,* 179–184.

Ferrari, P. F., Bonini, L., & Fogassi, L. (2009). From monkey mirror neurons to primate behaviours: Possible 'direct' and 'indirect' pathways. *Philosophical Transactions of the Royal Society: B, 364,* 2311–2323.

Ferrari, P. F., Visalberghi, E., Paukner, A., Fogassi, L., Ruggiero, A., & Suomi, S. J. (2006). Neonatal imitation in rhesus macaques. *Public Library of Science: Biology, 4,* e302.

Fragaszy, D., Deputte, B., Cooper, E. J., Colbert-White, E. N., & Hémery, C. (2011). When and how well can human-socialized capuchins match actions demonstrated by a familiar human? *American Journal of Primatology, 73,* 1–12.

Fragaszy, D., & Visalberghi, E. (2004). Socially biased learning in monkeys. *Learning & Behavior, 32,* 24–35.

Galef, B. G., Jr. (1988a). Communication of information concerning distant diets in a social, central-place foraging species: *Rattus norvegicus.* In T. R. Zentall, & B. G. Galef, Jr. (Eds.), *Social learning: Psychological and biological perspectives* (pp. 119–139). Hillsdale, NJ: Erlbaum.

Galef, B. G., Jr. (1988b). Imitation in animals: History, definition, and interpretation of data from the psychological laboratory. In T. R. Zentall, & B. G. Galef, Jr. (Eds.), *Social learning: Psychological and biological perspectives* (pp. 3–28). Hillsdale, NJ: Erlbaum.

Galef, B. G., Jr. (2010). The relationship between terminology and experiment in studies of social learning. *Paper presented at the meeting on social learning in humans and non-human animals: Theoretical and empirical dissections.* Kavli Royal Society International Centre, Chicheley Hall, Buckinghamshire, GB.

Galef, B. G., Jr., & Allen, C. (1995). A new model system for studying behavioural traditions in animals. *Animal Behaviour, 50,* 705–717.

Galef, B. G., Jr., Lim, T. C. W., & Gilbert, G. (2008). Evidence of mate-choice copying in Norway rats. *Animal Behaviour, 75,* 1117–1123.

Gallese, V. (2001). The 'shared manifold' hypothesis from mirror neurons to empathy. *Journal of Consciousness Studies, 8,* 33–50.

Gardner, E. L., & Engel, D. R. (1971). Imitational and social facilitatory aspects of observational learning in the laboratory rat. *Psychonomic Science, 25,* 5–6.

Gardner, H. (1993). *Creating minds.* New York, NY: Basic Books.

Gergely, G., Bekkering, H., & Király, I. (2002). Rational imitation in preverbal infants. *Nature, 415,* 755.

Gergely, G., Nadasdy, Z., Csibra, G., & Biro, S. (1995). Taking the intentional stance at 12 months of age. *Cognition, 56,* 165–193.

Goldman, A. I., & Sripada, C. S. (2005). Simulationist models of face-based emotion recognition. *Cognition, 94,* 193–213.

Guillaume, P. (1926/1971). *Imitation in children.* Chicago, IL: University of Chicago Press.

Haruki, Y., & Tsuzuki, T. (1967). Learning of imitation and learning through imitation in the white rat. *Annual of Animal Psychology, 17,* 57–63.

Herman, L. M., Matus, D. S., Herman, E. Y. K., Ivancic, M., & Pack, A. A. (2001). The bottlenosed dolphin's *(Tursiops truncatus)* understanding of gestures as symbolic representations of its body parts. *Animal Learning & Behavior, 29,* 250–264.

Heyes, C. M. (1994). Reflections on self-recognition in primates. *Animal Behavior, 47*, 909–919.

Heyes, C. M. (2002). Transformational and associative theories of imitation. In K. Dautenhahn, & C. L. Nehaniv (Eds.), *Imitation in animals and artifacts* (pp. 501–523). Cambridge, MA: MIT Press.

Heyes, C. M. (2010). Where do mirror neurons come from? *Neuroscience and Biobehavioral Reviews, 34*, 1527–1531.

Heyes, C. M., & Dawson, G. R. (1990). A demonstration of observational learning in rats using a bidirectional control. *Quarterly Journal of Experimental Psychology, 42B*, 59–71.

Heyes, C. M., Bird, G., Johnson, H., & Haggard, P. (2005). Experience modulates automatic imitation. *Cognitive Brain Research, 22*, 233–240.

Heyes, C. M., & Ray, E. D. (2000). What is the significance of imitation in animals? *Advances in the Study of Behavior, 29*, 215–245.

Heyes, C. M., Jaldow, E., & Dawson, G. R. (1994). Imitation in rats: Conditions of occurrence in a bidirectional control procedure. *Learning and Motivation, 25*, 276–287.

Hinde, R. A. (Ed.), (1969). *Bird vocalizations.* Cambridge, England: Cambridge University Press.

Hogan, D. E. (1988). Learned imitation by pigeons. In T. R. Zentall, & B. G. Galef, Jr. (Eds.), *Social learning: Psychological and biological perspectives* (pp. 225–238). Hillsdale, NJ: Erlbaum.

Hopper, L. M. (2010). 'Ghost' experiments and the dissection of social learning in humans and animals. *Biological Reviews, 85*, 685–701.

Hopper, L. M., Lambeth, S. P., Schapiro, S. J., & Whiten, A. (2008). Observational learning in chimpanzees and children studied through 'ghost' conditions. *Proceedings of the Royal Society of London: B, 275*, 835–840.

Hoppitt, W. J. E., & Laland, K. N. (2008). Social processes influencing learning in animals: A review of the evidence. *Advances in the Study of Behavior, 38*, 105–165.

Horne, P. J., & Erjavec, M. (2007). Do infants show generalized imitation of gestures? *Journal of the Experimental Analysis of Behavior, 87*, 63–87.

Horner, V., & Whiten, A. (2005). Causal knowledge and imitation/emulation switching in chimpanzees (*Pan troglodytes*) and children (*Homo sapiens*). *Animal Cognition, 8*, 164–181.

Huang, C.-T., Heyes, C., & Charman, T. (2006). Preschoolers' behavioural reenactment of "failed attempts": The roles of intention-reading, emulation and mimicry. *Cognitive Development, 21*, 36–45.

Huber, L., Range, B., Voelkl, B., Szucsich, A., Virányi, Z., & Miklosi, A. (2009). The evolution of imitation: What do the capacities of non-human animals tell us about the mechanisms of imitation?. *Philosophical Transactions of the Royal Society of London. Series B, Biological Sciences, 364*, 2299–2309.

John, E. R., Chesler, P., Bartlett, F., & Victor, I. (1968). Observational learning in cats. *Science, 159*, 1489–1491.

Kaiser, D. H., Zentall, T. R., & Galef, B. G., Jr. (1997). Can imitation in pigeons be explained by local enhancement together with trial and error learning? *Psychological Science, 8*, 459–465.

Kaminski, J., Nitzschner, M., Wobber, V., Tennie, C., Braeuer, J., Call, J., et al. (2011). Do dogs distinguish rational from irrational acts? *Animal Behaviour, 81*, 195–203.

Kawai, M. (1965). Newly acquired pre-cultural behavior of the natural troop of Japanese monkeys on Kashima Islet. *Primates, 6*, 1–30.

Klein, E. D., & Zentall, T. R. (2003). Imitation and affordance learning by pigeons (*Columba livia*). *Journal of Comparative Psychology, 117*, 414–419.

Kurdek, L., & Rodgon, M. (1975). Perceptual, cognitive, and affective perspective taking in kindergarten through sixth-grade children. *Developmental Psychology, 11*, 643–650.

III. THE STRUGGLE FOR CREATIVITY

Laland, K. N. (2004). Social learning strategies. *Learning & Behavior, 32*, 4−14.

Laland, K. N., & Galef, B. G. (Eds.), (2009). *The question of animal culture.* Cambridge, MA: Harvard University Press.

Laland, K. N., & Williams, K. (1997). Shoaling generates social learning of foraging information in guppies. *Animal Behaviour, 53*, 1161−1169.

Langergraber, K. E., Boesch, C., Inoue, E., Inoue-Murayama, M., Mitani, J., Nishida, T., et al. (2010). Genetic and "cultural" similarity in wild chimpanzees. *Proceedings of the Royal Society of London: B, 278*, 408−416.

Lefebvre, L., & Palameta, B. (1988). Mechanisms, ecology, and population diffusion of socially learned food-finding behavior in feral pigeons. In T. R. Zentall, & B. G. Galef, Jr. (Eds.), *Social learning: Psychological and biological perspectives* (pp. 141−164). Hillsdale, NJ: Erlbaum.

Lepage, J. F., & Théoret, H. (2007). The mirror neuron system: Grasping others' actions from birth? *Developmental Science, 10*, 513−523.

Lorenz, K. (1935). Der kumpanin der umvelt des vogels: Die artgenosse als ausloesendesmoment socialer verhaltensweisen. *Journal für Ornithologie, 83*, 137−213 289−413

Marler, P. (1970). A comparative approach to vocal learning: Song development in white-crowned sparrows. *Journal of Comparative and Physiological Psychology, 71*, 1−25.

McGregor, A., Saggerson, A., Pearce, J., & Heyes, C. (2006). Blind imitation in pigeons (*Columba livia*). *Animal Behaviour, 72*, 287−296.

Meltzoff, A. N. (1995). Understanding the intentions of others: Re-enactment of intended acts by 18-month-old children. *Developmental Psychology, 31*, 838−850.

Meltzoff, A. N., & Moore, M. K. (1977). Imitation of facial and manual gestures by human neonates. *Science, 198*, 75−78.

Mesoudi, A., & White, A. (2008). The multiple roles of cultural transmission experiments in understanding human cultural evolution. *Philosophical Transactions of the Royal Society of London. Series B, Biological Sciences, 363*, 3489−3501.

Miles, H. L., Mitchell, R. W., & Harper, S. E. (1996). Simon says: The development of imitation in an enculturated orangutan. In A. E. Russon, K. A. Bard, & S. T. Parker (Eds.), *Reaching into thought* (pp. 278−299). New York, NY: Cambridge University Press.

Miller, H. C., Rayburn-Reeves, R., & Zentall, T. R. (2009). Imitation and emulation by dogs using the bidirectional control procedure. *Behavioural Processes, 80*, 109−114.

Mineka, S., & Cook, M. (1988). Social learning and the acquisition of snake fear in monkeys. In T. R. Zentall, & B. G. Galef, Jr. (Eds.), *Social learning: Psychological and biological perspectives* (pp. 51−75). Hillsdale, NJ: Erlbaum.

Mitchell, R. W. (1987). A comparative-developmental approach to understanding imitation. In P. P. G. Bateson, & P. H. Klopfer (Eds.), *Perspectives in ethology* (Vol. 7, pp. 183−215). New York, NY: Plenum Press.

Mitchell, R. W. (1997a). A comparison of the self-awareness and kinesthetic-visual matching theories of self-recognition: Autistic children and others. *New York Academy of Sciences, 818*, 39−62.

Mitchell, R. W. (1997b). Kinesthetic-visual matching and the self-concept as explanations of mirror-self-recognition. *Journal for the Theory of Social Behavior, 27*, 101−123.

Moore, D. L., Byers, D. A., & Baron, R. S. (1981). Socially mediated fear reduction in rodents: Distraction, communication, or mere presence? *Journal of Experimental Social Psychology, 17*, 485−505.

Müller, C. A., & Cant, M. A. (2010). Imitation and traditions in wild banded mongooses. *Current Biology, 20*, 1171−1175.

Myowa-Yamakoshi, M., & Matsuzawa, T. (1999). Factors influencing imitation of manipulatory actions in chimpanzees (*Pan troglodytes*). *Journal of Comparative Psychology, 113*, 128−136.

Nagell, K., Olguin, R. S., & Tomasello, M. (1993). Processes of social learning in the tool use of chimpanzees (*Pan troglodytes*) and human children (*Homo sapiens*). *Journal of Comparative Psychology, 107*, 174—186.

Nguyen, N. H., Klein, E. D., & Zentall, T. R. (2005). Imitation of two-action sequences by pigeons. *Psychonomic Bulletin & Review, 12*, 514—518.

Nielsen, M. (2008). The social motivation for social learning. *Behavioral and Brain Sciences, 31*, 33.

Nottebohm, F. (1970). Ontogeny of bird song. *Science, 167*, 950—956.

Palameta, B., & Lefebvre, L. (1985). The social transmission of a food-finding technique in pigeons: What is learned?. *Animal Behaviour, 33*, 892—896.

Pepperberg, I. M. (1986). Acquisition of anomalous communicatory systems: Implications for studies on interspecies communication. In R. Schusterman, J. Thomas, & F. Wood (Eds.), *Dolphin behavior and cognition: Comparative and ethological aspects* (pp. 289—302). Hillsdale, NJ: Erlbaum.

Pepperberg, I. M. (1988). The importance of social interaction and observation in the acquisition of communicative competence: Possible parallels between avian and human learning. In T. R. Zentall, & B. G. Galef, Jr. (Eds.), *Social learning: Psychological and biological perspectives* (pp. 279—299). Hillsdale, NJ: Erlbaum.

Pepperberg, I. M. (2002). Allospecific referential speech acquisition in Grey parrots: Evidence for multiple levels of avian vocal imitation. In K. Dautenhahn, & C. Nehaniv (Eds.), *Imitation in animals and artifacts* (pp. 109—131). Cambridge, MA: MIT Press.

Piaget, J. (1936). *The origins of intelligence in children.* New York, NY: W.W. Norton & Company.

Piaget, J. (1962). *Play, dreams and imitation in childhood.* New York, NY: W.W. Norton & Company.

Pongrácz, P., Miklósi, A., Kubiny, E., Gurobi, K., Topál, J., & Csányi, V. (2001). Social learning in dogs: The effect of a human demonstrator on the performance of dogs in a detour task. *Animal Behaviour, 62*, 1109—1117.

Prather, J. F., Peters, S., Nowicki, S., & Mooney, R. (2008). Precise auditory—vocal mirroring in neurons for learned vocal communication. *Nature, 451*, 305—310.

Range, F., Viranyi, Z., & Huber, L. (2007). Selective imitation in domestic dogs. *Current Biology, 17*, 868—872.

Ray, E., & Heyes, C. (2011). Imitation in infancy: The wealth of the stimulus. *Developmental Science, 14*, 92—105.

Rendell, L., Fogarty, L., Hoppitt, W. J. E., Morgan, T. J. H., Webster, M. M., & Laland, K. N. (2011). Cognitive culture: Theoretical and empirical insights into social learning strategies. *Trends in Cognitive Science, 15*, 68—76.

Rizzolatti, G., Fadiga, L., Fogassi, L., & Gallese, V. (2002). From mirror neurons to imitation: Facts and speculations. In A. Meltzoff, & W. Prinz (Eds.), *The imitative mind: Development, evolution, and brain bases* (pp. 247—266). Cambridge, UK: Cambridge University Press.

Roberts, D. (1941). Imitation and suggestion in animals. *Bulletin of Animal Behaviour, 1*, 11—19.

Russon, A. E., & Galdikas, B. M. F. (1993). Imitation in free-ranging rehabiliatant orangutans. *Journal of Comparative Psychology, 107*, 147—161.

Russon, A. E., & Galdikas, B. M. F. (1995). Constraints on great apes' imitation: Model and action selectivity in rehabilitant orangutan (*Pongo pygmaeus*) imitation. *Journal of Comparative Psychology, 109*, 5—17.

Saar, S., Mitra, P. P., Derégnaucourt, S., & Tchernichovski, O. (2008). Developmental song learning in the zebra finch. In H. P. Ziegler, & P. Marler (Eds.), *Neuroscience of birdsong.* Cambridge, UK: Cambridge University Press.

III. THE STRUGGLE FOR CREATIVITY

Searle, J. R. (1983). *Intentionality, an essay in the philosophy of mind.* New York, NY: Cambridge University Press.

Selman, R. L. (1980). *The growth of interpersonal understanding.* New York, NY: Academic Press.

Sternberg, R. J., & Lubart, T. I. (1991). An investment theory of creativity and its development. *Human Development, 34*, 1–31.

Strupp, B. J., & Levitsky, D. A. (1984). Social transmission of food preferences in adult hooded rats (*Rattus norvegicus*). *Journal of Comparative Psychology, 98*, 257–266.

Tanji, J., & Hoshi, E. (2008). Role of the lateral prefrontal cortex in executive behavioral control. *Physiological Review, 88*, 37–57.

Tennie, C., Call, J., & Tomasello, M. (2006). Push or pull: Imitation vs. emulation in great apes and human children. *Ethology, 112*, 1159–1169.

Tennie, C., Call, J., & Tomasello, M. (2009). Ratcheting up the ratchet: On the evolution of cumulative culture. *Philosophical Transactions of the Royal Society of London. Series B, Biological Sciences, 364*, 2405–2415. Available from http://dx.doi.org/10.1098/rstb.2009.0052.

Tennie, C., Call, J., & Tomasello, M. (2010). Evidence for emulation in chimpanzees in social settings using the floating peanut task. *PLoS One, 5*(5), e10544. Available from http://dx.doi.org/10.1371/journal.pone.0010544.

Tennie, C., Hedwig, D., Call, J., & Tomasello, M. (2008). An experimental study of nettle feeding in captive gorillas. *American Journal of Primatology, 70*, 584–593.

Thornton, A., & Malapert, A. (2009). Experimental evidence for social transmission of food acquisition techniques in wild meerkats. *Animal Behaviour, 78*, 255–264.

Thorpe, W. H. (1961). *Bird song: The biology of vocal communication and expression in birds.* Cambridge, MA: Harvard University Press.

Thorpe, W. H. (1963). *Learning and instinct in animals* (2nd ed.). Cambridge, MA: Harvard University Press.

Thorpe, W. H. (1967). Vocal imitation and antiphonal song and its implications. In D. W. Snow (Ed.), *Proceedings of the XVI international ornithological congress* (pp. 245–263). Oxford, England: Blackwell.

Tolman, C. W. (1964). Social facilitation of feeding behaviour in the domestic chick. *Animal Behaviour, 12*, 245–251.

Tomasello, M. (1990). Cultural transmission in the tool use and communicatory signaling of chimpanzees? In S. Parker, & K. Gibson (Eds.), *"Language" and intelligence in monkeys and apes: Comparative developmental perspectives* (pp. 271–311). Cambridge, UK: Cambridge University Press.

Tomasello, M. (1994). The question of chimpanzee culture. In R. W. Wrangham, W. C. McGrew, F. B. M. de Waal, & P. G. Heltne (Eds.), *Chimpanzee cultures* (pp. 301–317). Cambridge, MA: Harvard University Press.

Tomasello, M., & Call, J. (2004). The role of humans in the cognitive development of apes revisited. *Animal Cognition, 7*, 213–215.

Tomasello, M., Call, J., Warren, J., Frost, G. T., Carpenter, M., & Nagell, K. (1997). The ontogeny of chimpanzee gestural signals: A comparison across groups and generations. *Evolution of Communication, 1*, 223–259.

Tomasello, M., Gust, D., & Frost, T. (1989). A longitudinal investigation of gestural communication in young chimpanzees. *Primates, 30*, 35–50.

Tomasello, M., Savage-Rumbaugh, S., & Kruger, A. C. (1993). Imitative learning of actions on objects by children, chimpanzees, and enculturated chimpanzees. *Child Development, 64*, 1688–1705.

Topal, J., Byrne, R. W., Miklosi, A., & Csanyi, V. (2006). Reproducing human actions and action sequences: 'Do as I do!' in a dog. *Animal Cognition, 9*, 355–367.

Tulving, E. (2004). Episodic memory and autonoesis: Uniquely human? In H. S. Terrace, & J. Metcalfe (Eds.), *The missing link in cognition* (pp. 3–56). New York, NY: Oxford University Press.

Uzgiris, I. C. (1981). Two functions of imitation during infancy. *International Journal of Behavioral Development, 4*, 1–12.

Vallery-Radot, R. (1916). *The life of Pasteur (R. L. Devonshire, Trans.)*. New York, NY: Doubleday.

van Schaik, C. P., Ancrenaz, M., Borgen, G., Galdikas, B., Knott, C. D., Singleton, I., et al. (2003). Orangutan cultures and the evolution of material culture. *Science, 299*, 102–105.

Visalberghi, E., & Fragaszy, D. (1990). Do monkeys ape? In S. Parker, & K. Gibson (Eds.), *Language and intelligence in monkeys and apes: Comparative developmental perspectives* (pp. 247–273). Cambridge, UK: Cambridge University Press.

Voelkl, B., & Huber, L. (2000). True imitation in marmosets. *Animal Behaviour, 60*, 195–202.

Warden, C. J., & Jackson, T. A. (1935). Imitative behavior in the rhesus monkey. *Journal of Genetic Psychology, 46*, 103–125.

Weigle, P. D., & Hanson, E. V. (1980). Observation learning and the feeding behavior of the red squirrel (*Tamiasciurus hudsonicus*): The ontogeny of optimization. *Ecology, 61*, 213–218.

White, D. J., Gros-Louis, J., King, A. P., Papakhian, M. A., & West, M. J. (2007). Constructing culture in cowbirds (*Molothrus ater*). *Journal of Comparative Psychology, 121*, 113–122.

Whiten, A. (1993). Human enculturation, chimpanzee enculturation, and the nature of imitation: Commentary on cultural learning, by Tomasello et al. *Behavioral and Brain Sciences, 16*, 538–539.

Whiten, A. (1998). Imitation of the sequential structure of actions by chimpanzees (*Pan troglodytes*). *Journal of Comparative Psychology, 112*, 270–281.

Whiten, A. (2000). Primate culture and social learning. *Cognitive Science, 24*, 477–508.

Whiten, A. (2005). The imitative correspondence problem: Solved or sidestepped? In S. Hurley, & N. Chater (Eds.), *Perspectives on imitation: From neuroscience to social science* (Vol. 1, pp. 220–222). Cambridge, MA: MIT Press.

Whiten, A., & Custance, D. (1996). Studies of imitation in chimpanzees and children. In C. M. Heyes, & B. G. Galef (Eds.), *Social learning in animals: The roots of culture* (pp. 291–318). San Diego, CA: Academic Press.

Whiten, A., & Ham, R. (1992). On the nature and evolution of imitation in the animal kingdom: Reappraisal of a century of research. In P. J. B. Slater, J. S. Rosenblatt, C. Beer, & M. Milinski (Eds.), *Advances in the study of behavior* (Vol. 21, pp. 239–283). New York, NY: Academic Press.

Whiten, A., Horner, V., Litchfield, C., & Marshall-Pescini, S. (2004). How do apes ape? *Learning & Behavior, 32*, 36–52.

Whiten, A., McGuigan, N., Marshall-Pescini, S., & Hopper, L. M. (2009). Emulation, imitation, over-imitation and the scope of culture for child and chimpanzee. *Philosophical Transactions of the Royal Society of London. Series B, Biological Sciences, 364*, 2417–2428.

Zajonc, R. B. (1965) Social facilitation. *Science, 149*, 269–274.

Zentall, T. R. (1996). An analysis of imitative learning in animals. In C. M. Heyes, & B. G. Galef, Jr. (Eds.), *Social learning and tradition in animals* (pp. 211–243). Hillsdale, NJ: Erlbaum.

Zentall, T. R., Edwards, C. A., & Hogan, D. E. (1983). Pigeons' use of identity. In M. L. Commons, R. J. Herrnstein, & A. Wagner (Eds.), *The quantitative analyses of behavior: Vol. 4. Discrimination processes* (pp. 273–293). Cambridge, MA: Ballinger.

Zentall, T. R., & Hogan, D. E. (1975). Key pecking in pigeons produced by pairing key light with inaccessible grain. *Journal of the Experimental Analysis of Behavior, 23*, 199–206.

III. THE STRUGGLE FOR CREATIVITY

Zentall, T. R., Sutton, J. E., & Sherburne, L. M. (1996). True imitative learning in pigeons. *Psychological Science, 7*, 343–346.

Zukow-Goldring, P., & Arbib, M. A. (2007). Affordances, effectivities, and assisted imitation: Caregivers and the directing of attention. *Neurocomputing, 70*, 2181–2193.

Commentary on Chapter 12: Imitation and Creativity

Creativity is often thought of as a no-holds-barred, anything-goes, outside-the-box process in which rules are set aside and thinking is unconstrained in any way. That is a romantic, but I think very mistaken, way of thinking about creativity. As poet McClatchy (2014) said at a recent reading (in response to a question from fellow Yale professor Penelope Laurans about structure in his poetry), "I can't write without some form in mind."

Creativity often needs some structure and constraints, as shown in the work of psychologist Haught (2013, in press) (see also Haught & Johnson-Laird, 2003), who gave subjects writing tasks either with or without constraints, and found that those who have constraints of various kinds tended to produce more creative products. Of course, there is no denying that constraints can be *too* constraining (e.g., paint-by-numbers), but they can also be too lax, leaving the situation wide open. And while creativity may indeed sometimes take us "outside the box," even then the box from which one is escaping still plays a key role: its part of the context responsible for producing the creative product.

Imitation is often thought of as exactly the opposite of such a wild, untamed, out-of-the-clear-blue-sky process; it seems like a rote, completely constrained, unimaginative, and almost unthinking process. How interesting could mere copying be? I love the way Zentall's chapter shows the incredible complexity of the kinds of cognition involved in many kinds of imitation. This made me think of Beghetto and Kaufman's (2007; Kaufman & Beghetto, 2009) 4-C model of creativity. The model begins with mini-c, "the personal insights that are part of the learning process. These subjective self-discoveries are meaningful to the person, even if other people may not recognize the ideas as being creative" (Beghetto, Kaufman, & Baer, in press).

Acquiring a skill typically requires practice, and some of that practice may be essentially imitation of expert work. In the same talk by McClatchy mentioned earlier, the poet described how, as an apprentice poet, he tried to write like other poets he admired (and subsequently, he added, spent years trying to overcome that purposefully acquired influence). Writers often do this—it's a common practice in creative writing courses to intentionally mimic the styles of others—and this conscious (and often self-conscious) imitation probably results in temporary declines in creativity as writers copy, then adapt, new styles of writing. Some creativity commentators have even reported across-the-board declines in creativity at certain developmental stages. Probably the most famous claim of this kind came from Paul Torrance, who (based on scores on his divergent thinking tests) thought there was a major "slump" in the upper elementary grades during which creativity decreased for a period of time (the result perhaps of an internalization of school-based values of what are the right and wrong ways to do things and a resulting loss of freedom to experiment; Runco, 1999; Torrance, 1968). Other studies have contested this pattern with data that show a more linear development, with creativity increasing fairly regularly with age (Baer, 1996; Lau & Cheung, 2010; Mullineaux & Dilalla, 2009). Still others have suggested a J-shaped trajectory (Smolucha & Smolucha, 1985) or U-shaped trajectories of various kinds (Besancon & Lubart, 2008; Daugherty, 1993; Gardner, 1987; Rosenblatt & Winner, 1988).

It seems likely that the different results reported by this diverse group of researchers may be the result of their looking at creativity in different domains. Creativity in painting, for example, might indeed show temporary declines as children (or perhaps older art students) study and practice—and sometimes directly copy—the work of others in order to gain technical skill, whereas creativity in chemistry might show no such decline (and if it did, this would probably happen at different ages). As evidence of this domain-specific effect on the developmental trajectory of creativity, Gardner and Davis (2013) found that the development of creativity in adolescents moved in two totally opposite directions in the domains of creative writing and art during the historical period they studied. (It should be noted that Gardner and Davis did not propose that this data necessarily represented regular patterns of adolescent creativity development, but rather suggested that these differences might be the result of societal trends in the 21-year period they examined.)

Imitation as a way to develop a skill necessary for creative performance may be part of the normal development of creativity in some (and probably many) domains; something that may or may not occur at a particular age but which, whenever it occurs, is an almost unavoidable aspect of the development of one's creative thinking skills. Imitation may not be the antithesis or the enemy of creativity. It may, in fact, be an essential part of creativity.

Zentall's chapter helped me to appreciate the cognitive complexity of imitative behavior in both animals and humans and also the complexity of creativity, which may include and often even require, not only constraints, but also imitation.

References

Baer, J. (1996). Does artistic creativity decline during elementary school years? *Psychological Reports, 78*, 927–930.

Beghetto, R. A., & Kaufman, J. C. (2007). Toward a broader conception of creativity: A case for mini-c creativity. *Psychology of Aesthetics, Creativity, and the Arts, 1*, 73–79.

Beghetto, R. A., Kaufman, J. C., & Baer, J. (in press). *Creativity and the common core.* New York, NY: Teachers College Press.

Besancon, M., & Lubart, T. I. (2008). Differences in the development of creative competencies in children schooled in diverse learning environments. *Learning and Individual Differences, 18*, 381–389.

Daugherty, M. (1993). Creativity and private speech: Developmental trends. *Creativity Research Journal, 6*, 287–296.

Gardner, H. (1987). *Art, mind, and brain: A cognitive approach to creativity.* New York, NY: Basic Books.

Gardner, H., & Davis, K. (2013). *The app generation.* New Haven, CT: Yale University Press.

Haught, C. (2013). Role of constraints in creativity. *Paper presented at the annual meeting of the American Psychological Association.* Honolulu, HI, August 2013.

Haught, C. (in press). Constraints that foster creativity. In J. C. Kaufman, & J. Baer (Eds.), *The Cambridge companion to creativity and reason in cognitive development.* Cambridge University Press.

Haught, C., & Johnson-Laird, P. N. (2003). Creativity and constraints: The production of novel sentences. In: *Proceedings of the 25th annual meeting of the Cognitive Science Society* (pp. 528–532).

Kaufman, J. C., & Beghetto, R. A. (2009). Beyond big and little: The four C model of creativity. *Review of General Psychology, 13*, 1–12.

Lau, S., & Cheung, P. C. (2010). Developmental trends of creativity: What twists of turn do boys and girls take at different grades? *Creativity Research Journal, 22*, 329–336.

McClatchy, J. D. (2014). *Yale University's Jonathan Edwards College Presents: Master's Tea with J.D. McClatchy. Talk and poetry reading.* New Haven, CT, April 10, 2014.

Mullineaux, P. Y., & Dilalla, L. F. (2009). Preschool pretend play behaviors and early adolescent creativity. *Journal of Creative Behavior, 43*, 41–57.

Rosenblatt, E., & Winner, E. (1988). The art of children's drawings. *Journal of Aesthetic Education, 22*, 3–15.

Runco, M. A. (1999). Fourth grade slump. *The Encyclopedia of Creativity, 1*, 743–744.

Smolucha, L. W., & Smolucha, F. C. (1985). A fifth Piagetian stage: The collaboration between analogical and logical thinking in artistic creativity. *Visual Arts Research, 11*, 90–99.

Torrance, E. P. (1968). A longitudinal examination of the fourth grade slump in creativity. *Gifted Child Quarterly, 12*(4), 195–199.

13

Of Course Animals Are Creative: Insights from Generativity Theory

Robert Epstein

American Institute for Behavioral Research and Technology, Vista, CA, USA

Commentary on Chapter 13: Defining Animal Creativity: Little-C, Often; Big-C, Sometimes

Dean Keith Simonton

Department of Psychology, University of California, Davis, Davis, CA, USA

Of course animals are creative, especially if one defines "creative" advantageously. That is easy to do, given that the word comes from the vernacular. That gives us a great deal of latitude in constructing our definition.

My favorite definition of creative, which works pretty well in most natural uses of the term but is by no means definitive, is as follows: "Creative" is how we label someone's behavior or the product of that behavior if that behavior or its product is both *new* to some degree and also *of value* to a pertinent community. Sometimes, if someone behaves frequently and consistently in this way, we might even label the individual him- or herself "creative."

375

So if a child builds the same new and amazing block structure over and over again, we might say she was creative the first time she built it but probably not the second, and if the very first time she built it, she was simply copying someone else, again, it is unlikely we would label her behavior or the structure creative. If the structure is simple and boring, the language of creativity also won't usually be applied. Notable people in that child's community—a teacher, a parent, or a grandparent, typically—have to recognize the structure as intricate or interesting or esthetically pleasing. In other words, they have to *value* it before they will call it creative—generally speaking, the younger the child, the more relaxed the requirement.

The same criteria are generally applied to adults, even to accomplished adults such as artists and inventors. The main difference between judgments made regarding the creativeness of children and the creativeness of adults is the nature of the community. Mom or Dad have all the authority it takes to pronounce their child's artwork creative (but perhaps not the artwork of another child), whereas an adult writer or sculptor or inventor is at the mercy of a larger, more discerning panel of judges: critics, editors, art collectors, and so on—a panel that might even change from time to time, meaning that something that is judged creative by one community or in one generation might not be by another community or the next generation.

Novelty is still critical, however, no matter how generous the judges. No matter how squiggly Jackson Pollack's squiggles, if a dozen other artists had adopted his style before he did, his work would probably have been ignored—dismissed, perhaps, as copycat art, even if Pollack knew nothing of those other artists. So novelty is critical, and so is the community. In fact, one of the oddest things about the language of creativity is that one cannot credibly apply it to oneself. "Look, everybody— see how creative I am" does not impress and might even bring ridicule.

With these elements in mind, are animals creative? In this brief essay I will not only show that all animals are creative to some extent, I will also offer what I believe is an evolutionarily sound model for understanding why *both* animals and people behave creatively. Specifically, I will offer evidence that the same neural mechanisms, which can be expressed in formal terms using a theory I introduced in the 1980s called "generativity theory" (Epstein, 1985a, 1990, 1991, 1996a, 1999, 2014)— underlie the emergence of novel behavior in both animals and people.

THOSE AMAZING ANIMALS

In 1984, my colleagues and I reported that pigeons with appropriate training could solve Köhler's (1925) classic box-and-banana problem in a human-like, or, if you prefer, chimp-like manner, when faced with the problem for the first time. Like bright children faced with similar problems, our pigeons first looked confused for a while and then fairly suddenly solved the problem, pushing a box under a toy banana, climbing onto the box, and pecking the banana (Epstein, Kirshnit, Lanza, & Rubin, 1984). This kind of performance is not only novel—we could guarantee that, after all, by controlling and monitoring the behavioral histories of our pigeons—it also has obvious value. With animals, there is no community of peers available to judge a performance creative, but there is an obvious situation in which novel behavior demonstrates value: that is, when it solves a problem.

People have witnessed novel problem-solving behavior in animals for as long as they have been observing them, no doubt. Our family dog, Tiny Bryan, drove us all crazy for nearly a year by finding new ways to get through whatever gate we installed to keep her (yes, "Bryan" was female) out of our dining room, where she often left mementos on the carpet. Even more remarkably, none of us ever *saw* her get through a gate—never! So she not only was devising new ways to get through, she also had the good sense to practice her art only after determining that no one was observing her, just like Andy Dufresne in "The Shawshank Redemption." Brilliant!

Sometimes it was obvious that Tiny Bryan was using brute force because a gate was damaged, but most of the time we had no idea how she did it. When I installed an especially strong and tall gate, we were truly perplexed. Had she grown wings? My point is that with each new gate, our dog had to devise one or more new ways to get through.

In both laboratory and field settings, notable examples of novel problem-solving performances have been observed in many species, including dogs (e.g., Bräuer, Bös, Call, & Tomasello, 2013); all non-human primate species, to my knowledge, that have been studied (e.g., Avdagic, Jensen, Altschul, & Terrace, 2013; Boinski, 1988; Burkart, Strasser, & Foglia, 2009; Call & Tomasello, 1995; Capitanio & Mason, 2000; Fujita, Sato, & Kuroshima, 2010; Inoue-Nakamura & Matsuzawa, 1997; Kappeler, 1987; Menzel, Savage-Rumbaugh, & Lawson, 1985; Santos, Ericson, & Hauser, 1999; Tecwyn, Thorpe, & Chappell, 2012; Watson & Ward, 1996); elephants (Hart, Hart, McCoy, & Sarath, 2001), who are even said to be capable of exhibiting "insight" (Foerder, Galloway, Barthel, Moore, & Reiss, 2011); rodents (Tokimoto & Okanoya, 2004; Wass et al., 2012); marine mammals (e.g., Hille, Dehnhardt, & Mauck, 2006; Mann et al., 2008; Scholtyssek, Kelber, Hanke, & Dehnhardt, 2013; Schusterman, Thomas, & Wood, 1986); a wide variety of birds (e.g., Funk, 2002; Liker & Bokony, 2009; Pepperberg, 2004; Webster & Lefebvre, 2001); and even, to some extent, fish (Bshary, Wickler, & Fricke, 2002; Reader & Laland, 2000).

It is one of the great mysteries of comparative psychology that birds of the corvid family, particularly crows and ravens, are especially adept in this regard (e.g., Bird & Emery, 2009; Heinrich, 1995; Heinrich & Bugnyar, 2005; Hunt, Corballis, & Gray, 2001; Taylor, Elliffe, Hunt, & Gray, 2010; Taylor, Medina, et al., 2010; Von Bayern, Heathcote, Rutz, & Kacelnik, 2009; Wimpenny, Weir, Clayton, Rutz, & Kacelnik, 2009). It has even been argued that crows "understand their physical and social worlds" in much the same way that apes do, suggesting that convergent evolutionary processes are responsible for the extraordinary intelligence of both (Emery & Clayton, 2004, p. 1903; cf. Kirsch, Güntürkün, & Rose, 2008).

In one of the more ambitious studies of this sort, seven New Caledonian crows (*Corvus moneduloides*) were presented with various tasks requiring the novel use of small sticks to retrieve food from one of two types of containers (Wimpenny et al., 2009). The tasks were varied, the authors stated, in order to determine the extent to which crows "understand" (p. 1) what they are doing and to shed light on the "cognitive mechanisms" (p. 1) underlying the performances. In what was clearly the most spectacular of the performances—and indeed the only one of its sort reported in the study—a crow dubbed "Betty" (i) immediately tried to reach a long stick (20 cm) that was out of reach in a transparent tube (this was spurious behavior, not relevant to the solution), (ii) picked up (with its beak) a short stick (6 cm) that was within reach, (iii) used the short stick to retrieve a medium-length stick (10 cm) from a different transparent tube, (iv) briefly tried to retrieve food from another transparent tube using the medium-length stick (also spurious behavior, given that the stick was not long enough to reach the food), (v) used the medium-length stick to retrieve the 20 cm stick from the first tube, and then (vi) used the long stick to retrieve the food. The entire performance, which was labeled "the first observation of spontaneous three-tool sequential tool use in a non-human animal" (p. 6), took about 53 s (cf. Epstein, 1985c, 1987; Epstein & Medalie, 1983).

The fact that Betty *immediately* tried to reach the long object upon flying into the test area suggests a long history of experience with respect to the sticks, and indeed, this was actually Betty's fourth trial in somewhat similar test situations, and she also had extensive experience with various sticks and tubes in a "pre-testing procedure" ("a minimum of 30 trials" plus "six familiarization trials" (p. 13)). The fact that Betty tried to reach the food with the medium-length stick also indicates that this was not a fully "reasoned" performance in the human sense.

Nevertheless, it was a novel performance that solved a problem, and in that sense it could be called creative. It was also unique among the other birds tested, meaning that Betty is very much like Sultan, Köhler's (1925) precocious chimpanzee that was able to solve the famous box-and-banana problem in an "insightful" fashion when five other chimpanzees, standing nearby, could not. Unlike Betty, however, Sultan was fairly inactive and looked pensive for several minutes before suddenly solving the problem. That period of quiescence is important, if only because it can be so easily interpreted in human terms.

Like other recent studies of this sort with corvids (e.g., Taylor, Elliffe, et al., 2010), the Wimpenny et al. (2009) study was designed to shed light on cognitive mechanisms, specifically by making predictions along the following lines:

a. *A lack of spurious actions would suggest that the birds were planning or reasoning.* But *all* of the birds tried at times to reach the food with the shortest stick, and even Betty, in her most impressive performance, reached at the wrong times for both the long stick and the food. Collectively, the birds made spurious probes with the short stick on at least 16 of 29 trials in the authors' first experiment.

b. *A correct choice of sticks would suggest that the birds were planning or reasoning.* However, generally speaking, "subjects did not appear to choose in advance the tools they required" (p. 5).

c. *When presented with the sticks but no food, a bird shouldn't bother manipulating the sticks, which would suggest that successful performances were truly "goal-directed."* But when a bird that had learned to retrieve sticks from tubes was confronted with this situation, although it never probed the empty food-tube with the shortest tool, it did use that tool "to extract further tools on all trials and used these to probe into [the food-tube]" (p. 6). In a subsequent procedure, "all subjects did insert an extracted tool into the (empty) food-frame on at least one trial" (p. 9).

d. *If birds without relevant experiences with sticks could solve the problem, that would suggest they were capable of advanced reasoning abilities.* But in the authors' first experiment, all three of the birds that lacked experiences with sticks "failed in all tasks where tools had to be extracted" (p. 6). In a second experiment, one bird "used a tool to extract another tool on his very first trial" (p. 12), but it wasn't able to retrieve food with any tool. Moreover, one bird "frequently took the extracted tools to other parts of the aviary, suggesting that, although crows responded to some extent appropriately to the contingencies of the task, they were motivated to extract tools *per se*" (p. 9).

Although the authors argued that the number of errors they observed in some situations was less than one might expect by chance, overall, they offered little evidence of any substantive cognitive activity in their birds, and, indeed, they were forced to conclude that "we do not implicate reasoning (or a lack of it) as an explanation for our crows' behavior" (p. 12).

III. THE STRUGGLE FOR CREATIVITY

Other researchers, reviving the old debate about behaviorism that began more than a century ago (see King, 1930; Watson & McDougall, 1928), have taken various stands on this issue, some insisting that problem-solving performances in animals demonstrate clear signs of human-like cognitive activity (e.g., Taylor, Elliffe, et al., 2010), others taking a more conservative stand (e.g., Shettleworth, 2009, 2010). This debate has never yielded a winner and never will; it is pointless, in my view (Epstein, 1984, 1996a, 2014).

ANIMAL ART

In judging the creativeness of animal behavior, we needn't limit ourselves to problem solving. Novel human behavior can be judged to be creative, after all, if it has virtually *any* kind of value—esthetic (think Michelangelo), economic (the first Apple desktop computer), or even just bizarre (Lady Gaga). Non-human animals lack a linguistic community capable of judging their behavior to be creative, but that doesn't mean humans can't step up to fill the void. Oddly enough, even pigeons can be trained to discriminate good human art from bad human art (Watanabe, 2010). It should be a simple matter for us to make similar discriminations regarding animal creativity.

Are there examples of novel animal behavior that humans might reasonably judge to be creative? A YouTube video that has had nearly 10 million views shows an elephant named Paya, a resident at the Maesa Elephant Camp in Chiang Mai, Thailand, painting a fairly simple but beautiful image of an elephant holding a flower.[1] Unfortunately, this performance will not satisfy our definition of creative, because Paya was painstakingly *trained* to paint this image over and over again every day to entertain tourists, and he was also probably coached during the making of the video.[2]

Other elephants, however, given more of a free trunk (so to speak), have occasionally covered canvasses with images impressive enough to draw large sums from human art collectors (Bourgeois, 2005; Suttle & The, 1997). A 2012 exhibition at the University College London's Grant Museum of Zoology displayed paintings by elephants, gorillas, orangutans, chimpanzees, and even a bowerbird which many people would consider to be indistinguishable from comparable human art (Guerke, 2012).

Not every painting by a particularly talented elephant is praiseworthy, but the same can be said of the paintings of human artists. Years ago, I had the pleasure of viewing Andrew Wyeth's "Helga" exhibit at

[1]http://www.youtube.com/watch?v=He7Ge7Sogrk.

[2]http://urbanlegends.about.com/library/bl_elephant_painting.htm.

a museum in Boston. This was one of the most unique art exhibits of its day, because the collector who bought the completed paintings insisted on buying and having the right to preserve and show more than 200 preliminary drawings and studies that preceded the finished works, some so crude that a young child—or chimp—could have made them. We are willing, in short, to judge someone creative if he or she behaves creatively only occasionally. If we give the same latitude to animals, we will almost certainly judge many animals to be creative.

GENERATIVITY THEORY

Whether novel animal behavior has value in the same sense that novel human behavior has value is debatable, just as the significance of computer-generated art and music is debatable, or even the significance, for that matter, of any computer behavior that seems to mimic human intelligence (Epstein, Roberts, & Beber, 2008; Turing, 1950). But the fact that animals do new things is indisputable. Where does novel behavior come from, and is it possible that the same basic processes are responsible for the novel behavior of both animals and people?

In the early 1980s, I proposed a theory to try to explain the emergence of novel behavior in multiple species. According to generativity theory, (i) new behavior emerges as previously established repertoires of behavior become interconnected over time, and (ii) the interconnection process is both orderly and predictable. Combinatorial theories of creativity have been around for a long time (consider Hull, 1935; Poincaré, 1946; Rothenberg, 1971); the main thing I added to the picture was an assertion of orderliness, as well as a way of quantifying and measuring this orderliness. Specifically, I speculated that multiple processes operate simultaneously on the probabilities of multiple behaviors, and, by implication, on their underlying neurological systems.

This is a sensible and, in some respects, simple theory. It is also predictive and parsimonious. Most important for present purposes, generativity theory explains why novel behavior—a small portion of which will inevitably have value in specific situations—is ubiquitous in the animal kingdom. The orderly and rapid interconnection of repertoires of behavior is *adaptive*; it is about as adaptive a process as nature could possibly devise because it guarantees that all previously established behaviors, whether learned or programmed by genes, will be available to tackle new challenges as they arise moment to moment in time in the environment. Generativity theory shows how this process works without making any assumptions about speculative cognitive mechanisms—or denying their existence.

The same mechanisms underlie the emergence of novel behavior in both animals and people because novelty-yielding behavioral variability has value, just as novelty-yielding genotypic and phenotypic variability have value in evolution. Species survive and sometime diverge because of phenotypic variants. The range of variability in traits yielded by sexual reproduction is so large that genuinely new traits inevitably emerge in every generation—a phenomenon that helps to protect a species from extinction when environmental conditions change and that ensures that superior traits and even superior species will ultimately triumph over time. Similarly, novelty-yielding behavioral variability in an individual organism helps to guarantee that that individual's behavior will be effective under changing conditions, occasionally even producing behavior so new that it can change the organism's environment in significant ways.

These two types of variability might even overlap. Genes might occasionally produce individual organisms so creative that they find effective new ways to compete against other species for limited resources, eventually dislodging those species from the gene pool.

Two ubiquitous situations in the natural environment assure that behavioral competition occurs almost continuously. First, the real world surrounds us constantly with multiple, novel, and vague stimuli which set multiple behaviors in motion. Second, when behavior is ineffective—which it is in small ways hundreds of times a day—a process called "resurgence" occurs: previously established behaviors that were effective under conditions similar to the current ones are activated (Epstein, 1983, 1985b, 1996a; cf. Epstein & Skinner, 1980). Generative processes thus assure not only that behavior will *vary* but that it will vary in ways that are especially likely to be adaptive in the current situation. In this sense the variability that occurs in individual organisms is even more adaptive than the variability that drives the evolution of species; the latter is blind, but the former guarantees the emergence of new behavior that is specifically relevant to the properties of the new situation. Generativity theory also has immediate practical value, because when you identify variables and parameters that contribute to the emergence of novel behavior, you can manipulate those variables and parameters for useful ends (Epstein, 1996b, 2000, 2011; Epstein, Kaminaka, Phan, & Uda, 2013; Epstein & Phan, 2012; Epstein, Schmidt, & Warfel, 2008).

To explore the potential power of generative competition, I expressed the theory in formal terms and modeled it with a computer algorithm. Representing just four long-studied behavioral processes in a series of equations I called "transformation functions" (Figure 13.1), I was able to model a number of creative performances I had studied in laboratory settings with both pigeons and people, such as the manner in which

(1) *Extinction:* $y_{n+1} = y_n - y_n * \varepsilon$

(2) *Reinforcement:* $y_{n+1} = y_n + (1-y_n) * \alpha$

(3) *Resurgence:* for $\lambda_{yy'} < 0$ and $y'_n - y'_{n-1} < 0$,

$y_{n+1} = y_n + (1-y_n) * (-\lambda_{yy'}) * y'_n$

(4) *Automatic* for $\lambda_{yy'} > 0$ and $y'_n - y'_{n-1} > 0$,

Chaining: $y_{n+1} = y_n + (1-y_n) * \lambda_{yy'} * y'_n$

FIGURE 13.1 **The transformation functions of generativity theory.** According to generativity theory, multiple behavioral processes operate simultaneously on the probabilities of multiple behaviors. In one possible instantiation of the theory, four basic behavioral processes are represented (above). y_n is the probability of behavior y at cycle n of the algorithm, y'_n is the probability of behavior y' at cycle n of the algorithm, ε is a constant for extinction (it determines the rate at which the probability of behavior y decreases over cycles of the algorithm), α is a constant for reinforcement (it determines the rate at which the probability of behavior y increases over cycles of the algorithm as a result of certain environmental events), and $\lambda_{yy'}$ is the constant of interaction between behaviors y and y'.

people solved Maier's (1931) classic "two-string problem" (Figure 13.2). Running simultaneously in a "state" algorithm, the transformation functions generate a "probability profile," in which overlapping probability curves show how behavior changes over time, producing novel behavior almost continuously as other behaviors become interconnected over time (Figure 13.3). I also devised a way to show the orderliness in the novel performance of a single subject using a graphical technique that generates a "frequency profile." The frequency profile yields overlapping curves that look very much like probability curves (Figure 13.4).

This methodology proved effective in predicting novel performances in both pigeons and people in laboratory performances. The differences between the two species seem largely to be parametric. The general principles—that new behavior emerges as old behaviors merge and that multiple behavioral processes operate simultaneously on the probabilities of different behaviors—seem valid for both species and perhaps for many others as well.

The formal representation and quantification of the creative process in individual organisms is no small feat, but with few exceptions the existing literature on animal creativity and problem solving shows little or no awareness of any of the main implications of generativity theory and its supporting research. Instead researchers are (and I say this respectfully) wasting their time debating about what animals may or may not "understand" regarding their performances, as if that information—were it even possible to determine with any degree of confidence, which it is not—would add anything important to our own "understanding" of the performances.

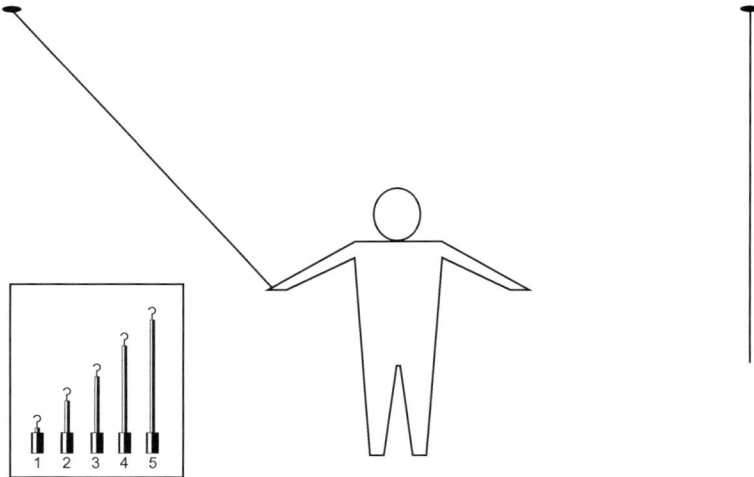

FIGURE 13.2 **Maier's (1931) Two-string problem.** Subjects are instructed to tie the two ends of the strings together, but they quickly learn that they can't reach both strings at once. They learn this by pulling one string toward the other and reaching. Most people then try pulling the second string toward the first, which makes little sense. When provided with a long heavy object (#5 in inset), a subject is highly likely to use it to extend his or her reach, but the object that is provided is not long enough to reach the other string. When provided with a short heavy object (#1), a subject is much more likely to solve the problem, which requires tying the object, short or tall, to one string and swinging it, then pulling the other string toward the swinging string and catching it when it comes near. Appropriately, the problem is sometimes called "the pendulum problem." Provided with a long object, if a subject is able to solve the problem at all, automatic chaining is usually involved. The person ties the long object to the end of a string and then pulls the object toward the second string; this is one way of using the object to extend one's reach. When that fails, the subject often lets go of the object, which causes the attached string to swing in a pendulum motion. The solution follows rapidly. Objects of intermediate lengths produce predictable outcomes according to those lengths.

When you speculate about what an animal "understands" regarding an arrangement of stimuli or the manner in which it solved a problem, you are simply talking about *more behavior*. You are asking whether the animal can not only solve the problem but can also *state a principle* or *visualize a causal flow diagram* or perhaps even *perform mathematical calculations*. The problem here—which has existed as long as comparative psychology has existed—is that *as long as you can conceive of a way for the animal to have solved the problem without engaging in these so-called "high-level" cognitive manipulations, you must make the parsimonious assumption that the animal is not doing so.*

Even if we were able to show definitely that an animal had formulated the (human) verbal equivalent of a formal principle, I have little doubt that the processes giving rise to the emergence of such a

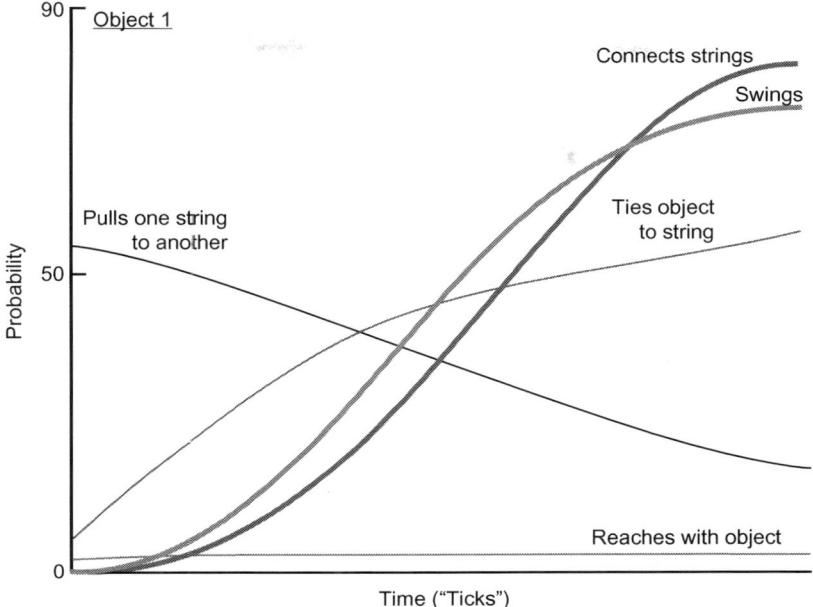

FIGURE 13.3   Probability profile for Maier's (1931) two-string problem. A probability profile generated by the transformation functions shown in Figure 13.1, generated for five behaviors relevant to Maier's (1931) two-string problem. The abscissa is labeled "ticks," which are cycles of the computer algorithm, each a scalable moment of unspecified duration. The profile was generated with parameters for a short object (#1 in Figure 13.2), which generally produced rapid solutions to the problem and no irrelevant reaching. Note that pulling one string toward the other decreases steadily in probability and that other behaviors increase in probability in an orderly sequence. Tying the object to the string makes swinging more likely, which, in turn, makes connecting the strings more likely. The computer model that generates the curve uses discrete state methodology, running a set of initial probabilities through all four equations to generate a new set of probabilities, then running those through the equations again, and so on.

principle—again, this is just more behavior, after all—would be similar if not identical to the processes that led to the emergence of the problem-solving performance itself. In footnote 5 of an early paper I published on the principle of resurgence, I explained how generative processes—in particular, the process of resurgence itself—could account for my formulation of the formal principle of resurgence itself (Epstein, 1985b, p. 151; also see Epstein, 1996a, p. 145). A statement of a relationship among variables—in other words, "reasoning"—is, first and foremost, a *statement*; again, it is just more behavior, presumably amenable to the same sort of analysis that can be applied to all behavior.

Without exception, all of the studies of animal problem solving or creativity I have seen in recent years lend themselves to a rigorous

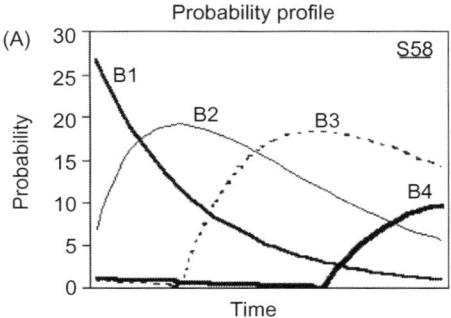

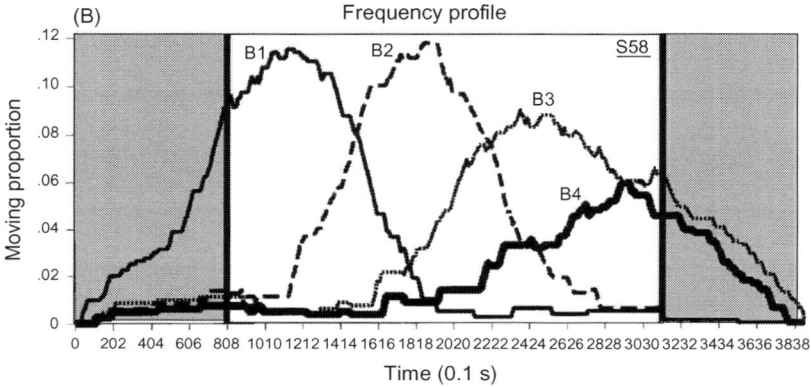

FIGURE 13.4 **Predicting individual behavior moment to moment in time.** (A) This probability profile, produced by the transformation functions of the generativity model, predicts the behavior of a human subject on a touch-screen task. The subject has been instructed to move a spot across the screen into a goal area. Tapping three patches on the screen (B1, B2, and B3) will move the spot in various directions and at varying speeds; tapping a fourth patch (B4) has no effect. The model predicts that the subject will begin tapping B1, then gradually shift to B2, then gradually shift to B3, with responses alternating among the three choices along the way (where the curves overlap). It also predicts that toward the end of the session, the subject will begin tapping B4, even though doing so has no effect. (B) This frequency profile shows actual data obtained from one subject (S58) during a 5 min session. The pattern of responding is predicted well by the probability profile, including the shift to B4.

analysis using the tools of generativity theory. Video recordings can be examined frame by frame, coded, and represented by frequency profiles, which instantly reveal orderliness which is almost entirely invisible to the naked eye. Instead, researchers are still relying on crude verbal descriptions of the performances, or, at most, rough tabulations of "percentage correct" and other data aggregated across organisms or trials. Problem situations, which by definition, are bounded in specific ways, can easily be represented in formal terms using transformation functions, and those functions can then be used to model and predict individual performances. In short, the orderliness in animal creativity can be quantified and studied rigorously using advanced tools of the natural sciences.

Whether generativity theory is correct in its particulars is beside the point. The predictive power of this type of theory is so great that some form of it almost certainly must be correct. Almost certainly, multiple processes must be acting simultaneously on the probabilities of multiple behaviors and their counterparts in the nervous system, and the net result generates a wide range of behavior continuously in time—everything from mundane grooming to profound "insight."

Meanwhile, many researchers in psychology and biology who are rightfully fascinated by some of the extraordinary human-like capabilities of crows, chimpanzees, and other animals, are barking up the wrong tree. Like the naturalists of the 1800s, they continue to anthropomorphize, insinuating that human-like performances by animals are interesting only if an animal's cognitive world is like a human's.

Meanwhile, I wonder, as many have before me, why we continue to bother speculating about the cognitive world of *humans*; I have long seen this as a dilemma in which our consciousness interferes with our scientific objectivity (Epstein, 1982, 2008).

Acknowledgments

I thank Adam Peng for suggesting relevant articles on animal creativity and Misti Vaughn for help in the preparation of the manuscript. I can be reached at re@aibrt.org.

References

Avdagic, E., Jensen, G., Altschul, D., & Terrace, H. S. (2013). Rapid cognitive flexibility of rhesus macaques performing psychophysical task-switching. *Animal Cognition*, 1—13. Available from: http://dx.doi.org/10.1007/s10071-013-0693-0.

Bird, C. D., & Emery, N. J. (2009). Insightful problem solving and creative tool modification by captive nontool-using rooks. *Proceedings of the National Academy of Sciences USA*, *106*, 10370—10375.

Boinski, S. (1988). Use of a club by a wild white-faced capucin (*Cebus capucinus*) to attack a venomous snake (*Bothrops asper*). *American Journal of Primatology*, *14*, 177—179.

Bourgeois, P. (2005). *Elephant's painting raises $7,000 for tsunami relief on eBay*. Knight Ridder Tribune News Service, Retrieved from: http://search.proquest.com/docview/456580893?accountid=28103.

Bräuer, J., Bös, M., Call, J., & Tomasello, M. (2013). Domestic dogs (*Canis familiaris*) coordinate their actions in a problem-solving task. *Animal Cognition*, *16*, 273—285. Available from: http://dx.doi.org/10.1007/s10071-012-0571-1.

Bshary, R., Wickler, W., & Fricke, H. (2002). Fish cognition: A primate's eye view. *Animal Cognition*, *5*, 1—13.

Burkart, J. M., Strasser, A., & Foglia, M. (2009). Trade-offs between social learning and individual innovativeness in common marmosets (*Callithrix jacchus*). *Animal Behaviour*, *77*, 1291−1301.

Call, J., & Tomasello, M. (1995). Use of social information in the problem solving of orangutans (*Pongo pygmaeus*) and human children (*Homo sapiens*). *Journal of Comparative Psychology*, *109*, 308−320.

Capitanio, J. P., & Mason, W. A. (2000). Cognitive style: Problem solving by *Rhesus macaques* (*Macaca mulatta*) reared with living or inanimate substitute mothers. *Journal of Comparative Psychology*, *114*, 115−125.

Emery, N. J., & Clayton, N. S. (2004). The mentality of crows: Convergent evolution of intelligence in corvids and apes. *Science*, *306*, 1903−1907.

Epstein, R. (1982). The mythological character of categorization research in psychology. *The Journal of Mind and Behavior*, *3*, 161−169.

Epstein, R. (1983). Resurgence of previously reinforced behavior during extinction. *Behaviour Analysis Letters*, *3*, 391−397.

Epstein, R. (1984). The case for praxics. *The Behavior Analyst*, *7*, 101−119.

Epstein, R. (1985a). Animal cognition as the praxist views it. *Neuroscience and Biobehavioral Reviews*, *9*, 623−630.

Epstein, R. (1985b). Extinction-induced resurgence: Preliminary investigations and possible applications. *Psychological Record*, *35*, 143−153.

Epstein, R. (1985c). The spontaneous interconnection of three repertoires. *Psychological Record*, *35*, 131−141.

Epstein, R. (1987). The spontaneous interconnection of four repertoires of behavior in a pigeon (*Columba livia*). *Journal of Comparative Psychology*, *101*, 197−201.

Epstein, R. (1990). Generativity theory and creativity. In M. A. Runco, & R. S. Albert (Eds.), *Theories of creativity* (pp. 116−140). Newbury Park, CA: Sage.

Epstein, R. (1991). Skinner, creativity, and the problem of spontaneous behavior. *Psychological Science*, *2*, 362−370.

Epstein, R. (1996a). *Cognition, creativity, and behavior: Selected essays*. Westport, CT: Praeger.

Epstein, R. (1996b). *Creativity games for trainers*. New York, NY: McGraw-Hill.

Epstein, R. (1999). Generativity theory. In M. A. Runco, & S. Pritzker (Eds.), *Encyclopedia of creativity* (pp. 759−766). New York, NY: Academic Press.

Epstein, R. (2000). *The big book of creativity games*. New York, NY: McGraw-Hill.

Epstein, R. (2008). Why private events are associative: Automatic chaining and associationism. *Journal of Mind and Behavior*, *29*, 267−280.

Epstein, R. (2011). Exercises. In M. A. Runco, & S. Pritzker (Eds.), *Encyclopedia of creativity* (2nd ed., pp. 480−487). San Diego, CA: Academic Press.

Epstein, R. (2014). On the orderliness of behavioral variability: Insights from generativity theory. *Journal of Contextual Behavioral Science*, *3*, 279−290.

Epstein, R., Kaminaka, K., Phan, V., & Uda, R. (2013). How is creativity best managed? *Creativity and Innovation Management*, *22*, 359−374.

Epstein, R., Kirshnit, C., Lanza, R., & Rubin, L. (1984). "Insight" in the pigeon: Antecedents and determinants of an intelligent performance. *Nature*, *308*, 61−62.

Epstein, R., & Medalie, S. D. (1983). The spontaneous use of a tool by a pigeon. *Behaviour Analysis Letters*, *3*, 241−247.

Epstein, R., & Phan, V. (2012). Which competencies are most important for creative expression? *Creativity Research Journal*, *24*, 278−282.

Epstein, R., Roberts, G., & Beber, G. (Eds.), (2008). *Parsing the turing test: Methodological and philosophical issues in the quest for the thinking computer*. Dordrecht, The Netherlands: Springer.

Epstein, R., Schmidt, S. M., & Warfel, R. (2008). Measuring and training creativity competencies: Validation of a new test. *Creativity Research Journal*, *20*, 7−12.

Epstein, R., & Skinner, B. F. (1980). Resurgence of responding after the cessation of response-independent reinforcement. *Proceedings of the National Academy of Sciences USA, 77*, 6251−6253.

Foerder, P., Galloway, M., Barthel, T., Moore, D. E., & Reiss, D. (2011). Insightful problem solving in an Asian elephant. *PLoS One, 6*, e23251.

Fujita, K., Sato, Y., & Kuroshima, H. (2010). Learning and generalization of tool use by tufted capuchin monkey (*Cebus paella*) in tasks involving three factors: Reward, tool, and hindrance. *Journal of Experimental Psychology, 37*, 10−19.

Funk, M. (2002). Problem solving skills in young yellow-crowned parakeets (*Cyanoramphus auriceps*). *Animal Cognition, 5*, 167−176.

Guerke, B. (2012). *Paintings by animals raise questions about nature of art.* McClatchy—Tribune Business News, Retrieved from: http://search.proquest.com/docview/919822545?accountid=28103.

Hart, B. L., Hart, L. A., McCoy, M., & Sarath, C. R. (2001). Cognitive behavior in Asian elephants: Use and modifications of branches for fly switching. *Animal Behaviour, 62*, 839−847.

Heinrich, B. (1995). An experimental investigation of insight in common ravens (*Corvus corax*). *Auk, 112*, 994−1003.

Heinrich, B., & Bugnyar, T. (2005). Testing problem solving in ravens: String pulling to reach food. *Ethology, 111*, 962−976.

Hille, P., Dehnhardt, G., & Mauck, B. (2006). An analysis of visual oddity concept learning in a California sea lion (*Zalophus californianus*). *Learning & Behavior, 34*, 144−153.

Hull, C. (1935). The mechanism of the assembly of behavior segments in novel combinations suitable for problem solving. *Psychological Review, 42*, 219−245.

Hunt, G. R., Corballis, M. C., & Gray, R. D. (2001). Laterality in tool manufacture by crows. *Nature, 414*, 707.

Inoue-Nakamura, N., & Matsuzawa, T. (1997). Development of stone use by wild chimpanzees (*Pan troglodytes*). *Journal of Comparative Psychology, 111*, 159−173.

Kappeler, P. M. (1987). The acquisition process of a novel behavior pattern in a group of ring-tailed lemurs (*Lemur catta*). *Primates, 28*, 225−228.

King, W. P. (Ed.), (1930). *Behaviorism: A battle line!* New York, NY: Macmillan.

Kirsch, J., Gúntúrkúr, R., & Rose, J. (2008). Insight without cortex: Lessons from the avian brain. *Conscious Cognition, 17*, 475−483.

Köhler, W. (1925). *The mentality of apes.* London: Routledge & Kegan Paul.

Liker, A., & Bokony, V. (2009). Larger groups are more successful in innovation problem solving in house sparrows. *Proceedings of the National Academy of Sciences USA, 109*, 7893−7898.

Maier, N. R. F. (1931). Reasoning in humans. II. The solution of a problem and its appearance in consciousness. *Journal of Comparative Psychology, 12*, 181−194.

Mann, J., Sargeant, J. L., Watson-Capps, J. J., Gibson, Q. A., Heithaus, M. R., Connor, R. C., et al. (2008). Why do dolphins carry sponges? *PLoS One, 3*, 1−7.

Menzel, E. W., Jr., Savage-Rumbaugh, E. S., & Lawson, J. (1985). Chimpanzee (*Pan troglodytes*) spatial problem solving with the use of mirrors and televised equivalents of mirrors. *Journal of Comparative Psychology, 99*, 211−217.

Pepperberg, I. M. (2004). "Insightful" string-pulling in Grey parrots (*Psittacus erithacus*) is affected by vocal competence. *Animal Cognition, 7*, 263−266.

Poincaré, H. (1946). *The foundations of science.* Lancaster, PA: Science Press.

Reader, S. M., & Laland, K. N. (2000). Diffusion of foraging innovations in the guppy. *Animal Behaviour. 60*, 175−180.

Rothenberg, A. (1971). The process of Janusian thinking in creativity. *Archives of General Psychology, 24*, 195−205.

Santos, L. R., Ericson, B. N., & Hauser, M. D. (1999). Constraints on problem solving and inhibition: Object retrieval in cotton-top tamarins (*Saguinus oedipus oedipus*). *Journal of Comparative Psychology*, 113, 186—193.

Scholtyssek, C., Kelber, A., Hanke, F. D., & Dehnhardt, G. (2013). A harbor seal can transfer the same/different concept to new stimulus dimensions. *Animal Cognition*, 16, 915—925.

Schusterman, R. J., Thomas, J. A., & Wood, F. G. (Eds.), (1986). *Dolphin cognition and behavior: A comparative approach*. Hillsdale, NJ: Lawrence Erlbaum Associates.

Shettleworth, S. J. (2009). Animal cognition: Deconstructing avian insight. *Current Biology*, 19, R1039—R1041.

Shettleworth, S. J. (2010). Clever animals and killjoy explanations in comparative psychology. *Trends in Cognitive Science*, 14, 477—481.

Suttle, G., & The, N.T. (1997). Cindy the elephant painting brings $13,000. The News Tribune. Retrieved from: http://search.proquest.com/docview/264722963?accountid=28103.

Taylor, A. H., Elliffe, D., Hunt, G. R., & Gray, R. D. (2010). Complex cognition and behavioural innovation in New Caledonian crows. *Proceedings of the Royal Society Biological Sciences*, 277, 2637—2643.

Taylor, A. H., Medina, F. S., Holzhaider, J. C., Hearne, L. J., Hunt, G. R., & Gray, R. D. (2010). An investigation into the cognition behind spontaneous string pulling in New Caledonian crows. *PLoS One*, 5, e9345.

Tecwyn, E. C., Thorpe, S. K. S., & Chappell, J. (2012). What cognitive strategies do orangutans (*Pongo pygmaeus*) use to solve a trial-unique puzzle-tube task incorporating multiple obstacles?. *Animal Cognition*, 15, 121—133.

Tokimoto, N., & Okanoya, K. (2004). Spontaneous construction of "Chinese boxes" by Degus (*Octodon degu*): A rudiment of recursive intelligence? *Japanese Psychological Research*, 46, 255—261.

Turing, A. M. (1950). Computing machinery and intelligence. *Mind*, 59, 433—460.

Von Bayern, A. M. P., Heathcote, R. J. P., Rutz, C., & Kacelnik, A. (2009). The role of experience in problem solving and innovative tool use in crows. *Current Biology*, 19, 1965—1968.

Wass, C., Denman-Brice, A., Rios, C., Light, K. R., Kolata, S., Smith, A. M., et al. (2012). Covariation of learning and "reasoning" abilities in mice: Evolutionary conservation of the operations of intelligence. *Animal Behavior*, 38, 109—124.

Watanabe, S. (2010). Pigeons can discriminate good and bad paintings by children. *Animal Cognition*, 13, 75—85.

Watson, J. B., & McDougall, W. (1928). *The battle of behaviorism: An exposition and an exposure*. London: Kegan Paul, Trench, Trubner & Co.

Watson, S. L., & Ward, J. P. (1996). Temperament and problem solving in the small-eared bushbaby (*Otolemur garnettii*). *Journal of Comparative Psychology*, 110, 377—385.

Webster, S. J., & Lefebvre, L. (2001). Problem solving and neophobia in a columbiform-passeriform assemblage in Barbados. *Animal Behaviour*, 62, 23—32.

Wimpenny, J. H., Weir, A. A. S., Clayton, L., Rutz, C., & Kacelnik, A. (2009). Cognitive processes associated with sequential tool use in New Caledonian crows. *PLoS One*, 4, e6471.

Commentary on Chapter 13: Defining Animal Creativity: Little-C, Often; Big-C, Sometimes

Epstein's (this issue) main argument has already been emphatically endorsed in my *Origins of Genius: Darwinian Perspectives on Creativity*

(Simonton, 1999). More specifically, I indicated more than a dozen years ago how generativity theory can be translated into the terms of Campbell's (1960) blind variation and selective retention theory of creativity. By this I mean that the only way to generate creative ideas is to produce ideational or behavioral "invariants" that must be generated and tested, for the very simple reason that the organism does not know in advance whether the idea or act will work (Simonton, 2011). That is, the "blindness" consists in the lack of prior knowledge of the variation's utility or value. The testing may take place either externally or internally (cf. the "Skinnerian" versus "Popperian creatures" in Dennett, 1995). Indeed, it can be shown through both formal mathematics and Monte Carlo simulations that it is impossible to produce a creative variant if its effectiveness already known prior to the generation and test (Simonton, 2012, 2013a). Furthermore, I argued way back in 1999 that animals can display creativity, and that humans can be likewise creative according to processes used by animals.

That much said, I have one objection to Epstein's definition of creativity. To me it seems that he has conflated two sets of creativity criteria, one operating on the individual level and the other operating on the group level. In short, he does not distinguish between "little-c" creativity and "Big-C" Creativity (Simonton, 2013b). The former is a personal experience, the latter a consensual assessment. At the personal level, an organism can no doubt generate behaviors that are both novel and valuable to that organism at a particular time in that creature's life. Indeed, Epstein provides examples in his article. Yet at the consensual level, other organisms may not view that very same behavior as creative. The behavior is either already known at the group level or the behavior is not seen as adaptive by other members of the group.

This distinction is critical for comprehending the creative process. According to Epstein, if an organism comes up with a behavior that is novel to it, but not novel to other organisms in the group, then it cannot be creative. Yet the creative process by which novelty is generated does not incorporate the group evaluation directly. The latter is only added later, after the behavior has already been generated. Moreover, Epstein's definition implies that an organism cannot be creative as an individual operating in isolation of the social world. Yet this implication cannot handle the following two scenarios.

In the first scenario, a person is stranded in Antarctica and thus compelled to survive by her wits. The individual comes up with a number of personally novel and valuable adaptations that allows her to survive until the rescue team arrives. When that team finally shows up, and the person reports to them the ingenious ways she devised to ensure survival, the techniques are greatly admired, and soon become integrated into manuals for survival under such extreme conditions.

In the second scenario, we have the same stranded person, who devises precisely the same survival techniques, and using the exact same cognitive processes. Yet in this case, when the survivor reveals her newly discovered secrets of success, the rescuers just laugh, and tell the naive soul that the techniques just described can be found in any well-worn survival manual. In short, in the second scenario, the person just re-invented the wheel.

The contradiction should now be clear. The survival behaviors are exactly the same in the two scenarios, and the processes leading to those behaviors are exactly the same as well, yet the first is deemed "creative" while the latter is not according to Epstein's definition. In other words, I am arguing that singular animals adapting to their environments can display creativity without any reference to their conspecifics. If an adaption is novel and valuable to a given organism, then that organism is creative, regardless to whether or not conspecifics already have that adaptation in their behavioral repertoires.

Thus, Epstein may have sold animal creativity short. To make our judgments fair, creativity really has to be defined in terms of the life history of a singular organism, ignoring the life histories of other organisms. Little-c creativity is all that matters, and generativity theory applies regardless of what others can do or know. Did this specific creator generate a novel and valuable behavior?

In the human species, Big-C Creativity only obtains when other humans also judge an individual's idea or behavior to be both novel and valuable (see, e.g., the systems theory of Csikszentmihalyi, 2014). Without that consensual assessment, the personal judgment remains subjective, but still completely valid at the personal level (Simonton, 2013b). Yet it is very rare in the animal world for Big-C Creativity events to take place. Such events require that conspecifics implicitly sanction the personal assessment by adopting the novel adaptive behavior. When the personal act of creativity thus becomes widely disseminated, then the group consensus emerges that that act is creative as well. Stated differently, conspecifics must have the capacity and willingness to engage in social learning, adopting novel and adaptive practices on the basis of observing others display creativity.

Of course, we know that animal Big-C Creativity actually takes place. Perhaps the best known examples come from Japanese Macaque troops in which individual monkeys have devised new and adaptive behaviors that are later adopted by other monkeys, and in time become a part of troop culture. However, if each member troop had independently invented the same adaptation, then these separate events would count as little-c (personal) but not Big-C Creativity (consensual). In human cultures it sometimes happens that two or more inventors to come up with the same idea independently of each other, but only one wins priority for the invention. A classic example is the telephone. Although Bell and Gray got to the patent office on the same day, Bell arrived first, and thus we ended up with Bell Telephone rather than Gray Telephone. Both Bell and Gray had been engaged in personal creativity, but only Bell received credit for consensual creativity.

References

Campbell, D. T. (1960). Blind variation and selective retention in creative thought as in other knowledge processes. *Psychological Review, 67*, 380–400.

Csikszentmihalyi, M. (2014). The systems model of creativity and its applications. In D. K. Simonton (Ed.), *The Wiley handbook of genius* (pp. 533–545). Oxford: Wiley.

Dennett, D. C. (1995). *Darwin's dangerous idea: Evolution and the meanings of life.* New York: Simon & Schuster.

Simonton, D. K. (1999). *Origins of genius: Darwinian perspectives on creativity.* New York, NY: Oxford University Press.

Simonton, D. K. (2011). Creativity and discovery as blind variation: Campbell's (1960) BVSR model after the half-century mark. *Review of General Psychology, 15*, 158–174.

Simonton, D. K. (2012). Combinatorial creativity and sightedness: Monte Carlo simulations using three-criterion definitions. *International Journal of Creativity & Problem Solving, 22* (2), 5–17.

Simonton, D. K. (2013a). Creative thought as blind variation and selective retention: Why sightedness is inversely related to creativity. *Journal of Theoretical and Philosophical Psychology, 33*, 253–266.

Simonton, D. K. (2013b). What is a creative idea? Little-c versus Big-C creativity. In J. Chan, & K. Thomas (Eds.), *Handbook of research on creativity* (pp. 69–83). Cheltenham, UK: Edward Elgar.

PART IV

PUSHING THE BOUNDARIES OF CREATIVITY

Conservatism Versus Innovation: The Great Ape Story

Josep Call[1,2]

[1]School of Psychology and Neuroscience, University of St Andrews,
St Andrews, UK [2]Max Planck Institute for Evolutionary Anthropology,
Leipzig, Germany

Commentary on Chapter 14: Conservatism Versus Innovation: The Great Ape Story

Weihua Niu

Pace University, Dyson College of Arts and Sciences, New York, NY, USA

Dictionary.com defines innovation as the "introduction of new things or methods" and it lists "tradition" as one of its antonyms. For those of us interested in ape behavior and cognition, this poses an interesting puzzle. Chimpanzees and orangutans are well known for their social traditions which persist over generations (Whiten et al., 1999; van Schaik et al., 2003) and yet, they have been traditionally considered some of the most innovative creatures in the planet. Can one be tradition-bound and yet a naturally-born innovator?

While some analyses suggest that this may indeed be possible, as evidenced by the correlation that exists between social learning and innovation across species (Reader & Laland, 2002), others have argued

Animal Creativity and Innovation.
DOI: http://dx.doi.org/10.1016/B978-0-12-800648-1.00014-0

that innovation is either not as common and robust as previously thought (Hrubesch, Preuschoft, & van Schaik, 2009) or it may easily be curtailed by social learning and social influence (Price, Lambeth, Schapiro, & Whiten, 2009; van de Waal, Borgeaud, & Whiten, 2013). The currently popular characterization of chimpanzees as conservative and conformists contrasts with the traditional notion of chimpanzees as flexible problem solvers and innovators.

The main goal of this chapter is to resolve this apparent contradiction at both the empirical and theoretical levels. To do so, I will review the evidence that has recently accumulated in problem solving and innovation on the one hand, and social learning and tradition on the other hand. Additionally, I will consider individual and contextual variables that may contribute to resolve the existing apparent contradiction. I will end the chapter with some reflections about the question of flexible cognition and cumulative culture.

PROBLEM SOLVING AND INHIBITORY CONTROL

Innovation is a crucial component of problem solving that can adopt different forms including applying an old solution to a new problem or producing a new solution to an old problem (Kummer & Goodall, 1985). Manrique, Voelter, and Call (2013) have argued that these two types of innovation may be quite different from a cognitive standpoint. Whereas using an old solution to a new problem requires transferring motor sequences and functional knowledge about object–object relations to a new situation, producing a new solution to an old problem also requires the inhibition of behaviors that were successful in the past. There is ample evidence that great apes display both types of innovation but I will primarily focus on the second type of innovation.

Inhibitory control has been documented in several primate species in a variety of contexts including object search, reaching, reversal learning, and even in temporal discounting. Amici, Aureli, and Call (2008) compared the performance of monkeys and great apes in a battery of five tasks that included all of these contexts. Results revealed two clusters of species with chimpanzees, orangutans, bonobos, and spider monkeys belonging to the high performance cluster and gorillas, capuchin monkeys, and long-tailed macaques forming the second low performance cluster. Interestingly, this distribution would not be predicted by phylogeny alone but it correlates well with some socio-ecological variables such as a social group's likelihood to split and merge into multiple subgroups of different sizes and compositions over the course of multiple days (so-called fusion–fission dynamics). Interestingly, Aureli, Schaffner,

Boesch, Bearder, and Call (2008) had hypothesized that those species with high levels of fission—fusion dynamics (chimpanzees, orangutans, bonobos, and spider monkeys) should perform better in inhibitory control tasks than those species with low levels of fission—fusion dynamics (gorillas, capuchin monkeys, and long-tailed macaques).

Although most of the tasks used by Amici et al. (2008) only required relatively simple motor responses such as touching or pointing to one of two objects, one of the tasks (i.e., swing door) required a more complex reaching response. More specifically, subjects faced a panel with two square holes next to each other and covered by two transparent "doors" attached to the top part of each hole by a set of hinges that could only be opened by pushing them forward. A piece of banana was placed behind one of the doors so that directly reaching for the banana (and pushing the door forward in the process) invariably produced the undesirable outcome of displacing the banana away from the subject's reach. To solve the task, subjects had to refrain from directly reaching for the banana and instead had to reach into the non-baited door to grab the banana from behind. Orangutans performed better than chimpanzees, gorillas, and 4-year-old children in this task. This task is reminiscent to some of the experiments that Köhler (1925) did with chimpanzees who were capable of taking indirect travel routes to a goal when the most direct route was blocked.

Puzzle boxes typically require even more complex manipulations than detour reaching (or traveling) tasks which in some cases even involve using tools. One particularly interesting case is when individuals have been using a solution for some time, with the possibility that it may have become fixated, and they must upgrade to a more efficient solution. Some studies have shown that apes are capable of sequentially innovating multiple solutions (Lehner, Burkart, & van Schaik, 2011; Manrique et al., 2012). For instance, Lehner et al. (2011) found that orangutans produced multiple solutions to extract liquid from a container and quickly abandoned those solutions that became ineffective after the conditions of the task were altered. Manrique et al. (2012) reported similar findings in all great ape species in a non-tool-using task.

Sequential innovation is not the sole purview of the great apes. Auersperg, von Bayern, Gajdon, Huber, and Kacelnik (2011) presented the so-called Multiple Access Box to New Caledonian crows and keas. Subjects could use four different methods (open a door, insert a stick, drop a ball, pull a string) to extract a piece of food placed inside the box. Once subjects became proficient with one of the methods, that method blocked and subjects had to find a new way to get the food with one of the remaining methods. Note that unlike the previous studies, subjects were prevented from entirely using a previously successful method, e.g., by completely removing the string from the box, which might have eliminated some of the difficulties inherent to refraining from using a previously effective method if it was still available but no longer effective. In general, keas showed a preference for dropping the ball whereas crows preferred using the stick. Moreover, although one individual in each species discovered all methods, keas tended to discover more methods and do this faster than New Caledonian crows.

There are some studies, however, that have reported a failure to innovate in great apes. For instance, Hanus, Mendes, Tennie, and Call (2011) reported a case of innovation failure in chimpanzees probably caused by functional fixedness (Duncker, 1945). More specifically, Hanus et al. (2011) found that some chimpanzees that were unable to solve the floating peanut task (i.e., spitting water into a vertically-oriented tube to extract a peanut lying at the bottom of it) succeeded as soon as their familiar water dispensing device was replaced by a new one. Hanus et al. (2011) suggested that the familiar water dispensing device may have acquired a fixed function of supplying water that the new device lacked, thus hindering innovation for a new use of water.

Other studies that have reported individuals failing to abandon solutions that no longer worked have been interpreted as evidence of conservatism (Bonnie et al., 2012; Hrubesch et al., 2009). Bonnie et al. (2012) observed that chimpanzees continued to fish at an artificial termite mound after the food became gradually scarcer (fewer holes remained productive) over a period of several months. However, food was never completely exhausted because some holes continued to produce food for the entire period. In fact, there was a notable reduction in the use of the termite mound that precisely matched the amount of food available—something that strongly suggests that chimpanzees did abandon some of their original practices. Note that there was no alternative available when some of the holes became exhausted other than the other holes that remained productive. For an individual exploiting the still productive holes it may be a good idea (and nearly cost-free) to check those sources that were productive in the past in case they became productive again—as occurred in the second part of the study. If subjects had completely abandoned the occasional inspection of previously rewarded holes, they would have never been able to re-exploit them once they became productive again.

Hrubesch et al. (2009) also found persistence in chimpanzees in the face of failure. Namely, chimpanzees persisted in using sticks to get out-of-reach food from a platform instead of adopting a technique based on rattling the platform which produced more food than the stick technique. However, those individuals who persisted in using tools were also those who were the most efficient ones at using the tools. This means that they had less pressure to change to the new alternative. In fact, those chimpanzees who switched to rattling the platform were mostly adult males, which some studies have characterized as less persistent than females at using tools for extended periods of time (Lonsdorf, 2005). When platform rattling was made ineffective, adult males who had used that technique continued to do so despite its low probability of success and their persistence was interpreted as conservatism. However, note that for those chimpanzees the option of changing to sticks may not have been a viable option, which is why they invented rattling in the first place. Moreover, trying that technique may not be an indication of conservatism but a case of probing as noted above. What would be needed is two techniques matched in difficulty so that one can assess whether individuals abandon one technique in favor of the other as soon as the latter provides a greater payoff. Would individuals adopt the technique offering the higher payoffs regardless of which of the two techniques they acquired first?

Another issue to consider is that suboptimal solutions (both in terms of motor efficiency and/or payoffs) may persist under an intermittent reinforcement regime but may disappear once the payoff becomes zero—although as we have seen they may occasionally reappear as a way of probing old solutions. In other words, it may be easy to mistake behavioral conservatism for responses under a partial reinforcement regime. This explanation is supported by data on reverse reward contingency tasks (RRC) in which subjects have to pick the smaller of two quantities to receive the larger quantity—a procedure similar to reversal learning. Several studies have shown that learning to suppress the choice of the larger quantity is slower when subjects receive non-zero quantities for incorrect choices than when they receive nothing (Silberberg & Fujita, 1995; Vlamings, Hare, & Call, 2010). In fact, one of the most effective correction procedures to train subjects to select the smallest quantity in RRC tasks consists of giving them no reward upon making the wrong choice (Silberberg & Fujita, 1995).

Although the zero payoff effect provides a good explanation for why apes were more likely to innovate when a solution becomes completely ineffective compared to when it still provides intermittent reinforcement, there are other studies showing that apes do upgrade to solutions better solution even under partial reinforcement conditions. For instance, Manrique and Call (2011) reported that chimpanzees, bonobos, and orangutans extracted juice from a box by dipping and licking a piece of electric cable (three copper filaments covered by plastic and held tightly together by a plastic shaft). Interestingly, all orangutans tested and one chimpanzee discovered the use of the cable as a straw to extract the liquid—a technique which was much more efficient than the dipping technique in terms of both speed and quantity of juice extracted.

In summary, the reviewed evidence suggests that ape conservatism in most problem solving tasks may be more apparent than real. Apes' persistence may occur under partial reinforcement or because other options are not possible. Re-trying a previously successful solution as a way to probe whether conditions have changed is quite adaptive and it may represent an alternative explanation to conservatism. Furthermore, some studies have documented both innovation and even repeated innovation of more efficient solutions even though the existing ones produced non-zero payoffs. Future studies could investigate whether individuals trained to obtain food with one technique abandon it for another one matched in terms of difficulty but producing a higher payoff.

SOCIAL LEARNING AND SOCIAL TRADITIONS

It is well known that different wild populations of chimpanzees and orangutans display different behavioral repertoires that some authors consider cultural variants, mainly related to the use of tools to extract food. For instance, using stones to crack open nuts, sticks to fish for termites or ants, or leaves to extract liquids are present in some populations but not others despite the fact that the potential for the appearance and spread of those behaviors are available in multiple sites. Consequently, the behavioral tool repertoires of the different populations vary considerably, not just in terms of the type of tool or the resource exploited but also in the techniques used to explait the resources. For instance, western chimpanzees in Goualougo and Tai display 22 and 21 types of tool use, respectively, whereas eastern chimpanzees in Budongo only use eight types of tools (Sanz & Morgan, 2007; Whiten et al., 1999). The contrast is even more striking between populations of eastern chimpanzees, since Gombe and Mahale chimpanzees display 22 and 16 tool types, respectively, many more than those found among the chimpanzees of Budongo forest.

But perhaps even more striking than the large differences between populations is that some of these populations seem extremely conservative in the tools that they use and the way they use them. Gruber, Muller, Strimling, Wrangham, and Zuberbühler (2009) conducted a field experiment that consisted of presenting chimpanzees with a log filled with honey. The honey could be extracted by inserting tools through a set of holes. Chimpanzees in Kanyawara used sticks to extract the honey, which is not surprising since they had been observed using sticks for similar tasks. In contrast, Sonso chimpanzees (who live only 180 km away from Kanyawara) used their fingers and leaves (but not sticks) to extract the honey—in fact leaf sponging is a technique that they commonly use to extract water from crevices in their natural habitat. More importantly, when leaves became ineffective because the honey that remained was too deep inside the log, they completely failed to innovate a solution to continue exploiting this resource. They were unable to use sticks to extract the honey—a behavior that is quite widespread among other wild and captive chimpanzee populations. Even pre-inserting sticks into the holes filled with honey so that when chimpanzees would realize about their utility when they pull them out produced no positive results. Chimpanzees extracted the honey dipped tools, licked them, and then discarded them, or stripped the leaves from the sticks and inserted them into the hole while discarding the now leafless stick!

Mesoudi (2011) has argued that conservatism, like the one exemplified above, may be the reason why there is so little evidence of cumulative culture in nonhuman great apes or nonhuman animals more generally. Furthermore, Hrubesch et al. (2009, see also Whiten, Horner, & de Waal, 2005) have pointed out that both conservatism and social conformity may contribute to make cumulative culture in nonhuman animals a rare phenomenon.

The possibility that chimpanzees are behavioral conformists, i.e., once they have learned a technique and used it repeatedly they are incapable of abandoning when it becomes obsolete. This explanation, however, is seriously undermined by the studies reviewed in the previous section. Alternatively, it is conceivable that chimpanzees are conservative because they are social conformists—they are either unable to change what they have learned socially or unwilling to deviate from what others are doing. We explore these two possibilities next.

One hypothesis to explain the putative social conformity is that social learning leaves an indelible mark, much stronger and difficult to change than individual learning. In other words, socially acquired information is more resilient to change than individually acquired information. For instance, Price et al. (2009) found that subjects who learned to use a tool by observation had greater difficulty inventing a new technique compared to those individuals who learned the use of the tool individually. Marshall-Pescini and Whiten (2008) also suggested that chimpanzees became "stuck" on less efficient solutions that they learned socially. The authors confronted young chimpanzees with an artificial fruit that could be opened with a tool in two ways. One way was simpler (but less efficient) than the other. A human demonstrated first the simple method to the chimpanzees and three subjects learned it (two other chimpanzees learned this method on their own during a baseline period). Then the experimenter demonstrated the second more complex and efficient method but none of the chimpanzees adopted it, presumably as the authors argued because they became stuck on the first method they (socially) learned.

Although this evidence is suggestive, it is hard to attribute the fixation effect to social learning per se, given that two other chimpanzees who learned on their own also failed to benefit from the human demonstration (of the second technique). Moreover, it is unclear whether individuals actually learned socially in the first place given that two chimpanzees learned individually. An additional control group with additional exposure (and without demonstrations) could have helped tease apart these alternatives. Age might have also been an issue in this study because the median age of the chimpanzees tested was only 3 years and in fact, no chimpanzee younger than 3 years of age learned how to operate the apparatus. Nevertheless, it is still possible that social learning has a dampening effect on the probability of innovation, and that such an effect might contribute to explain why chimpanzees who learned socially failed to adopt more efficient techniques (Gruber, Muller, Reynolds, Wrangham, & Zuberbühler, 2011; Gruber et al., 2009). However, better data are needed to confirm this possibility. Also, note that Hrubesch et al. (2009) found the opposite effect, i.e., individually acquired behavior was immune to social influence even when the alternative solution was more profitable.

But as mentioned before, it is possible that individuals did not abandon their current technique because it was profitable enough. Even if social learning were to have the dampening effect once the behavior has been acquired and consolidated, this does not necessarily mean that when they acquire social information they do it in an inflexible manner. On the contrary, there is some evidence showing that chimpanzees are selective about the information that they adopt depending on the situational constraints. For instance, Horner and Whiten (2005) found that chimpanzees observing a human model using a sequence of actions (some needed to actually open and others completely unnecessary and ineffective) to open an artificial fruit copied the demonstrated techniques depending on whether the artificial fruit was clear or opaque. Chimpanzees who witnessed the actions on the opaque box were more likely to copy all the demonstrated actions, whereas chimpanzees witnessing the demonstration on the clear artificial fruit only copied those actions that were actually needed to open the box (they ignored unnecessary actions such as inserting a stick that had no causal effect because its path to the food was blocked by a barrier). Buttelmann, Carpenter, Call, and Tomasello (2008) also reported selectivity in copying the body part executing an action depending on the contextual constraints faced by a human demonstrator touching a box that produced music and lights. More specifically, enculturated chimpanzees who observed a human using an unusual body part to activate the box, used the same unusual body part more often when there was no physical constraint that it would have impeded the human to use a usual body part to activate the box (e.g., activating a night light by touching it with the forehead rather than with the hand). These results suggest that chimpanzees are not completely tied to what they see when they are learning a new technique—but the data are not incompatible with the idea that chimpanzees become less flexible once they have already learned the technique by observation. In fact, it is an open question whether chimpanzees who learned to open an opaque artificial fruit compared to those who learned from the clear box would be equally likely to abandon their technique as soon as it became obsolete.

Another possible explanation for why chimpanzees may give priority to social compared to individual information is the influence that group majorities may have on individuals. That is, social learning per se is not the key, but majorities in general bias subjects' behavior.

The effect of group majorities on the individual's behavior is well documented in several domains in various species (Day, MacDonald, Brown, Laland, & Reader, 2001; Galef & Whiskin, 2008; Pike & Laland, 2010). Moreover, several studies have shown that after chimpanzees acquire a solution using social information and later on they discover an alternative solution individually, they typically revert back to the initial socially learned solution both in a token selection (Bonnie, Horner, Whiten, & de Waal, 2007) or a tool-using task (Whiten et al., 2005). Some authors have interpreted these findings as an indication of social conformity (Claidiere & Whiten, 2012; Whiten et al., 2005). However, Van Leeuwen and Haun (2013) pointed out that conformity in those studies is conceptualized as preserving the initial solution rather than abandoning it in favor of an alternative solution used by the majority of individuals. In fact, Van Leeuwen, Cronin, Schütte, Call, and Haun (2013) found that chimpanzees did not abandon their initial solution when confronted with a majority of individuals using an alternative but equally profitable solution. However, chimpanzees did abandon their original solution when they observed (and experienced) that a new solution provided a more profitable payoff. Similarly, Yamamoto, Humle, and Tanaka (2013) found that chimpanzees that learned a "straw-dipping" technique abandoned it for a "straw-sucking" one that was more efficient after observing conspecifics using it. These two studies corroborate previous findings reviewed in the previous section showing that chimpanzees are perfectly capable of abandoning techniques in favor of more efficient or profitable ones.

Another aspect that ties with the literature on problem solving is that regardless of the way that subjects learned to solve a task (either social or individual learning), it is clear that individuals try new things even when their current solution is already profitable. Dean, Kendal, Schapiro, Thierry, and Laland (2012) also observed that chimpanzees and capuchin monkeys continued exploring the properties of an apparatus after they succeeded but unlike children they failed to acquire more complex (and profitable) solutions by observation. Thus, once a solution is found, chimpanzees seem to engage in either further exploration or a "testing-the-waters" approach that may reveal new possibilities. What is less clear and still a matter of debate is whether (and if yes, why) the newly discovered options are ultimately adopted by the individual. In any case, what this suggests is that social learning does not in general (although there may be some exceptions, e.g., Price et al., 2009) make the adoption of other strategies less likely.

In summary, the idea that social learning makes change more difficult is intriguing but the evidence supporting it is mixed, with some studies potentially showing a greater resilience for behavior learned via social influence while others show no such effect. The effect of the social environment, rather than the social learning mechanisms per se, is another tantalizing possibility but once again the evidence is rather mixed. The following section will attempt to throw some light onto these results.

INDIVIDUAL AND CONTEXTUAL DIFFERENCES

The two previous sections have revealed two important things. First, there is a set of mixed results with regard to the innovative tendencies of the great apes. Second, and perhaps more importantly, there seem to be a variety of factors that play a role in the observed differences. Next, I will briefly explore both individual attributes (personality, age, and species) and contextual variables (task, setting) that may be contribute to clarifying the relation between innovation and conservatism.

Recent years have witnessed an unprecedented research interest in the question of individual differences and personality in nonhuman animals. Boldness-shyness is one of the personality dimensions has been most intensively investigated. Typically, this dimension is assessed by presenting individuals with novel foods and/or novel objects and researchers measure the latency to taste or to approach those novel items. This test has revealed both species and individual differences. At the species level, bonobos are shier than chimpanzees and orangutans—a result that has been linked to the higher degree of tool use and varied foraging repertoire in chimpanzees and orangutans, compared to bonobos (Herrmann, Hare, Cissewski, & Tomasello, 2011). At the individual level, bolder chimpanzees perform better than shier chimpanzees in physical (but not social) cognition tasks (Herrmann, Call, Hare, Hernandez-Lloreda, & Tomasello, 2007). In contrast, no relation between those two variables was found for 2.5-year-old children.

Thus, bolder individuals are more likely to try new things and therefore they would be the prime candidates for innovation. However, the relationship between boldness and innovation may not be as straightforward as it may appear. Günther, Brust, Dersen, and Trillmich (2014) found that although shier cavies (*Cavia aperea*) are slower than bolder individuals to explore and learn an object discrimination task, they are faster than bolder individuals at reversal learning when the previously rewarded option is no longer rewarded. Thus, bolder individuals may be quicker to innovate but they are slower to abandon a previously profitable option, thus hindering their adoption of alternative solutions. This is important because it supports the idea that innovation and inhibitory control are related but may not be same phenomenon. Additionally, it is tantalizing to link boldness and impulsivity even though these two constructs do not appear to be one and the same, at least not in humans (Cross, Copping, & Campbell, 2011).

Whereas the relation between boldness and innovation has received some empirical support, it is an open question whether one could also find personality traits that identify individuals that are more prone to adopt the behavior and products of the innovators once they have become familiar with them. Groups formed by a mixture of innovators and copiers would be capable of sustaining cultural evolution. The faithful adoption of others' behavior or their products (without further innovation) could be facilitated by social learning itself (although this will depend on the task), or perhaps those individuals who rely on social learning are also less prone to innovate than those who are bolder and rely on individual learning. Herrmann et al. (2007) could have potentially discovered the relationship between boldness and social learning in chimpanzees. Unfortunately, the data on social learning showed a floor effect, meaning that the tasks (or the procedure) may not have been sensitive enough to detect social learning. Alternatively, this may mean that chimpanzees (which are bolder than 2.5-year-old children) tend to rely more on individual than social learning, especially when the tasks are not opaque—an issue that we will return to when we discuss the effect of task on performance.

Age, sex, and species are other attributes of the individual that may impact on innovation and conservatism. Manrique and Call (2015) found that task persistence in the face of failure was age dependent in the great apes. They found that the youngest and the oldest individuals included in the study were the most persistent ones whereas the middle aged subjects adapted quicker to the new conditions and modified their behavior accordingly. Marshall-Pescini and Whiten (2008) found that chimpanzees younger than 3 years of age failed to adopt any of the demonstrated methods. However, as Marshall-Pescini and Whiten (2008) pointed out, the chimpanzees they tested may have been too young to benefit from demonstrations. With regard to sex differences, Hrubesch et al. (2009) found that adult male chimpanzees were mainly responsible for inventing a platform rattling technique, presumably because they were more efficient than with the alternative tool-using technique—something that fits with the idea that female chimpanzees are more prone than male chimpanzees to use tools for extended periods of time (Lonsdorf, 2005). There are also species differences in several tasks. For instance, Manrique, Voelter, and Call (2013) found that chimpanzees, bonobos, and gorillas invented one solution based on forcefully projecting the reward out of the apparatus while only one of the orangutans did. It is conceivable that banging actions are more prominent in the repertoire of African apes than in orangutans. In contrast, orangutans were more likely than chimpanzees and bonobos to discover the use of a piece of electric cable as a straw to drink from a container (Manrique & Call, 2011).

Contextual factors may also play an important role in explaining the observed differences between studies. One such factor is the type of task, more specifically, the nature of the relation between the elements of the task. Puzzle box tasks are grounded on physical (causal) relations between their object components and food delivery. For instance, a tool is used to displace a piece of food or some box "defense" needs to be removed before the subjects can access the food. In contrast, token tasks are grounded on an arbitrary relation between objects and food delivery. For instance, depositing a token inside a particular container or choosing one of two tokens is what determines the delivery of the reward. There is no physical rule governing the reward delivery but an arbitrary rule set by the experimenter. It is conceivable that these arbitrary rules are more opaque than those grounded on physical relations but once acquired they may be more easily transmitted faithfully down a chain of individuals. In fact, Horner and Whiten (2005) found that when the physical relations in a puzzle box were made opaque, chimpanzees were more likely to copy the behavior of a demonstrator including actions that were irrelevant to solve the task. This could explain why studies using tokens might be more likely to produce uniformity in groups compared with studies not based on tokens. An individual confronted with a task in which she cannot deploy her individual knowledge might be more likely to rely on the behavior of others (Hopper, Schapiro, Lambeth, & Brosnan, 2011). Note that tokens *per se* are not causing this effect because when individuals have acquired knowledge about tokens on their own they can become immune to the influence of social information (van Leeuwen et al., 2013).

The setting where the task takes place is another important consideration. Some authors have argued that innovation may be more likely to occur in the captive setting because it is a "safe" environment basically devoid of dangerous creatures and stimuli—clearly a case that it is not true in the field. As a consequence, experimenting in the field, especially for young individuals, may be more costly than in the laboratory. In a similar vein, Lehner et al. (2011) pointed out that captive (or rehabilitant) orangutans were more prone to innovation that wild orangutans (Russon, Kuncoro, Ferisa, & Handayani, 2010) and suggested that innovation in captivity may have been associated with positive reinforcement (e.g., finding food rewards under novel containers). However, this should not be taken as an indication that innovation does not occur in the wild. On the contrary, innovation has been reported in the field, although it may not persist (Kummer & Goodall, 1985) and it still may be less prominent than in the laboratory. Unfortunately, with a few exceptions (Gruber et al., 2009; Matsuzawa, 1994) there has been comparatively little field experimentation in the area of ape innovation.

CONSERVATISM, SOCIAL CONFORMITY, AND CUMULATIVE CULTURE

Neither conservatism nor social conformity provides a compelling argument to explain the putative lack of cumulative culture among non-human apes. Conservatism is not present in every study and when it appears it depends on the return benefits and individual variables such as boldness, age, and species. In fact, most studies show non-conservatism in one way or another. Moreover, some studies show repeated innovation and even those that do not indicate that subjects can revert to alternative solutions not just when current solutions cease to produce benefits but also even when their current solutions continue to be rewarded. Those findings paired with data on inhibitory control and reversal learning suggest that apes in general should not be characterized as conservative. This conclusion fits with large scale studies that have shown that primates, and apes in particular, score high both on innovation and inhibitory control (Amici et al., 2008; MacLean et al., 2014; Reader & Laland, 2002).

Social conformity is an equally unsatisfactory explanation for the lack of cumulative culture since it is not observed in every study when complex tasks are involved. More robust is the evidence based on token tasks (or opaque puzzle boxes type tasks), but most of this evidence is based on reverting to the original solution rather than abandoning the original solution for another one that is equally beneficial. Additionally, some studies have shown that apes can abandon their current solutions for other more profitable ones even when social pressure would predict the opposite. Interestingly, the most compelling case for social conformity has been reported in vervet monkeys, not apes. Van de Waal et al. (2013) observed that migrating vervet monkey males abandoned (not just reverted) their original preference for one type of corn and adopted the preference displayed by most members of the group where they had just immigrated.

Although this finding may qualify as a case of social conformity, which is one of the pillars of human culture, it is still hard to consider it as cumulative culture because the acquired technique is still rather simple—a choice between two types of food. Tennie, Call, and Tomasello (2009) have argued that one key aspect of cumulative culture is that the practice or product derived from it cannot entirely be invented by a single individual, no matter how innovative. In our culture, a computer, the wheel, or even a mundane paper clip are just some examples of massively concentrated knowledge distilled and accumulated over multiple generations which could not have been invented by a single individual. Social institutions would also represent cases of collective knowledge accumulated over multiple generations that could not be invented by one individual alone without a proper starting point. Incidentally, this complexity criterion would also disqualify dropping a token inside a container to obtain a food reward since individuals could learn this on their own.

Since neither conservatism nor social conformity appear to be a satisfactory explanation for the putative lack of cumulative culture in nonhuman animals, we must search elsewhere. I will provide a sketch of an answer here. Let's begin our search by returning to the original conundrum that motivated the present chapter. Innovation and tradition are antonyms and yet, they work in conjunction, at least in humans, to create cumulative culture. Apes display both innovation and traditions but no cumulative culture. What else may be needed to develop cumulative culture? Technical sophistication is one aspect that has already been mentioned. But technical sophistication develops when the best solutions are preserved and for that to happen two things are necessary. First, subjects have to face a problem that requires new solutions, and second, individuals have to be able to benefit from the behavior of others. This means that social learning mechanisms such as imitative learning and teaching need to be in place so that observers/apprentices can copy the existing solutions which in turn they can further improve on their own. Note that once solutions reach a certain level of complexity or arbitrariness, imitative learning and instruction may be the only way to acquire them from and transmit them to others, respectively. This is particularly true for behaviors that are not supported on objects or artifacts but are both arbitrary and conventionalized among members of a particular group. For instance, think about the hand gesture consisting of extending the index and middle fingers while keeping the thumb and the two other fingers retracted used to indicate "victory." Individual invention would be highly unlikely to produce this exact same pattern and even more unlikely would be to associate this particular finger topography with its referent. Note that arbitrary conventions are often developed within human groups as a marker of group identity, and consequently, they should be things that could only be acquired and transmitted socially.

There is another consideration that may help explain cumulative culture in humans beyond social learning mechanisms. Humans like doing things *with* others and *like* others in their groups. The intrinsic motivation that compels humans to be and behave like others paired with our extended mental temporal horizon, that is, our ability to think about the past and into the distant future, has enabled humans to transfer knowledge not only to those present here and now but also to those that will be present in future generations. In a sense, humans have become knowledge "hoarders" for the benefit of future generations with traditional songs, libraries, and the internet as our knowledge repositories. This extended society encompassing past, present, and future generations bound together by a cumulative common knowledge has become one of the key features of humankind.

TWO DIRECTIONS FOR FUTURE RESEARCH

But let's leave our species aside and return to nonhuman apes by putting on the table two important outstanding issues that will allow us to close the chapter. First, there is the relation between innovation and tradition. At the species level it seems that there is a good relation between the two (Reader & Laland, 2002), at least in primates and provided we take social learning as a proxy for tradition. Whether innovation and tradition are also related at the individual level or instead constitute two different factors is still an open and fascinating question. Recent analyses on physical cognition in apes have suggested that at the individual level things like learning, inhibitory control, and inference may be segregated not part of a single G factor (Herrmann & Call, 2012). Although Herrmann et al. (2007) also explored the relation between social learning and other aspects of cognition, we did not find a relation, although as I mentioned earlier, a floor effect may have prevented us from uncovering any relation. It is conceivable that the high social learning and innovation is a result of different individuals contributing to each of those two aspects or alternatively, they may also be also related at the individual level. Future studies are needed to elucidate this question. If the answer is that different individuals contribute it may help us reconcile the terms innovation and tradition antonyms;

they work together to produce cultural progress and when that progress is preserved along the lines that I suggested earlier, cumulative culture emerges.

The second point is the puzzling conservatism displayed by the Sonso chimpanzees. Does this represent a case of social conformity, cultural override, functional fixedness, or perhaps a combination of the three? I was extremely puzzled, and I still am, when I read that study. I would have thought that chimpanzees could have overcome their bias. Is this special about this population? Is this perhaps typical of wild populations? It is imperative to test other populations using the same methodology to find out whether this phenomenon is restricted to this population/behavior or whether it is something more general in the field. It is conceivable that this conservatism does not apply to other populations. After all, other populations adapt to artificial provisioning devices (van Lawick-Goodall, 1971) and respond well to field experiments of various sorts (Matsuzawa, 1994). However, the extent of their flexibility, especially compared to captive populations is unknown. It is conceivable that wild chimpanzees in general are not as conservative as suggested by this study. Field experimentation is therefore sorely needed to put lab and field results on an equal footing.

References

Amici, F., Aureli, F., & Call, J. (2008). Fission–fusion dynamics, behavioral flexibility and inhibitory control in primates. *Current Biology, 18*, 1415–1419.

Auersperg, A. M. I., von Bayern, A. M. P., Gajdon, G. K., Huber, L., & Kacelnik, A. (2011). Flexibility in problem solving and tool use of kea and New Caledonian crows in a Multi Access Box paradigm. *PLoS One, 6*, e20231.

Aureli, F., Schaffner, C. M., Boesch, C., Bearder, S. K., Call, J., et al. (2008). Fission–fusion dynamics: New frameworks for comparative research. *Current Anthropology, 49*, 627–654.

Bonnie, K. E., Horner, V., Whiten, A., & de Waal, F. B. M. (2007). Spread of arbitrary conventions among chimpanzees: A controlled experiment. *Proceedings of the Royal Society B, 274*, 367e372.

Bonnie, K. E., Milstein, M. S., Calcutt, S. E., Ross, S. R., Wagner, K. E., & Lonsdorf, E. V. (2012). Flexibility and persistence of chimpanzee (*Pan troglodytes*) foraging behavior in a captive environment. *American Journal of Primatology, 74*, 661–668.

Buttelmann, D., Carpenter, M., Call, J., & Tomasello, M. (2008). Rational tool use and tool choice in human infants and great apes. *Child Development, 79*, 609–626.

Claidiere, N., & Whiten, A. (2012). Integrating the study of conformity and culture in humans and nonhuman animals. *Psychological Bulletin, 138*, 126–145.

Cross, C. P., Copping, L. T., & Campbell, A. (2011). Sex differences in impulsivity: A meta-analysis. *Psychological Bulletin, 137*, 97–130.

Day, R. L., MacDonald, T., Brown, C., Laland, K. N., & Reader, S. M. (2001). Interactions between shoal size and conformity in guppy social foraging. *Animal Behaviour, 62*, 917–925.

Dean, L. G., Kendal, R. L., Schapiro, S. J., Thierry, B., & Laland, K. N. (2012). Identification of the social and cognitive processes underlying human cumulative culture. *Science, 335,* 1114—1118.

Duncker, K. (1945). *On problem-solving.* Washington, DC: The American Psychological Association, Inc.

Galef, B. G., & Whiskin, E. E. (2008). Conformity" in Norway rats? *Animal Behaviour, 75,* 2035—2039.

Gruber, T., Muller, M. N., Reynolds, V., Wrangham, R. W., & Zuberbühler, K. (2011). Community-specific evaluation of tool affordances in wild chimpanzees. *Scientific Reports, 1.* Available from: http://dx.doi.org/10.1038/srep00128.

Gruber, T., Muller, M. N., Strimling, P., Wrangham, R. W., & Zuberbühler, K. (2009). Wild chimpanzees rely on cultural knowledge to solve an experimental honey acquisition task. *Current Biology, 19,* 1806—1810.

Günther, A., Brust, V., Dersen, M., & Trillmich, F. (2014). Learning and personality types are related in cavies *(Cavia aperea). Journal of Comparative Psychology, 128,* 74—81.

Hanus, D., Mendes, N., Tennie, C., & Call, J. (2011). Comparing the performances of apes *(Gorilla gorilla, Pan troglodytes, Pongo pygmaeus)* and human children *(Homo sapiens)* in the floating peanut task. *PLoS One, 6,* e19555.

Herrmann, E., & Call, J. (2012). Are there geniuses among the apes? *Philosophical Transactions of the Royal Society of London. Series B, Biological Sciences, 367*(1603), 2753—2761. Available from: http://dx.doi.org/10.1098/rstb.2012.0191.

Herrmann, E., Call, J., Hare, B., Hernandez-Lloreda, M. V., & Tomasello, M. (2007). Humans have evolved specialized skills of social cognition: The cultural intelligence hypothesis. *Science, 317,* 1360—1366.

Herrmann, E., Hare, B., Cissewski, J., & Tomasello, M. (2011). A comparison of temperament in nonhuman apes and human infants. *Developmental Science, 14*(6), 1393—1405. Available from: http://dx.doi.org/10.1111/j.1467-7687.2011.01082.x.

Hopper, L. M., Schapiro, S. J., Lambeth, S. P., & Brosnan, S. F. (2011). Chimpanzees' socially maintained food preferences indicate both conservatism and conformity. *Animal Behavior, 81,* 1195—1202. Available from: http://dx.doi.org/10.1016/j.anbehav.2011.03.002.

Horner, V., & Whiten, A. (2005). Causal knowledge and imitation/emulation switching in chimpanzees *(Pan troglodytes)* and children *(Homo sapiens). Animal Cognition, 8,* 164—181. Available from: http://dx.doi.org/10.1007/s10071-004-0239-6.

Hrubesch, C., Preuschoft, S., & van Schaik, C. (2009). Skill mastery inhibits adoption of observed alternative solutions among chimpanzees *(Pan troglodytes). Animal Cognition, 12,* 209—216. Available from: http://dx.doi.org/10.1007/s10071-008-0183-y.

Köhler, W. (1925). *The mentality of apes.* London: Routledge and Kegan Paul.

Kummer, H., & Goodall, J. (1985). Conditions of innovative behavior in primates. *Philosophical Transactions of the Royal Society of London Series B-Biological Sciences, 308,* 203—214.

Lehner, S. R., Burkart, J. M., & van Schaik, C. P. (2011). Can captive orangutans *(Pongo pygmaeus abelii)* be coaxed into cumulative build-up of techniques? *Journal of Comparative Psychology, 125,* 446—455.

Lonsdorf, E. V. (2005). Sex differences in the development of termite-fishing skills in wild chimpanzees *(Pan troglodytes schweinfurthii)* of Gombe National Park, Tanzania. *Animal Behavior, 70,* 673—683.

MacLean, E. L., Hare, B., Nunn, C. L., et al. (2014). The evolution of self-control. *PNAS, 111,* E2140—E2148.

Manrique, H. M., & Call, J. (2011). Spontaneous use of tools as straws in great apes. *Animal Cognition, 14,* 213—226.

Manrique, H. M., & Call, J (2015). Age-dependent cognitive inflexibility in great apes. *Animal Behaviour, 102,* 1—6. Available from: http://dx.doi.org/10.1016/j.anbehav.2015.01.002.

Manrique, H. M., Voelter, C. J., & Call, J. (2013). Repeated innovation in great apes. *Animal Behavior, 85,* 195–202. Available from: http://dx.doi.org/10.1016/j.anbehav. 2012.10.026.

Marshall-Pescini, S., & Whiten, A. (2008). Chimpanzees (*Pan troglodytes*) and the question of cumulative culture: An experimental approach. *Animal Cognition, 11,* 449–456. Available from: http://dx.doi.org/10.1007/s10071-007-0135-y.

Matsuzawa, T. (1994). Field experiments on use of stone tools in the wild. In R. W. Wrangham, W. C. McGrew, F. B. M. de Waal, & P. G. Heltne (Eds.), *Chimpanzee cultures* (pp. 351–370). Cambridge, MA: Harvard University Press.

Mesoudi, A. (2011). *Cultural evolution: How darwinian theory can explain human culture & synthesize the social sciences.* London: University of Chicago Press.

Pike, T. W., & Laland, K. N. (2010). Conformist learning in nine-spined sticklebacks' foraging decisions. *Biology Letters, 6,* 466–468.

Price, E. E., Lambeth, S. P., Schapiro, S. J., & Whiten, A. (2009). A potent effect of observational learning on chimpanzee tool construction. *Proceedings of the Royal Society B: Biological Sciences, 276*(1671), 3377–3383.

Reader, S., & Laland, K. (2002). Social intelligence, innovation, and enhanced brain size in primates. *Proceedings of the National Academy of Sciences, 99,* 4436–4441.

Russon, A. E., Kuncoro, P., Ferisa, A., & Handayani, D. P. (2010). How orangutans (*Pongo pygmaeus*) innovate for water. *Journal of Comparative Psychology, 124,* 14–28.

Sanz, C. M., & Morgan, D. B. (2007). Chimpanzee tool technology in the Goualougo Triangle, Republic of Congo. *Journal of Human Evolution, 52,* 420–433.

Silberberg, A., & Fujita, K. (1995). Pointing at smaller food amounts in an analogue of Boysen and Berntson's (1995) procedure. *Journal of the Experimental Analysis of Behavior, 66,* 143–147.

Tennie, C., Call, J., & Tomasello, M. (2009). Ratcheting up the ratchet: On the evolution of cumulative culture. *Philosophical Transactions of the Royal Society B, 364,* 2405–2415.

van de Waal, E., Borgeaud, C., & Whiten, A. (2013). Potent social learning and conformity shape a wild primate's foraging decisions. *Science, 340,* 483–485.

Van Lawick-Goodall, J. (1971). *In the shadow of man.* London: Houghton Mifflin.

Van Leeuwen, E. J., & Haun, D. B. M. (2013). Conformity in nonhuman primates: Fad or fact? *Evolution and Human Behavior, 43,* 1–7. Available from: http://dx.doi.org/ 10.1016/j.evolhumbehav.2012.07.005.

Van Leeuwen, E. J. C., Cronin, K. A., Schütte, S., Call, J., & Haun, D. B. M. (2013). Chimpanzees flexibly adjust their behaviour in order to maximize payoffs, not to conform to majorities. *PLoS One, 8,* e80945. Available from: http://dx.doi.org/10.1371/ journal.pone.0080945.

van Schaik, C. P., Ancrenaz, M., Borgen, G., Galdikas, B., Knott, C. D., Singleton, I., et al. (2003). Orangutan cultures and the evolution of material culture. *Science, 299,* 102–105.

Vlamings, P., Hare, B., & Call, J. (2010). Reaching around barriers: The performance of the great apes and 3- to 5-year-old children. *Animal Cognition, 13,* 273–285.

Whiten, A., Goodall, J., McGrew, W. C., Nishida, T., Reynolds, V., Sugiyama, Y., et al. (1999). Cultures in chimpanzees. *Nature, 399,* 682–685. Available from: http://dx.doi. org/10.1038/21415.

Whiten, A., Horner, V., & de Waal, F. B. M. (2005). Conformity to cultural norms of tool use in chimpanzees. *Nature, 437,* 737–740.

Yamamoto, S., Humle, T., & Tanaka, M. (2013). Basis for cumulative cultural evolution in chimpanzees: Social learning of a more efficient tool-use technique. *PLoS One, 8,* e5576.

Commentary on Chapter 14: Conservatism Versus Innovation: The Great Ape Story

Can tradition and innovation go hand in hand? This is the question posited by Josep Call in his chapter on *Conservatism Versus Innovation: The Great Ape Story*. As someone who is interested in studying culture and human creativity, I am naturally drawn to the thesis of this paper. It is almost a known fact that certain characteristics of individuals, including being traditional or conservative, are negatively associated with creativity. When comparing the creativity of members from different cultures, a general belief is that a collectivist-orientated culture, where individuals are driven to maintain the harmonious state of a society and conform to the cultural norm, may have a more stifling effect on the development of creativity than an individualist-orientated culture, where individual autonomy and self-directed learning is encouraged. However, such a belief has been challenged by recent evidence making the relationship between tradition and creativity far more complex than once thought.

But let's put aside the debate on whether or not tradition hampers the development of human creativity and focus on the thesis of this chapter, that is, to examine the relationship between tradition and innovation displayed in the nonhuman world. I want to emphasize that creativity and innovation are not synonymous. Creativity is defined as the ability to generate original and appropriate ideas in any domain and innovation is the successful implementation of creative ideas. Whereas creativity is often measured through open-ended tasks that independent raters can evaluate at a later date, innovation is assessed through problem solving, either by applying an old solution to a new problem or producing a new solution to an old problem. This chapter adopts the concept of innovation.

This chapter begins by posting an interesting puzzle that chimpanzees and orangutans (or apes) are conservative and conformist species. Nevertheless, they are also believed to be the most flexible problem solvers and innovators on Earth. In exploring this puzzle, the author first carefully examines the empirical evidence on animal problem solving, inhibitory control, and innovation across different ape species. His conclusions: conservatism and innovation are not necessarily contradicted by one another. They are both adaptive mechanisms that apes developed for pursuing a better payoff. In fact, there are mixed results regarding apes' conservatism, some of which suggest that apes are not as conservative as people have generally believed. The observed conservatism (e.g., persistently using existing tools or methods acquired either through social learning and individual exploration) can be explained by partial reinforcement. In other words, apes choose to stick to tradition when old solutions still work. When a solution becomes completely ineffective, apes are more likely to innovate. Therefore, depending on the payoff, apes are capable of choosing between repeating previously successful solutions and innovating new solutions when the old ones fail.

To further examine the relationship between apes' conservatism and innovation, the author explored the evidence on social learning, social influence, and conformity, as well as their impact on innovation across different several subgroups of apes. Again, results are mixed. There are evidences that socially acquired information (by observing another individual's successful solution) has a dampening effect on the probability of innovation, such that apes choose to abandon their own individually gained solution to conform to the majority. However, other studies have shown that apes are capable of abandoning techniques in favor of more efficient or profitable solutions regardless of how techniques are acquired (either through social observation or individual exploration). In other words, innovation can take place in the face of majority influence to pursue better outcomes.

So what influences apes' likelihood to innovate or conform? The answer: it depends. Similar to humans, individual factors such as age, gender, personalities (e.g., boldness-shyness) and contextual variables (such as task and setting), can contribute to the likelihood and degree to which individuals or species conform or innovate.

To me, the most intriguing discussion of this chapter is on conservatism, social conformity, and cumulative culture. Apes preserve tradition through social learning and innovation, but have yet to develop cumulative culture. It is tempting to blame conservatism and social conformity on the lack of cumulative culture among apes. However, conservatism and social conformity were not present in all studies examined by Call's chapter. Most importantly, human beings also display a great deal of conservatism and innovation to establish, preserve, and evolve their civilization.

Back to the initial question proposed by this chapter: "Can tradition and innovation go hand in hand?" Certainly the question can be examined by applying it to human beings. Preserving tradition and taking risky steps for innovation seem to be on two opposite ends of a spectrum, yet both are needed to make progress in any field. On one hand, it is important to actively acquire and master the existing knowledge, norms, and skills accumulated and passed on over generations; on the other hand, it is also important to take an occasional "gamble" and shift away from tradition in order to make progress in a field.

There are two common approaches in psychology to examine the relationship between tradition and innovation/creativity of human beings. One is by studying the cognitive process and examining the relationship between fixation and creativity or insight problem solving. Fixation refers to knowledge about an existing paradigm. In fixation, knowledge or an obvious solution is automatically activated which leads to an inability to generate new solutions. Tradition takes the form of knowledge or skill fixation, which may undermine innovation and creativity. There seems to be overwhelming evidence from cognitive research that establishes a negative relationship between fixation and creativity.

Another approach is to study individual differences and to find some characteristics that promote creativity. These characteristics include personality traits (openness to experience, extraversion), gender, religious affiliation, and culture (collectivist versus individualist).

However, new evidence from both approaches suggests that the relationship between preserving tradition and innovation/creativity is more complex than people used to think. For example, recent evidence demonstrates that creativity can benefit from tasks that are highly structured (in a way rather fixed), especially for those individuals who have a higher personal need for structure (Rietzschel, Slijkhuis, & Van Yperen, 2014). This suggests that fixation and structure can sometimes help innovation when individuals use structured knowledge as a base for further exploration.

Similarly, closely examining the Confucian ideology and creativity, Niu (2013) claimed that the notion of tradition does not necessarily contradict with creativity, at least in the Chinese context. Compared to the Western conception of creativity, creativity in the Chinese context puts greater emphasis on the continuation between past, present, and future. Therefore, to be creative, a person should actively acquire knowledge accumulated in the past and practice certain stills to reach perfection. Some other scholars also examined the relationship between Confucianism and innovation within organizations and concluded that maintaining a harmonious relationship in fact could help innovation at the organizational level. For example, a study of Chinese employees in a business setting found that Zhongyong (the doctrine of mean), a core value of Confucianism prescribing the extent to which a person puts priority on traditional propriety and interpersonal harmoniousness by following the doctrine of mean (often viewed as an important characteristics of conservatism), was found to have a positive effect on organizational innovation, especially when employees realize the importance of innovation to the organization (Du, Ran, & Cao, 2014).

Evidence like the two listed above suggest that tradition does not necessarily hamper creativity at the individual and group levels. When making structure and tradition meaningful to individuals, this structure and tradition can even stimulate creativity and facilitate the process of innovation.

So what can we learn from the studies on conservatism and innovation in the animal world? Conservatism is not always apparent. Preserving tradition and striving for innovation are both important adaptive mechanisms that are not necessarily on opposite ends of a spectrum; rather, they can work hand in hand in facilitating cultural progress. The findings from animal research seem to echo what we have learned from research with human beings, and eventually help us better understand the progress of cultural evolution.

References

Du, J., Ran, M., & Cao, P. (2014). Context contingent effect of Zhongyong on employee innovation behavior. *Acta Psychologica Sinica*, 46(1), 113–124.

Niu, W. (2013). Confucian ideology and creativity. *Journal of Creative Behavior*, 46(4), 274–284.

Rietzschel, E. F., Slijkhuis, J. M., & Yperen, N. W. V. (2014). Task structure, need for structure, and creativity. *European Journal of Social Psychology*, 44, 386–399.

Tools for the Trees: Orangutan Arboreal Tool Use and Creativity

Anne E. Russon[1], Purwo Kuncoro[2] and Agnes Ferisa[2]

[1]Psychology Department, Glendon College, York University, Toronto, ON, Canada [2]Orangutan Kutai Project, Kutai National Park, E. Kalimantan, Indonesia

Commentary on Chapter 15: Tools for the Trees: Orangutan Arboreal Tool Use and Creativity

David H. Cropley

Department of Engineering, Defence and Systems Institute (DASI), University of South Australia, Mawson Lakes, SA, Australia

INTRODUCTION

This paper assesses orangutan tool use in arboreal travel and positioning in terms of its characteristics and the role of creativity. Tools have long been a topic of research interest because of their implications for cognition (Goodall, 1963) and, more recently, for innovation (Lefebvre, 2013). Orangutans have contributed relatively little to tool-related issues because, until recently, they appeared to make little use of tools in the wild (Yamakoshi, 2004). It is now worth reconsidering orangutan tool use because recent revisions to the tool definition, notably lifting the requirement that tool objects be unattached, mean some of their forest canopy manipulations now quality as tool use (Shumaker, Walkup, & Beck, 2011). Their well-known tree swaying to cross canopy gaps is a prime example. Studying orangutans' tool use in arboreal travel may also offer new insights into creativity because it favors improvisation and innovation. Their arboreal habitat is highly unstable and unpredictable, given mature orangutans' heavy body weight, so they must adjust their travel on a moment-to-moment, flexible, calculating, context-contingent basis (Hunt, 2004; Povinelli & Cant, 1995). This paper takes advantage of the new tool definition to explore creativity in orangutans' tool use in arboreal travel and positioning. Our basis for exploring creativity includes the tools that orangutans have been observed using, identified in our observational records of rehabilitant orangutan behavior and recent reviews of primate and orangutan tool use (e.g., Meulman & van Schaik, 2013; Shumaker et al., 2011). Our aims are to attempt to identify and characterize the role of creativity, in terms of innovation and improvisation, in generating these tools.

BACKGROUND

Tool Use

We highlight major issues in the history of studying tool use as they have affected primatology, especially orangutans. Shumaker et al. (2011) review this history in detail.

Early Approaches

Tool use came to be important in the study of primate behavior with Goodall's (1963) report of chimpanzees using and making tools. This was a landmark discovery because of the prevailing belief that tools are uniquely human and critical in the evolution of human intelligence. The interest in nonhuman tool use that followed led to Beck's formalizing the definition of a tool, especially the nature of tool items and how the tool item-goal connection is established, as "the external employment of an *unattached* environmental object to alter more efficiently the form, position, or condition of another object, another organism, or the user itself, when the user holds or carries the tool during or just prior to use and is responsible for the proper and effective orientation of the tool" (Beck, 1980, p. 10). Until recently, Beck's definition served as the standard in the study of tool use. At least in primatology, it has had several important consequences.

Beck's tools were pivotal to Parker and Gibson's (1977) influential reconstruction of the evolution of great ape intelligence. They proposed that ancestral apes' seasonal reliance on embedded foods (foods protected from predators by embedding matrices) favored the evolution of flexible extractive foraging techniques assisted by "intelligent" tool use (tool use in which users understand the causal dynamics involved; Parker & Poti', 1990). Beck's definition also led to defining "proto" tools as attached objects (e.g., fixed substrates) used to alter target objects to distinguish them from "true" tools in Beck's sense, even when they were used in functionally similar fashion (Parker & Gibson, 1977; Yamakoshi, 2004). In result, studies of primate tool use focused on obtaining foods, especially embedded ones, more than other uses (e.g., communication, travel).

From this perspective, true tools are relatively rarely used by wild orangutans. Yamakoshi's (2004) review of tool use in feeding found wild orangutans using only five tools while wild chimpanzees use 27. Recent reviews covering a wider range of behavior identified 38 wild orangutan true tools (Meulman & van Schaik, 2013), still fewer than wild chimpanzees' 43 tools (Sanz & Morgan, 2007), and only 2/38 are used in arboreal travel (*snag ride, branch hook*). Some have suggested that arboreality is what limits orangutans' tool use in the wild because they are exceptional tool users in captivity (Meulman, Sanz, Visalberghi, & van Schaik, 2012; Meulman & van Schaik, 2013). One argument is that arboreality leaves few manipulators available for tool use because most (typically three) are occupied with supporting the body (Cant, 1987).

Changing Views

Beck's definition has met criticisms over the years that recently culminated in several scholars recommending its revision, Beck included (Bentley-Condit & Smith, 2010; Shumaker et al., 2011; St. Amant & Horton, 2008). Critical to this paper is that revisions drop the *unattached* criterion: the consensus is that manipulation of the tool object, not its detachment, is the essential property of tool use (Shumaker et al., 2011; St. Amant & Horton, 2008), because detachment per se does not change the relation of the mediating object to the target. In hammer-and-anvil nut cracking, for example, stones that are not manipulated or moved before being used as anvils are functionally similar to those that are (Boinski, Quatrone, & Swartz, 2000; Panger, 1998, 1999; Parker & Gibson, 1977). The two new definitions are very similar; we adopted the Shumaker et al. (2011) version because it was the basis for a comprehensive reassessment of animal tool use: tool use is (i) employing, externally, an unattached or manipulable-attached environmental object (ii) to alter more efficiently the form, position, or condition of a target item (another object or organism, or the user itself) when the user

(iii) holds or directly manipulates the tool during or prior to use and (iv) is responsible for its proper and effective orientation. To a large degree, the proto-true distinction disappears.

Alternatives to the extractive foraging model also appeared, several of which singled out arboreal travel as a critical pressure in primate cognitive evolution. Milton (1981) argued that *finding* foods is a major task for frugivorous primates in seasonal tropical rainforest habitats because of variation in where and when they are available; most primate fruits grow in the canopy, so finding food entails arboreal travel. This brought research attention to orangutan arboreal travel, especially gap crossing, and some concluded that it may be orangutans' greatest cognitive challenge and the context in which their cognitive achievements are most evident (e.g., Chevalier-Skolnikoff, Galdikas, & Skolnikoff, 1982; MacKinnon, 1974; Povinelli & Cant, 1995; Russon, 1998). These views stimulated study of orangutan arboreal locomotion, but mostly its biomechanics (Thorpe & Crompton, 2009). Beyond Povinelli and Cant (1995) and Hunt (2004), little of this study has concerned cognitive or tool-using features of arboreal travel. Appreciating why and how tools may enter into orangutan arboreal travel requires understanding the challenges of arboreal travel for orangutans.

Orangutan Arboreal Travel and Tool Use

Habitat Structure

Habitat structures define the general challenges that orangutans face as arboreal travelers. Three stand out: (i) the canopy's discontinuous nature (notably gaps between trees), (ii) its compliance (deformation of vegetation under an individual's weight), and (iii) its unpredictability (instability, change over time) (Povinelli & Cant, 1995; Thorpe & Crompton, 2009). Arboreal travel may be especially complex for orangutans because their extremely heavy bodies make it especially risky and energetically costly (Cant, 1987; Elton, Foley & Ulijaszek, 1998; Hanna, Schmidt, & Griffin, 2008; Warren & Crompton, 1998). Body weight is a critical factor because difficulties associated with the strength, stability, and deformation of arboreal substrates all increase with increasing weight (Cant, 1987). Gaps effectively become wider because outward-extending branches normally taper outwards and become weaker, less stable, and more easily deformable downward under weight, so they cannot support heavy weight at the periphery; deformation and risks of breaking supports increase so multiple versus single supports are often needed and instabilities effectively increase (Cant, 1987, 1998). Heavier individuals are relatively more fragile than lighter ones, so are at greater risk of serious injury or death from falls

(Povinelli & Cant, 1995). How orangutans deal with these challenges is then central to understanding their arboreal travel.

Orangutan Canopy Use

Orangutan canopy use further defines their arboreal challenges. As frugivores they have three major tasks: vertical travel (climb trees or lianas), horizontal travel (cross gaps), and positioning for feeding (to access/obtain and manipulate foods, especially fruits found mostly at thin ends of terminal branches) (Povinelli & Cant, 1995). Tree trunks are continuous and stable vertically but separated horizontally; branches extending laterally between trees normally taper outwards so they become weaker, less stable, and more compliant downwards under weight. So gaps between trees, even narrow ones, typically exist between ends of thin branches. Lianas typically extend vertically from the ground into the canopy and some provide relatively horizontal connections across gaps, but their strength and stability are highly variable (Cant, Finlinson, Mendel, & Povinelli, 1990). For these reasons, Povinelli and Cant (1995) concluded that gap crossing is the single most important and severe problem facing heavier bodied (>10 kg) arboreal travelers; mature orangutans' very heavy bodies (>35 kg) then strongly influence every solution to their arboreal tasks.

What Orangutans Do

Prominent among orangutans' responses to these arboreal travel challenges are (i) using compliance, that is, manipulating versus adjusting to or ignoring it, and (ii) given their heavy body weight, making great use of suspension and multiple supports (Povinelli & Cant, 1995; Thorpe, Crompton, & Alexander, 2007). In gap crossing they have several ways of manipulating vegetation compliance, for example, lunging, pulling, bending, and oscillating (e.g., Cant, 1987; Thorpe & Crompton, 2005, 2006; Thorpe et al., 2007; van Schaik, Noordwijk, & Wich, 2006). They make and use several forms of reachers to cope with gaps and hangers for suspension. Securing multiple supports can entail modifying the form (join several slender lianas into a "rope") and relative position of each. Very few of these mediating devices have qualified as tools because they remain attached; given the manipulations involved, some are now recognized as tools under the new definition (Shumaker et al., 2011).

An additional arboreal travel challenge very little considered is water. Preferred orangutan habitats are water dominated (Delgado & van Schaik, 2000) and water can create forest gaps. Some orangutan methods for crossing water gaps are water-specific (e.g., check depth with a stick, boats) but others resemble methods used for crossing arboreal gaps (e.g., various forms of bridging). Water and the canopy pose some similar travel challenges, so orangutan water travel may derive from their arboreal travel. Water's buoyancy resembles the canopy's compliance in that both support some weight but give way, and applying force to both generates propulsion. A few rehabilitants that "swim" short distances do so in a *lunge-and-glide* fashion very similar to the *lunge transfer* orangutans often use to cross arboreal gaps (Hunt et al., 1996; Russon, Kuncoro, Ferisa, & Handayani, 2010).

Important in positioning for feeding is securing a position that is stable and frees some manipulators for obtaining food. Both features are critical when force is needed to obtain foods, for example, some embedded foods. A rehabilitant once applied so much force to crack a termite nest that he almost fell off his branch (Russon, 2002). Orangutans typically use three manipulators in securing an arboreal feeding position but favor sitting while feeding (Cant, 1986, 1987), so securing stable, hand-freeing feeding positions could involve modifying canopy structures if no natural sitting places are available.

Creativity in Orangutan Arboreal Travel Tool Use

Creativity should be relatively common in orangutans' arboreal travel and positioning given the low predictability and instability of their arboreal habitat and their known reliance on non-stereotyped, figure-it-out-as-you-go forms of locomotion (Hunt, 2004; Povinelli & Cant, 1995; Thorpe et al., 2007). By creativity, we mean generating approaches or solutions to problems that are novel and useful relative to a particular context (Kaufman, Butt, & Kaufman, 2011; Sternberg, 1999). The Kaufman et al. (2011) model of animal creativity is helpful in framing our exploration of orangutan creativity because it proposes three foundational contributors (novelty recognition and seeking, observational learning, innovation), two of which we can explore in orangutan arboreal positioning: innovations, and novelty recognition and seeking as manifest in improvisations (which show recognizing and engaging with versus avoiding novelty).

Innovations, here, are new or modified individually learned behaviors whose acquisition is contingent on appropriate experience but not simply a product of social learning or environmental induction (Reader & Laland, 2003; van Schaik et al., 2006). This resembles definitions used in most systematic studies of orangutan innovation (Bastian, van Noordwijk, & van Schaik, 2012; Russon et al., 2010; Russon et al., 2009; van Schaik et al., 2006). Improvisations are on the spot, spontaneous solutions to a task facing the actor, that is, invented in their performance, not planned or practiced in advance, and not necessarily repeated (Bertinetto, 2012).

Improvisation should be relatively common, given the low predictability of the arboreal supports orangutans use. It may be critical for supports that prove more or less compliant or fragile than expected because locomotion must be adjusted on the spot using the materials and configurations at hand. Thorpe et al. (2007) offer several examples using tools. Improvisations may not be worth re-using if the situational configuration is very unlikely to recur. They suggest the inventiveness upon which orangutans can draw, however, and some may serve as a basis for innovations. Some tools that orangutans have used in arboreal travel qualify as innovative, for example, *snag ride* (ride a dead tree as it breaks and falls), *bite vine* (bite off vines restricting the sway of a tree or liana), and *branch hook* (use a branch to catch and pull in out-of-reach vegetation) (Bastian et al., 2012; Russon et al., 2010; Russon et al., 2009; van Schaik et al., 2006). The new tool definition could well turn up more tools used for these purposes.

THIS STUDY

We review available published and unpublished reports of orangutans' tool use in arboreal travel and positioning, using the newly revised definition of a tool, with the aims of assessing its extent and the role of improvisation and innovation. This assessment is exploratory, given the very limited systematic evidence available at this time. We made the following predictions to guide assessment:

1. Under the new tool definition, more of the techniques orangutans use in arboreal travel and positioning will qualify as tool use than currently recognized.
2. Since gap crossing is considered the most important travel challenge for large arboreal animals, most orangutan tool use will serve horizontal travel.
3. Because multiple supports are much used by orangutans and highly important given their heavy body weight, the use of several tools in combination should be relatively common.
4. Some orangutan tools used in arboreal travel and positioning will qualify as improvisational or innovative.

METHODS

We compiled reports of orangutans' tool use in arboreal travel and positioning for feeding based on revised tool definitions that remove the detachment criterion. Our compilation relies on data-mining, that is, including all credible reports, however obtained (Bates & Byrne, 2007; Lefebvre, Whittle, Lascaris, & Finkelstein, 1997).

We included all events accepted as tool use in previous reviews covering tools in wild, rehabilitant, or captive orangutans (Bastian et al., 2012; Herzfeld & Lestel, 2005; Meulman & van Schaik, 2013; Russon et al., 2009; Shumaker et al., 2011; van Schaik et al., 2006). Given our interests concern arboreal travel and positioning, we focused on tools used by wild and rehabilitant orangutans in native forest; tools used by captives offer additional insights into generating processes. The Shumaker et al. (2011) review used the revised tool definition; all others used the traditional tool definition. We added any additional published reports found by searching electronic bibliographic services (e.g., Scopus, Web of Science).

We also included events that qualified as tool use from our observational studies of rehabilitant orangutans. Rehabilitant data were collected at six sites: Tanjung Puting (1989–1992), Wanariset caging facilities (1994), Sungai Wain (1995–2000), Meratus (1999–2003), Kaja Island (2004–2006), Palas I and II islands (*ad hoc*, 2004 and 2007), and Samboja Lestari (2006–2008). Rehabilitants living in cages were treated as captives due to the behavioral opportunities and constraints involved.

Measures

Tools

We used the Shumaker et al. (2011) tool definition that drops the unattached criterion for tool objects. Operationally, tools were (i) freely manipulable external objects (unattached or manipulable-attached) used to (ii) alter the physical properties of a target item (form, position, condition) when the user (iii) held or directly manipulated the tool prior to use and (iv) was responsible for its proper and effective orientation.

Arboreal Tools

We included tools used in the three standard tasks in orangutan arboreal travel and positioning: *vertical travel* (climbing up and down), *horizontal travel* (travel between vertical structures, mostly crossing gaps between trees and including gaps defined by water), and *positioning for feeding* (positioning for accessing or manipulating food). We classified tools used in several tasks as *multipurpose* (e.g., *reacher, hanger*).

Distinguishing and Classifying Tools

We used three levels: *type* (defining tool-target relationship), *mode* (broad use/purpose), and *task* (the arboreal task it addressed).

1. *Tool types* were differentiated by the physical relationship created between tool and target objects (e.g., *rake, hammer, hanger*). "Tool" is essentially a relational term, referring to objects only when they are used in particular kinds of physical relationships with target objects (Beck, 1980; Reynolds, 1982; Russon, Mitchell, Lefebvre, & Abravanel, 1998). Objects modified to serve a specific tool-target relationship have been named for that relationship (e.g., *rake, sponge, wiper*) but other objects can substitute for them and the tool-named object can be used in other tool-target relationships (e.g., a *rake* to *poke, hammer*, or *scratch*). We noted tool types used in combination to solve a given task, as a basis for indexing complexity (Byrne & Russon, 1998; Case, 1985; Russon, 1998; Yamakoshi, 2004) and to suggest how new tool forms are generated.
2. *Tool modes* grouped tool types by common purpose, for example, *reacher* groups tools that extend the user's reach (e.g., *rake, rope, hanger*), *ladder* groups tools ladder-climbed to reach objects overhead (e.g., standard *ladder, pole, stacked tower, conspecific*).
3. *Tool tasks.* We classified tool modes and types by the arboreal task they served (positioning for feeding, horizontal travel, vertical travel, multipurpose).

Tool Use Prevalence

We used all prevalence evidence available. Minimally, we reported the total number of wild, rehabilitant, and captive sites reporting each tool type. We recorded within-site prevalence estimates where available for wild and rehabilitant sites (typically A—absent, R—rare, H—habitual (several individuals), C—customary (most individuals), E—absent for ecological reasons: van Schaik et al., 2006). In the absence of other evidence, we treated tool types listed as standardized primate positional modes (Hunt et al., 1996) as customary at all wild and rehabilitant sites.

Creativity

We explored creativity for wild and rehabilitant orangutans using two of the three factors in the Kaufman et al. (2011) animal creativity model: novelty recognition and seeking, and innovation. Improvisation, likely to be relatively common in orangutan arboreal travel and positioning, requires recognizing and engaging with novelty.

1. *Innovation.* We accepted a tool type as innovative if it met at least one of three criteria. (i) The newest systematic studies identified it as such, based on the within-site prevalence estimates mentioned above (Bastian et al., 2012; Russon et al., 2010; Russon et al., 2009;

van Schaik et al., 2006). (ii) Tool types reported elsewhere were known at less than 25% of the wild or rehabilitant sites where they are ecologically possible. (iii) Tool types reported in only one or very few orangutans were performed with a level of skill suggesting extensive practice-based refinement.

2. *Improvisation.* We accepted tool use as improvisational if it was described as a one-time spontaneous act that dealt with an unusual and apparently unforeseen situation.

Coding

We coded all reports of orangutan tool use in these arboreal travel and positioning tasks that we located in the literature and in our own observational data. For each event we coded tool type, mode, and task, where it occurred (site, wild-rehabilitant-captive orangutans), the reporting source, a descriptive narrative of the event, and any clarifying comments. We coded orangutans as *wild* if they had been living freely in native habitat since birth, *rehabilitant* if they were ex-captives living free or semi-free in native forest life when the event was observed, and *captive* if they were living in cages when the event was observed because cage life alters environmental and behavioral possibilities. Sites at which the event was observed were field research sites, rehabilitation/reintroduction forests, and captive facilities (cages at rehabilitation projects, zoos, research laboratories, or private homes). Reporting sources were listed as the identifying published review or, for events not covered in published reviews, the reporting publication, and/or observers.

RESULTS

Beyond the reports of orangutans using tools for these arboreal tasks identified in reviews of orangutan tool use or innovation, most published reports we found were single events, described for illustrative purposes, observed while collecting data for studies of arboreal positioning or cognition. The tool use database compiled here derives from 12 wild and seven rehabilitant field sites in Borneo and Sumatra and a large (uncounted) number of captive facilities. Unless stated otherwise, the findings reported represent orangutans living in native habitat (wild, rehabilitants). In this database we identified 15 tool modes and 42 tool types used in arboreal positioning in native habitat (46 tool types including captives) (see definitions and catalog in Appendix A).

Tool Use Frequency and Distribution

Table 15.1 shows the distribution of these tools (modes and types) by the arboreal positioning task they addressed: vertical travel, horizontal travel, positioning for feeding, multipurpose (see Table 15.1). These classes are not mutually exclusive but are useful as an initial basis for exploring relationships. As predicted, we found more tools used in arboreal positioning in native habitat under the revised than under the standard definition of a tool, and more used in horizontal travel than in vertical travel or positioning for feeding.

Creativity

We tentatively identified 19 of the tool types used in native habitat as innovations and seven as improvisations. Their distribution by arboreal task is shown in Table 15.1. Two more innovations were tentatively identified in rehabilitants (both multipurpose).

We attempted to order tool types within each mode and class so as to suggest the behavioral pathway that may have led to each, especially those that qualify as innovations or improvisations. We used two common ordering indices, prevalence (common precedes rare) and complexity (less precedes more complex); both can suggest relationships between types.

Well-founded prevalence estimates (intra- and inter-population) are available for only the few tool types that have been studied

TABLE 15.1 Distribution of Tool Use in Arboreal Positioning, by Task and Living Conditions

Arboreal task	Tool modes	Tool types					Innov	Improv
		Total	Wild	Rehab	Native	Captive		
Position for feeding	2	4	2	3	4	0	2	1
Horizontal travel	7	17	8	13	17	2	7	3
Vertical travel	3	12	3	11	12	4	6	3
Multipurpose	3	13	6	10	10	7	6	0
Total	15	46	19	37	42	13	19	7

Cell values: total number of tool modes and types per category; Tool types: number of tool types identified in the orangutan populations identified (wild, rehabilitants living in native habitat, captives, total—all three populations, native—wild and rehabilitant); Innov: number of innovations tentatively identified; Improv: number of improvisations tentatively identified.

systematically (Bastian et al., 2012; Herzfeld & Lestel, 2005; Meulman & van Schaik, 2013; Russon et al., 2010; Russon et al., 2009; van Schaik et al., 2006). Both arboreal positioning and cognitive studies probably selected tool-related events because they were unusual but neither assessed prevalence. For wild and rehabilitant orangutans, we therefore estimated the prevalence of each tool type as: (i) the number of wild, rehabilitant, and captive sites at which it is reported, (ii) rare, if no prevalence information was available, or (iii) customary (performed by most community members), if the behavior constitutes a standardized primate locomotor or positional mode (Hunt et al., 1996; van Schaik et al., 2006). For captives, we estimated prevalence as the number of facilities at which a tool type has been reported, using the newest systematic reviews (e.g., Herzfeld & Lestel, 2005; Shumaker et al., 2011). We do not consider the captive estimate a serious limitation because our main interest is forest-living orangutans, so captives mainly suggest whether and how tools used in forest contexts are expressed in artificial ones.

It was possible to assess complexity for the majority of events because most reports described the context and form of tool use in considerable detail. We estimated complexity by (i) the number of object-object relations intrinsic to the tool or tool composite (several tools used simultaneously to achieve an outcome where component tools typically differ in type, e.g., stone hammer-anvil nut cracking; Shumaker et al., 2011) and (ii) the number of different tools types used in combination.

Findings are presented below by the arboreal task addressed.

Arboreal Positioning for Feeding

Table 15.2 shows tool modes and types reported in arboreal positioning for feeding. Two may be innovations, one an improvisation.

Chairs address the problem that arboreality leaves orangutans few manipulators for handling tools or food items. With the exception of *sit*, their arboreal feeding positions use three manipulators for support greater than 60% of the time; *sit* uses no manipulators for support but often one as an anchor for stability (Cant, 1987). *Sit* is the single most common feeding posture (Cant, 1987) and it is preferred over suspend for bark feeding, perhaps because bark feeding requires applying force and that favors stable supports (Myatt & Thorpe, 2011). We used the term *chair* to refer to the arboreal substrate on which an orangutan sits. Normally, *chairs* are suitable naturally existing structures (e.g., branch, liana).

We identified three tool chairs (*bent branch, cherry picker, X-chair*), presumably used when no natural chairs are available at the location needed. Reports indicate all three are rare but we suspect *bent branch* is

TABLE 15.2 Positioning for Feeding: Tool Modes and Types

Tools	Creativity	Wild (12)	Rehabilitant (7)	Captive	Sources
P1 CHAIR 0. sit (P1)	—	C	C	C	Hunt et al. (1996)
1. bent branch	?	1	4	E?	Kuncoro (unpublished data), Russon et al. (2010), Shumaker et al. (2011)
2. cherry picker	IN	0	2	E?	Russon (unpublished data)
3. X-chair	IN	0	1	E?	Shumaker et al. (2011)
P2 STICK PUSH SPINY FRUIT	IM	1	0	E?	Meulman and van Schaik (2013), Shumaker et al. (2011)
Tool types: 4		2	3	0	

Tools: Mode (P1, P2, etc.) and type (P1.1, P1.2, etc.) as defined in Appendix A. Types: 0, non-tool technique (to suggest potentially related non-tool solutions); #, suggested ordering by prevalence/complexity; a, b, etc., combinations with other tool types.
Creativity: —, not relevant (not a tool); C, customary; ?, uncertain; IN, possible innovation; IM, possible improvisation (Russon et al., 2009).
Wild sites (n = 12): Suaq Balimbing, Ketambe, Bohorok, Gunung Palung, Tanjung Puting, Sabangau, Tuanan, Sungai Lading, Kutai, Lower Kinabatangan, Lokan, Ulu Segama.
Rehabilitant sites (n = 7): Ketambe, Meratus, Sungai Wain, Samboja Lestari, Kaja Island, Palas I and II islands, Tanjung Puting.
Captive sites (total number unknown): zoos, research laboratories, private homes.
Cell values: ##, number of sites at which the mode has been reported; C, standardized locomotor mode (Hunt et al., 1996), inferred customary; E/E?, unsuitable/possibly unsuitable ecological conditions.

relatively common and under-reported: structures that could serve as *chairs* to access foods on terminal branches are likely to be relatively slender and therefore bend under an orangutan's weight. Making a *bent branch* chair merely involves using this compliance, that is, manipulating and orienting bending. *Cherry picker* appears to extend *bent branch*: an orangutan, once positioned on a *bent branch chair*, pulls adjacent vegetation to raise or lower the chair, thus accessing different parts of the food tree. Both *cherry picker* users were also *bent branch* users. *X-chair* was used by a single orangutan to access vertically climbing rattans (to extract the "heart" tissue); she used it multiple times in skillful fashion so it probably constitutes her innovation to a relatively common local feeding problem. Prevalence indices are consistent with this ordering and interpretation.

We treated *stick push spiny fruit* as an arboreal positioning tool because it frees manipulators for obtaining or processing a food item: in this case, it frees them from holding the fruit (and also protects their hands from the fruit's spines).

Horizontal Travel

Table 15.3 shows tool modes in horizontal arboreal travel, including travel across gaps created by water. As predicted, we found more tool modes and types in horizontal travel than in the other arboreal tasks considered.

TABLE 15.3 Horizontal Travel: Tool Modes and Types

Tools	Creativity	Wild (12)	Rehabilitant (7)	Captive	Sources
H1 LUNGE TRANSFER (L9f)	–	C	C	E	Hunt et al. (1996)
H2 BRIDGE					
1. body (P14)	–	C	–	E?	Hunt et al. (1996)
2. held-vegetation	–	3	–	E?	Campbell (1992), Kuncoro (unpublished data), MacKinnon (1974)
0. log bridge natural	–	1	4	E?	MacKinnon (1974), Russon et al. (2010), Russon (personal observation)
3. log bridge-built	?	0	1	E?	Shumaker et al. (2011)
4. bent-tree bridge	IN	0	1	E?	Russon et al. (2010), Shumaker et al. (2011)
H3 TREE SWAY					
1. bend tree (L16)	–	C	C	E	Hunt et al. (1996)
a. + *bite vine*	[IN]	0	(1)	E	Russon (personal observation)
b. + *hanger*	IM	0	1	E	Russon (personal observation)
2. oscillating (L16)	–	C	C	E?	Hunt et al. (1996)
a. + *rake*	IM	1	0	E	Shumaker et al. (2011)
b. + *hanger*	IM	0	1	E	Shumaker et al. (2011)

(Continued)

TABLE 15.3 (Continued)

Tools	Creativity	Wild (12)	Rehabilitant (7)	Captive	Sources
3. pole vault	IN	0	3	(1)	Kuncoro (unpublished data), Russon (personal observation), Shumaker et al. (2011)
4. block tree sway	IM	0	2	E	Kuncoro (unpublished data), Shumaker et al. (2011)
H4 LIANA SWING					
1. tarzan swing	—	4	3	E?	Campbell (1992), Cant (1987), Russon (personal observation), Shumaker et al. (2011), Thorpe and Crompton (2006)
a. + bite vine	[IN]	3	(1)	0	Russon et al. (2009), Shumaker et al. (2011), van Schaik et al. (2006)
b. + make liana rope	IM	0	1	0	Russon (personal observation)
2. oscillating	IN	1	0	0	Manduell, (2008)
H5 CATAPULT					
1. liana stretch	IM	1	0	E	Thorpe et al. (2007)
2. oscillating	IN	0	1	0	Russon (personal observation)
3. bungee jump	IN	0	1	E?	Smith (unpublished data)
H6 BOAT					
1. float	IM→IN	0	3	2	Kuncoro (unpublished data), Russon et al. (2010), Russon (personal observation), Shumaker et al. (2011)
a. + untie	IM	0	2	E	Russon (personal observation), Shumaker et al. (2011)
2. paddle	IN-HI	0	2	E	Kuncoro (unpublished data), Shumaker et al. (2011)
b. + untie	IM	0	1	E	Shumaker et al. (2011)

(Continued)

IV. PUSHING THE BOUNDARIES OF CREATIVITY

TABLE 15.3 (Continued)

Tools	Creativity	Wild (12)	Rehabilitant (7)	Captive	Sources
H7 SWIM	IN	0	1	1	Bender and Bender (2013), Russon et al. (2010)
Tool types: 15		**8**	**13**	**2**	

Tools: Mode (H1, H2, etc.) and type (H1.1, H1.2, etc.) as defined in Appendix A. Types: 0, non-tool technique (to suggest potentially related non-tool solutions); #, suggested ordering by prevalence/complexity; a, b, etc., combinations with other tool types.
Creativity: −, not relevant (not a tool, not possible in native habitat); C, customary; ?, uncertain; IN, possible innovation; IM, possible improvisation; [IN], innovation doubtful because the behavior may be (i) rarely needed or (ii) under-reported (Russon et al., 2009), HI, human influenced.
Wild sites (*n* = 12): Suaq Balimbing, Ketambe, Bohorok, Gunung Palung, Tanjung Puting, Sabangau, Tuanan, Sungai Lading, Kutai, Kinabatangan, Lokan, Ulu Segama.
Rehabilitant sites (*n* = 7): Ketambe, Meratus, Sungai Wain, Samboja Lestari, Kaja Island, Palas I and II Islands, Tanjung Puting.
Captive sites (total number unknown): zoos, research laboratories, private homes.
Cell values: ##, number of sites at which the mode has been reported; C, standardized locomotor mode (Hunt et al., 1996), inferred customary; E/E?, unsuitable/possibly unsuitable ecological conditions.

The standardized gap crossing mode is *lunge transfer* (L9f, Hunt et al., 1996). It qualifies as tool use because it manipulates compliance to narrow the gap: lunging deforms the support tree, and pulling the target tree bends it closer. We estimated it to be customary across forest sites because it is a standardized locomotor mode. Modes H2−H5 could all derive from *lunge transfer*. *Bridge* involves creating various physical structures across the gap (own body, hold vegetation on both sides close together, place logs across, bend trees across) (Figure 15.1). *Body bridge*, for example, could originate in simply pausing during a *lunge transfer* after deforming and at the point of holding both support and target vegetation, before moving into the target tree. *Tree sway, liana swing,* and *catapult* all involve more complex ways of deforming the support so that it swings close enough to the target and/or generates sufficient propulsion (bend, oscillate, stretch) for the rider to transfer across (clamber, leap). *Pole vault* was reported in situations where no trees or lianas were available for crossing a gap, so it probably originated as an improvisation to "recreate," to the degree possible, the natural tool object used in swaying/swinging across gaps. Reports suggest skilled performance, so we considered it an innovation. In some of these modes additional tools are used to improve effectiveness (Figure 15.2).

The three reports of *catapulting* show differences in complexity, suggesting progressive refinement and elaboration with practice. Summaries below illustrate.

FIGURE 15.1   *Make bridges across water.* An adolescent male orangutan used tree sway on a slender tree to cross a water gap then overbalanced it to ride down to the ground. Trees that orangutans manipulate in this fashion tend to stay bent, eventually providing 'permanent' bridges across the body of water. In this photo, the orangutan bent one tree (right) and descended onto a second tree that others had bent horizontal (left). © A. Russon (2005)

FIGURE15.2   *Tree sway combined with hanger.* An adolescent female rehabilitant climbs on a slender pole tree with its top broken off while holding a liana hanger (upper left photo). She sways the tree towards a coconut floating beyond her reach—still holding the liana hanger, probably to help control and to limit the sway (upper right photo). Leaning out from the bent pole tree and the liana hanger, she grabs the coconut (lower left photo). Pulling back on the liana, she puts her weight on or pulls the liana hanger to help her clamber, with the coconut, back to shore (lower right photo). © A. Russon (2005).

Liana stretch: An adolescent male, holding a liana angled 45° from vertical, wanted to transfer into a target tree at a higher level than his current position. He grabbed a lower terminal branch of another nearby tree, pulled himself down into the core of that tree—thereby stretching the liana, and then launched himself upwards towards the target tree. The recoil of the stretched liana added sufficient momentum to catapult him close enough to the terminal tree to grab one of its terminal branches (Thorpe et al., 2007).

Oscillating catapult. A juvenile male rehabilitant climbed to near the top of a slender vehicle tree and then pumped it back and forth repeatedly, like a swing, to amplify its oscillation. When he had achieved high oscillation and at the forward extreme of the vehicle tree's swing, he let go and was catapulted 5–10 m forward through space. He landed in and grabbed a target tree, so the direction of his catapulting was well aimed for transferring across a gap. As an adolescent, the same male continued catapulting and extended it to horizontal rubber cable (Russon, unpublished data, 2006, 2009).

Bungee jump. An adolescent female rehabilitant "bungee-jumped" to obtain grasses across a stream that were out of her reach arboreally. She avoided ground travel because wild pigs and cobras visited the stream. She positioned herself on a horizontal rubber cable above the grass area, then used her body weight to oscillate it vertically; once oscillating vertically, she shifted her position and pumped so that the cable oscillated diagonally (out-down, back-up). When she had sufficient oscillation she shifted position so that she was hanging from the cable upside-down (by her feet), then at the extreme of each out-down oscillation grabbed grasses on either side of the stream by hand. When she had collected enough grass she let the cable's oscillations subside and ate her grass hanging from the cable. She used this technique regularly over several months' observation (Smith, unpublished data, 2006), and was still using it several years later (Russon, unpublished data, 2012).

The two water-related modes, *boat* and *swim*, could derive from *bridge* and *lunge transfer* respectively. *Boat* could emerge from walking onto what looks like a *log bridge* (i.e., a log fallen across shallow water that is floating but looks to be lying on the ground). One rehabilitant climbed on such a log and became so uneasy when it tipped and rolled under his weight that he backed offer after crawling very carefully only a few cm. Others crossed the same floating log successfully, balancing and moving more skillfully along it. The form of *swimming* described in rehabilitant orangutans closely resembled *lunge transfer* in the positioning and movements involved (Figure 15.3). More complex boating is reported only in rehabilitants that interact often with humans and their

FIGURE 15.3 *"Lunge and glide" swim.* An adolescent female rehabilitant enters a pond (upper left photo), crouches in preparation for lunging (upper right photo), lunges towards the other side of the pond (lower left photo), and glides until she can grab vegetation on the edge of the pond (lower right photo) © A. Russon (2005).

boats, so while it is rare it almost certainly emerged partly because of these additional opportunities.

Vertical Travel

Table 15.4 shows tools used in vertical arboreal travel.

Wild orangutans travel up and down ladder-like vegetation to reach items out of reach above or below, that is, vegetation offering relatively horizontal supports for climbing (Figure 15.4). It is presumably relatively common because *ladder-climb*, climbing vertical supports with a diagonal gait like a person climbing a ladder, is a standard locomotor mode (L8b: Hunt et al., 1996). The *ladder* tool mode includes two tool types that manipulate natural vegetation to create ladder-climbable structures (*two-tree*, *zipper*), two that position a detached pole (*prop*, *balance*) and then climb it like a ladder, and two that use other climbable objects available in captivity (*stool*, *stacked tower*).

Two-tree and *zipper* both suggested improvisation because they rely on very unusual vegetation configurations (two saplings close together below a target tree, strong branches diverging outward from the tree's trunk). Both were used tentatively but with considerable skill,

TABLE 15.4 Vertical Travel: Tool Modes and Types

Tools	Creativity	Wild	Rehabilitant	Captive	Sources
V1 LADDER					
0. ladder-climb (L8b)	–	C	C	E?	Hunt et al. (1996)
1. two-tree	IM	0	1	E	Russon (2003)
2. zipper	IM	0	1	E	Russon (2003)
3. prop pole	IN	0	3	2	Shumaker et al. (2011)
4. balance pole	IN	0	1	3	Russon (personal observation), Shumaker et al. (2011)
a. + join poles	–	E	E	1	Shumaker et al. (2011)
5. stool	–	E	E	7	Shumaker et al. (2011)
a. + rake	–	E	E	1	Shumaker et al. (2011)
6. stack tower	–	E	E	4	Shumaker et al. (2011)
V2 RIDE					
1. basic (L17)	–	C	C	E	Hunt et al. (1996)
2. snag	IN	5/9	(4 similar)	E	Meulman and van Schaik (2013), Russon et al. (2009), Shumaker et al. (2011), van Schaik et al. (2006)
3. ski	IM	0	1	E	Russon (2003)
4. play	[IN]	0	2	E	Bastian et al. (2012), Russon (personal observation)
V3 RAPPEL					
1. basic	[IN]	1	0	0	Campbell (1992)
2. rope-loop	IN	0	1	0	Russon (personal observation), Smith (unpublished data)
a. + tie rope	IM	0	1	0	Russon (personal observation)
Tool types: 12		**3**	**11**	**4**	

Tools: Mode (V1, V2, etc.) and type (V1.1, V1.2, etc.) as defined in Appendix A. Types: 0, non-tool technique (to suggest potentially related non-tool solutions); #, suggested ordering by prevalence/complexity; a, b, etc., combinations with other tool types.

Creativity: –, not relevant (not a tool, not possible in native habitat); C, customary; ?, uncertain; IN, possible innovation; IM, possible improvisation; [IN], innovation doubtful because the behavior may be (i) rarely needed or (ii) under-reported (Russon et al., 2009).

Wild sites ($n = 12$): Suaq Balimbing, Ketambe, Bohorok, Gunung Palung, Tanjung Puting, Sabangau, Tuanan, Sungai Lading, Kutai, Kinabatangan, Lokan, Ulu Segama.

Rehabilitant sites ($n = 7$): Ketambe, Meratus, Sungai Wain, Samboja Lestari, Kaja Island, Palas I and II islands, Tanjung Puting.

Captive sites (total number unknown): zoos, research laboratories, private homes.

Cell values: ##, number of sites at which the mode has been reported; C, standardized locomotor mode (Hunt et al., 1996), inferred customary; E/E?, unsuitable/possibly unsuitable ecological conditions.

FIGURE 15.4 *Natural ladder.* A juvenile orangutan sits near the base of a natural liana ladder, about to climb it to enter the crown of the giant fig tree supporting it. © A. Russon (2001).

suggesting the orangutan applied existing expertise (ladder-climbing) but also went beyond it to handle the unusual configuration (manipulated them to enable their use as a ladder). *Prop* and *balance* were reported where no natural ladders were available, and both recreated the essential ladder properties (tool objects were tall enough, rigid, climbable) and ladder-target configuration (beside, below) using the objects available. The same holds true for ladder tools reported in captivity using stools, tires, barrels, and friends.

Ride, as a standard locomotor mode, resembles *tree sway* but is used for vertical descent. The tool forms we identified are probably derived from it (*snag ride, ski ride, play ride*). The latter two were reported in juvenile-adolescent rehabilitants, so both may involve motor skill practice. *Play ride* occurred during social play in the form of repeatedly breaking trees or branches and riding them down through the canopy. *Ski ride* may represent an exploratory improvisation (it was seen once, in a juvenile male rehabilitant).

Rappel, the controlled descent down (or ascent up) a near-vertical surface using a rope, was reported three times in two forms: *basic* in wild orangutans, *rope-loop* (more complex) in rehabilitants. *Basic rappel* was reported at only one wild orangutan site but it could be more common, especially where lianas are important in arboreal travel, because orangutans prefer multiple supports in arboreal positioning (Myatt & Thorpe, 2011) so pushing feet against a vertical surface is a probable accompaniment to climbing up or down a liana. *Rope-loop* rappel, seen in one rehabilitant on a small forested island, showed well-honed, sophisticated skill (Smith, personal observation, 2006; Russon, personal observation, 2012). It is described below:

> *Rope-loop rappel.* One adolescent female rehabilitant, ca 8 years old when first observed, rappelled to descend from a horizontal cable (slung ca 3 m above the ground between vertical posts) to the ground and to ascend from ground to cable. Positioned on the horizontal cable, she held a second rubber cable ca 5−6 m long; one of its ends was permanently fastened near the top of a post but the other end was free. She looped the free end of her rubber cable over the horizontal cable and pulled it enough to create a u-shaped loop below the horizontal cable, with the rest of the free portion of the cable dangling down to the ground. To rappel, she sat or stood on the u-shaped loop while holding the free portion of the cable firmly so that the loop supported her body weight. Then she gradually loosened the free part of the cable, thereby lengthening the u-shaped loop which in turn lowered her. She typically descended on the loop to 0.5−1.0 m above ground, where she could pick up fruit provisions that staff had placed there while still sitting in her loop seat. Fruits in hand, she pulled the free section of the cable to shorten the u-shaped loop, thereby lifting her back up to the horizontal cable. Smith observed her rappel this way at least four times over the course of a 4 month observation period; Russon observed this twice several years later. She did not need to rappel to descend or ascent; she could (and others did) simply climb down and up nearby poles or vertically hanging ropes.

Multipurpose Tools

Table 15.5 lists multipurpose tools reported in arboreal travel and positioning.

Reacher, the broadest mode, included seven tool types (*branch pull, rake, hanger, rope, slave, pull water, ruler*). The most derived are probably *ruler* and *pull water*, the most complex a multi-support *rope-rake* combination used in trying to obtain a coconut floating out of reach (Figure 15.5) and a *rope-hook* combination (probably a composite) used to snare and pull in out of reach items.

Ruler may derive from *rake*: a wild male tried repeatedly to *rake* in target vegetation but failed because the branch he used was too short,

TABLE 15.5 Multipurpose: Tool Modes and Types

Tools	Creativity	Wild (12)	Rehabilitant (7)	Captive	Sources
M1 REACHER					
1. branch pull (L9f, L11b)	–	C	C	E?	Hunt et al. (1996)
2. rake	–	6	4	~15	Meulman and van Schaik (2013), Russon et al. (2009), Russon et al. (2010), Shumaker et al. (2011), van Schaik et al. (2006)
3. hanger					
a. + make rope (join)	–	C?	2	0	Davenport (1967), Russon et al. (2010), Russon (unpublished data)
b. + tree sway	IM	0	2	0	Russon (1998), Russon et al. (2010)
c. + make rope (join) + rake	IM	0	1	0	Russon (unpublished data)
4. rope	IN	0	2	3	Davenport (1967), Kuncoro (unpublished data), Shumaker et al. (2011)
5. rope-hook composite + make rope (join)	IN	0	0	2	Russon (unpublished data), Shumaker et al. (2011)
6. slave reacher	IN	0	0	1	Shumaker et al. (2011)
7. pull water	IN	0	1	0	Russon et al. (2010)
8. ruler	IN	1	3	0	Fox and Bin'Muhammad (2002), Russon et al. (2010), Shumaker et al. (2011)
M2 ATTACH					
1. join	–	C?	2	~11	Davenport (1967), Russon et al. (2010), Russon (unpublished data), Shumaker et al. (2011), van Schaik et al. (2006)
2. tie or knot	IN	0	2	7	

(Continued)

TABLE 15.5   (Continued)

Tools	Creativity	Wild (12)	Rehabilitant (7)	Captive	Sources
					Herzfeld and Lestel (2005), Shumaker et al. (2011)
M3 UNBLOCK					
1. *bite vine*	[IN]	4	3	E?	Russon et al. (2009), Shumaker et al. (2011), van Schaik et al. (2006)
2. *untie knots*	IN	E?	2	7	Herzfeld and Lestel (2005), Shumaker et al. (2011)
Tool types: 12		**6**	**10**	**7**	

Tools: Mode (M1, M2, etc.) and type (M1.1, M1.2, etc.) as defined in Appendix A. Types: 0, non-tool technique (to suggest potentially related non-tool solutions); #, suggested ordering by prevalence/complexity; a, b, etc., combinations with other tool types.
Creativity: —, not relevant (not a tool, not possible in native habitat); C, customary; ?, uncertain; IN, possible innovation; IM, possible improvisation; [IN], innovation doubtful because the behavior may be (i) rarely needed or (ii) under reported (Russon et al., 2009).
Wild sites (*n* = 12): Suaq Balimbing, Ketambe, Bohorok, Gunung Palung, Tanjung Puting, Sabangau, Tuanan, Sungai Lading, Kutai, Kinabatangan, Lokan, Ulu Segama.
Rehabilitant sites (*n* = 7): Ketambe, Meratus, Sungai Wain, Samboja Lestari, Kaja Island, Palas I and II islands, Tanjung Puting.
Captive sites (total number unknown): zoos, laboratories, private homes.
Cell values: ##, number of sites at which the mode has been reported; C, standardized locomotor mode (Hunt et al., 1996), inferred customary; E/E?, unsuitable/possibly unsuitable ecological conditions.

so eventually he found a longer branch and used it successfully (Fox & Bin'Muhammad, 2002). The first branch may have helped him assess the distance to the target vegetation, opening the door to using sticks deliberately to assess distance. Russon et al. (2010) reconstructed behavioral pathways within one rehabilitant site which suggested that *pull water* derives from failed attempts to lunge and grab or *rake* objects that were floating on still water out of reach. Such failures result in the hand or rake plunging into water in front of the target object, which in turn result in *pull water* as unintended by-products of pulling back from the lunge or pulling in the rake. Deliberate *pull water* could develop from noticing this effect and then actively creating it. The *rope-hook* tool use is described below.

> *Rope-hook composite tool.* An adolescent female rehabilitant in a large socialization cage tried to obtain rambutan fruits that staff had placed outside a section of the cage housing a new orangutan, to

FIGURE 15.5 *Multi-support rope-rake combination.* An adolescent female tried repeatedly to grab a coconut floating on a pond but could not reach it, so she suspended herself from two overhead supports (holding three slender lianas in her left foot as a 'rope' hanger and a strong liana in her left hand), reached into the water with her right hand and foot, and lifted out a long branch (left photo). She repositioned her right hand so she was suspended from three liana supports and then, holding the branch with her right foot, tried to retrieve the coconut with the branch rake (right photo). She manipulated multiple arboreal supports (the liana rope she made combined with two other lianas) to create a stable suspensory position near the coconut, then added the use of a rake. © A. Russon (2005).

tempt him to eat. After several failed attempts to reach the rambutan by hand she left, retrieved an empty rice sack (provided for behavioral enrichment), unraveled one of the long strings from which it was woven, and collected a cup-shaped piece of orange peel. She carried both items back to the side of the cage near the rambutan. She poked a hole in the center of the orange peel with a finger then threaded her string through the hole in the orange peel. Holding the two ends of the string together in one hand and the orange peel in the other, she pulled gently on both to slide the orange peel to the center of her doubled string "rope." Holding the two ends of the rope together between her first and second finger, she reached through the cage bars as far as possible and cast its orange peel end towards the rambutan. After a few casts she succeeded in landing the orange peel behind a rambutan, and then gently pulled the rope back to drag the fruit closer. She retrieved several rambutan fruits using this tool. She had to remake her tool once because another orangutan stole her first one, including finding new and somewhat different materials. She did so in ways suggesting well-honed skills and good understanding of the qualities needed in her tool materials and the relationships to be achieved.

Attach is ostensibly rare in arboreal life if we exclude behaviors that bring vegetation together for travel (*lunge transfer, branch pull, rake, tree/liana sway*), feeding positions (*X-chair, stick push spiny fruit*), and nesting (weaving nest materials together, bending and locking branches of multiple trees together then building a nest on them: Prasetyo et al., 2009; van Schaik et al., 2006). Modes recognized are *join* and *tie or knot;* most *handles* involve one or both (Maple, 1980).

DISCUSSION

Our review found that orangutans use (and make) a substantial number of tools in arboreal positioning, under the new broader definition of a tool. It is consistent what is known about orangutans' major arboreal positioning challenges in native habitat (canopy compliance, discontinuity, unpredictability, heavy body weight) and their solutions to them (manipulating compliance, connecting discontinuous forest structures, improvising in unexpected situations, preference for multiple arboreal supports). All are prominent in the tool use we identified. Creating connections, for example, is central to 24/27 tool types that orangutans have been reported to use in arboreal positioning and travel (3/4 tools used in positioning for feeding, 13/13 in horizontal travel, 8/10 in vertical travel; water-related tool types excluded). Some of these tools would not in fact work well, or at all, if the tool object was detached: *liana swing* and *tree sway* are prime examples. Something like *tree sway* is possible using a detached pole (*pole vault*), but it entails first planting the pole firmly into the ground (i.e., connecting it). For these reasons, we consider that our findings offer a credible first impression of the importance of tools in orangutans' arboreal positioning and travel.

We considered creativity in terms of innovation and improvisation. Our findings suggest substantial creativity but are clearly inconclusive. Systematic evidence is currently negligible and we know of no good basis for comparison (wild orangutan tools were identified using the standard tool definition and innovations using systematic prevalence data that are unavailable here). Our findings are worth following up, however, because they are in line with the importance of unpredictability in orangutan arboreal life. Beyond the typical unpredictabilities in arboreal habitat, orangutans face variation in forest type, damage, and heterogeneity between and within sites, and differ in body weight as a function of age, sex and reproductive condition. For all these reasons, improvisation and innovation should occur relatively often.

Our descriptive evidence also suggests the processes involved in generating tool innovations or improvisations. We discuss four that stand out: deliberately recruiting accidental occurrences, recreating (to the degree possible) naturally occurring structures or configurations, substituting alternative for original tool objects, and combining multiple objects to solve one problem.

Recruiting Accidents

Great apes are known to be especially astute at noticing accidental solutions to problems, understanding the cause-effect relations involved, and recruiting them into their own voluntary repertoire—where recruiting almost undoubtedly involves modifying or refining them (Byrne, 1995, 1999; Russon, 2003; Russon et al., 2009). This is reflected in orangutans' using versus simply experiencing compliance, and compliance-based tools could arise this way. (i) Experiencing accidental bending in a slender support tree caused by leaning or reaching to grab a target tree could lead to deliberate leaning to bend it (*tree sway-bend*). (ii) Experiencing a vehicle tree's recoil from repeated failed attempts to *bend* could lead to deliberate *oscillating tree sway*. (iii) Unintentionally over-oscillating a vehicle tree could overbalance it and carry the rider to the ground, leading to deliberately overbalancing as in *ride*, or deliberately breaking the tree as in *play ride*. As another example, we suggested earlier that *pull water* to draw floating objects into reach may arise from accidentally pulling water when attempts to lunge and grab them or rake them in fail.

Recreating Natural Structures

Many of the tools we identified are physically and functionally similar to natural structures in orangutans' native habitat, for example, *chairs*, *bridges*, *ladders*, *ropes*, vaulting *poles*. Orangutans use such tools when the natural structures they normally use are unavailable, so natural structures probably serve as sensory models for finding and/or making a suitable tool. This is consistent with great apes' strong capacity for understanding and manipulating physical cause-effect relations and the likelihood that they create "novel" solutions by gradually building more complex structures upon pre-existing simpler ones (Russon et al., 2009; Yamakoshi, 2004). Orangutans' causal understanding should be especially refined concerning *weight* (Povinelli & Cant, 1995), *distance* (Russon, 1998; Russon et al., 2010; Thorpe et al., 2007), and *multiple supports*. Tool use we identified as improvisational similarly suggests using what is already known and then going beyond it to cope with novel or unforeseen challenges.

Substituting Tool Objects

Strong understanding of cause-effect relations probably enables substituting the materials used to make tools. Orangutans have made and used *hangers*, for instance, from lianas, branches (that they cracked deliberately), string or other fibrous materials, wire, and sheets: important are functional properties, not specific forms.

Combining Tools

We found tool combinations (two more tools used in sequence to solve one problem) but did not treat them as distinct tool types. They were used in *hanger, bend tree sway, oscillate tree sway, tarzan swing*, and *rope-loop rappel* to extend the distance reached or traveled (*hanger, rake, bite vine*) and to attach a rappel rope overhead (*tie rope*). In the most complex case, a female trying to reach a coconut floating on water made a hanger by joining three slender lianas, combined this with two stronger lianas to make a three-point suspensory support as close as possible to the coconut; suspended from this three-point support, she then used a rake to try to reach a coconut floating on water. One tool combination, *rope-hook* (a rope with a hook attached to catch and retrieve items out of reach), qualified as a tool composite, two or more tools used simultaneously to achieve a single outcome (Shumaker et al., 2011). *Oscillating tree sway* combined with *hanger*, where the *hanger* served as a metatool (Shumaker et al., 2011), may also qualify as a tool composite. Combining tools could relate to orangutans' tendency to use multiple supports and/or to refine and build new skills out of simpler ones. The latter suggests the potential for cumulative build-up of increasingly complex tools over time.

CONCLUSION

Our review of orangutans' tool use in arboreal travel and feeding postures only scratches the surface since it has not been the focus of systematic study in the field. At minimum, it promises to be a fruitful topic for follow-up systematic studies. With more systematic study, we predict that some of the tool use that we identified as innovative will prove to be relatively common but also that an even broader range of tools will appear. We also predict that our impression that creativity is relatively common will be confirmed by systematic studies, given the everyday uncertainties inherent in orangutan arboreality.

APPENDIX A: CATALOG: TOOLS ORANGUTANS USE IN ARBOREAL TRAVEL AND POSITIONING

This catalog includes all tools that orangutans have been reported to use in arboreal positioning, where tools were identified using the Shumaker et al. (2011) definition. Reports of orangutan tool use were obtained from recent systematic reviews of orangutan tool use (e.g., Meulman & van Schaik, 2013; Shumaker et al., 2011) and innovation (e.g., Russon et al., 2009; van Schaik et al., 2006), our cumulative data base on rehabilitant orangutan behavior, and additional reports of orangutan tool use and innovation found by searching the published literature (e.g., Bastian et al., 2012; Manduell, 2008; Russon et al., 2010; Thorpe et al., 2007). Tools are distinguished by type (the tool-target relationship manipulated), and grouped by mode (tool function) and arboreal task (positioning for feeding, horizontal travel, vertical travel, multipurpose).

Tool types are numbered within modes to suggest complexity (0—no tools used, 1 or higher—one or more tool types used). For types recognized as standard locomotor or postural modes, we note the identifying code in Hunt et al. (1996). We ordered types to suggest possible relations between them.

Positioning for Feeding

P1. *chair*: vegetation or other items used for stable substrate support in the canopy, typically to obtain or manipulate food (e.g., extract embedded foods) or to play.

1. *sit*: the ischia bear a substantial portion of the body weight (usually >50%), the torso is relatively orthograde (P1, Hunt et al., 1996); assumption is positioned on existing natural structures (e.g., branch, liana).

2. *bent branch*: deliberately bend and position a branch (or palm petiole or slender tree) horizontal beside a target food source, then sit on it to obtain the food items.

3. *cherry picker*: make a *bent branch* chair, sit (or stand) on it, then pull on nearby vegetation to move it up or down; typically used while obtaining plant foods to access different parts of the target food source.

4. *X-chair*: bend two very slender trees into an "X" form adjacent to a food source, sit on the outer side of one lower leg of the "X," and seated there obtain food items (Russon, 2003).

P2. *stick push spiny fruit*: push a spiny fruit into a crevice with a stick (Rijksen, 1978).

Horizontal Travel Modes

H1. *lunge transfer*: to cross a gap between trees, take a position on a support tree, lunge towards a target tree and grasp one of its small branches, pull the branch, and gradually move into the target tree. Lunging deforms branches and narrows or closes the gap (see L9f, Hunt et al., 1996). A lunge is an incomplete leap; formally, it is a brachiation-like motion in which the user grasps a small support in the target tree with one hand, then hand over hand or hand over foot pulls one of its branches closer.

H2. *bridge*: travel on a structure providing a pathway across an obstacle (e.g., gap, river).

1. *body bridge*: a mother takes a *postural bridge position* in a gap so her infant can cross it on her body (Hunt et al., 1996). In a postural bridge, feet and hands grasp support(s) on either side of a gap so the body spans the gap in tension (P14, Hunt et al., 1996).

2. *held-vegetation bridge*: to help youngsters cross wide gaps, mothers narrow the gap by holding branches closer together so they can cross (MacKinnon, 1974).

3. *existing log bridge*: travel on trees or other vegetation that have fallen across streams to cross the streams (MacKinnon, 1974).

4. *make log bridge*: drag logs or vines to a river or pond bank, place them as bridges across water, then walk/crawl on them to cross the water (Galdikas, 1982).

5. *bent-tree bridge*: *tree sway* (H3) to cross a body of water and then overbalance the tree to *ride* (V3) it down to the ground on the other side. Trees repeatedly bent this way stay horizontal and become permanent bridges that others use (Russon et al., 2010).

H3. *tree sway*: to cross a gap between trees, the rider uses their body weight to deform the vehicle tree on which they are positioned far enough so they can reach a nearby target tree (see *liana swing* for a similar technique using a liana) (L16, Hunt et al., 1996).

1. *bend tree*: deform a vehicle tree by leaning in the intended travel direction, so body weight bends the tree closer to the destination (L16, Hunt et al., 1996).

2. *oscillating*: pump a vehicle tree back and forth, like a swing, to amplify its sway in the intended travel direction (L16, Hunt et al., 1996; Sugardjito & van Hooff, 1986).

3. *pole vault*: plant a detached slender pole upright on/into the ground near a target location, climb/transfer onto the top of the pole, lean in the intended travel direction to bend it across a gap, and transfer into a target tree.
4. *block tree sway*: after having *tree swayed* and transferred into a target tree, retain hold of the vehicle tree to block a follower's pursuit along the same route (Russon, 2003).

H4. *liana swing*: swing on a liana to cross a gap (similar to *tree sway* with a liana vehicle).

1. *tarzan swing*: swing on a liana, tarzan-like, to cross a gap (van Schaik et al., 2006).
2. *oscillating*: pump a vehicle liana back and forth repeatedly to increase its swing, then cross a gap on it when it sways far enough (Manduell, 2008).

H5. *catapult*: use a device in which accumulated tension is suddenly released to propel an object some distance through space.

1. *liana stretch*: One orangutan, holding a liana angled at 45° to vertical, wanted to enter a target tree higher than his current position. He grabbed and pulled a lower terminal branch of another nearby tree, pulling himself down into the tree's core and stretching the liana. When he launched himself upwards towards the target tree the liana acted as a catapult, adding enough momentum for him to grab an outer branch of the target tree (Thorpe et al., 2007).
2. *oscillating*: pump a vehicle tree or liana back and forth repeatedly to create extreme sway, let go of the vehicle tree/liana at the forward extreme of its sway, and sail through space—typically toward nearby vegetation (bushes, target tree) where the rider lands.
3. *bungee jump*: use *oscillating catapult* on a horizontal (rubber) cable to sail across a gap to a target location; without letting go of the cable, grab food on the other side of the gap, and then ride back to the original location on the cable as it swings back, with the food (Smith, personal observation, 2006).

H7. *boat*: travel across/along waterways on floating wood, canoes, or rafts.

1. *float*: climb onto floating objects and ride them with the current.
2. *paddle*: propel a boat by paddling with sticks, boards, or dippers or, alternatively, pull vegetation growing near the water's surface or along the shoreline; can be combined with untying knots in the rope fastening the boat near shore and, after crossing a river, pulling the boat along by its rope while walking along the shoreline (Galdikas, 1982).

IV. PUSHING THE BOUNDARIES OF CREATIVITY

H6. *swim*: travel in water by floating, either with support (hold and pull on vegetation to propel the body) or without support (lunge-and-glide, or rudimentary paddling) (Russon et al., 2010).

Vertical Travel Modes

V1. *ladder*

1. *ladder-climb*: vertical travel on supports greater than 45°; supports are often relatively horizontal and never a single vertical support. The climber's limbs move in a diagonal sequence, like a person climbing a ladder with a diagonal gait (L8b, Hunt et al., 1996).
2. *two-tree ladder*: climb a pair of small saplings approximately 0.75 m apart, like a rung-less ladder, to enter a tree. From a bipedal position on the ground, grab an overhead support by hand to anchor and take some of the body weight, grasp one sapling in each foot, and inch up into the tree (Russon, 2003). See *ski ride* for similar travel down to the ground.
3. *zipper ladder*: an older adolescent male climbed down then up (out then in) two adjacent branches of an *Anthocephalus chinensis* tree to get fruit on their terminal branches; the two branches diverged from the trunk to the periphery. As he climbed out along the two branches he gradually pulled them closer together, like a zipper. This enabled him to reach fruit on the tree's terminal breaches. After eating, he reversed these actions to return to the center of the tree (Russon, 2003).
4. *prop pole*: reposition a detached pole (e.g., broken tree, stick), prop it against a vertical support, and then climb the pole (Shumaker et al., 2011).
5. *balance pole*: balance a detached pole upright and climb it fast (Shumaker et al., 2011).
6. *stool*: reposition a stool-like object beside or below a target, and then climb the stool to reach the target (Shumaker et al., 2011).
7. *stack tower*: stack several objects on top of each to make a tower beside or below a target, and then climb up the tower to reach the target (Shumaker et al., 2011).

V2. *ride*: similar to *tree sway* but for vertical descent (Cant, 1987; Hunt et al., 1996).

1. *basic*: climb onto a slender vertical support (typically a small tree) and bend or oscillate it to overbalance it, pulling it from vertical toward horizontal. The rider may shift to a suspensory posture as the support nears horizontal; after or during this process the rider releases hindlimb grip and drops, feet first, to the ground (L17, Hunt et al., 1996).

2. *snag ride*: ride on a snag (dead tree) as it falls, grabbing and shifting onto nearby vegetation before it crashes (Galdikas, 1983).
3. *ski ride*: ride on two adjacent slender saplings, staying in a quadrupedal position atop them until they near the ground, then step or jump onto the ground (Russon, unpublished data). See also *two-tree ladder* for a similar ascending mode.
4. *play ride*: climb on a small tree or a branch, bend or bite the tree or branch until it breaks, and ride as it crashes or slides down through the canopy (commonly, immatures).

V3. *rappel*: controlled descent of a near-vertical surface using a rope, either (i) slowly slide down the rope and push feet against the surface or (ii) slowly release the end of a doubled rope, looped around the body and fixed at a higher point, to lower the body and push feet against the surface for balance and partial support. Reverse for ascent.

1. *basic*: walk up/down a tree trunk while holding and pulling on a liana (Campbell, 1992).
2. *rope-loop rappel*: from an elevated position, descend to the ground by loosening a rope looped around the body and fastened overhead; in reverse, ascend from the ground back up by pulling on the rope (Smith, unpublished data, 2006). Has been combined with tie (loop and twist) the upper end of the rappel rope directly above the target ground location, then *rope-loop* rappel to the ground, pick up desired (food) items there, and *rope-loop* rappel back up with the food (Russon, unpublished data, 2012).

Multipurpose Arboreal Tool Modes

M1. *reacher*: an elongated object used to touch or retrieve an item the user cannot or does not want to contact directly (Shumaker et al., 2011).

1. *branch pull*: pull a branch to draw other vegetation within reach, typically a tree or liana (to enable transfer), or terminal branches (to get their fruit) (L9f in Hunt et al., 1996).
2. *rake*: use a rigid elongated object, detached, to hook and pull in a target object (Russon et al., 2010; Russon et al., 2009) (combines *branch hook* and *branch reach fruit* in Meulman & van Schaik, 2013; van Schaik et al., 2006).
3. *hanger*: hold an elongated object that is or has been attached to an overhead structure (branch, liana) so that it partially or completely supports the body, then lean or swing out from it to extend reach (Shumaker et al., 2011).
4. *rope*: use and sometimes make a rope (e.g., gather or twist together several thin branches, lianas, or other fibrous items) and then cast it to snag and then pull in a target object.

5. *rope with hook composite*: fasten a "hook" (e.g., orange or banana peel) to one end of a *rope*. Cast the rope, like a fishing line, towards a target item so that its hook catches the target, and then pull the rope in carefully to draw the target into reach.

6. *slave*. In a large socialization cage at a rehabilitation center, an adolescent male grabbed and led a female down to the grill floor of the cage. The cage had a double floor, grill work 1 m above tiles, to reduce orangutans' contact with debris. He pushed her down and held her while she reached through the grill and retrieved food fallen on the tiles, then took it from her (Russon, 2003).

7. *pull water*: to retrieve an item floating out of reach on still water, dip one hand into the water in front of the item and repeatedly pull the water towards oneself. This creates a small current that draws the item closer until it can be grabbed (Russon et al., 2010).

8. *ruler*: insert a long detached stick into water until it touches the bottom, apparently to test the water's depth (Russon et al., 2010).

M2. *attach*

1. *join*: connect several inflexible items (especially sticks, end to end), often to make reachers longer than any of the items used singly, or loop long, flexible items over a support then connect them (hold them together) to make a hanger or handle (Shumaker et al., 2011).

2. *tie or knot*: tie, twist or weave together fibrous items (e.g., straw, rope, lianas, cloth), often to make swinging or hanging devices (Herzfeld & Lestel, 2005). See also pulling and locking two or more trees together to form the base for building a nest (Prasetyo et al., 2009), *bridge nest* in van Schaik et al. (2006).

M3. *unblock*: remove an obstruction impeding the movement of other items.

1. *bite vine*: bite through a vine, typically to release a vehicle tree or liana being used to *tree sway* or *liana swing* (van Schaik et al., 2006).

2. *untie knots*: undo knots to release objects for use in travel, where the objects were tied up (Herzfeld & Lestel, 2005; Shumaker et al., 2011).

References

Bastian, M. L., van Noordwijk, M. A., & van Schaik, C. P. (2012). Innovative behaviors in wild Bornean orangutans revealed by targeted population comparison. *Behaviour, 149*, 275–297.

Bates, L. A., & Byrne, R. W. (2007). Creative or created: Using anecdotes to investigate animal cognition. *Methods, 42*, 12–21.

Beck, B. B. (1980). *Animal tool behavior: The use and manufacture of tools by animals*. New York, NY: Garland STPM.

Bender, R., & Bender, N. (2013). Swimming and diving behavior in apes (*Pan troglodytes* and *Pongo pygmaeus*): First documented report. *American Journal of Physical Anthropology, 152* (1), 156–162. Available from: http://dx.doi.org/10.1002/ajpa.22338.

Bertinetto, A. (2012). Performing the unexpected: Improvisation and artistic creativity. *Revista Internacional de Filosofía, 57*, 117–135.

Bentley-Condit, V. K., & Smith, E. O. (2010). Animal tool use: current definitions and an updated comprehensive catalog. *Behaviour, 147*, 185–221, Available from: http://dx. doi.org/10.1163/000579509X12512865686555.

Boinski, S., Quatrone, R. P., & Swartz, H. (2000). Substrate and tool use by brown capuchins in Suriname: Ecological contexts and cognitive bases. *American Anthropologist, 102*, 741–761.

Byrne, R. W. (1995). *The thinking ape*. Oxford, UK: Oxford University Press.

Byrne, R. W. (1999). Cognition in great ape ecology: Skill-learning ability opens up foraging opportunities. In H. O. Box, & K. R. Gibson (Eds.), *Mammalian social learning: Comparative and ecological perspectives* (pp. 333–350). Cambridge, UK: Cambridge University Press.

Byrne, R. W., & Russon, A. E. (1998). Learning by imitation: A hierarchical approach. *Behavioral and Brain Sciences, 21*, 667–721.

Campbell, J. L. (1992). *Ecology of Bornean orangutans (Pongo pygmaeus) in drought- and fire-affected lowland rainforest* (unpublished doctoral dissertation). USA: Anthropology, Pennsylvania State University.

Cant, J. G. H. (1987). Positional behavior of female Bornean orangutans (*Pongo pygmaeus*). *American Journal of Primatology, 12*, 71–90.

Cant, J. G. H., Finlinson, H. A., Mendel, F. C., & Povinelli, D. J. (1990). Stress tests of lianas to determine safety factors in the habitat of orangutans (Pongo pygmaeus) in northern Sumatra. *American Journal of Physical Anthropology, 81*, 203.

Case, R. (1985). *Intellectual development: Birth to adulthood*. New York, NY: Academic Press.

Chevalier-Skolnikoff, S., Galdikas, B. M. F., & Skolnikoff, A. (1982). The adaptive significance of higher intelligence in wild orangutans: A preliminary report. *Journal of Human Evolution, 11*, 639–652.

Davenport, R. K., Jr (1967). The orang-utan in Sabah. *Folia Primatologica, 5*, 247–263.

Delgado, R. A., Jr, & van Schaik, C. P. (2000). The behavioral ecology and conservation of the orangutan (Pongo pygmaeus): A tale of two islands. *Evolutionary Anthropology, 9*(5), 201–218.

Elton, S., Foley, R., & Ulijaszek, J (1998). Habitual energy expenditure of human climbing and clambering. *Annals of Human Biology, 25*(6), 523–531.

Fox, E. A., & Bin'Muhammad, I. (2002). New tool use by wild Sumatran orangutans (*Pongo pygmaeus abelii*). *American Journal of Physical Anthropology, 119*(2), 186–188.

Galdikas, B. M. F. (1982). Orangutan tool-use at Tanjung Puting Reserve, central Indonesian Borneo (Kalimantan Tengah). *Journal of Human Evolution, 11*(1), 19–33.

Galdikas, B. M. F. (1983). The orangutan long call and snag crashing at Tanjung Puting Reserve. *Primates, 24*(3), 371–384.

Goodall, J. M. (1963). My life among wild chimpanzees. *National Geographic Magazine, 124* (2), 272–308.

Hanna, J. B., Schmidt, D. L., & Griffin, T. M. (2008). The energetic cost of climbing in primates. *Science, 320*, 898.

Herzfeld, C., & Lestel, D. (2005). Knot tying in great apes: Etho-ethnology of an unusual tool behavior. *Social Science Information, 44*(4), 621–653. Available from: http://dx.doi. org/10.1177/0539018405058205.

Hunt, K. D. (2004). The special demands of great ape locomotion and posture. In A. Russon, & D. Begun (Eds.), *The evolution of thought: Evolution of great ape intelligence*. Cambridge, UK: Cambridge University Press (pp. 172–189).

Hunt, K. D., Cant, J. G. H., Gebo, D. L., Rose, M. D., Walker, S. E., & Youlatos, D. (1996). Standardized descriptions of primate locomotor and postural modes. *Primates, 37*, 363–387.

Kaufman, A. B., Butt, A. E., & Kaufman, J. C (2011). Towards a neurobiology of creativity in nonhuman animals. *Journal of Comparative Psychology, 125*(3), 255–272.

Lefebvre, L. (2013). Brains, innovations, tools and cultural transmission in birds, nonhuman primates, and fossil hominins. *Frontiers of Human Neuroscience, 7*, 1–10.

Lefebvre, L., Whittle, P., Lascaris, E., & Finkelstein, A. (1997). Feeding innovations and forebrain size in birds. *Animal Behaviour, 53*, 549–560.

MacKinnon, J. (1974). The behaviour and ecology of wild orangutans (*Pongo pygmaeus*). *Animal Behaviour, 22*, 3–74.

Manduell, K. L. (2008). Locomotor behaviour of wild orang-utans (*P. p. wurmbii*) in disturbed peat swamp forest, Sabangau, Central Kalimantan, Indonesia. MSc Thesis, Biology, Manchester Metropolitan University.

Maple, T. L. (1980). *Orang-utan Behavior*. New York: Van Nostrand Reinhold.

Meulman, E. J. M., Sanz, C. M., Visalberghi, E., & van Schaik, C. P. (2012). The role of terrestriality in promoting primate technology. *Evolutionary Anthropology, 21*, 58–68.

Meulman, E. J. M., & van Schaik, C. P. (2013). Orangutan tool use and the evolution of technology. In C. M. Sanz, J. Call, & C. Boesch (Eds.), *Tool use in animals: Cognition and ecology* (pp. 176–202). Cambridge, UK: Cambridge University Press.

Milton, K. (1981). Distribution patterns of tropical plant foods as an evolutionary stimulus to primate mental development. *American Anthropologist, 83*(3), 534–548.

Myatt, J. P., & Thorpe, S. K. S. (2011). Postural strategies employed by orangutans (*Pongo abelii*) during feeding in the terminal branch niche. *American Journal of Physical Anthropology, 146*, 73–82.

Panger, M. A. (1998). Object-use in free-ranging white-faced capuchins (*Cebus capucinus*) in Costa Rica. *American Journal of Physical Anthropology, 106*, 311–321.

Panger, M. A. (1999). Capuchin object manipulation. In P. Dolhinow, & A. Fuentes (Eds.), *The nonhuman primates* (pp. 115–120). Mountain View, CA: Mayfield.

Parker, S. T., & Gibson, K. R. (1977). Object manipulation, tool use and sensorimotor intelligence as feeding adaptations in *Cebus* monkeys and great apes. *Journal of Human Evolution, 6*, 623–641.

Parker, S. T., & Poti', P. (1990). The role of innate motor patterns in ontogenetic and experiential development of intelligent use of sticks in *Cebus* monkeys. In S. T. Parker, & K. R. Gibson (Eds.), *"Language" and intelligence in monkeys and apes* (pp. 219–246). New York, NY: Cambridge University Press.

Povinelli, D. J., & Cant, J. G. H. (1995). Arboreal clambering and the evolution of self-conception. *Quarterly Review of Biology, 70*(4), 393–421.

Prasetyo, D., Ancrenaz, M., Morrogh-Bernard, H. C., Utami Atmoko, S. S., Wich, S. A., & van Schaik, C. P. (2009). Nest building in orangutans. In S. A. Wich, S. S. Utami, T. Mitra´Setia, & C. P. van Schaik (Eds.), *Orangutans: Geographic variation in behavioral ecology and conservation* (pp. 269–277). Oxford: Oxford University Press.

Reader, S. M., & Laland, K. N. (Eds.), (2003). *Animal innovation*. Oxford: Oxford University Press.

Reynolds, P. C. (1982). The primate constructional system: The theory and description of instrumental tool use in humans and chimpanzees. In M. Van Cranach, & R. Harré (Eds.), *The analysis of action: Recent theoretical and empirical advances* (pp. 343–385). Cambridge: Cambridge University Press.

Rijksen, H. D. (1978). *A field study on Sumatran orangutans* (Pongo pygmaeus abelii *Lesson, 1827): Ecology, behaviour, and conservation*. Wageningen, The Netherlands: H. Veenman and Zonen.

Russon, A. E. (1998). The nature and evolution of intelligence in orangutans (*Pongo pygmaeus*). *Primates, 39*, 485–503.

Russon, A. E. (2002). Return of the native: Cognition and site-specific expertise in orangutan rehabilitation. *International Journal of Primatology*, 23(3), 461−478.

Russon, A. E. (2003). Innovation and creativity in forest-living rehabilitant orangutans. In S. M. Reader, & K. N. Laland (Eds.), *Animal innovation* (pp. 279−306). Oxford: Oxford University Press.

Russon, A. E., Kuncoro, P., Ferisa, A., & Handayani, D. P. (2010). How orangutans innovate for water. *Journal of Comparative Psychology*, 124(1), 14−28.

Russon, A. E., Mitchell, R. E., Lefebvre, L., & Abravanel, E. (1998). The evolution of imitation. In J. Langer, & M. Killen (Eds.), *Piaget, evolution and development* (pp. 103−143). Hillsdale, NJ: Lawrence Erlbaum Associates.

Russon, A. E., van Schaik, C. P., Kuncoro, P., Ferisa, A., Handayani, D. P., & van Noordwijk, M. A. (2009). Innovation and intelligence in orangutans. In S. A. Wich, S. S. Utami, T. Mitra Setia, & C. P. van Schaik (Eds.), *Orangutans: Geographic variation in behavioral ecology and conservation* (pp. 279−298). Oxford: Oxford University Press.

Sanz, C. M., & Morgan, D. B. (2007). Chimpanzee tool technology in the Goualougo Triangle, Republic of Congo. *Journal of Human Evolution*, 52(4), 420−433.

Shumaker, R. W., Walkup, K. R., & Beck, B. B. (2011). *Animal tool behavior: The use and manufacture of tools by animals, revised and updated edition*. Baltimore, MD: The Johns Hopkins University Press.

St. Amant & Horton (2008). Revisiting the definition of tool use. *Animal Behaviour, 75*, 1199−1208. Available from: http://dx.doi.org/10.1016/j.anbehav.2007.09.028.

Sternberg, R. J. (1999). A propulsion model of types of creative contributions. *Review of General Psychology, 3*, 83−100.

Sugardjito, J., & van Hooff, J. A. R. A. M. (1986). Age-sex class differences in the positional behavior of the Sumatran orangutan *Pongo pygmaeus abelii* in the Gunung Leuser National Park, Indonesia. *Folia Primatologica, 47*, 14−25.

Thorpe, S. K. S., & Crompton, R. H. (2006). Orangutan positional behavior and the nature of arboreal locomotion in Hominoidea. *American Journal of Physical Anthropology, 131*, 384−401.

Thorpe, S. K. S., & Crompton, R. H. (2009). Orangutan positional behavior. In S. A. Wich, S. S. Utami, T. Mitra Setia, & C. P. van Schaik (Eds.), *Orangutans: Geographic variation in behavioral ecology and conservation* (pp. 33−47). Oxford: Oxford University Press.

Thorpe, S. K. S., Crompton, R. H., & Alexander, R. McN. (2007). Orangutans use compliant branches to lower the energetic cost of locomotion. *Biology Letters, 3*, 253−356. Available from: http://dx.doi.org/10.1098/rsbl.2007.0049.

van Schaik, C. P., Noordwijk, M. A., & Wich, S. A. (2006). Innovation in wild orangutans (Pongo pygmaeus wurmbii). *Behaviour, 143*, 839−876.

Warren, R. D., & Crompton, R. H. (1998). Diet, body size and energy costs of locomotion in salutatory primates. *Folia Primatologica, 69*, 86−100.

Yamakoshi, G. (2004). Evolution of complex feeding techniques in primates: Is this the origin of great ape intelligence? In A. E. Russon, & D. R. Begun (Eds.), *The evolution of thought: Evolutionary origins of great ape intelligence* (pp. 140−171). Cambridge, UK: Cambridge University Press.

Commentary on Chapter 15: Tools for the Trees: Orangutan Arboreal Tool Use and Creativity

One of the problems, and joys, of creativity research is that the field is so broad. I like to characterize creativity as fundamentally concerned with solving problems, and creativity in any field can be seen in the same way. Thus, creativity in the context of medicine is concerned with the development of novel and effective solutions to problems of healthcare. Creativity in engineering is concerned with the development of novel and effective solutions to problems of structures, machines and technology. Russon, Kuncoro and Ferisa's chapter on orangutan arboreal tool use therefore adds another field to my list. Primatology, like many other fields, is as concerned with creativity—in this case in the sense of the development of novel and effective solutions to the day-to-day problems faced by orangutans—as any other field.

With hindsight, this link is rather obvious. However, it may offer some valuable and fresh insights into creativity closer to home, particularly perhaps because of the very limitations of the field. If we wish to study creativity in the activities of the orangutan, it almost goes without saying that we are, like Russon, Kuncoro and Ferisa, driven towards a focus on the "product"—the solutions that orangutans devise to address their problems of locomotion and feeding within their environment.

For an engineer interested in creativity, that is a logical and informative approach because it requires us to think very clearly about how we define the output of the creative process. What is the product, in other words? How do we recognize as creative the solutions that orangutans develop?

The authors tackle this issue first and quickly move to the necessity of a broader and more inclusive definition of "tools" and "tool use." They adopt more recent definitions that dispense with a rather restrictive condition of *unattached* environmental objects that not only allows more solutions to be incorporated into a study like theirs, but I think appropriately recognizes that we humans make similar use of many fixed objects in novel and effective ways, to solve problems that we face. Indeed, it would seem almost unfair to require among primates a higher standard of "tool use" than we do in ourselves. "What," the orangutans might ask, "do we have to do to convince you humans that we aren't stupid?"

Russon, Kuncoro and Ferisa's chapter highlighted, for me, the fact that the tools used by primates are, first and foremost, merely the means to an end. That end is the solution of a problem—of locomotion, or feeding, for example. Thus, any supplemental object (or process) that contributes to the solution of a problem—whether attached, unattached, natural or artificial—is a "tool" and may be judged novel and effective, or not.

The other vital element of product (or tool) creativity that strikes a chord with an engineer is the role of the orangutans' external environment. In engineering parlance, we might think of this as the set of constraints on the problem-solving task. The point is that the primate is not sitting around simply waiting for the muse to strike, but is highly influenced—or constrained—in its efforts to solve a given problem by the circumstances of that problem situation. Like humans attempting to solve a technological problem such as building a bridge across a river, the orangutan attempting to move horizontally through its habitat is constrained by factors such as body weight, gap size, the strength and availability of materials, and so on. Each problem becomes unique, at least at some level. Unlike humans, however, the orangutan doesn't have the benefit of records and technology—every gap that has to be crossed, while nominally the same as previous experiences, is a little different. Maybe this thin branch won't turn out to be quite as flexible as the one I used last time when I employed a "pole vault" solution?! Arguably, therefore, the primate is faced with uncertainty in every problem-solving situation, in a way that we humans are not, thanks to our repeatable and proven materials and methods.

This begs the question—which is the more creative? The human, who finds one highly novel and effective solution, and then "mass-produces" it, or the orangutan, who must reinvent the solution every time it is used, even if only to a small degree?

I've been interested in computational creativity for a while now—colloquially, the question of how we might automate creativity. It occurs to me, after reading Russon, Kuncoro and Ferisa's chapter that there may be valuable lessons to be learned by going back, so to speak, to first principles. Primates may provide creativity researchers with something approaching a cybernetic—that is, goal-directed—system through which we can study aspects of creativity, freer from some of the vagaries of human emotion and behavior. A broad definition of tool use that encompasses a wide range of primate activity opens up the opportunity to study nuances in tool use, creativity and problem solving that would not be possible under a restrictive definition that rules out all but a few activities.

IV. PUSHING THE BOUNDARIES OF CREATIVITY

I write this only a day after my own university conferred an honorary doctorate on the giant of primate research, Jane Goodall, and I was bitterly disappointed to learn that she was being invested with this award *in absentia*. I was disappointed, of course, because 48 h earlier it never occurred to me that I might have a great deal to discuss with her.

Russon, Kuncoro and Ferisa have broached an interesting aspect of creativity research that I am sure we will hear more about in due course. Not only has it opened up interesting perspectives on the nature of the creative product, but it has, ironically, confirmed that creativity is *not* monkey-business!

16

Insects as a Model System to Understand the Evolutionary Implications of Innovation

Emilie Snell-Rood, Eli Swanson and Sarah Jaumann

Department of Ecology, Evolution and Behavior,
University of Minnesota – Twin Cities, St. Paul, MN, USA

Commentary on Chapter 16: Insect Creativity as Applied to Human Organizational Behavior: A Form of Social Biomimicry?

Samuel T. Hunter

Department of Psychology, Pennsylvania State University, State College, PA, USA

INTRODUCTION

Research on animal innovation arose out of observations of complex and novel behavior in vertebrates, such as primates and birds. This focus on specific groups of vertebrates continues to largely define the field of innovation research. Such a focus is perfectly natural given the high innovation rates in these taxonomic groups, but still could result in a biased or overly narrow view of innovation. In this chapter, we argue that the basic elements of innovation, broadly defined, can apply to a range of behavioral traits in a variety of species, even species often regarded as "simple" or "pre-programmed," such as insects.

Animal Creativity and Innovation.
DOI: http://dx.doi.org/10.1016/B978-0-12-800648-1.00016-4

We apply concepts from the innovation literature to insect behavior, focusing largely on the extensive literature on how behavior can drive exposure to novel resources, thus leading to increased niche breadth and subsequent diversification. In doing so, we highlight how research on insect behavior and evolution can benefit from considering the literature on innovation in vertebrates, and vice versa.

DEFINING INNOVATION FROM AN EVOLUTIONARY PERSPECTIVE AND INCORPORATING "MISTAKES"

Defining Innovation

Innovation, broadly defined, describes the process of creation of new behaviors or modification of existing behaviors, and as such, represents an important way for animals to cope with ecological demands and novel stressors in appropriate ways (Reader, 2003 and references therein). One important consideration here is the meaning of the word "appropriate" when we say that an innovative behavior is an "appropriate" response. It's not immediately clear what constitutes an appropriate response. Does an appropriate response increase fitness? If a behavioral response decreases fitness, but is heritable, and later generations adapt to the negative consequences of the behavior, does it become an innovation later? Is a response appropriate if it does not immediately kill the innovator and thus may become beneficial in the future, or must it improve upon the previous state right away? Many potential innovations in terms of resource use, antipredator behavior, or habitat choice may in fact be "mistakes" occurring during exploration of different possibilities that do not improve fitness, or only do so under very specific conditions. Ramsey, Bastian, and van Schaik (2007) distinguish between the process and outcome of innovation—here, we focus on the *process that generates innovations, and argue that considering the mistakes* that occur through trial-and-error sampling is key to understanding the origin and consequences of innovations. Thus, our operational definition of an innovation does not require that a novel behavior is immediately "appropriate."

Some researchers suggest that innovations should be considered on a population level—a behavior is only an innovation if it's novel to the population as well as the individual (Reader, 2003). However, Ramsey et al. (2007) suggested that as long as the novel behavior does not arise through social learning or environmental induction of an innate behavior, then novelty must be due to innovation. In practice of course, it can be problematic to prove that a behavior arising in an individual is truly such an innovation when other members of the population

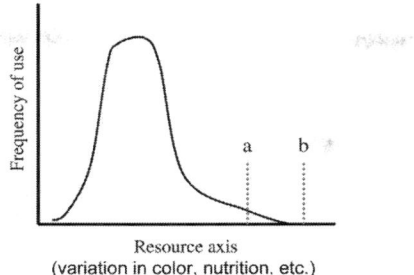

FIGURE 16.1 **A continuum of innovation.** Consider variation across a population or species in a behavior or set of resources used (*x*-axis). The use of a novel (b) or uncommonly used (a) resource might be considered an innovation relative to the frequency of resource use at either the individual or population level. By placing innovation on a continuum we can avoid much of the confusion associated with classifying a behavior as either an innovation or not.

already exhibit the behavior. Here we consider a continuum of novelty in defining an innovation. A population or species might use a wide range of resources. Furthermore, each resource may be used at different frequencies depending on ecological, behavioral and evolutionary factors such as availability, preference and suitability. A particular behavior may not be novel enough to be considered an innovation under one set of conditions, but the same behavior could be an innovation if the state of the environment changes. Thus, we consider innovations to be novel patterns of behavior or resource use relative to this distribution of potential responses based on ecological, behavioral, and evolutionary factors in the environment (Figure 16.1).

Where Do Innovations Come from?

Innovations ultimately arise from novel variation, whether generated through trial-and-error learning or at the genetic level through mutation. Although we focus on behavioral innovations, the developmental and genetic processes that lead to innovations are analogous, and both are important in evolutionary diversification; so throughout, we often consider both jointly. As with any process of exploration or selection, the likelihood of a novel, high-performing phenotype increases with an increase in selectable variation (Snell-Rood, 2012). For instance, the probability of developing an antibody that matches a novel antigen increases with the range of variant B cells produced by the immune system (Honjo & Habu, 1985; Hull, Langman, & Glenn, 2001). Likewise, the probability of stumbling upon a novel behavioral solution to a problem increases with the range of behavioral phenotypes sampled, from the range of resources and environments an individual interacts with to the range of motor patterns that individual expresses in response (Frank, 1996, 1997; Kaelbling, Littman, & Moore, 1996).

Thus, the process of innovating, at least through a trial-and-error process, necessitates "mistakes." This has long been recognized in the learning literature as the cost of being naïve (Dukas, 1998), and in the machine learning literature as the exploration-exploitation tradeoff (Kaelbling et al., 1996). For example, relative to a specialist on a given flower, generalist bumblebees must take time to try out a range of handling patterns (Laverty & Plowright, 1988). Interestingly, the process of innovation through insight and conscious thought may not necessitate mistakes and thus may have different evolutionary implications than those discussed here.

Throughout the remainder of this chapter, we consider novel behavior, both "appropriate" novel behaviors, along with less appropriate "mistake" behaviors, arguing that the latter are an important component of the process of innovation. In some cases, this process may occur over multiple generations, through a combination of first developmental, and later, genetic, innovation. Specifically, we apply these ideas to patterns of behavior and diversification in insects.

RESOURCE SEARCH IN INSECTS: THE ROLE OF TRIAL-AND-ERROR PROCESSES

How insects search for resources has been well-studied and may serve as a very basic model of the process by which innovation occurs. Most insects have an innate bias to search for particular colors or scents or an innate preference for one resource type versus another. However, many insects sample a range of different resources and gradually refine this search process through learning (Dukas, 2008; Papaj & Prokopy, 1989). For example, some butterflies are innately biased to look for green colors when searching for host plants (where they lay their eggs) versus blue or yellow colors when searching for nectar resources (Blackiston, Briscoe, & Weiss, 2011; Snell-Rood & Papaj, 2009; Weiss, 1997; Weiss & Papaj, 2003). These species will land on a range of possible resources and, through chemical feedback during landing, refine the resource search process (Hern, Edwards-Jones, & McKinlay, 1996). Feedback on the quality of a resource can come through tasting of a nectar reward during ingestion or tasting of host-plant specific chemicals on plant leaves through contact chemoreceptors in their feet (Smallegange, Everaarts, & Van Loon, 2006; Traynier, 1984, 1986). In this way, resource use in insects (as with behavior in most animals) is a complex mixture of both innate and learned components (Riffell et al., 2008).

To understand the origins and consequences of innovations, we can draw relevant points from the literature on how insects locate resources. First, *the probability of an innovation stems from the range of possible resources sampled during the search process*. This sampling range is partly influenced by genetics: for instance, some full sibling groups of butterflies are more likely to sample a broader range of host plants when naïve (Snell-Rood & Papaj, 2009). Nevertheless, this sampling range is also influenced by the environment: if an expected resource is not present, individuals may be induced to sample more broadly (Snell-Rood & Papaj, 2009). This observation recalls a broader literature on the role of environmental enrichment and stress in learning—more complex and novel environments can stimulate learning, brain development and exploration (Scotto Lomassese et al., 2000; van Praag, Kempermann, & Gage, 2000).

The second lesson we can take from the literature on insect resource use has to do with the *costs of searching, and thus the process that leads to innovation*. Searching for resources is incredibly costly in terms of time, energy and exposure to predators (Stephens & Krebs, 1986). Dividing attention among several possible resources increases the likelihood of predation and increases the amount of time to make a decision (Dukas, 2002; Dukas & Kamil, 2000). Data from butterflies has shown that genetic variation in the ability to search for novel colors (e.g., red host plants) is correlated with increased investment in costly neural tissue and delays in reproduction (Snell-Rood, Davidowitz, & Papaj, 2011; Snell-Rood, Papaj, & Gronenberg, 2009). If the process that results in innovation is so costly, there should be strong selection to minimize the costs of searching, for instance through the evolution of innate biases for common resources. For instance, species of butterflies that have shifted to use more red colored host plants have an innate bias to search for red, minimizing the costs of the search process (Bernard & Remington, 1991). Simple changes in genes associated with sensation and perception could result in major changes in the costs of searching (Briscoe, 2008; Kelber, 1999). Although the benefit of such innate biases that limit ranges of exploration is that search costs are mitigated, the drawback, given our first lesson, is that the probability of innovation can be reduced. The costs of searching suggest that the probability of innovation will vary quite a bit across species. Furthermore, these costs imply that after colonization of a novel environment there may be strong selection for reduced search and specialization, leading to diversification (as discussed further below).

Finally, the literature on insect resource search illustrates that *the opportunity for individual-level feedback can affect whether an exploratory process is more likely to lead to an appropriate novel behavior*, or an innovation. As an individual samples environments and resources, does it get feedback as to whether a resource increases its performance or fitness? For instance, in the process of searching for nectar resources, pollinators get immediate feedback on the sugar and amino acid content of nectar through chemoreceptors (Alm, Ohmeiss, Lanza, & Vriesenga, 1990; Cnaani, Thomson, & Papaj, 2006; Gardener & Gillman, 2002; Page, Erber, & Fondrk, 1998); insects can also learn from slightly more delayed post-ingestive feedback on other nutritional aspects of a food resource (Behmer, Elias, & Bernays, 1999). In contrast, for species that feed on different resources as adults and larvae, there may be limited opportunities for adults to gain direct information on the nutritional quality of a possible larval resource. Most species use cues that have evolved to be reliable indicators of host plant quality, such as plant size or color, which may be correlated with nitrogen content or chemical correlates of the plant family that a species is specialized to feed on (Myers, 1985). However, sometimes an increased range of resource sampling during the search process can lead to a novel resource that the species is not adapted to use. In this case, because the adult moves on before their offspring even hatches, there is no information on whether this novel behavior actually improves performance. In some cases, adult insects that live near their offspring may still be able to learn about differences in resource quality for their offspring (Xue, Egas, & Yang, 2007). Cases of search where direct feedback on performance is possible are more likely to lead to appropriate novel behaviors, or innovations, whereas lack of opportunities for direct feedback are more likely to lead to what is perceived initially as a mistake. Conversely, cases of search where such direct feedback on an individual's performance within its lifetime are not possible will be more strongly subject to natural or sexual selection. An insect that explores too widely and makes a series of poor egg-laying decisions will not get direct feedback, but will instead be selected against, providing a similar form of feedback, but on an evolutionary timescale.

Taken together, studies of resource searching in insects highlight the importance of considering *the process that leads to innovation*—in many cases, the mistakes during this process are just as important because they determine the costs associated with exploiting novel resources and thus the likelihood of innovation. In addition, mistakes broaden the range of behaviors that may one day lead to a true innovation. It is important to consider search costs and the fitness costs arising from possible mistakes. Such a simultaneous balance between the more proximate costs and benefits of innovations, coupled with the lineage-level costs and benefits of innovations, shapes variation in search processes that could lead to innovation.

HOST SHIFTS AND DIVERSIFICATION: BEHAVIORAL INNOVATION LEADS THE WAY

Insects present a remarkable opportunity for studying the evolutionary implications of innovation because they are resource specialized in a manner that allows the history of innovation to be read in extant species. In particular, by studying resource use across modern insects, we can reconstruct the history of resource shifts, and, along with studies of resource search patterns, we can infer how behavioral innovation may drive patterns of diversification.

Colonizing a Novel Environment

As discussed above, the process of resource search involves exploration, sampling and feedback, which can sometimes lead to an individual using a novel resource. In some cases, this can lead to an immediate increase in performance, for instance the bee that learns to climb into discarded cans of soda. Lab studies of hairstreak butterflies have shown that if caterpillars are forced to feed on novel legume host plants, a good number of species do just as well on the new host as on the host plant they normally feed on in the field (Pratt & Ballmer, 1991). Similarly, in some leaf miners, the larvae perform well on entirely novel hosts (Gratton & Welter, 1998). Adaptive maternal effects may also increase the chances of survival following an innovative behavior. For instance, in some beetles, maternal exposure to a novel resource increases offspring survival because females allocate more to individual eggs (Fox, Gordon, & Bojang, 2006; Fox & Savalli, 2000). These examples suggest the intriguing possibility that many innovative behaviors may be beneficial by chance. Thus, in some cases, exploring novel resources can lead to colonization and immediate "appropriate" use of these resources (Figures 16.2 and 16.3). This has been well documented in butterflies that have shifted to use novel invasive host plants (Graves & Shapiro, 2003) and parasitoid wasps that have learned to exploit invasive host species (Wei et al., 2013).

In contrast, in many cases, an innovative resource use behavior may lead to no immediate increase in performance, or possibly a decrease in performance. For instance, while tiger swallowtails have been observed to lay eggs on plants in the carrot family, a novel host for that species (but used by the related black swallowtail), their larvae do not survive on this host in the lab (Berenbaum, 1981). Repeated novel behavior may allow a species to escape what is initially an "ecological trap"—a resource or habitat a species is drawn to that has negative effects on performance and fitness (Schlaepfer, Runge, & Sherman, 2002).

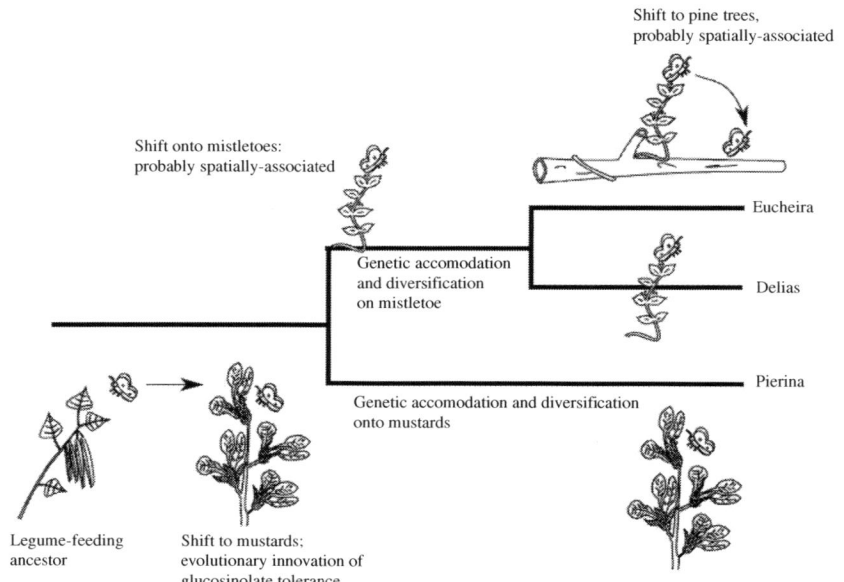

FIGURE 16.2   Innovations drive resource shifts which result in diversification. 'An initial shift from the ancestral legume host to mustards (Brassicaceae) and related capers was facilitated by a genetic innovation that allows detoxification of those plant defenses (Wheat et al., 2007). A secondary shift to a mistletoe parasite of capers was likely facilitated by spatial co-localization of the plants and exploratory host search behavior. The mistletoe shift led to an adaptive radiation of over 400 species (Braby & Trueman, 2006). Because mistletoe is found on a wide variety of different plants, the use of mistletoe led to the subsequent colonization of several different plant families, including pine and Ericaceae.

Repeated exposure to the same resource or environment increases the likelihood of selection on a key genetic (or developmental) innovation that allows the population to exploit a potential resource. Many species show pronounced genetic variation in performance on such resources (Ueno, Fujiyama, Irie, Sato, & Katakura, 1999). And many suites of detoxification genes show rapid adaptation in insects. For instance, variation in P450 enzymes allows many insects to detoxify novel chemicals—either of plant or anthropogenic origin, such as pesticides (Berenbaum, Favret, & Schuler, 1996; Li, Schuler, & Berenbaum, 2007; Schuler, 2011). What may appear to be a mistake may set the stage for an innovation and colonization of a new environment. For instance, many aquatic insects are drawn to shiny metal or paved surfaces that reflect polarized light and resemble water (Horvath, Kriska, Malik, & Robertson, 2009). While the majority of these instances currently result in high mortality, this could also represent the first step for many species colonizing anthropogenic environments. In butterflies, similarity of

FIGURE 16.3 **Innovation leading to diversification.** Mistakes leading to innovative resource shifts in several orders of insects have led to the colonization of novel environments and subsequent diversification. For example, a series of host shifts from capers to mistletoe to pine resulted in rapid diversification of pierid butterflies into many species, including the pine white ((A) image credit: Megan McCarty, creative commons). Kidney-spot ladybird beetles ((B) image credit: Gilles San Martin, creative commons) are one of many species of beetle that shifted from feeding on fungi to feeding on Hemipteran insects. The Euphorine wasps ((C) image credit: H Dumas, creative commons) historically parasitized larvae but radiated after they shifted to parasitizing adults. Apple-maggot flies ((D) image credit: Joseph Berger, creative commons), which originally fed on hawthorn, began feeding on apple when it was introduced to North America, and they are currently diversifying into several distinct host-races which either eat apple or hawthorn.

host plant chemical cues has led to rapid shifts to novel, non-native host plants (Renwick, 2002). Some of these novel host plants, such as garlic mustard, are initially toxic (Porter, 1994); however, *Pieris* butterflies with continued exposure to garlic mustard appear to be adapting to feed on it (Keeler & Chew, 2008).

Diversification in a Novel Environment

The road to diversification is open after a population colonizes a novel environment. This has been classically illustrated by adaptive radiations of species after colonizing an island, as in the case of Hawaiian *Drosophila* or Galapagos finches (Parent, Caccone, & Petren, 2008; Roderick & Gillespie, 1998). But it has also been well illustrated by insects colonizing new groups of resources. For instance, after a host switch, butterfly lineages show a burst of speciation (Fordyce, 2010; Janz, Nylin, & Wahlberg, 2006).

The signature of novel behavior leading to diversification can still be seen across a range of insect lineages. In many cases, resource shifts between spatially adjacent resources are quite common—this would be expected if the colonization of a novel resource is initiated by novel behavior, in this case sampling of a novel resource in the same vicinity as a more typical resource. For instance, in pierid butterflies, colonization of mistletoe hosts on ancestral caper hosts led to colonization of other plants that mistletoe parasitizes such as pine (Braby & Trueman, 2006; see Figures 16.2 and 16.3). Similar shifts between spatially co-located resources have been seen in several lineages of beetles (Giorgi et al., 2009; Leschen, 2000) and in parasitoid wasps (Shaw, 1988). Spatial associations between adult and larval resources may also fuel patterns of diversification through innovative behavior. It has long been recognized that nectar plants can influence patterns of adult egg-laying behavior in butterflies (Janz, Bergstrom, & Sjogren, 2005; Murphy, Menninger, & Ehrlich, 1984). But in at least one species of butterfly, it appears that nectar resources may even drive innovative egg-laying behavior on novel host plants (Forister, Nice, Fordyce, & Gompert, 2009).

There is even a signature of innovative behavior in the classic case of ecological speciation in apple maggot flies. Here, a host-race of apple-feeding flies emerged (and diverged) from the ancestral hawthorn-feeding population shortly after the introduction of apples in the 1800s (Dres & Mallet, 2002; Funk, Filchak, & Feder, 2002). These flies show evidence of learning in host search and preference behavior—they can learn to discriminate between different apple varieties and become more efficient at host searching through experience (Papaj & Prokopy, 1988; Prokopy, Bergweiler, Galarza, & Schwerin, 1994; Prokopy & Papaj, 1989). They also modify their acceptance of a novel host based on previous experience (Prokopy, Cooley, & Papaj, 1993; Prokopy, Papaj, Cooley, & Kallet, 1986). These behavioral data suggest that behavioral innovations in host use behavior may have initiated host-race formation (and speciation) in this species.

As reviewed above, the process that leads to innovation and colonization of novel resources necessitates mistakes, but, as reviewed in this section, what may initially be a mistake may eventually result in innovation. The mistakes during this process are equally important as the innovations themselves, in part because these mistakes are costly, thus leading to strong selection against broad search and exploration processes. Selection for a narrower search and more focused innate biases should then lead to more specialized behavior. Taken together, these processes can lead to evolutionary "oscillations" of generalist lineages that colonize new environments, followed by the evolution of specialists in those environments (Janz & Nylin, 2008; Nosil, 2002; Nylin & Janz, 2009; Nylin & Wahlberg, 2008). This process recalls that of genetic assimilation, where an initial plastic response results in the colonization of a novel environment, but the costs of plasticity eventually result in the loss of plasticity and genetic consolidation of the initially environmentally induced phenotype (Bateson & Gluckman, 2011; West-Eberhard, 2003). Overall, the processes linking both innovation and plasticity to diversification are quite similar.

THE SPREAD OF INNOVATIONS THROUGH SOCIAL INTERACTIONS

So far we have mostly discussed innovative behavior arising multiple times within individual insects. However, as has been seen in vertebrates, the use of social cues can result in the rapid spread of innovations in insect populations. Social insects often directly share information about resources with nest-mates, and evidence suggests that naïve or inexperienced individuals are likely to use and benefit from such information (Baude, Dajoz, & Danchin, 2008; Slaa, Wassenberg, & Biesmeijer, 2003). Bumblebees, for example, can learn nectar robbing behavior not just through individual trial and error learning, but also through cues left by conspecifics already performing this behavior (Leadbeater & Chittka, 2008). Honeybees use social cues embedded in the waggle dance to communicate the location of nectar resources to nest-mates in the hive. Similarly, ants leave pheromone trails leading nest-mates to rewarding resources. Some species of ants even "teach" nest-mates the location of food resources through a behavior known as tandem running. A "teacher" physically guides a naïve follower to the food source, and the follower learns the route much more quickly by following the teacher (Franks & Richardson, 2006). In any of these cases, if an individual forager successfully exploits a novel resource, the resource could potentially be exploited by many other foragers in the nest. Some social insects can also learn novel associations by paying attention to conspecifics. Bees have been shown to modify feeding preferences when conspecifics are paired with more rewarding flowers (Worden & Papaj, 2005).

Social information can also potentially play a big role in resource use in solitary insects. Many "non-social" insects, including some species of butterflies, flies, locusts, and cockroaches, are more likely to investigate or use a potential resource if conspecifics are present (Baur, Kostal, Patrian, & Stadler, 1996; Lihoreau & Rivault, 2011; Otis et al., 2006; Yu, Wang, Xiao, Shao, & Li, 2011). Such local enhancement in solitary insects, and even in social insects foraging individually, could promote the spread of an innovation by resulting in multiple individuals experiencing a novel resource. In ovipositing insects, benefits of larval aggregation, such as increased feeding rates (Inouye & Johnson, 2005) or protection against desiccation and predators (Clark & Faeth, 1997, 1998) may promote the use of conspecific cues in adult egg-laying behavior. In fact, the adults of several species of butterflies with gregarious larvae have been observed to lay eggs in groups (Reed, 2003; Sourakov, 1997).

One final point is that information about resources can be spread not just within, but also between species, in particular in social insects. Foraging bees have been shown to be attracted to flowers with hetero-specifics and learn associations between rewarding flowers and hetero-specific cues (Dawson & Chittka, 2012; Slaa et al., 2003). Multiple ant species also sometimes share pheromone trails leading to food (Wilson, 1965). If different species use different types of flowers, such cross-species behavior could expose individuals to new flower types, even potentially outside of their normal range of acceptable flowers. Thus, there are many opportunities for resource innovations to rapidly spread within and between insect species through the use of social information.

CONCLUSIONS AND IMPLICATIONS

Using the data on resource use and behavior in insects, we can begin to draw some general conclusions about when innovations may drive evolutionary patterns. First, we might predict that *species that exhibit more resource-related learning may be more likely to show more behavioral innovations and subsequent diversification*. While we have discussed examples of resource learning across many species of insects, there are just as many species where learning is not important. For instance, a number of butterflies show no evidence of learning in the context of host use (Parmesan, Singer, & Harris, 1995; Tabashnik, Wheelock, Rainbolt, & Watt, 1981; Thompson, 1988); thus, variation in resource use has a high heritability (Singer, Ng, & Thomas, 1988). Those species or populations where learning is more important may be more likely to display innovative foraging behavior, colonize a novel environment and fuel evolutionary diversification. This general prediction is consistent with findings in birds: species with larger brains tend to show more foraging innovations (Sol, Duncan, Blackburn, Cassey, & Lefebvre, 2005) which leads to subsequent survival in novel environments (Sol, Timmermans, & Lefebvre, 2002) and diversification (Nicolakakis, Sol, & Lefebvre, 2003; Sol & Price, 2008).

Second, the data from insect resource use emphasize *the importance of interactions between behavior and ecology in the emergence of innovation*. Behavior is important, but so is the abundance and location (both spatial and temporal) of some novel resource. In many cases, learning to exploit a novel resource that has recently increased in abundance is a favorable strategy even if performance on this resource is lower (West & Cunningham, 2002). Thus, plasticity in behavior, either through learning or relaxation of a preference, can result in the initial shift to using a novel resource (Nylin & Janz, 2009). Across many species of insects, it appears the initial use of a novel resource is driven through behavioral flexibility in a changing environment, followed later by genetic changes that lead to higher performance on that resource (Bowers, Stamp, & Collinge, 1992; Camara, 1997; Gassmann, Levy, Tran, & Futuyma, 2006).

Finally, while much of the present discussion has focused on innovations in the context of larval and adult resources, it's likely that these concepts apply to innovations in other contexts, such as mate choice and predator avoidance. However, resource use is likely much easier to study (in both the field and the lab) due to the frequency of occurrence. There is evidence that similar processes occur in the context of innovative mate choice. For instance, learning and experience can lead to altered mate preferences—in particular, a relaxation of a mate preference when the environment changes such that a preferred trait is no longer common (Bailey & Zuk, 2008; Fowler-Finn & Rodriguez, 2012).

As a concluding point, it appears there are many, many examples of behavioral innovation in resource use potentially leading to diversification in insects. Is it possible that this process is even more pronounced in insects than vertebrates? While the data are too limited to make such a conclusion, it could be that the high degree of specialization in insects, especially those that feed on plants (Jaenike, 1990), makes them highly likely to exhibit resource-based innovation and diversification. Regardless, overall these observations suggest that considering the innovation literature may shed light on resource diversification in insects, and considering the insect literature may broaden some of the principles generated in the innovation literature.

Acknowledgments

We appreciate the invitation to contribute this chapter to an exciting collection of ideas. During this time, EMS was supported on an NSF Postdoctoral Fellowship in Biology (No. 1306627) fellowship and ESR received support from NSF 1354737.

References

Alm, J., Ohmeiss, T. E., Lanza, J., & Vriesenga, L. (1990). Preference of cabbage white butterflies and honeybees for nectar that contains amino acids. *Oecologia, 84*(1), 53–57. Available from: http://dx.doi.org/10.1007/bf00665594.

Bailey, N. W., & Zuk, M. (2008). Acoustic experience shapes female mate choice in field crickets. *Proceedings of the Royal Society B—Biological Sciences, 275*(1651), 2645–2650. Available from: http://dx.doi.org/10.1098/rspb.2008.0859.

Bateson, P., & Gluckman, P. (2011). *Plasticity, robustness, development and evolution.* Cambridge, UK: Cambridge University Press.

Baude, M., Dajoz, I., & Danchin, E. (2008). Inadvertent social information in foraging bumblebees: Effects of flower distribution and implications for pollination. *Animal Behaviour, 76,* 1863–1873. Available from: http://dx.doi.org/10.1016/j.anbehav.2008.08.010.

Baur, R., Kostal, V., Patrian, B., & Stadler, E. (1996). Preference for plants damaged by conspecific larvae in ovipositing cabbage root flies: Influence of stimuli from leaf surface and roots. *Entomologia Experimentalis Et Applicata, 81*(3), 353–364. Available from: http://dx.doi.org/10.1046/j.1570-7458.1996.00106.x.

Behmer, S. T., Elias, D. O., & Bernays, E. A. (1999). Post-ingestive feedbacks and associative learning regulate the intake of unsuitable sterols in a generalist grasshopper. *Journal of Experimental Biology, 202*(6), 739–748.

Berenbaum, M. (1981). An oviposition "mistake" by *Papilio glaucus* (Papilionidae). *Journal of the Lepidopterists' Society, 35,* 75.

Berenbaum, M. R., Favret, C., & Schuler, M. A. (1996). On defining "key innovations" in an adaptive radiation: Cytochrome P450s and papilionidae. *American Naturalist, 148,* S139–S155. Available from: http://dx.doi.org/10.1086/285907.

Bernard, G. D., & Remington, C. L. (1991). Color vision in *Lycaena* butterflies—spectral tuning of receptor arrays in relation to behavioral ecology. *Proceedings of the National Academy of Sciences of the United States of America, 88*(7), 2783–2787. Available from: http://dx.doi.org/10.1073/pnas.88.7.2783.

Blackiston, D., Briscoe, A. D., & Weiss, M. R. (2011). Color vision and learning in the monarch butterfly, *Danaus plexippus* (Nymphalidae). *Journal of Experimental Biology, 214*(3), 509–520. Available from: http://dx.doi.org/10.1242/jeb.048728.

Bowers, M. D., Stamp, N. E., & Collinge, S. K. (1992). Early stage of host range expansion by a specialist herbivore, *Euphydryas phaeton* (Nymphalidae). *Ecology, 73*(2), 526–536. Available from: http://dx.doi.org/10.2307/1940758.

Braby, M. F., & Trueman, J. W. H. (2006). Evolution of larval host plant associations and adaptive radiation in pierid butterflies. *Journal of Evolutionary Biology, 19*(5), 1677–1690. Available from: http://dx.doi.org/10.1111/j.1420-9101.2006.01109.x.

Briscoe, A. D. (2008). Reconstructing the ancestral butterfly eye: Focus on the opsins. *Journal of Experimental Biology, 211*(11), 1805–1813. Available from: http://dx.doi.org/10.1242/jeb.013045.

Camara, M. D. (1997). A recent host range expansion in *Junonia coenia* Hubner (Nymphalidae): Oviposition preference, survival, growth, and chemical defense. *Evolution, 51*(3), 873–884. Available from: http://dx.doi.org/10.2307/2411162.

Clark, B. R., & Faeth, S. H. (1997). The consequences of larval aggregation in the butterfly *Chlosyne lacinia. Ecological Entomology, 22*(4), 408–415.

Clark, B. R., & Faeth, S. H. (1998). The evolution of egg clustering in butterflies: A test of the egg desiccation hypothesis. *Evolutionary Ecology, 12*(5), 543–552. Available from: http://dx.doi.org/10.1023/a:1006504725592.

Cnaani, J., Thomson, J. D., & Papaj, D. R. (2006). Flower choice and learning in foraging bumblebees: Effects of variation in nectar volume and concentration. *Ethology, 112*(3), 278–285. Available from: http://dx.doi.org/10.1111/j.1439-0310.2006.01174.x.

Dawson, E. H., & Chittka, L. (2012). Conspecific and heterospecific information use in bumblebees. *PLoS One, 7*(2). Available from: http://dx.doi.org/10.1371/journal.pone.0031444.

Dres, M., & Mallet, J. (2002). Host races in plant-feeding insects and their importance in sympatric speciation. *Philosophical Transactions of the Royal Society B—Biological Sciences, 357*(1420), 471–492. Available from: http://dx.doi.org/10.1098/rstb.2002.1059.

Dukas, R. (1998). Evolutionary ecology of learning. In R. Dukas (Ed.), *Cognitive ecology: The evolutionary ecology of information processing and decision making* (pp. 129–174). Chicago, IL: University of Chicago Press.

Dukas, R. (2002). Behavioural and ecological consequences of limited attention. *Philosophical Transactions of the Royal Society B—Biological Sciences, 357*(1427), 1539–1547. Available from: http://dx.doi.org/10.1098/rstb.2002.1063.

Dukas, R. (2008). Evolutionary biology of insect learning. *Annual Review of Entomology, 53,* 145–160.

Dukas, R., & Kamil, A. C. (2000). The cost of limited attention in blue jays. *Behavioral Ecology, 11*(5), 502–506. Available from: http://dx.doi.org/10.1093/beheco/11.5.502.

Fordyce, J. A. (2010). Host shifts and evolutionary radiations of butterflies. *Proceedings of the Royal Society B—Biological Sciences, 277*(1701), 3735–3743. Available from: http://dx.doi.org/10.1098/rspb.2010.0211.

Forister, M. L., Nice, C. C., Fordyce, J. A., & Gompert, Z. (2009). Host range evolution is not driven by the optimization of larval performance: The case of *Lycaeides melissa* (Lepidoptera: Lycaenidae) and the colonization of alfalfa. *Oecologia, 160*(3), 551–561. Available from: http://dx.doi.org/10.1007/s00442-009-1310-4.

Fowler-Finn, K. D., & Rodriguez, R. L. (2012). Experience mediated plasticity in mate preferences: Mating assurance in a variable environment. *Evolution, 66*(2), 459–468. Available from: http://dx.doi.org/10.1111/j.1558-5646.2011.01446.x.

Fox, C. W., & Savalli, U. M. (2000). Maternal effects mediate host expansion in a seed-feeding beetle. *Ecology, 81*(1), 3–7. Available from: http://dx.doi.org/10.1890/0012-9658(2000)081[0003:memhei]2.0.co;2.

Fox, C. W., Gordon, D. M., & Bojang, P. (2006). Genetic and environmental sources of variation in survival on nonnative host species in the generalist seed beetle, *Stator limbatus. Southwestern Naturalist, 51*(4), 490–501. Available from: http://dx.doi.org/10.1894/0038-4909(2006)51[490:gaesov]2.0.co;2.

Frank, S. A. (1996). The design of natural and artificial adaptive systems. In M. Rose, & G. Lauder (Eds.), *Adaptation* (pp. 451–505). New York, NY: Academic Press.

Frank, S. A. (1997). The design of adaptive systems: Optimal parameters for variation and selection in learning and development. *Journal of Theoretical Biology, 184*(1), 31–39. Available from: http://dx.doi.org/10.1006/jtbi.1996.0241.

Franks, N. R., & Richardson, T. (2006). Teaching in tandem-running ants. *Nature, 439* (7073), 153. Available from: http://dx.doi.org/10.1038/439153a.

Funk, D. J., Filchak, K. E., & Feder, J. L. (2002). Herbivorous insects: Model systems for the comparative study of speciation ecology. *Genetica, 116*(2–3), 251–267. Available from: http://dx.doi.org/10.1023/a:1021236510453.

Gardener, M. C., & Gillman, M. P. (2002). The taste of nectar—a neglected area of pollination ecology. *Oikos, 98*(3), 552–557. Available from: http://dx.doi.org/10.1034/j.1600-0706.2002.980322.x.

Gassmann, A. J., Levy, A., Tran, T., & Futuyma, D. J. (2006). Adaptations of an insect to a novel host plant: A phylogenetic approach. *Functional Ecology, 20*(3), 478–485. Available from: http://dx.doi.org/10.1111/j.1365-2435.2006.01118.x.

Giorgi, J. A., Vandenberg, N. J., McHugh, J. V., Forrester, J. A., Silpinski, S. A., Miller, K. B., et al. (2009). The evolution of food preferences in Coccinellidae. *Biological Control, 51*(2), 215–231. Available from: http://dx.doi.org/10.1016/j.biocontrol.2009.05.019.

Gratton, C., & Welter, S. C. (1998). Oviposition preference and larval performance of *Liriomyza helianthi* (Diptera: Agromyzidae) on normal and novel host plants. *Environmental Entomology, 27*(4), 926–935.

Graves, S. D., & Shapiro, A. M. (2003). Exotics as host plants of the California butterfly fauna. *Biological Conservation, 110*(3), 413–433. Available from: http://dx.doi.org/10.1016/s0006-3207(02)00233-1.

Hern, A., Edwards-Jones, G., & McKinlay, R. G. (1996). A review of the pre-oviposition behaviour of the small cabbage white butterfly, *Pieris rapae* (Lepidoptera: Pieridae).

Annals of Applied Biology, *128*(2), 349–371. Available from: http://dx.doi.org/10.1111/j.1744-7348.1996.tb07328.x.

Honjo, T., & Habu, S. (1985). Origin of immune diversity—genetic variation and selection. *Annual Review of Biochemistry*, *54*, 803–830. Available from: http://dx.doi.org/10.1146/annurev.bi.54.070185.004103.

Horvath, G., Kriska, G., Malik, P., & Robertson, B. (2009). Polarized light pollution: A new kind of ecological photopollution. *Frontiers in Ecology and the Environment*, *7*(6), 317–325. Available from: http://dx.doi.org/10.1890/080129.

Hull, D. L., Langman, R. E., & Glenn, S. S. (2001). A general account of selection: Biology, immunology, and behavior. *Behavioral and Brain Sciences*, *24*(3), 511.

Inouye, B. D., & Johnson, D. M. (2005). Larval aggregation affects feeding rate in *Chlosyne poecile* (Lepidoptera : Nymphalidae). *Florida Entomologist*, *88*(3), 247–252. Available from: http://dx.doi.org/10.1653/0015-4040(2005)088[0247:laafri]2.0.co;2.

Jaenike, J. (1990). Host specialization in phytophagous insects. *Annual Review of Ecology and Systematics*, *21*, 243–273. Available from: http://dx.doi.org/10.1146/annurev.ecolsys.21.1.243.

Janz, N., Bergstrom, A., & Sjogren, A. (2005). The role of nectar sources for oviposition decisions of the common blue butterfly *Polyommatus icarus*. *Oikos*, *109*(3), 535–538. Available from: http://dx.doi.org/10.1111/j.0030-1299.2005.13817.x.

Janz, N., & Nylin, S. (2008). The oscillation hypothesis of host-plant range and speciation. In K. Tilmon (Vol. Ed.), *Specialization, Speciation, and Radiation: The Evolutionary Biology of Herbivorous Insects*, 203–215, University of California Press, Oakland, CA.

Janz, N., Nylin, S., & Wahlberg, N. (2006). Diversity begets diversity: Host expansions and the diversification of plant-feeding insects. *BMC Evolutionary Biology*, *6*. Available from: http://dx.doi.org/10.1186/1471-2148-6-4.

Kaelbling, L. P., Littman, M. L., & Moore, A. W. (1996). Reinforcement learning: A survey. *Journal of Artificial Intelligence Research*, *4*, 237–285.

Keeler, M. S., & Chew, F. S. (2008). Escaping an evolutionary trap: Preference and performance of a native insect on an exotic invasive host. *Oecologia*, *156*(3), 559–568. Available from: http://dx.doi.org/10.1007/s00442-008-1005-2.

Kelber, A. (1999). Ovipositing butterflies use a red receptor to see green. *Journal of Experimental Biology*, *202*(19), 2619–2630.

Laverty, T. M., & Plowright, R. C. (1988). Flower handling by bumblebees: A comparison of specialists and generalists. *Animal Behaviour*, *36*, 733–740.

Leadbeater, E., & Chittka, L. (2008). Social transmission of nectar-robbing behaviour in bumble-bees. *Proceedings of the Royal Society B—Biological Sciences*, *275*(1643), 1669–1674. Available from: http://dx.doi.org/10.1098/rspb.2008.0270.

Leschen, R. A. B. (2000). Beetles feeding on bugs (Coleoptera, Hemiptera): Repeated shifts from mycophagous ancestors. *Invertebrate Taxonomy*, *14*(6), 917–929. Available from: http://dx.doi.org/10.1071/it00025.

Li, X. C., Schuler, M. A., & Berenbaum, M. R. (2007). Molecular mechanisms of metabolic resistance to synthetic and natural xenobiotics. *Annual Review of Entomology*, *52*, 231–253 Palo Alto: Annual Reviews

Lihoreau, M., & Rivault, C. (2011). Local enhancement promotes cockroach feeding aggregations. *PLoS One*, *6*(7). Available from: http://dx.doi.org/10.1371/journal.pone.0022048.

Murphy, D. D., Menninger, M. S., & Ehrlich, P. R. (1984). Nectar source distribution as a determinant of oviposition host species in *Euphydryas chalcedona*. *Oecologia*, *62*(2), 269–271. Available from: http://dx.doi.org/10.1007/bf00379025.

Myers, J. H. (1985). Effect of physiological condition of the host plant on the ovipositional choice of the cabbage white butterfly, *Pieris rapae*. *Journal of Animal Ecology*, *54*(1), 193–204. Available from: http://dx.doi.org/10.2307/4630.

Nicolakakis, N., Sol, D., & Lefebvre, L. (2003). Behavioural flexibility predicts species richness in birds, but not extinction risk. *Animal Behaviour, 65*, 445–452. Available from: http://dx.doi.org/10.1006/anbe.2003.2085.

Nosil, P. (2002). Transition rates between specialization and generalization in phytophagous insects. *Evolution, 56*(8), 1701–1706.

Nylin, S., & Janz, N. (2009). Butterfly host plant range: An example of plasticity as a promoter of speciation? *Evolutionary Ecology, 23*(1), 137–146. Available from: http://dx.doi.org/10.1007/s10682-007-9205-5.

Nylin, S., & Wahlberg, N. (2008). Does plasticity drive speciation? Host-plant shifts and diversification in nymphaline butterflies (Lepidoptera: Nymphalidae) during the tertiary. *Biological Journal of the Linnean Society, 94*(1), 115–130.

Otis, G. W., Locke, B., McKenzie, N. G., Cheung, D., MacLeod, E., Careless, P., et al. (2006). Local enhancement in mud-puddling swallowtail butterflies (*Battus philenor* and *Papilio glaucus*). *Journal of Insect Behavior, 19*(6), 685–698. Available from: http://dx.doi.org/10.1007/s10905-006-9049-9.

Page, R. E., Erber, J., & Fondrk, M. K. (1998). The effect of genotype on response thresholds to sucrose and foraging behavior of honey bees (*Apis mellifera* L.). *Journal of Comparative Physiology A—Neuroethology Sensory Neural and Behavioral Physiology, 182* (4), 489–500. Available from: http://dx.doi.org/10.1007/s003590050196.

Papaj, D. R., & Prokopy, R. J. (1988). The effect of prior adult experience on components of habitat preference in the apple maggot fly (*Rhagoletis-pomonella*). *Oecologia, 76*(4), 538–543.

Papaj, D. R., & Prokopy, R. J. (1989). Ecological and evolutionary aspects of learning in phytophagous insects. *Annual Review of Entomology, 34*, 315–350. Available from: http://dx.doi.org/10.1146/annurev.ento.34.1.315.

Parent, C. E., Caccone, A., & Petren, K. (2008). Colonization and diversification of Galapagos terrestrial fauna: A phylogenetic and biogeographical synthesis. *Philosophical Transactions of the Royal Society B—Biological Sciences, 363*(1508), 3347–3361. Available from: http://dx.doi.org/10.1098/rstb.2008.0118.

Parmesan, C., Singer, M. C., & Harris, I. (1995). Absence of adaptive learning from the oviposition foraging behavior of a checkerspot butterfly. *Animal Behaviour, 50*, 161–175. Available from: http://dx.doi.org/10.1006/anbe.1995.0229.

Porter, A. (1994). Implications of introduced garlic mustard (*Alliaria petiolata*) in the habitat of *Pieris virginiensis* (Pieridae). *Journal of the Lepidopterists' Society, 48*, 171–172.

Pratt, G., & Ballmer, G. (1991). Acceptance of *Lotus scoparius* (Fabaceae) by larvae of Lycaenidae. *Journal of the Lepidopterists' Society, 45*, 188–196.

Prokopy, R. J., Bergweiler, C., Galarza, L., & Schwerin, J. (1994). Prior experience affects the visual ability of *Rhagoletis pomonella* flies (Diptera, Tephritidae) to find host fruit. *Journal of Insect Behavior, 7*(5), 663–677. Available from: http://dx.doi.org/10.1007/bf01997438.

Prokopy, R. J., Cooley, S. S., & Papaj, D. R. (1993). How well can relative specialist *Rhagoletis* flies learn to discriminate fruit for oviposition. *Journal of Insect Behavior, 6*(2), 167–176.

Prokopy, R. J., & Papaj, D. R. (1989). Can ovipositing *Rhagoletis pomonella* females (Diptera Tephritidae) learn to discriminate among different ripeness stages of the same host biotype *Florida Entomologist, 72*(3), 489–494. Available from: http://dx.doi.org/10.2307/3495187.

Prokopy, R. J., Papaj, D. R., Cooley, S. S., & Kallet, C. (1986). On the nature of learning in oviposition site acceptance by apple maggot flies. *Animal Behaviour, 34*, 98–107. Available from: http://dx.doi.org/10.1016/0003-3472(86)90011-4.

Ramsey, G., Bastian, M. L., & van Schaik, C. (2007). Animal innovation defined and operationalized. *Behavioral and Brain Sciences, 30*(4), 393. Available from: http://dx.doi.org/10.1017/s0140525x07002373.

Reader, S. M. (2003). Innovation and social learning: Individual variation and brain evolution. *Animal Biology*, *53*(2), 147–158. Available from: http://dx.doi.org/10.1163/157075603769700340.

Reed, R. D. (2003). Gregarious oviposition and clutch size adjustment by a *Heliconius* butterfly. *Biotropica*, *35*(4), 555–559.

Renwick, J. A. A. (2002). The chemical world of crucivores: Lures, treats and traps. *Entomologia Experimentalis Et Applicata*, *104*(1), 35–42.

Riffell, J. A., Alarcon, R., Abrell, L., Davidowitz, G., Bronstein, J. L., & Hildebrand, J. G. (2008). Behavioral consequences of innate preferences and olfactory learning in hawkmoth–flower interactions. *Proceedings of the National Academy of Sciences of the United States of America*, *105*(9), 3404–3409. Available from: http://dx.doi.org/10.1073/pnas.0709811105.

Roderick, G. K., & Gillespie, R. G. (1998). Speciation and phylogeography of Hawaiian terrestrial arthropods. *Molecular Ecology*, *7*(4), 519–531. Available from: http://dx.doi.org/10.1046/j.1365-294x.1998.00309.x.

Schlaepfer, M. A., Runge, M. C., & Sherman, P. W. (2002). Ecological and evolutionary traps. *Trends in Ecology & Evolution*, *17*(10), 474–480. Available from: http://dx.doi.org/10.1016/s0169-5347(02)02580-6.

Schuler, M. A. (2011). P450s in plant–insect interactions. *Biochimica Et Biophysica Acta— Proteins and Proteomics*, *1814*(1), 36–45. Available from: http://dx.doi.org/10.1016/j.bbapap.2010.09.012.

Scotto Lomassese, S., Strambi, C., Strambi, A., Charpin, P., Augier, R., Aouane, A., et al. (2000). Influence of environmental stimulation on neurogenesis in the adult insect brain. *Journal of Neurobiology*, *45*(3), 162–171.

Shaw, S. R. (1988). Euphorine phylogeny—the evolution of diversity in host utilization by parasitoid wasps (Hymenoptera, Braconidae). *Ecological Entomology*, *13*(3), 323–335. Available from: http://dx.doi.org/10.1111/j.1365-2311.1988.tb00363.x.

Singer, M. C., Ng, D., & Thomas, C. D. (1988). Heritability of oviposition preference and its relationship to offspring performance within a single insect population. *Evolution*, *42*(5), 977–985. Available from: http://dx.doi.org/10.2307/2408913.

Slaa, E. J., Wassenberg, J., & Biesmeijer, J. C. (2003). The use of field-based social information in eusocial foragers: Local enhancement among nestmates and heterospecifics in stingless bees. *Ecological Entomology*, *28*(3), 369–379. Available from: http://dx.doi.org/10.1046/j.1365-2311.2003.00512.x.

Smallegange, R. C., Everaarts, T. C., & Van Loon, J. J. A. (2006). Associative learning of visual and gustatory cues in the large cabbage white butterfly, *Pieris brassicae*. *Animal Biology*, *56*(2), 157–172. Available from: http://dx.doi.org/10.1163/157075606777304159.

Snell-Rood, E. (2012). Selective processes in development: Implications for the costs and benefits of phenotypic plasticity. *Integrative and Comparative Biology*, *52*(1), 31–42. Available from: http://dx.doi.org/10.1093/icb/ics067.

Snell-Rood, E. C., Davidowitz, G., & Papaj, D. R. (2011). Reproductive tradeoffs of learning in a butterfly. *Behavioral Ecology*, *22*(2), 291–302. Available from: http://dx.doi.org/10.1093/beheco/arq169.

Snell-Rood, E. C., & Papaj, D. R. (2009). Patterns of phenotypic plasticity in common and rare environments: A study of host use and color learning in the cabbage white butterfly *Pieris rapae*. *American Naturalist*, *173*(5), 615–631. Available from: http://dx.doi.org/10.1086/597609.

Snell-Rood, E. C., Papaj, D. R., & Gronenberg, W. (2009). Brain size: A global or induced cost of learning? *Brain Behavior and Evolution*, *73*(2), 111–128. Available from: http://dx.doi.org/10.1159/000213647.

Sol, D., Duncan, R. P., Blackburn, T. M., Cassey, P., & Lefebvre, L. (2005). Big brains, enhanced cognition, and response of birds to novel environments. *Proceedings of the National Academy of Sciences of the United States of America*, 102(15), 5460–5465. Available from: http://dx.doi.org/10.1073/pnas.0408145102.

Sol, D., & Price, T. D. (2008). Brain size and the diversification of body size in birds. *American Naturalist*, 172(2), 170–177. Available from: http://dx.doi.org/10.1086/589461.

Sol, D., Timmermans. S., & Lefebvre, L. (2002). Behavioural flexibility and invasion success in birds. *Animal Behaviour*, 63, 495–502. Available from: http://dx.doi.org/10.1006/anbe.2001.1953.

Sourakov, A. (1997). Social oviposition behavior and life history of *Aglais cashmirensis* from Nepal. *Holarctic Lepidoptera*, 4, 75–76.

Stephens, D., & Krebs, J. (1986). *Foraging theory*. Princeton, NJ: Princeton University Press.

Tabashnik, B. E., Wheelock, H., Rainbolt, J. D., & Watt, W. B. (1981). Individual variation in oviposition preference in the butterfly, *Colias eurytheme*. *Oecologia*, 50(2), 225–230. Available from: http://dx.doi.org/10.1007/bf00348042.

Thompson, J. N. (1988). Variation in preference and specificity in monophagous and oligophagous swallowtail butterflies. *Evolution*, 42(1), 118–128. Available from: http://dx.doi.org/10.2307/2409120.

Traynier, R. M. M. (1984). Associative learning in the ovipositional behaviour of the cabbage butterfly, *Pieris rapae*. *Physiological Entomology*, 9(4), 465–472.

Traynier, R. M. M. (1986). Visual learning in assays of sinigrin solution as an oviposition releaser for the cabbage butterfly, *Pieris rapae*. *Entomologia Experimentalis Et Applicata*, 40(1), 25–33.

Ueno, H., Fujiyama, N., Irie, K., Sato, Y., & Katakura, H. (1999). Genetic basis for established and novel host plant use in a herbivorous ladybird beetle, *Epilachna vigintioctomaculata*. *Entomclogia Experimentalis Et Applicata*, 91(1), 245–250. Available from: http://dx.doi.org/10.1046/j.1570-7458.1999.00490.x.

van Praag, H., Kempermann, G., & Gage, F. H. (2000). Neural consequences of environmental enrichment. *Nature Reviews Neuroscience*, 1(3), 191–198.

Wei, K., Tang, Y. L., Wang, X. Y., Yang, Z. Q., Cao, L. M., Lu, J. F., et al. (2013). Effects of learning experience on behaviour of the generalist parasitoid *Sclerodermus pupariae* to novel hosts. *Journal of Applied Entomology*, 137(6), 469–475. Available from: http://dx.doi.org/10.1111/jen.12031.

Weiss, M. R. (1997). Innate colour preferences and flexible colour learning in the pipevine swallowtail. *Animal Behaviour*, 53, 1043–1052. Available from: http://dx.doi.org/10.1006/anbe.1996.0357.

Weiss, M. R., & Papaj, D. R. (2003). Colour learning in two behavioural contexts: How much can a butterfly keep in mind? *Animal Behaviour*, 65, 425–434. Available from: http://dx.doi.org/10.1006/anbe.2003.2084.

West, S. A., & Cunningham, J. P. (2002). A general model for host plant selection in phytophagous insects. *Journal of Theoretical Biology*, 214(3), 499–513. Available from: http://dx.doi.org/10.1006/jtbi.2001.2475.

West-Eberhard, M. J. (2003). *Developmental plasticity and evolution*. New York, NY: Oxford University Press.

Wheat, C. W., Vogel, H., Wittstock, U., Braby, M. F., Underwood, D., & Mitchell-Olds, T (2007). The genetic basis of a plant-insect coevolutionary key innovation. *Proceedings of the National Accdemy of Sciences of the United States of America*, 104(51), 20427–20431. Available from: http://dx.doi.org/10.1073/pnas.0706229104.

Wilson, E. (1965). Trail sharing in ants. *Pysche*, 72, 2–7.

Worden, B. D., & Papaj, D. R. (2005). Flower choice copying in bumblebees. *Biology Letters*, 1(4), 504–507. Available from: http://dx.doi.org/10.1098/rsbl.2005.0368.

Xue, H. J., Egas, M., & Yang, X. K. (2007). Development of a positive preference-performance relationship in an oligophagous beetle: Adaptive learning? *Entomologia Experimentalis Et Applicata, 125*(2), 119–124. Available from: http://dx.doi.org/10.1111/j.1570-7458.2007.00605.x.

Yu, H.-P., Wang, Z.-T., Xiao, K., Shao, L., & Li, G.-Q. (2011). The presence of conspecific decoys enhances the attractiveness of an NaCl resource to the yellow-spined locust, *Ceracris kiangsu. Journal of Insect Science, 11*.

Commentary on Chapter 16: Insect Creativity as Applied to Human Organizational Behavior: A Form of Social Biomimicry?

Across the fields of engineering, architecture, and industrial design, there has been a growing interest in what is known as biomimicry, or biologically inspired design (e.g., Bosner, 2006; Bosner & Vincent, 2006; Helms, Vattam, & Goel, 2009). Oversimplifying somewhat, biomimicry essentially involves taking cues or analogies from nature to inject novel design ideas in man-made products and processes. As legend goes, for example, Velcro is a result of a Swiss engineer noticing burred seeds strongly attached to his dog after a walk—a design he then mimicked in the lab (Benyus, 2002). One of my favorite examples is the legend of the bullet train, said to be inspired by a Japanese design team ruminating on the sonic booms that occurred as the early versions of the bullet train entered various tunnels around Japan. Using the observation that a Kingfisher could dive for fish without causing a great disturbance in the water due to its uniquely tapered beak, the bullet train was redesigned with the Kingfisher in mind and sonic booms were subsequently silenced.

In a twist on the notion of products inspired from nature, Snell-Rood, Swanson, and Jaumann (this volume) have seemingly introduced a form of social biomimicry, whereby we might consider creative processes from insect communities and explore analogical relationships to human creative activities. Akin to the architect turning to beehive cooling mechanisms as a means to guide the design of an efficient and sustainable office complex (Fragkou & Stevenson, 2012), I think we too can learn a great deal from scholars investing in efforts uncovering creative methods in nonhuman communities. As an organizational psychologist, I find this notion more fascinating than Velcro, Gecko Tape and even Bullet Trains.

PARALLELS WITH ORGANIZATIONAL PSYCHOLOGY

Upon first read, I was overwhelmingly struck by the sheer number of parallels between a discussion on insects and human creative behavior and, in particular, with the work my research lab does on innovation in organizations. Perhaps most strikingly was the broad reaffirmation that creativity is well conceptualized as a form of complex problem solving (Mumford, Whetzel, & Reiter-Palmon, 1997). A group of insects faced with increasingly aggressive predators or limited food sources can be linked fairly well to organizations facing similar competitors looking to increase market share, or an economic downturn that limits funds available for R&D spending. Indeed, necessity is the mother of invention for all of us.

Throughout the chapter, Snell-Rood and colleagues return to a notion of resource availability and its complex role in creative and innovative processes—a second striking parallel. In particular, through their discussion on tiger swallowtails, it seems as if there may be form of complacency (via lack of diversification) that can occur when a set of resources are too plentiful. Many organizations tasked with innovating also worry about such complacency and often refer longingly to their start-up days, when there was an energy within the organization or group that is difficult to replicate and sustain as an organization grows (Estrin, 2009). Moreover, the authors highlight the complex nature of resources, suggesting that while complacency is to be avoided, so too is the lack of resources which limits a group's ability to take risks and explore novel approaches to survival. This complex and potentially curvilinear relationship with resources and success was fascinating to see in a discussion of insect behavior.

The authors also comment extensively on errors and mistakes as important impetuses for novel behavior. A number of products, ranging from the Post-It note, Teflon, and even Viagra were the result of error—albeit serendipitous ones (Pina e Cunha, Clegg, & Mendonca, 2010). Moreover, Snell-Rood and colleagues nearly rip a page from innovation pioneer David Kelley's book on design and creativity, noting that mistakes early in the process are helpful for outlining and detailing the problem to be solved. The founder of IDEO and Stanford D-School professor is quoted as saying "we need to fail faster to succeed sooner" (Kelley, 2001, p. 36).

The final strong parallel that emerged, though there are certainly more (e.g., reward structures, social hierarchies, communication), was how inherently social creativity and innovation are. The great artist or designer is often depicted as a bit of a loner, operating in the wee hours of the night, toiling away in their workshop or studio. In reality, organizational innovation is the result of a collective hive, functioning in dynamic but often fluid ways. I found myself mentally comparing images of honeybees doing their "waggle dance" with the project kick-offs and expertise laden lunch meetings occurring at Google and Pixar.

LEARNING FROM ORGANIZATIONAL PSYCHOLOGY

In addition to these parallels, I also found myself asking if the field of organizational psychology and organizational behavior could humbly offer insight, or possible paths of insight, into nonhuman creative behavior. The first possible path was the critical role of leadership in facilitating innovation. Doing things differently, the very hallmark of creative endeavors, means breaking from established norms and operating procedures. This form of deviance is quite difficult in organizational settings and it is frequently the *leader* that gives permission for such approaches (Hunter & Cushenbery, 2011; Mumford, Hunter, Eubanks, Bedell-Avers, & Murphy, 2007). It may be worth considering the mechanisms by which non-human "leaders" encourage novel approaches—or even if this occurs in insect communities.

A second path that organizational psychology might offer is in the process approach to conceptualizing creativity and innovation (Baughman & Mumford, 1995). Within the social sciences, creativity is generally defined as the generation of solutions that are novel and useful, while innovation is defined as the implementation of creative ideas (Mumford & Gustafson, 1988) suggesting a natural progression from generation to implementation. Despite this depiction as a qualitatively distinct two-part process, in reality there are a number of complex sub-processes comprising and linking these two broader activities. Through this more complete lens, for example, we find that organizations able to define their problems effectively (i.e., perform well in an early stage process) are in a better position to *develop* novel solutions to those problems. Moreover, organizations that effectively prototype and test their ideas (a later stage process) are better poised to select those that are likely to succeed. To be clear, such notions are certainly touched on in the chapter by Snell-Rood and colleagues, but the literature surrounding such processes may be useful in framing and extending future investigations.

WHAT ORGANIZATIONAL PSYCHOLOGY MIGHT LEARN

I found it to be a fascinating thought experiment to ask what my field might learn from scholars studying animal and insect communities. Although there are many, the first that emerged was the notion of more solitary insects, such as butterflies, learning from other insects via indirect mechanisms. The link to entrepreneurs as the more solitary innovators to other entrepreneurs, entrepreneurial groups, and even larger organizations is an interesting one, suggesting to me that we may need to spend more time understanding how entrepreneurs learn from the larger, more complex innovation system.

Additionally, although we have begun to explore the notion a bit (e.g., Lovelace & Hunter, 2013; Mumford & Hunter, 2005), Snell-Rood and colleagues offer paths for further exploration in the realm of mistakes and creativity. The authors successfully argue that mistakes are an essential part of the innovation process. So successful is their argument that it begs the question as to whether organizations would benefit from active pursuit of mistakes as a means to speed up and enhance the learning process. Anecdotally, we have seen this in some organizations; Hewlett Packard, for example, would provide $1000.00 rewards for the biggest failure in a creative endeavor thereby encouraging the mistake and learning process (Berkun, 2007; Estrin, 2009). Despite such anecdotes, however, empirical work has been limited.

Finally, a central thesis of the chapter was the notion of oscillation between colonization and subsequent diversification. This broader and more macro concept is certainly linked to work by Benner and Tushman (2003) and March (1991) who explore cycles of exploiting an innovation and exploring a new ideas, yet underscores the criticality of considering the broader backdrop of activities occurring as the more micro creative and innovative processes are engaged in. As Apple moved into smartphones, for example, they began a colonization effort—the first iPhone was quite original in the cell phone space. Once colonized, however, the organization began specialization activities, expanding into a range of touchscreen, tablet, and handheld devices. The processes required for each were certainly impacted by the broader innovation activities. Thus, it seems that organizational psychology would be well served to also consider such macro processes as a key shaper of more localized innovation efforts. Even more generally, it is clear we have much to learn from other disciplines utilizing their own unique and innovative lenses to understand the fascinating phenomenon of creativity.

References

Baughman, W. A., & Mumford, M. D. (1995). Process-analytic models of creative capacities: Operations influencing the combination-and-reorganization process. *Creativity Research Journal. 8*, 37–62.

Benner, M. J., & Tushman, M. L. (2003). Exploitation, exploration, and process management: The productivity dilemma revisited. *Academy of Management Review, 28*, 238–256.

Benyus (2002). *Biomimicry: Innovation inspired by nature*. New York, NY: HarperCollins.

Berkun, S. (2007). *The myths of innovation*. Sebastopol, CA: O'Reilly Media.

Bosner, R. (2006). Patented biologically-inspired technologogical innovations: A twenty year view. *Journal of Bionic Engineering, 3*, 39–41.

Bosner, R., & Vincent, J. (2006). Technology trajectories, innovation, and the growth of bio-mimetics. *Journal of Mechanical Engineering Science, 221*, 1177–1180.

Estrin, J. (2009). *Closing the innovation gap*. New York, NY: McGraw-Hill.

Fragkou, D., & Stevenson, V. (2012). Study of beehive and its potential biomimicry application on capsule hotels in Tokyo, Japan. *Proceedings of second conference on people and buildings*. London, UK.

Helms, M., Vattam, S. S., & Goel, A. K. (2009). Biologically inspired design: Process and products. *Design Studies, 30*, 606–622.

Hunter, S. T., & Cushenbery, L. (2011). Leading for innovation: Direct and indirect influences. *Advances in Developing Human Resources, 13*, 248–265.

Kelley, T. (2001). *The art of innovation: Lessons in creativity from IDEO, America's leading design firm*. New York, NY: Currency/Doubleday.

Lovelace, J. B., & Hunter, S. T. (2013). Charismatic, ideological, and pragmatic leaders' influence on subordinate creative performance across the creative process. *Creativity Research Journal*.

March, J. G. (1991). Exploration and exploitation in organizational learning. *Organization Science, 2*, 71–87.

Mumford, M., & Gustafson, S. (1988). Creativity syndrome: Integration, application, and innovation. *Psychological Bulletin, 103*, 27–43.

Mumford, M. D., & Hunter, S. T. (2005). The creativity paradox: Sources, resolutions, and directions. In F. J. Yammarino, & F. Dansereau (Eds.), *Research in multi-level issues* (Vol. IV). Oxford, England: Elsevier.

Mumford, M. D., Hunter, S. T., Eubanks, D. L., Bedell, K. E., & Murphy, S. T. (2007). Developing leaders for creative efforts: A domain-based approach to leadership development. *Human Resource Management Review, 17*, 402–417.

Mumford, M. D., Whetzel, D., & Reiter-Palmon, R. (1997). Thinking creatively at work: Organizational influences on creative problem solving. *Journal of Creative Behavior, 31*, 7–17.

Pina e Cunha, M., Clegg, S. R., & Mendonca, S. (2010). On serendipity and organizing. *European Management Journal, 28*, 319–330.

Creating Creative Animals[1]

Karen Pryor
Karen Pryor Academy, Watertown, MA, USA

Commentary on Chapter 17: Creating Creative Animals

James C. Kaufman
University of Connecticut, Neag School of Education, Storrs, CT, USA

I met my first creative animals in 1963, when I became head dolphin trainer at Sea Life Park, a new oceanarium and adjoining research institute in Hawaii. There were no professional dolphin trainers at that time, just a few circus trainers trying their hand with these new animals. I had no preparation. My background was in marine invertebrate zoology. I had trained one dog and one horse using traditional correction-based methods, which was apparently sufficient to recommend me for the position.

Fortunately a Sea Life Park consultant had provided a brief manuscript on training a dolphin with reinforcement-based training procedures based on principles of behavior acquisition first defined by Harvard researcher B.F. Skinner. This information enabled me and three additional trainers to train the multiple and unusual species of dolphins found in Hawaiian waters; to develop shows for the public at Sea Life Park in time for opening day; and to undertake a number of research projects in the next few years.

[1]Pryor (2015).

Animal Creativity and Innovation.
DOI: http://dx.doi.org/10.1016/B978-0-12-800648-1.00017-6

483

My second-in-command, and in due course my replacement as head trainer, was a Swedish ethologist, Ingrid Kang Shallenberger, whose background in animal behavior was invaluable. Ingrid and I, working together, were probably the first to demonstrate experimentally that one can teach an animal to produce new, innovative behaviors deliberately, repeatedly, and on request. We did it initially by accident while training a rough-toothed dolphin (*Steno bredanensis*) named Malia in the shows at Sea Life Park in Hawaii. Malia, when cued to show us new, previously unreinforced responses, came up with some astonishing responses, produced spontaneously and full-blown, including coasting with her tail in the air, jumping upside down, and drawing patterns in the silt on the floor of the tank with the tip of her dorsal fin.

The US Navy, interested in dolphin cognition for military purposes, funded a replication of this event. We used a naïve young female rough-toothed dolphin named Hou. Using two extra observers we taped, filmed, and diagramed each training session. Hou was neither as experienced nor as bold as Malia, and easily became discouraged, but eventually she did come to meet the criterion that might be described as, "Show me something that has not been reinforced before." Her innovations were not nearly as original as Malia's, being normal dolphin actions such clapping the jaws, spitting water, breaching (a sort of sideways splashing jump) and resting her head on the edge of the training platform. But she got the point.

For video of the experiment go to http://www.reachingtheanimal-mind.com/chapter_05.html and click on Creative Porpoise Experiment.

The resulting paper was published in the *Journal of Experimental Analysis of Behavior* (Pryor, Haag, & O'Reilly, 1969). It has been cited upwards of 800 times over the years, and assigned to students in multitudes of psychology departments. However, as far as I know, no scientists repeated the experiment, with the exception of one study of walrus vocalizations (Shusterman & Reichmuth, 2008). The general attitude seemed to be, "Oh, those smart dolphins, aren't they fun; but I don't have any dolphins, so it doesn't apply to my work." The research that did get done, mostly with rats and birds, differed greatly from the Hou study, both in goals and in procedures, and generally produced evidence of increased variability of behavior, but limited to single response types such as key pressing (Pryor & Chase, 2014).

ABOUT "ANECDOTAL" OBSERVATIONS

Farmers, zoo keepers, and animal trainers are aware that animals can sometimes not just problem-solve but innovate, and in very creative ways. I had a golden retriever that learned to escape the back yard by climbing a ladder over an eight foot wall; my other dog, a Great Dane, thus learned to do it too. My father's cat spent 2 days making a large and careful pattern under the dining-room table with a ball of yarn; then she yowled until the household came and admired it. Noel Perrin, a Dartmouth professor who owned a pig, discovered one day that a pile of shingles in the corner of the pigpen had been neatly laid out in rows around the base of the pen, an art form that took many hours of careful work.

Orangutans are natural innovators, even in the wild (Russon, Purwo, Ferisa, & Handayani, 2010). I once met an old lady orangutan at the National Zoo in Washington, DC, who had discovered that by patting her head or waving her arms she could get the children outside her cage to play an orang version of "Simon Says." Washoe, the famous signing chimp, tasting a piece of watermelon for the first time, gave it a new name by merging two previously learned signs: candy-fruit. These events are regarded as rarities, even among oh-so-smart great apes: sporadic observations, interesting, even glamorous contributions to our views of animal cognition, but essentially anecdotal, not, you know, really science.

"Anecdotal" is a slightly contemptuous word. It has come to mean something you merely SAW. Current conventional wisdom holds that nothing is proven until you can reproduce it in the lab. Flashes of insight or instances of creativity in an animal (if that is indeed what is happening) are one-time events, not exactly good fodder for replication. So, these are treated as pretty stories, indicators, maybe, but that's it. For many years I bought into that view. Erich Klinghammer, founder of Wolf Park in Indiana, an ethologist and protégé of Konrad Lorenz, set me straight. Erich had come across a book I'd written about training dolphins with operant conditioning (Pryor, 1973) and invited me to Wolf Park to teach the staff to work with the wolves in the same way. One day I was telling him about something or other I had seen a dolphin do, and apologized for the description as being merely an anecdote. "Don't say anecdote," Erich declared. "An anecdote is something told to others, usually accompanied by an explanation that may or may not be true. What you saw your dolphin DO is an observation. A qualified observer reporting on what they have observed is reliable information. Observation is the fundamental basis for all science." Oh. Thank you, Erich.

Ingrid Shallenberger conveyed a similar insight about one-time or unusual events we had observed in wild dolphins. Ingrid said, sensibly, "If you have seen an animal do something surprising, it is really more parsimonious to say that this kind of animal apparently sometimes does this behavior, than to assume that you have seen the one and only time it ever happened in the history of the species."

So: novel or innovative behavior. Perhaps it is not as rare, nor as trivial, as people might assume.

SPONTANEOUS CREATIVITY

The waters around Hawaii teem with little-known deep ocean species of cetaceans. During my tenure at Sea Life Park we held nine different species of dolphins and small whales in our tanks. Some species, usually those with opportunistic foraging behaviors such as the rough-toothed dolphins and Pacific bottlenose dolphins (*Tursiops gilli*), could be spontaneously creative. The bottlenoses in our training tanks developed games, such as greeting people not by popping their heads out of the water but by turning upside down and laying their tails on the rim of the tank, a disconcerting sight to guests. They also developed a game of coming out of the water and balancing on the top of the waist-high tank wall, tails in the air. Now and then one fell out, a nuisance because (i) they could get hurt and (ii) they are heavy and it took a stretcher and three or four people to hoist them back in.

Rough-toothed dolphins, nicknamed *Stenos*, were particularly given to experimental and exploratory behavior. Back in the 1960s, there were no restrictions on collecting dolphins, and we were in fact pioneering in the handling and care of many little-known species. Some were shy but some were extremely bold. One day a recently captured pair of *Stenos* needed to be moved so their tank could be cleaned. Assuming that they would not swim into an adjoining tank voluntarily, the trainers put a net across the tank in order to herd them. One *Steno* immediately swam to the bottom and picked up the lower edge of the net in its jaws. The other one swam under the net, and then turned around and lifted the net so the first one could swim under it too. They played like this, back and forth, several times, while the confused trainers watched; the two then swam briskly into the adjoining tank on their own.

Other creative dolphin games at Sea Life Park included a fad for wearing "jewelry," capturing a bit of plant material that had fallen or been blown into the tank and carrying it over a fin or in the beak for hours; chasing sea birds away by spitting at them; and developing underwater patterns of bubbles, such as loops, lines, and circles, and then swimming through them.

Perhaps the most neophobic, conservative species of cetaceans we worked with were the Pacific spotted dolphins, *Stenella attenuata*. They avoid strange objects, adapt to new situations slowly, and prefer to form long-term close associations with a select group of conspecifics only. It was possible to establish trained behavior in these animals, but we saw very little initiative or even play behaviors in the species. They were however not incapable of creating new behavior in the wild, as Ingrid and I were to learn a few years after I left Sea Life Park (see the section "Creativity in the Wild").

Spontaneous innovative behavior is not of course limited to dolphins. We kept a pair of Malaysian river otters at Sea Life Park for a while, using them in the shows and walking them on leashes and harnesses around the grounds. They were docile and easy to handle most of the time. The trouble with training them was that no matter what you marked and reinforced, the otters would do it once or twice and then do something different. I complained about this over lunch to a professor who was doing some research at our facilities. He said nonsense, that's not the way reinforcement works. So I took him to the training area. I put an otter into one of the training tanks and held a small ring made of garden hose in the water. The otter swam through the hoop. I marked the move with a whistle and gave him a fish. The professor nodded. I put the hoop in the tank again. The otter circled around and swam through the hoop again, but he left his tail dangling on the other side which brought him to a halt. Do I pay him or not? I chose to pay him. On the third pass, he swam through the hoop, grabbed it with one hind foot as he passed, and carried it away.

See? The professor saw. He thought about it for a moment and then said "It takes me three years to teach graduate students to think like that."

MODERN, MARKER-BASED TRAINING

Modern training is defined by the use of the event marker, reinforcement, shaping, discriminative stimuli or cues, and absence of correction and punishment (Pryor & Ramirez, 2014). In spite of the best efforts of marine mammal trainers to share their procedures, modern, marker-based training continued to be thought of as a special technique solely for cetaceans for about 30 years. A breakthrough finally occurred in 1992, due at least partly to the confluence of three different events. First, I had published a book describing positive reinforcement training for people, which the publisher, over my objections, titled *Don't Shoot the Dog!* (Pryor, 1999). While the parents I had hoped to reach avoided the book because of the title, the dog community began buying it and adapting the technology for their dogs. Secondly, because dog trainers already used a whistle for cueing behavior, trainer Gary Wilkes came up with a different event marker that was ideal for many kinds of terrestrial animals: a toy noisemaker, the clicker, which was neutral, distinct, brief, loud, sturdy, and cheap. Thirdly, the Internet arrived, making real-time discussion lists available for the early-adopter "clicker training" enthusiasts, so the technology could grow and be shared.

Behavior analysts and many other researchers tend to use the conditioned reinforcer purely as a substitute for primary reinforcement, and may continue to use it that way, without following it up with a primary reinforcer, to maintain learned behavior, at least until the value of the conditioned reinforcer extinguishes. When used as an event marker, in contrast, the marker's feedback is no longer needed once the behavior has been established and brought under stimulus control. An already-learned behavior can henceforth be maintained by a variable schedule of reinforcement and by many types of reinforcers including normal life experiences.

Marker-based training has gradually transformed dog training in the United States and abroad (Pryor, 2009). The modern training has also spread widely through the zoo community. Sometimes the techniques are brought into the zoo by marine mammal trainers; sometimes keepers who own dogs and are already using clicker training transfer their skills to their zoo animals. Sometimes instruction is provided from outside consultants or keeper organizations and sometimes keepers self-teach through books and videos. In many zoos in the United States and elsewhere keepers can now provide vaccinations, medicine, hoof trimming, and other physical care to their animals, whether large (tigers, hippos) or small (marmosets, birds, small reptiles) without resorting to force such as chemical or physical restraints.

Once an animal becomes aware that its own actions make the marker happen, it is not a big step for the animal to start exploring the possibilities of other actions, not through whining or pawing or other submissive displays of infancy, but through calculated innovative action designed to trigger human activity. In 1979 as a consultant to the National Zoo I held one session of marker-based training with a young female Indian elephant named Shanthi. She quickly discovered several ways to earn the marker and ensuing food treats. She then used eye contact and her trunk to indicate something she wanted me to do: go around the corner of her cage. There I discovered she had reached through the bars and was holding the padlock to the keeper's entrance door in the tip of her trunk. The message seemed clear: how about letting me out? (Her keeper and I agreed we couldn't open that door, but we could open the door to her play yard, and out she went, at a trot, to chase pigeons.)

In my dolphin training days I was, I confess, a species chauvinist. We had created creative dolphins. I felt sure one could follow the same procedures and elicit innovations from other animals known to be intelligent, such as elephants and the great apes. I assumed however that animals are not supposed to be highly cognitive would not be able to grasp such an abstract concept.

Well, I was wrong. For example, the modern dog training community has made wide use of creative training. Training for novel behavior increases exploratory behavior and behavioral variability in general, useful in preparing to develop the large repertoires of service dogs (such as guide dogs for the blind) and working dogs (such as search and rescue animals). A useful game for dogs, nicknamed "101 things to do with a box," consists of presenting the dog with a cardboard box and then marking (usually with the clicker) and reinforcing (with food treats) any interactions that ensue, such as glancing at the box, climbing in it, knocking it over, or dragging it around. Besides being amusing for onlookers the game has real benefits. One can use the marker to select and strengthen a single behavior from the dog's offerings, such as getting into a box, that can be then brought under stimulus control for other uses (get in the car, get in a shipping crate, get on the veterinary's wobbly scales). People training dogs to perform on stage or in competitions may use the game to encourage the dog itself to come up with new behaviors that can be woven into performance routines. The game is beneficial in rehabilitating military and police dogs with a long history of aversive control, when they are to be retired to family life. These veteran animals are often reluctant, in a training situation, to initiate any behavior on their own. Short bouts of the creative game offer a high rate of reinforcement for initiating behavior of any sort, building confidence and increasing healthy social interactions.

By 2010, zoo keepers who were using marker-based training for husbandry and medical care were also developing innovative behavior in many species. Emily Insalaco, curator of behavioral husbandry at the Denver Zoo in Colorado, told me that an African Painted Dog (*Lycaon pictus*) came up with over 40 novel behaviors, many of them involving jumps and aerial maneuvers. Melissa Nelson Slater, curator for behavior and enrichment at the Wildlife Conservation Society's Bronx Zoo in New York, NY, reports that as of December, 2013, keepers had established creative behavior games with Western lowland gorillas, a dromedary camel, a Nubian goat, silvery langurs, white cheeked gibbons, California sea lions, and Pacific walrus.

Training for innovation can relieve stress and boredom and increase activity levels such as social interactions and play. Zoo keepers in the United States frequently use creative behavior games to provide mental exercise for potentially troublesome male gorillas. In this application the game is called "Show me something new." The cue is waggling both hands in the air, a version of the American Sign Language sign for play (Pryor & Ramirez, 2014). Gorillas are often capable of coming up with at least one new behavior a day, including making comical faces and gestures (video at http://www.reachingtheanimalmind.com/chapter_05.html).

CREATIVE TRAINING PRECURSORS

Modern training procedures for innovative behavior are much the same, whatever the species involved. The procedures can be undertaken successfully with a naïve animal, but are facilitated by preparation. Useful precursors for the learner include:

- Multiple experiences of shaping with an event marker
- An existing repertoire of numerous behaviors under stimulus control
- Exposure to varied environments
- Freedom from corrective or punitive experiences during previous training.

Useful precursor skills for the trainer or teacher include:

- Good observational skills for noticing small shifts in behavior
- Good timing with the event marker
- Experience in breaking behavior down into small increments
- Ability to deliver a high rate of reinforcement (not just a high rate of treats, but of contingent reinforcement of increasingly challenging criteria)
- Experience in observing and recognizing emotional signals, i.e., evidence of internal states of affect
- Avoidance of punishment or correction of errors.

The trainer's goal in training for innovation may be to develop an animal that can create a new behavior on a daily basis, like gorillas; or to identify new behaviors for other purposes, as with performing dogs; or simply to enrich the animal's life. The game can be varied in many ways; one can for example establish the criterion, "show me old behaviors but change to a different behavior after every click." While not requiring entirely new motor patterns this version definitely requires some cognitive work and can be very satisfying for the animal. (For slides of a dog playing this version go to http://www. reachingtheanimalmind.com/chapter_05.html and click on "Peggy plays the box game.")

"Concept training" is also a popular cognitive challenge among trainers of dogs and of some other species. Examples include matching to sample; sorting by attributes (large/small, black/white, wood/metal); responding to sequenced cues or "sentences" (nose-ball-bump; paw-big-box-tap); imitation (of the action of another animal or of a person); and of course, "show me something new."

With or without selecting specifically for innovation, this kind of training, with its varied but always high rate of reinforcement, rapidly shifting criteria, and absence of punishment for errors, produces a global change in the learner. Both Malia and Hou, the original "creative" dolphins, continued to exhibit increased activity and surprising new behaviors and initiatives for the rest of their lives. Hou started jumping into her neighbors' tanks. Malia actually came out of the water and skidded across the training platform to get attention by gently tapping a trainer on the ankle; it became necessary to give her an opportunity to do this occasionally on cue, to prevent her from stranding herself frequently and perhaps putting herself at risk. Regardless of species, this global change, from passive to active, from timid to bold and confident, from rigid and neophobic to flexible and novelty-seeking, is usually permanent in the individual animal, greatly improving its general well-being in captivity.

CREATIVITY IN THE WILD

Over the past decades the resistance to anything that might be considered "anthropomorphism" led to the opposite extreme: the view that animals are automatons, instinct-driven machines without emotions or even conscious awareness, a misconception that naturalist Joseph Wood Krutch labeled "mechanomorphism." Field scientists, rather than laboratory scientists, have supplied most of the evidence to the contrary: the macaque that first washed the sand off her food, the British titmice who opened milk bottles to drink the cream, and Jane Goodall's chimpanzees that used tools. As some aspects of innovative behavior were backed up with laboratory evidence, such as the New Caledonian crows that not only use tools but make them, the very concept of looking for innovative behavior in the wild became a little more acceptable.

A valuable contribution to this area of research has been a collection of book chapters under the title of *Animal Innovation* (Reader & Laland, 2003) surveying and discussing a wide range of studies and topics in this area. Innovation in the natural habitat has been recorded in many species of primates and birds. Innovative foraging techniques have been displayed by female guppies (Laland & Reader, 1999).

And what brings about the development of innovation in the wild? Innovation is more frequent in some species of than in others, perhaps related to general level of plasticity of behavior and opportunistic foraging habits. For example killer whales (*Orcinus orca*) and Atlantic bottlenose dolphins (*Tursiops truncatus*) are capable of innovating new hunting techniques in reaction to prey scarcity, new food resources, or other environmental changes (Bel'kovich, Ivanova, Yefremenkova, Koverovitsky, & Kharitonov, 1991; Sol, 2003), while some other species of cetaceans are neophobic and limited in their behavioral variability.

In social groups, innovation may arise more frequently in low-status animals, perhaps driven by having less access to resources; or, it has also been suggested, by having more idle time than high status animals that must defend their territories, protect mating opportunities, and deflect competitors (Box, 2003). However, environmental changes or pressures may force the development and spread of novel behavior, even in the most conservative species (Lee, 2003).

Aboard the Queen Mary

In 1979, Ingrid Kang Shallenberger and I had a remarkable opportunity to observe three kinds of innovative behavior in the very neophobic and conservative Pacific spotted dolphin, *Stenella attenuata*. As Principal Investigator I traveled with Ingrid aboard the tuna fishing vessel *Queen Mary*, during one of a government-sponsored series of research trips related to dolphin mortality in purse-seining for tuna in the Eastern Tropical Pacific. The two species that are most frequently accompanied by schools of commercially valuable tuna are the Pacific spotted dolphin (*Stenella attenuata*) and the spinner dolphin (*Stenella longirostris*). Dolphin mortality in this fishery had become a major concern. Sea Life Park was the only facility that had maintained these fragile pelagic animals in captivity, and Ingrid and I were among the few that had extensive experience with their behavior and training. They were extremely neophobic. As open ocean dwellers they had trouble dealing with barriers. One would presume they could simply jump over the net, once surrounded; other species of dolphins do, but Pacific spotted dolphins just couldn't. By the time of our research expedition, mortality rates had been considerably mitigated by modifications in both the fishing procedures and the nets; but populations were not rebounding as anticipated, and further research was needed (Figures 17.1 and 17.2).

The fishing procedure is efficient. A purse-seiner tuna fishing vessel uses a large, deep net to surround the dolphin schools and any tuna that may be swimming beneath them. When dolphins are spotted, the ship approaches the school, which often flees; the ship then chases the school until the dolphins can be herded by small speedboats back toward the ship, and enclosed by the purse net.

The net, which may be 600 feet deep and half a mile long, is lowered from the stern of the vessel in a circle around the dolphins and tuna, if present. The bottom rim of the net is "pursed" or drawn shut underneath the dolphins, tuna, sharks, and whatever else may have been accidentally confined. The net is gradually drawn aboard, reducing the diameter. The dolphins, waiting in a group, are released from a narrow

FIGURE 17.1   The author's colleague Ingrid Shallenberger taking behavioral data among a school of pacific spotted dolphins inside a tuna fishing vessel's net. The animals, unafraid of the scientist in the water, swim or rest quietly while waiting to be released.

FIGURE 17.2   The net, half a mile long and 600 feet deep, is closed at the bottom and the majority of the webbing is pulled back onto the ship, leaving a single channel from which the dolphins can be safely released. The ship reverses, pulling the net out from under the dolphins; the splashes beyond the net are made by freed dolphins leaping.

channel opening created in the far part of the net, and the tuna, if present, are brought aboard and frozen.

Each time a school or schools of spotted dolphins were surrounded, Ingrid and I traveled into the net in a rubber boat and snorkeled next to or among the animals, taking Focal Animal Samples of behavior of individuals. Our data showed that spotted dolphins are organized in clan-like schools of from 30 to 300 or so animals. Each such school is comprised of specific types of small closely associated subgroups, consisting of females and young, juveniles, young adults, and, the most conspicuous, darkly marked senior male subgroups, swimming in close formation like fighter jets (Pryor & Shallenberger, 1991). We saw sets with one to up to three such schools, separated by 150 feet or more of empty water, inside the net, associating only with their own group members, each school containing its own one to four subgroups of senior males.

Female and younger subgroups generally avoided the senior males, moving aside to let the male subgroups pass through the school. However, in one set of the net, involving a small school with a single senior male subgroup, we saw the exception to this behavior. Ingrid and I were observing from underwater, when a loud speedboat engine started up just outside the net. All the animals in that small school physically plastered themselves against the senior male subgroup. A-ha! Now we could surmise what happens in the chase. Where the subgroups of dominant males go, everyone else will go, too. The males are probably not the leaders, exactly, but the followed.

It is not my purpose here to discuss the politics or economics of this fishery, nor to teach you about dolphin behavior. However, the behavioral evidence suggested that many schools of dolphins in the fishery have been encircled multiple times. Thus senior male subgroups had an opportunity to innovate some specific method of escape for themselves and their school. I shall go into some detail, here, to help readers appreciate what remarkable courage and initiative some of these senior male groups displayed in order to evade capture successfully.

The captain, first mate, and navigator of the *Queen Mary* described for us six strategies spotted dolphin schools have developed to escape the net. We witnessed the first three of these maneuvers.

1. Hide in a rain cloud.
   Small puffy cumulus clouds are a common sight over tropical oceans. Sometimes one of these small clouds holds enough moisture to produce rain. Ingrid and I were watching the chase from the pilot house of the *Queen Mary* when a cloud about 30 degrees to port began to just pour rain. Instantly the dolphin school fleeing ahead of the ship veered leftward and raced into the little rainstorm. They disappeared completely. The captain had no way of telling whether or where they would come out. The chase was aborted.
2. Schools in an aggregation disperse during the chase.
   Ingrid and I witnessed one chase in which, suddenly, the big aggregation of dolphins racing away in the distance broke up into at least four separate schools, still traveling in tight groups but headed in four separate directions. There was no way to tell which school, if

any, had the tuna with them, and no way to round them up again. A perfect escape.

3. Beat the boat.

As the dolphins are being herded back toward the ship by the speedboats the ship itself must slow down in order to lay the net. In one set as Ingrid and I watched from the port rail, the dolphins were clumped inside the area encircled by the net, as far from the ship and its engine noise as possible. However the bow of the ship had not yet reached the other end of the net to close the circle. Abruptly a group of dolphins sprinted for the bow of the *Queen Mary* and were able to cross to the starboard side of the ship. In moments a major part of the school had succeeded in following them across the bows. The dolphins were free. They knew it, and they were expressing that by stotting, like sheep or antelopes, jumping high in the air, a behavior that (in my opinion) expresses excitement or elation but also signals "this way to safety" for animals still behind. The engines stopped; there was nothing the ship could do about this except halt and pull the net back in.

4. Swim through the bubbles.

During the period when the dolphins are encircled but the circle is not yet closed, speedboats can go back and forth over the gap at high speed, creating motor noise and a wall of bubbles impenetrable both to vision and to echolocation. The bubbles are a harmless barricade, but the dolphins can neither see nor sense by sonar the extent of the barrier; and the noise created by the speedboat motors generating the bubbles is alarming. Nevertheless, tuna vessel captains reported to us that some schools do escape by swimming right through this obstacle. The courage it took to take this path the first time had to be enormous.

5. Dive under the net before it is pursed.

Once the dolphins are surrounded, they cannot escape by jumping, due to lack of experience; and the net can be hanging down 600 feet, a long dive for this surface-feeding species, and especially for calves. However, pulling the purse line closed around the base of the net takes time. Some schools dive to the bottom during that time period, cross under the lower hem of the net before it is closed, and thus go free; and the tuna, if present, always go with them.

6. Schools scatter into individuals

This unusual maneuver has evolved only in coastal waters where the purse-seining fishery first began. In these waters captains reported to me having seen schools that instead of fleeing from ship and speedboats, scatter into single individuals or small groups all around the ship over a wide area, making no effort to flee and becoming impossible to herd. Since the tuna may be anywhere or nowhere under this acreage of dolphins, there is no point in setting the net.

Two of these six maneuvers, hiding in under a rain cloud or dispersing into different schools, may have developed as methods for escaping from dolphin-predatory killer whales. However, pressure from the tuna purse seining fleet seems to be the sole pressure that pushed these non-innovators into trying the rest of these daring and successful maneuvers: going through the bubbles, diving under the net, and escaping across the ship's bows—plus the method of just spreading out widely and refusing to be chased. Pressed hard enough, even these stability-loving neophobes could indeed come up with something new.

CONCLUSIONS

While traditional trainers often regarded their methods as secret and proprietary, modern training developments, based on principles rather than methods, are widely shared in the training community, through the internet and through organizations and conferences. However, many modern training tools, such as the use of the conditioned reinforcer as an event marker, have been either absent from or poorly represented in the peer-reviewed literature.

Modern marker-based training is based on underlying science, but it has developed largely as a technology, proven in the use rather than in the laboratory. Practitioners are interested in outcomes, not in antecedents, and have little motivation to collect or publish data; and until very recently very few university psychology departments have focused on this potential area for research.

Furthermore the specific disciplines of behavior analysis and of modern animal training, while both based on the same scientific discoveries, have diverged somewhat along separate paths (Pryor, 2015, in press). For example, one current practice, presented in some student textbooks as the *only* way to shape behavior, is to develop a reinforced behavior and then shut off the reinforcement abruptly, putting the learner into an extinction curve. In this condition a burst of varied activity may occur from which new or improved behavior may be selected. The drawback, of course, is that extinction is highly aversive. The burst of activity is highly likely to be aggression-related or flight-related, rather than exploratory or inventive. The accompanying emotional and hormonal states are detrimental to learning and extremely detrimental to the learner/teacher relationship. Modern animal trainers utilize much more benign procedures both for shaping new versions of existing behavior and for eliciting new behavior.

Where do we go next? Research is underway at several institutions on the efficacy of the event marker in a wide range of conditions, and on detrimental effects of correction, extinction, and deprivation on learning. There appears to be a strong link between play and creativity which might warrant further study (Pellis & Pellis, 2009). Modern training technology may make possible more efficient experimental designs for accessing and investigating cognitive skills in animals, and, in fact, for any experiment in which efficiency would be improved if the animal understood what it was supposed to be doing. Finally, we should not overlook the value of marker-based training techniques and protocols in developing behavioral skills, including creative behavior, in human beings (see www.tagteach.com for current applications and research).

References

Bel'kovich, V. M., Ivanova, E. E., Yefremenkova, O. V., Koverovitsky, L. B., & Kharitonov, S. P. (1991). Searching and hunting behavior in the bottlenose dolphin (*Tursiops truncatus*) in the Black Sea. In K. Pryor, & K. S. Norris (Eds.), *Dolphin societies, discoveries and puzzles*. Los Angeles, CA: University of California Press.

Box, H. O. (2003). Characteristic and propensities of marmosets and tamarins: Implications for studies of innovation. In S. M. Reader, & K. N. Laland (Eds.), *Animal innovation*. Oxford, UK: Oxford University Press.

Laland, K. N., & Reader, S. M. (1999). Foraging innovation in the guppy. *Animal Behavior*, 57, 331–340.

Lee, P. C. (2003). Innovation as a behavioral response to environmental challenges: A cost and benefit approach. In S. M. Reader, & K. N. Laland (Eds.), *Animal innovation*. Oxford, UK: Oxford University Press.

Pellis, S., & Pellis, V. (2009). *The playful brain: Venturing to the limits of neuroscience*. Oxford, UK: Oneworld Press.

Pryor, K. (1973). *Lads before the wind: Adventures in dolphin training*. New York, NY: Harper & Row.

Pryor, K. (1999). *Don't shoot the dog! The new art of teaching and training* (Rev. ed.). New York, NY: Bantam Books.

Pryor, K. (2009). *Reaching the animal mind*. New York, NY: Scribner, pp. 82–87, 90–95.

Pryor, K. (2015). In A. Kaufman & J. Kaufman (Eds.), *Creativity and innovation in animals*.

Pryor, K. (in press). *Inside and outside behavior analysis*. Cambridge Center for Behavioral Science, Boston, MA.

Pryor, K., & Chase, S. (2014). Training for variable and innovative behavior. *International Journal of Comparative Psychology*, 27(2), 361–368.

Pryor, K., & Ramirez, K. R. (2014). Modern animal training: A transformative technology In F. McSweeney, & E. Murphy (Eds.), *A handbook of operant and classical conditioning* New York, NY: Wiley and Blackwell.

Pryor, K., & Shallenberger, I. K. (1991). Social structure in spotted dolphins (*Stenella attenuata*) in the tuna purse seine fishery in the Eastern Tropical Pacific. In K. Pryor, & K. S. Norris (Eds.), *Dolphin societies, discoveries and puzzles*. Los Angeles, CA: University of California Press.

Pryor, K. W., Haag, R., & O'Reilly, J. (1969). The creative porpoise: Training for novel behavior. *Journal of the Experimental Analysis of Behavior*, 12, 653–661.

Reader, S. M., & Laland, K. N. (2003). Animal Innovation: An introduction. In S. M. Reader, & K. N. Laland (Eds.), *Animal innovation*. Oxford, UK: Oxford University Press.

Russon, A. E., Purwo, K., Ferisa, A., & Handayani, D. P. (2010). How orangutans (*Pongo pygmaeus*) innovate for water. *Journal of Comparative Psychology, 124,* 14–28.

Schusterman, R. J., & Reichmuth, C. (2008). Novel sound production through contingency learning in the Pacific walrus (*Odebenus rosmarus divergens*). *Animal Cognition, 11,* 319–327.

Sol, D. (2003). Behavioral innovation: A neglected issue in the ecological and evolutionary literature? In S. M. Reader, & K. N. Leland (Eds.), *Animal innovation*. Oxford, UK: Oxford University Press.

Commentary on Chapter 17: Creating Creative Animals

There are many reasons why I was so excited to tackle the topic of creativity and animals in an edited book (one was the chance to work with my wife Allison). Foremost among them was that studying creativity and studying animals are surprisingly similar experiences. As a creativity researcher, I've seen my entire field frequently dismissed as not being real science. In the question-and-answer session of a job talk I had one rather rude professor ask whether the topic of creativity really belonged within education, alongside core topics such as reading and motivation. I've seen comparable incidents happen in Allison's career—wanting to devote your life to studying something that the average person actually finds interesting means that you also spent your career defending your work from those who refuse to believe that anything interesting or fun can also be real science.

In reading Karen Pryor's excellent chapter I was struck by the commonalities and big picture themes that can emerge from working with dolphins, gorillas, or people. One big one is Pryor's articulation of the difference between an anecdote and an observation. She remembers being told by Erich Klinghammer that a "qualified observer reporting on what they have observed is reliable information. Observation is the fundamental basis for all science." Such a distinction is at the core of what I believe is the most sophisticated way of assessing creativity.

The idea of asking people evaluate artistic work has been around for nearly a century (Cattell, Glascock, & Washburn, 1918). Teresa Amabile (1982, 1996) used this concept to form the Consensual Assessment Technique (CAT), in which qualified experts assign ratings to creative work (which could be anything from artwork to a scientific theory). Numerous studies have demonstrated that experts agree with each other at a strikingly high rate about which products are creative (Baer, Kaufman, & Gertile, 2004; Hennessey, Kim, Guomin, & Weiwei, 2008; Kaufman & Baer, 2012). The raters must have a certain amount of expertise in the domain area in order to show agreement; pure novices will not agree with experts (Kaufman, Baer, & Cole, 2009; Kaufman, Baer, Cropley, Reiter-Palmon, & Sinnett, 2013).

An anecdote becomes a professional observation with rigor, patterns, statistical evidence, and appropriate expertise on the part of the observer. Indeed, Kaufman and Rosenthal (2009) have applied the same tenets of the CAT methodology to how animal behavior is studied. Karen Pryor has been removing the stigma from using observation as an essential tool for the last 45 years, much as Teresa Amabile has been removing the stigma from using subjective opinions about creativity as a key way of measuring creativity.

Pryor also covers another concept dear to my heart, which is the notion that creativity is both a means and an end. She discusses how training for novelty can help accomplish very practical things, such as helping police dogs adjust to life-after-service or helping enormous creatures be weighed at a veterinarian's office. Yet she also covers the way that training for innovation can lead to play behaviors that can keep an animal mentally alert and socially active. Most pressingly, dolphin schools have used their creativity to avoid capture in tuna nets. Animals can use their creativity for recreation or to save their lives—much as humans can use their creativity to play (Russ, 2013) or to heal from life-shattering trauma (Forgeard, 2013).

Within the field of creativity, the idea of training is quite salient; many people (typically not researchers) have made a small fortune training organizations or people to be more creative. Most corporate training has a spotty track record and somewhat dubious reputation; if only the systematic work that has been done on training creativity in animals that Pryor has documented could be applied to humans!

References

Amabile, T. M. (1982). Social psychology of creativity: A consensual assessment technique *Journal of Personality and Social Psychology, 43*, 997–1013.

Amabile, T. M. (1996). *Creativity in context: Update to the social psychology of creativity.* Boulder, CO: Westview.

Baer, J., Kaufman, J. C., & Gentile, C. A. (2004). Extension of the consensual assessment technique to nonparallel creative products. *Creativity Research Journal, 16*, 113–117.

Cattell, J., Glascock, J., & Washburn, M. F. (1918). Experiments on a possible test of aesthetic judgment of pictures. *American Journal of Psychology, 29*, 333–336.

Forgeard, M. J. C. (2013). Perceiving benefits after adversity: The relationship between self-reported posttraumatic growth and creativity. *Psychology of Aesthetics, Creativity, and the Arts, 7*, 245–266.

Hennessey, B. A., Kim, G., Guomin, Z., & Weiwei, S. (2008). A multicultural application of the Consensual Assessment Technique. *International Journal of Creativity and Problem Solving, 18*, 87–100.

Kaufman, A. B., & Rosenthal, R. (2009). Can you believe my eyes? The importance of interobserver reliability statistics in observations of animal behaviour. *Animal Behaviour, 78*(6), 1487–1491.

Kaufman, J. C., & Baer, J. (2012). Beyond new and appropriate: Who decides what is creative? *Creativity Research Journal, 24*, 83–91.

Kaufman, J. C., Baer, J., & Cole, J. C. (2009). Expertise, domains, and the Consensual Assessment Technique. *Journal of Creative Behavior, 43*, 223–233.

Kaufman, J. C., Baer, J., Cropley, D. H., Reiter-Palmon, R., & Sinnett, S. (2013). Furious activity vs. understanding: How much expertise is needed to evaluate creative work? *Psychology of Aesthetics, Creativity, and the Arts, 7*, 332–340.

Russ, S. W. (2013). *Pretend play in childhood: Foundation of adult creativity*. Washington, DC: American Psychological Association.

18

Animal Creativity and Innovation: An Integrated Look at the Field

William J. O'Hearn, Allison B. Kaufman and James C. Kaufman

Department of Ecology and Evolutionary Biology, University of Connecticut, Field Technician, Storrs, CT, USA

Creativity can be an immensely powerful tool, whether in human hands or nonhuman paws, claws, or flippers (A.B. Kaufman & Kaufman, 2014). It can entail finding a new problem to solve or creating a solution to an existing problem. Creativity can be found across many different domains, from art to science to business to everyday life (J.C. Kaufman, 2012; J.C. Kaufman & Baer, 2004). In many ways, the material presented in this book simply represents another domain of creativity, nonhuman creativity, and it can be found in anything from Indonesian orangutans (Russon) manipulating their environment to solve unique problems to an African Grey Parrot intentionally using phonemes to request a break or a reward (Pepperberg).

Creativity in nonhumans is significant because it is an adaptive force that provides solutions to unforeseen problems. For example, in a scenario in which a critical food source becomes unavailable, as discussed in apes the chapter by Call, individuals in a population who are unable to find an alternate food source will die. Yet how does an animal find a solution to this problem? There needs to be a mechanism that allows animals to find alternate resources and form new behaviors to utilize them; this is creativity. Creativity here is like a behavioral mutation that allows animals to climb an environmental wall. Onward in time, if others members of the population see the new behavior and the benefits it provides, they can imitate it. This phenomenon is known as social learning. It is responsible for the spread of adaptive behavior to the community level, a prime example being sponging in dolphins as discussed by Patterson and Mann.

As with human creativity, creativity of nonhumans comes with natural variation. Animals of varying cognitive abilities (such as working memory and cognitive control, as noted by Vartanian) show a spectrum of creative complexity concurrent with their brain capacity (discussed in Navarrete and Laland) and environmental needs (discussed in Lee and Moura). As proposed in a model of animal creativity (A.B. Kaufman, Butt, Colbert-White, & Kaufman, 2011; J.C. Kaufman & Kaufman, 2004), these behaviors can begin with something as simple as neophilia, or attraction to new stimuli, which in turn may provide a possible evolutionary advantage. The pinnacle can be full-blown innovation and tool use, where an individual can identify and manipulate parts of the environment. In this way, creativity is a tool that animals use to help them cope with their surroundings. It is a part of the adaptive strategies that allow animals to survive in new environments as well as utilize new and available resources. Call's chapter provides an example, discussing how the traditional method of acquiring food in chimps fishing for termites was readily used until an innovative method was presented with a greater pay off. The use of the novel method was based on the adaptive significance it provided to the individual. Similarly, this demonstrates what Sol discusses in his chapter, that an innovative animal is capable of shaping its own evolution. Creativity is an adaptive process; it is behavioral evolution allowing change on the individual time scale instead of the species time scale. Ward takes this further when he remarks on the human propensity to accumulate, store, and remember novel ideas.

As we develop the concept of creativity as an adaptive force, more and more of the actual mechanisms are uncovered; many of them are named in these chapters. Snell-Rood presents the exploration-exploitation trade off that shows us the costs of being naïve in a harsh environment. Zentall demonstrates the role of social learning in spreading a novel adaptive behavior within a population (with an expansion on the role of imitation by Baer), and Epstein provides the Theory of Generativity to explain how animals form solutions to problems they've never seen from behaviors they've seldom used (with an expansion on the perspective of the individual by Simonton). The diversity of the mechanisms described in these and many more studies illustrate the need for flexibility in our understanding of the inner workings of non-human creativity. This is where the interdisciplinary nature of creativity is exceedingly advantageous, because it allows us to draw parallels to complimentary processes in any other domain of creativity. Reiter-

Palmon and Hunter, both industrial/organizational psychologists, see parallels between the behavior of business organizations and groups of animals. As Cropley writes in his commentary, "creativity is fundamentally concerned with solving problems, and creativity in any field can be viewed the same way." Be it engineers or apes, the problem and problem solving processes are the same. Indeed, it is not just creativity that has the same basic mechanisms across humans and animals; play, often thought of as a precursor for creativity, also shows the same fundamental principles, as explored in the Burghardt/Russ and Michell/Hoffman dyads.

Similarity in process does not, of course, mean similarity in experimental design. One common test of human creativity involves asking a question and measuring the number of answers, the types of answers, and their elaboration and originality (Torrance, 1974). However, the study of nonhuman creativity cannot use this approach. Animals cannot give a single answer to our questions, let alone enough answers to provide a measure of creativity. So instead the tools of ecology are applied to nonhuman creativity. Researchers and trainers look at behaviors animals display which we deem "creative," and attempt to work out how they arrived at those behaviors, as well as what social or environmental factors made them adaptive. As Lee demonstrates in her chapter, ecological staples of energy acquisition, time costs and mortality risks can be used to assess the costs and benefits of creativity. And environmental factors are key in how creativity arises, as detailed by several commentators (e.g., Gabora and Ranjan, Forgeard and Jayawickreme). Creative behaviors can be replicated and encouraged, as demonstrated by Pryor's training of Malia the dolphin as well as other animals at Sea Life Park. Ultimately however, the best way to study creativity in nonhumans is to look the external conditions that caused the innovation and speculate the incentives which led the animal to the behavior; or what Chappell would call studying "the Gap." This disconnect between investigative techniques in the two domains (psychology and biology) is indicative of the myriad forms of creativity; there is no one right way to study it.

The field of nonhuman creativity faces a problem of clarity. Though the terms creativity and innovation are thrown around regularly in creativity literature, the definitions of these words are inconsistent and often times absent. Within the study of human and nonhuman creativity innovation and creativity can be defined and used in vastly different ways, and as Niu notes, they are often used (erroneously) as synonyms. Definitions range from the simple, "something new and different" (Chappell, this book) to the complex "the amalgamation of existing forms of something novel, to novel ways of using existent forms, or even of inferring novel solutions to problems based in circumstantial information" (Pepperberg, this book). As pointed out by Hennessey and Stathis, without a clear definition of creativity and innovation it becomes difficult to compare experimental findings between academics. Plucker, Beghetto, and Dow (2004) refer to this the "definition problem" and argue it is responsible for much of the confusion found in the field today. There are some areas of agreement; in human creativity research, much of the definitions, even when differing, still center around the core concepts of novelty and task appropriateness (Barron, 1955; Guilford, 1950; Stein, 1953).

This problem extends farther than simply the language we use to describe creativity; it is also an issue of how we consider creativity. Most creativity theories tend to breaks down creativity into components, such as Big C versus little c, the process versus the product, the 4 Cs or the 5 As (Glăveanu, 2013; J.C. Kaufman, 2009, in press; J.C. Kaufman & Beghetto, 2009). Here, Glăveanu even proposes a "proto-c." The difficulty arises when we try to have a conversation between each theory, because each view of creativity is right but in a slightly different way, and from a slightly different perspective. In order to speak on creativity with any depth or reach, a commonality must be found on what it is we are discussing. The definition needs to be flexible enough to bring all the minds in this rapidly expanding field together; as the current situation is akin to discussing the colors of a painting when everyone in the room is a different type of colorblind. Similarly, the terms imagination and innovation, although conveying nuanced differences from creativity, cover the same broad content. It can be useful to focus on similarities as opposed to the more common focus on differences.

Nonhuman creativity as a field is expanding; studies are exploring creativity in previously unexplored animals like butterflies (Snell-Rood) and novel models are being developed and used (Auersperg). Preconceived notions such as the definition of tool use are being questioned (Russon), and unexpected areas of the brain are being examined as playing important roles in the creative process (Petrosini et al., Benedek, and Knudsen et al.). As more and more people with a variety of research backgrounds realize the importance of creativity, the concept itself grows and expands in new directions. The studies and ideas presented in this book do not represent the whole of the field of nonhuman creativity (nor was this volume intended to be a complete handbook); they are merely a descriptive snapshot of the work being done. Presently the field of nonhuman creativity lacks recognition from the larger creativity community. Nonetheless, if testimonials like that of Beghetto's commentary are any indication, it is simply a matter of time.

References

Barron, F. (1955). The disposition toward originality. *Journal of Abnormal and Social Psychology, 51*, 478–485.

Glăveanu, V. P. (2013). Rewriting the language of creativity: The Five A's framework. *Review of General Psychology, 17*, 69.

Guilford, J. P. (1950). Creativity. *American Psychologist, 5*, 444–454.

Kaufman, A. B., Butt, A. B., Colbert-White, E. N., & Kaufman, J. C. (2011). Towards a neurobiological model of creativity in nonhuman animals. *Journal of Comparative Psychology, 125*, 255–272.

Kaufman, A. B., & Kaufman, J. C. (2014). Applying theoretical models on human creativity to animal studies. *Animal Behavior and Cognition, 1*, 77–89.

Kaufman, J. C. (2009). *Creativity 101*. New York, NY: Springer.

Kaufman, J. C. (2012). Counting the muses: Development of the Kaufman-Domains of Creativity Scale (K-DOCS). *Psychology of Aesthetics, Creativity, and the Arts, 6*, 298–308.

Kaufman, J. C. (in press). *Creativity 101* (2nd ed.). New York, NY: Springer.

Kaufman, J. C., & Baer, J. (2004). The Amusement Park Theoretical (APT) model of creativity. *Korean Journal of Thinking and Problem Solving, 14*, 15–25.

Kaufman, J. C., & Beghetto, R. A. (2009). Beyond big and little: The Four C model of creativity. *Review of General Psychology, 13*, 1–12.

Kaufman, J. C., & Kaufman, A. B. (2004). Applying a creativity framework to animal cognition. *New Ideas in Psychology, 22*, 143–155.

Plucker, J. A., Beghetto, R. A., & Dow, G. T. (2004). Why isn't creativity more important to educational psychologists? Potentials, pitfalls, and future directions in creativity research. *Educational Psychologist, 39*, 83–96.

Stein, M. I. (1953). Creativity and culture. *Journal of Psychology, 36*, 311–322.

Torrance, E. P. (1974). *Torrance tests of creative thinking: Norms—technical manual*. Lexington, MA: Ginn.

Index

Note: Page numbers followed by "*f*" and "*t*" refer to figures and tables, respectively.

Zeitfracht Medien GmbH
Ferdinand-Jühlke-Straße 7
99095 Erfurt, Deutschland
produktsicherheit@kolibri360.de